physica status solidi (a)

ISSN 0031-8965 * VOL. 85 · NO. 1 · SEPTEMBER 1984

AKADEMIE-VERLAG · BERLIN

THEORIA CUM PRAXI

Classification Scheme

1. Structure of Crystalline Solids
   1.1 Perfectly Periodic Structure
   1.2 Solid-State Phase Transformations
   1.3 Alloys. Metallurgy
   1.4 Microstructure (Magnetic Domains See 18; Ferroelectric Domains See 14.4.1)
   1.5 Films
   1.6 Surfaces
2. Non-Crystalline State
3. Crystal Growth
4. Bonding Properties
5. Mössbauer Spectroscopy
6. Lattice Dynamics. Phonons
7. Acoustic Properties
8. Thermal Properties
9. Diffusion
10. Defect Properties (Irradiation Defects See 11)
    10.1 Metals
    10.2 Non-Metals
11. Irradiation Effects (X-Ray Diffraction Investigations See 1 and 10)
12. Mechanical Properties (Plastic Deformations See 10)
    12.1 Metals
    12.2 Non-Metals
13. Electron States
    13.1 Band Structure
    13.2 Fermi Surfaces
    13.3 Surface and Interface States
    13.4 Impurity and Defect States
    13.5 Elementary Excitations (Phonons See 6)
        13.5.1 Excitons
        13.5.2 Plasmons
        13.5.3 Polarons
        13.5.4 Magnons
14. Electrical Properties. Transport Phenomena
    14.1 Metals. Semi-Metals
    14.2 Superconductivity. Superconducting Materials and Devices
    14.3 Semiconductors
        14.3.1 Films
        14.3.2 Surfaces and Interfaces
        14.3.3 Devices. Junctions (Contact Problems See 14.3.4)
        14.3.4 High-Field Phenomena, Space-Charge Effects, Inhomogeneities, Injected Carriers (Electroluminescence See 20.3; Junctions See 14.3.3)
    14.4 Dielectrics
        14.4.1 Ferroelectrics
15. Thermoelectric and Thermomagnetic Properties
16. Photoconductivity. Photovoltaic Effects
17. Emission of Electrons and Ions
    17.1 Field Emission Microscope Investigations
18. Magnetic Properties
    18.1 Paramagnetic Properties
    18.2 Ferromagnetic Properties
        18.2.1 Ferromagnetic Films
    18.3 Ferrimagnetic Properties
    18.4 Antiferromagnetic Properties

(Continued on cover three)

Author Index

Author Index

physica status solidi (a)

applied research

Volume 85 · Number 1 · Pages 1 to 304, K1 to K96, and A1 to A8

September 16, 1984

PSSA 85(1) 1—304, K1—K96, A1—A8 (1984)

ISSN 0031-8965

AKADEMIE-VERLAG · BERLIN

Subscriptions and orders for single copies should be directed
in the GDR:

to the Postzeitungsvertrieb or
to the Akademie-Verlag, DDR-1086 Berlin, Leipziger Str. 3 – 4:

in the other socialist countries:

to a book-shop for foreign language literature or
to the competent news-distributing agency;

in the FRG and BERLIN (WEST):

to a book-shop or
to the wholesale distributing agency Kunst und Wissen, Erich Bieber, OHG, D-7000 Stuttgart 1, Wilhelmstr. 4 – 6;

in the other Western European countries:

to Kunst und Wissen, Erich Bieber GmbH, CH-8008 Zürich, Dufourstr. 51;

in USA and CANADA:

to Verlag Chemie International, Inc., 303 N.W. 12th Avenue, Deerfield Beach, FL 33441, USA;

in other countries:

to the international book- and journal-selling trade,
to Buchexport, Volkseigener Außenhandelsbetrieb der Deutschen Demokratischen Republik, DDR-7010 Leipzig, Postfach 160, or
to the Akademie-Verlag, DDR-1086 Berlin, Leipziger Str. 3 – 4.

Editorial Note:

"physica status solidi (a)" undertakes that an original paper *accepted for publication* before the 23rd of any month will be published within 50 days of this date unless the author requests a postponement. In special cases there may be some delay between *receipt* and *acceptance* of a paper due to the review and, if necessary, revision of the paper.

Schriftleiter und verantwortlich für den Inhalt:
Professor Dr. Dr. h. c. P. Görlich, DDR-1020 Berlin, Leipziger Straße 3 – 4, Postfach 1233 bzw. DDR-6900 Jena, Schillbachstr. 24.

Verlag:
Akademie-Verlag, DDR-1086 Berlin, Leipziger Str. 3 – 4; Fernruf: 2236221 und 2236229; Telex-Nr.: 114420; Bank: Staatsbank der DDR, Berlin, Kto.-Nr.: 6836-26-20712.

Chefredakteur:
Dr. H.-J. Hänsch.

Redaktionskollegium:
Prof. Dr. E. Gutsche, Dr. H.-J. Hänsch, Dr. H. Lange, Dr. S. Oberländer.

Anschrift der Redaktion:
DDR-1086 Berlin, Leipziger Straße 3 – 4, Postfach 1233; Fernruf: 2236279.

Veröffentlicht unter der Lizenznummer 1620 des Presseamtes beim Vorsitzenden des Ministerrates der Deutschen Demokratischen Republik.

Gesamtherstellung:
VEB Druckerei „Thomas Müntzer", DDR-5820 Bad Langensalza.

Erscheinungsweise:
Die Zeitschrift „physica status solidi (a)" erscheint jeweils am 16. eines jeden Monats. Jährlich erscheinen 6 Bände zu je 2 Heften.

Bezugspreis:
Je Band 200,— M zuzüglich Versandspesen (Preis für die DDR: 130,— M).

Bestellnummer dieses Bandes: 1085/85.

Printed in the German Democratic Republic.

AN EDV 20735

Contents

Review Article

Original Papers and Short Notes

Structure

Lattice properties

Defects, atomistic aspects

Magnetism

Extended electronic states and transitions

Localized electronic states and transitions

Electric transport

Device-related phenomena

Pre-Printed Titles

Systematic List

Contents of Volume 85 Continued on Page 307

Review Article

phys. stat. sol. (a) **85**, 11 (1984)

Subject classification: 10.1; 9; 21

Institute of Metal Physics,
Academy of Sciences of the USSR, Ural Scientific Centre, Sverdlovsk[1])

Effect of Dislocation Structure on Creep and Fracture of Metals and Alloys

I. Normal Deformation Conditions[2])

By

V. A. Pavlov, V. I. Shalaev, and V. M. Alyabiev

Contents

1. Introduction

With the rapid advance of nuclear engineering in the past few years need has arisen for materials capable of satisfactory performance under irradiation. The major features peculiar to the behaviour of materials under load when irradiated by high-energy particles are: 1. high acceleration of creep in the 0.4 to $0.5T_{\text{melt}}$ temperature range (T_{melt} is the melting point) as compared to normal deformation conditions, 2. considerable increase in volume swelling, 3. embrittlement, 4. heterogenization of solid solutions and the emergence of segregations on point defect sink surfaces.

In the final analysis, these phenomena stem from the fact that during irradiation a considerable point defect (vacancy and interstitial atom) concentration is produced which is largely in excess over the equilibrium concentration determined by the temperature conditions of deformation. Point defects promote diffusion processes and, by interacting with dislocations, affect the development of plastic deformation, void formation, and embrittlement. The interaction of point defects with atoms of solid solution elements may lead to the development of a fast diffusion of point defect complexes "defect–atom", which largely affect the rate of void formation, the rate creep, etc.

---

[1]) GSP-170, 620219 Sverdlovsk, USSR.

[2]) Part II see phys. stat. sol. (a) **85**, No. 2 (1984).

To approach correctly the problem of selecting reactor engineering materials, consideration should be given to the peculiarities of the mechanism of plastic deformation and fracture under irradiation. It would be expeditious to draw a thorough comparison of the behaviour of materials when subject to load under normal deformation conditions and when exposed to irradiation. Investigations should probably be carried out in the temperature range between 0.4 and $0.6T_{\text{melt}}$ where the acceleration of creep and the swelling of reactor materials are most pronounced and the creep under normal deformation conditions is controlled by the climb of edge dislocations due to point defects migrating to them.

2. Dislocation Structure

Let us examine the structure of dislocations, since the character of the interaction of point defects with them depends on the structural peculiarities of their nuclei.

Dislocations are subdivided into edge and screw dislocations. In the nucleus of an edge dislocation there is an edge of an incomplete plane of atoms, and the motion of that dislocation is associated with a particular crystallographic slip plane. Screw dislocations contain no extra plane and can, as they move, pass from one slip plane to another by way of cross slip. During plastic deformation, dislocation pile-ups are produced near obstacles, the elastic stress field of the pile-ups hindering plastic deformation. The deformation can develop further only provided the dislocations are capable of surmounting these obstacles in some way or another.

The motion of edge dislocations becomes possible only when the deformation temperature is so high as to allow point defects to migrate with sufficient velocity to dislocations. By the action of an external force and due to the diffusion process of the "growth" or "dissolution" of the incomplete plane, the leading pile-up dislocation will then climb over to a higher-lying (or lower-lying) slip plane. The dislocation that has become free will move over the new plane up to the next obstacle, and the dislocation source will be able to come into play again. These features of the motion of an edge dislocation are responsible for the shearing mechanism of the deformation which, however, is governed by the dislocation climb diffusion process. It stands to reason that the deformation in this case may develop only at sufficiently high temperatures capable of assuring the requisite rate of migration of point defects to dislocations and an adequate point defect concentration.

Screw dislocations surmount obstacles by cross slip. Under the effect of an external force and thermal energy, part of the dislocations present in the pile-up head pass over to another slip plane and from the latter, again by cross slip, to a higher-lying or lower-lying slip plane parallel to the first, and move further over a new plane. Thus, an obstacle in the way of screw dislocations is surmounted by double cross slip. In this case the shearing mechanism of plastic deformation is governed by a thermally activated, but diffusionless cross slip process. Point defects do not interact with a screw dislocation. Plastic deformation can easily develop at low temperatures if, of course, the mobility of screw dislocations does not lower with decrease in temperature, as is the case with b.c.c. metals [1].

The motion of dislocations becomes more complicated in the case of their splitting (Fig. 1b). A dislocation splits up into two partial dislocations with a stacking fault between them. The degree of splitting, i.e. the width of an extended (splitting) dislocation is determined by the stacking fault energy γ. The dislocation width is $h \approx \approx Gb^2/\gamma$, where G is the shear modulus and b the Burgers vector. In order that an edge dislocation may climb several vacancies (or interstitial atoms) at a time should approach one and the same dislocation site. The climbing event in this case assumes cooperative character. The cross slip of extended dislocations becomes difficult because

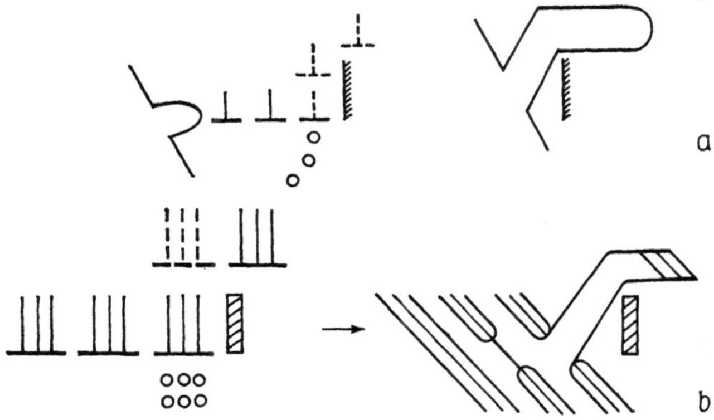

Fig. 1. Diagram of motion of edge and screw dislocations: a) no splitting, b) splitting

preliminarily, additional energy is required to convert an extended dislocation to a contraction of stacking faults. In the place where cross slip will occur the dislocation is compressed by external stresses and then the dislocation splits up in a new plane.

3. Dislocation and Grain Boundary Diffusion

Studying the parameters of dislocation and grain boundary diffusion is of great importance in many aspects. Primarily, these studies have general scientific significance in terms of establishing regularities of diffusion for the most frequently occurring crystal lattice defects. The evidence obtained permits judgement as to the state of the lattice in these defets. Secondly, the dislocation diffusion plays an important part in the behaviour of metals during creep, when plastic deformation is controlled by the climb of edge dislocations. Thirdly, dislocations and grain boundaries are good point defect sinks. Dislocation and grain boundary diffusion ensures the sinking of excessive vacancies. Therefore, the diffusing power of the dislocation structure of a material predetermines, in the final analysis, the swelling and embrittlement of materials during irradiation.

The most trustworthy evidence for the effect of defects being formed during plastic deformation on the diffusion characteristics was probably first obtained in [2]. The investigations were carried out on zinc single crystals with mercury serving as a diffusing element. Three crystals of different orientation were subjected to plastie deformation by extension and cut into portions along the basal plane, on whose thoroughly cleaned surface a mercury drop was placed. The diffusion was studied in the temperature range between 287 and 363 K. Fig. 2 presents curves, where x is the

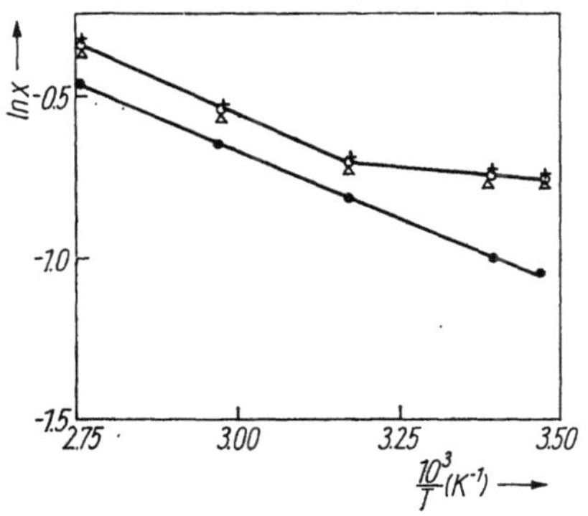

Fig. 2. $\ln x$ vs. $1/T$ curves for diffusion of mercury in zinc; non-deformed crystal (●), crystal deformed by 11 (○), 23 (△), and 46% (+) [2]

penetration depth of mercury, for a non-distorted crystal and three specimens of different degrees of preliminary deformation: 11, 23, and 46% change in length. In the range of low temperatures, below 314 K, the diffusion in deformed specimens proceeds preferentially along dislocations produced by plastic deformation, and the activation energy is very low, about 1.5×10^{-20} J. At higher temperatures, due to partial lattice defect relaxation and the volume diffusion coming into play, the activation energy rises up to about 5.0×10^{-20} J, which coincides approximately with the quantity Q of 5.5×10^{-20} J for a non-distorted crystal. The ratio of the dislocation diffusion activation energy Q_d to the mercury-in-zinc diffusion activation energy thus amounts to $Q_v \approx 0.3$. Besides, distorted crystals exhibit high pre-exponential D_0 values.

Recently, comprehensive investigations of both the individual isolated dislocation directional self-diffusion and the small-angle crystal boundary directional self-diffusion have been performed for different metals using radiative isotopes [3 to 15], local X-ray analysis [16], internal friction [17], and computer simulation [18] methods.

For small-angle boundary self-diffusion experiments specimens with small-angle boundaries were prepared (Fig. 3). The self-diffusion coefficient depends on: 1. the Burgers vector, 2. the direction of diffusion, 3. the type of dislocation (edge or screw dislocation), 4. the degree of dislocation splitting, which is determined by the magnitude of the stacking fault energy γ. To suit the purpose, crystals with a small-angle slope boundary formed by a series of edge dislocations (Fig. 3 a) or extended dislocations (Fig. 3b) were prepared. In that case, partial dislocations with a stacking fault ribbon between them lay along the small-angle boundary. To study screw dislocations small-angle torsion boundaries were used. By changing the crystallographic planes that form small-angle boundaries, dislocations with different Burgers vectors could be produced.

The results of those investigations have been analysed in detail in a review [11]. Table 1 furnishes dislocation diffusion data for metals with a f.c.c. lattice. The diffusion parameters were obtained straightforwardly from experimental data, without resorting to an assumption concerning the effective values of the cross-section of a dislocation pipe. For material volume diffusion the standard expression

$$D_v = D_{0v} \exp(-Q_v/kT) \tag{1}$$

was employed. Similarly, for dislocation diffusion

$$D_d = D_{0d} \exp(-Q_d/kT) \,. \tag{2}$$

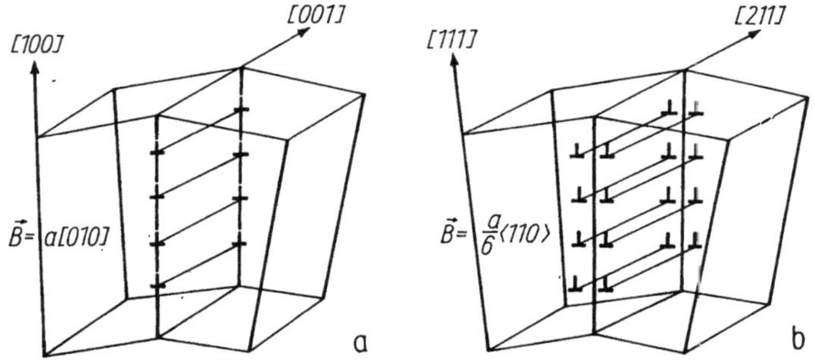

Fig. 3. Diagram of small-angle boundaries; a) boundary formed by a series of non-dissociated edge dislocations aligned in ⟨001⟩ direction with Burgers vector $\boldsymbol{b} = a\langle 010\rangle$; b) boundary formed by dissociated edge dislocations aligned with [211] direction with partial Burgers vector $\boldsymbol{b} = a/6\langle 110\rangle$

Table 1
Dislocation diffusion in f.c.c. metals

No.	material	diffusing element	D_d ($cm^4 s^{-1}$)	Q_d (eV)	Burgers vector	t	D_{0v} ($cm^2 s^{-1}$)	Q_v (eV)	Q_d/Q_v	diffusion conditions
1	Al	^{110}Al	5.2×10^{-17}	0.74	$a\langle 110\rangle$ or $(a/2)\langle 110\rangle$	[110]	0.44	1.92	0.4	directional diffusion along low-angle slope boundary edge dislocations [3]
2	Al	^{110}Al	—	—	—	—	0.44	1.92	—	fast high-angle boundary diffusion, but not enough data for low-angle boundary [4]
3	Ni	^{63}Ni	3.1×10^{-15}	1.76	$(a/2)\langle 110\rangle$	[211]	1.9	2.95	0.6	directional diffusion along small-angle slope boundary edge dislocations [5]
4	Ni	^{63}Ni	6.6×10^{-15}	1.95	$(a/2)\langle 110\rangle$	[110]	1.9	2.95	0.7	directional diffusion along dissociated screw dislocations in low-angle torsion boundary [5]
5	Ni	^{63}Ni	9.3×10^{-18}	1.08	$a\langle 100\rangle$ or $(a/2)\langle 110\rangle$	[110]	1.9	2.95	0.4	directional diffusion along low-angle slope boundary edge dislocations [6]
6	Ni	^{63}Ni	5.6×10^{-12}	1.6	$(a/2)\langle 110\rangle$	[211]	1.9	2.95	0.5	directional diffusion through thin crystal along edge dislocations [7]
7	Al	^{110}Al	7.0×10^{-17}	0.85	$(a/2)\langle 110\rangle$	—	0.18	1.31	0.6	diffusion transports of mass along dislocations due to capillarity [8]
8	NiCo	^{63}Ni	5.6×10^{-11}	1.31	$(a/2)\langle 110\rangle$	—	0.44	2.78	0.5	directional diffusion through thin crystal along edge dislocations [9]
9	Cu	Ge		1.26	$(a/2)\langle 110\rangle$	[211]	0.20	189	0.7	measurements of internal friction for the fastening of edge dislocations by germanium atoms [10]

The results tabulated in Table 1 provide conclusive evidence for fast dislocation diffusion. The dislocation activation energy Q_d varies from 0.4 to 0.7 of the crystal lattice self-diffusion activation energy Q_v, the quantity D_{0d} ranging between 10^{-11} and 10^{-17} cm^4 s^{-1}.

Noteworthily, non-dissociated dislocations with a large Burgers vector $\boldsymbol{b} = a\langle 100\rangle$ or $\boldsymbol{b} = a/2\langle 110\rangle$ for aluminium and nickel possess the property that the diffusion along them proceeds with a very small activation energy. The ratio $Q_d/Q_v \approx 0.4$. This is so because the lattice of the nucleus of dislocations with a large Burgers vector is highly distorted and the distortions are not capable of relaxation by splitting.

In the dislocation diffusion case, where the dislocations with the Burgers vector $\boldsymbol{b} = a/2\ \langle 110\rangle$ lie along different directions in the $\{111\}$ plane, the distortions may relax by splitting up into partial dislocations with smaller Burgers vectors $\boldsymbol{b} = a/6\ \langle 112\rangle$. For these dislocations the diffusion proceeds more slowly and the ratio of the dislocation activation energy to the crystal volume diffusion activation energy ranges between 0.5 and 0.7.

It should be noted that the diffusion coefficient tends to decrease with increasing degree of dislocation splitting which is determined by the stacking fault energy γ. The smaller the stacking fault energy the larger the width of dislocation splitting. Table 2 presents data on the stacking fault energy for different elements. It is seen

Table 2
Values of stacking fault energies [1, 19]

non-transition group elements	type of lattice	γ (mJ/m^2)	transition group elements	type of lattice	γ (mJ/m^2)
Cu	f.c.c.	70	Zr	h.c.p.	220
Ag	f.c.c.	25	Th	h.c.p.	220
Au	f.c.c.	10	V	b.c.c.	100
Zn	h.c.p.	250	Nb	b.c.c.	220
Cd	h.c.p.	150	Ta	b.c.c.	1100
Al	f.c.c.	250	Cr	b.c.c.	300
Si	diamond lattice	50	Mo	b.c.c.	300
Ge	diamond lattice	90	W	b.c.c.	300–400
Pb	f.c.c.	50	Re	b.c.c.	1800
Ca	f.c.c. above 250 °C,	15	Fe	b.c.c.	140
			Rh	f.c.c.	180
			Ir	f.c.c.	190
			Ni	f.c.c.	150, 40
	h.c.p. above 450 °C				above Curie point
C	h.c.p. graphite	0.1	Pd	f.c.c.	90
			Pt	f.c.c.	1120
			Co	h.c.p., at 450 °C f.c.c.	30
			Ti	h.c.p., at 882 °C b.c.c.	10
			Se	h.c.p. and f.c.c. at 23 °C, normally h.c.p.	10

that zinc, aluminium, and nickel which possess a high stacking fault energy of 150 to 250 mJ/m², exhibit a high dislocation diffusion velocity, and the ratio Q_d/Q_v lies within the range from 0.5 to 0.6. In copper, which has a low stacking fault energy $\gamma = 70$ mJ/m², the diffusion coefficient is low and the ratio of activation energies Q_d/Q_v is about 0.7. This important property was noted earlier by Weertman [20]. He indicated that the dislocation diffusion velocity rises with increasing Burgers vector and in the case of dislocations with a Burgers vector $\boldsymbol{b} = a/2\ \langle 110 \rangle$ the diffusion will slow down as disloactions dissociate into partial dislocations with a smaller vector $\boldsymbol{b} = a/2\langle 112 \rangle$.

The above analysis of experimental dislocation diffusion data for dislocations with different Burgers vectors and different degrees of dislocation splitting does not take into account the structural changes in the nucleus of dislocations of different types. The data presented in Table 1 for pure nickel (No. 3 and 4) suggest that extended edge dislocation diffusion proceeds faster than extended screw dislocation diffusion. The dislocation diffusion coefficients for a Burgers vector $\boldsymbol{b} = a/2\ \langle 110 \rangle$ dislocation lying along the $1/\sqrt{6}\ \langle 211 \rangle$ direction (edge orientation) are approximately a factor of 5 higher than those of a $\boldsymbol{b} = a/2\ \langle 110 \rangle$ dislocation lying approximately along the $a/2\ [211]$ direction (approximately screw orientation), the ratio Q_d/Q_v is about 0.6 for edge dislocations and about 0.7 for screw dislocations.

Later studies of the diffusion of individual dislocations confirm the data given in Table 1. Thus, in [11] the self-diffusion of gold was investigated. The activation energy Q_d turned out to be equal to 1.24 eV, $Q_v = 1.75$ eV, and $Q_d/Q_v = 0.7$. Investigations of the dislocation diffusion have resulted in the conclusion that there exists a vacancy mechanism of diffusion [13]. In [14] the data of earlier investigations [8] have been confirmed for the Burgers vector $\boldsymbol{b} = a/2\ \langle 110 \rangle$ dislocation diffusion in aluminium. Interesting data about the dependence of the dislocation diffusion parameters on the valence of diffusing elements such as zinc, germanium, and arsenic are furnished

Table 3
Grain boundary diffusion

metal	diffusing element	Q_v (kcal/mol)	Q_g (kcal/mol)	Q_g/Q_v	ref.
Au	Au	221.9	96.3	0.43	[24]
	Au	282.6	126.4	0.45	[25]
Ni	Ge	254.5	123.7	0.48	[26]
	Nd	250.4	107.1	0.42	
Co	Co	242.8	117.2	0.47	[27]
Ni + Fe 315.%	Cr	250.3	148.2	0.59	[28]
+ Cr 1.35%	Ni	262.0	156.6	0.6	
Ni	Ni	284.6	129.3	0.45	[29]
Cu	Zn	198.8	60.28	0.3	[23]
	Ag	176.2	57.70	0.3	
Pb	Pb	98.79	17.16 to 49.81	0.17 to 0.44	[36]
Al	Zn	128.9	39.22	0.29 to 0.42	[37]
Au	Ag		67.39	0.4	
	Ag		63.63	0.35	
	Cd		64.46	0.36	[38]
Ag	In		75.77	0.37	
	Sb		57.77	0.35	

in [15]. It has turned out that the ratio Q_d/Q_v increases with increasing ion charge and is equal to 0.29, 0.34, and 0,37, respectively.

At present there is sample experimental evidence concerning the study of the grain boundary diffusion parameters [21 to 37]. Some investigation results are consolidated in Table 3 which presents a fairly broad spectrum of different pure metals and alloys with f.c.c., b.c.c., and h.c.p. lattices. It can be seen that the diffusion along grain boundaries, just as along dislocations, proceeds much more easily than that in the bulk of a material. The ratio of the grain boundary activation energy to the material volume activation energy, Q_g/Q_v, lies between 0.2 and 0.6. In [35] interesting evidence is presented concerning the effect of stacking fault energy on the value of the pre-exponential factor D_0 in the diffusion equation. The ratio D_{0g}/D_{0v} decreases with decrease in stacking fault energy.

4. Effect of Stacking Fault Energy on Creep Rate

During plastic deformation under creep conditions, when the flow of a material is controlled by the climb due to the migration of point defects to dislocations and due to the dislocation diffusion of those point defects, the creep and fracture characteristics may be expected to depend substantially on the dislocation diffusion parameters. Since the latter are determined by the dislocation structure which is associated with the stacking fault energy, it should be cassumed that the creep characteristics depend strongly on dislocation structure and, specifically, stacking fault energy [39]. In different form, this idea was expounded in a number of papers [40 to 42]. For metals with f.c.c. and b.c.c. lattices, special investigations were carried out to ascertain the dependence of the creep rate on the stacking fault energy of dislocations [43 to 52]. Let us dwell in more detail upon some papers of interest [46, 47]. They appeared somewhat later than [43, 44], but the investigations performed are more complete and take account of previous results. The metals that were tested are as follows: Al, Pt, Pd, Ni, Ag, Mo, V. Their characteristics are specified in Table 4.

To investigate the steady-state creep rate f.c.c. metals were selected of which the stacking fault energy varied within 25 to 250 mJ/m². The tests were carried out at temperatures corresponding to the assigned values of the volume diffusion coefficient $D_v = 10^{-11}$ and 10^{-12} cm² · s⁻¹ and also at stresses with respect to the elasticity modulus $\sigma/E = 5 \times 10^{-5}$ and $\sigma/E = 8.1 \times 10^{-15}$. For comparison we here also present data for b.c.c. meals, borrowed from [52 to 55], as well as data for vanadium [46]. The characteristics of these materials are also given in Table 4. For these materials the stacking fault energy varies in the limits from 100 to 400 mJ/m².

Table 4

metal	lattice	γ (mJ/m²)	purity (%)	ref.
aluminium	f.c.c.	250	99.99	[46, 47]
platinum	f.c.c.	120	99.9	[46, 47]
palladium	f.c.c.	90	99.8	[46, 47]
nickel	f.c.c.	150	99.92	[46, 47]
silver	f.c.c.	25	99.99	[46, 47]
vanadium	b.c.c.	100	99.99	[53]
molybdenum	b.c.c.	300	99.9967	[52]
niobium	b.c.c.	220	99.995	[52]
tungsten	b.c.c.	400	99.999	[52, 54]
tantalum	b.c.c.	100	99.997	[52, 55]

For convenience in analysis of experimental data, the dependence of the creep rate on deformation conditions may be written as

$$\dot{\varepsilon} = A\gamma^q D_v \left(\frac{\sigma}{E}\right)^n, \tag{3}$$

where A is a constant, D_v the volume self-diffusion coefficient, σ the stress, E the elasticity modulus.

Fig. 4 presents the stacking fault energy dependence of referred steady-state creep rate $\dot{\varepsilon}/D_v$. The behaviour of the curve is qualitatively similar to that established in [43, 44]. However, the creep tests here have been carried out using two values of the diffusion coefficient and referred stress. This allows the influence of the variation of the quantities D_v and σ on the creep rate to be considered separately. Comparison of the data obtained shows that an increase of D_v or $(\sigma/E)^5$ by a factor of ten leads to an increase of $\dot{\varepsilon}$ by one order of magnitude. In addition, other conditions being equal, the referred rate of creep $(\dot{\varepsilon}/D_v)$ of different metals increases systematically with the magnitude of their stacking fault energy.

However, some discrepancies from the results of [44] have been noticed. In experiments [46, 47] the stacking fault energy dependence of the creep rate is less sharp $q = 2.4 \pm 0.2$ at $n = 4.7$ than that established in [43, 44] where $q = 3.5$ has been obtained with $n = 5$. A similar conclusion may be drawn from [48 to 50] where, from measurements on pure metals with f.c.c. lattice, values from 2.4 to 3.5 are indicated for the quantity q. Nonetheless, investigations of the creep in a number of copper-based solid solutions [45] may be mentioned where it is asserted that the dependence of $\dot{\varepsilon}/D_v$ on γ is exponential in character. The above discrepancy of experimental data apparently may not be attributed to only random errors in experiment or errors in the choice of the initial elasticity modulus values and in allowance for their actual variation with temperature [46, 47].

Thus, as a fact, the occurrence of a sharp stacking fault energy dependence of steady-state creep rate is confirmed by many investigations and evokes no doubts. The variation of the stacking fault energy γ by a factor of ten results in that the steady-state creeep rate $\dot{\varepsilon}$ varies by more than two orders of magnitude in metals with f.c.c. lattice. For b.c.c. metals the $\dot{\varepsilon} = f(\gamma)$ dependence behaves similarly. In some papers, the adopted value of the stacking fault energy for metals with b.c.c. lattice is, in our view, too high [54]. In this case the $\dot{\varepsilon} = f(\gamma)$ dependence in the region of high creep rates decviates from the linear dependence towards high γ values.

The decrease of the creep rate $\dot{\varepsilon}$ with decrease in stacking fault energy γ can probably be explained, as already indicated above, by the fact that the dislocation diffusion velocity decreases and the formation of jogs is difficult.

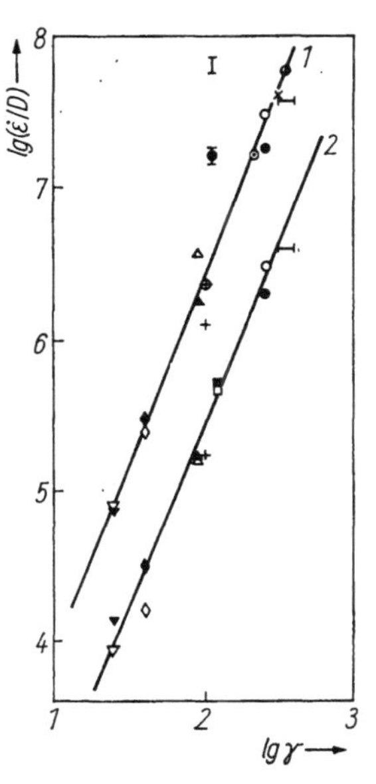

Fig. 4. Dependence of steady-state creep rate of pure metals $(\dot{\varepsilon}/D_v)$ on magnitude of stacking fault energy γ [47, 52 to 55]; (1) $\sigma/E = 8.1 \times 10^{-5}$, (2) $\sigma/E = 5 \times 10^{-5}$: f.c.c. metals (empty symbols $D_v = 10^{-11}$ cm²/s, full symbols 10^{-12} cm²/s: Al (○, ●), Pt (□, ■), Pd (△, ▲), Ni (◇, ◆), Ag (▽, ▼); b.c.c. metals $\sigma/E = 8.1 \times 10^{-5}$: W ⊢⊣, Mo ×, Nb ⊙, Ta I, V +

Mention should be made of the peculiarities in the behaviour of nickel which displays a creep rate corresponding to metals with low stacking fault energy ($\gamma = 40$ mJ/m^2). In their work Barret and Sherbi [43] set γ equal to 225 mJ/m^2 which is apparently too high. It has been shown in [56, 57] that when passing through the Curie point (above 633 K) nickel has low stacking fault energy.

5. Effect of Stacking Fault Energy on Fracture

Investigations of the creep under the deformatioon conditions mentioned above have also permitted detection of another important regularity. It has turned out that the time to fracture τ as a function of stacking fault value γ may be described by the empirical equation [46, 50]

$$\tau = B\gamma^{-p}\,, \tag{4}$$

where B and p are constants at the σ/E and D_v values selected. The regularity is graphically represented in Fig. 5. It is worth noting that B depends on the quantity σ/E, and p on D_v. When $D_v = 10^{-11}$ cm^2 s^{-1} $p \approx 1.5$, and when $D_v = 10^{-12}$ cm^2 s^{-1} $p \approx 1.0$.

In creep tests of metals and alloys, the well-known relationship between the time to fracture τ and the steady-state creep rate $\dot{\varepsilon}$ has been established [59, 60],

$$\tau\dot{\varepsilon}^m = \varepsilon_0\,, \tag{5}$$

where ε_0 is a constant. The exponent m is close to unity in most of the f.c.c. metals and may take on values between 0.8 and 1.2 for different groups of metals [61]. The quantity ε_0, known as the "plasticity margin", determines the plasticity of materials and has, for each material, a characteristic value which depends weakly on test conditions. The product $\tau\dot{\varepsilon}$ coincides in most cases with the magnitude of deformation accumulated by a specimen on the equilibrium creep area. Clearly, ascertaining the nature of the constant ε_0 is an important problem. However, all the expressions proposed earlier, on the basis of which it could be calculated, involve parameters not always capable of direct measurement and explain only the constant ε_0 for a given material [63, 64]. From this point of view, [46, 58] are of interest.

In virtue of (4) and (5) it is possible to write

$$\varepsilon_0 = \tau\dot{\varepsilon} = A'\gamma^{q-p}\,. \tag{6}$$

It should be noted that (6) with $q - p = 1$ satisfies a large body of experimental evidence obtained by different authors. Davis and Wilshire [64] have gathered numerical values of the quantity ε_0 for f.c.c. metals in which the exponent m is equal to unity. These data are presented in Table 5 (second column) in which the mean values of the product measured in [46] and the known values of stacking fault energies of metals are also given.

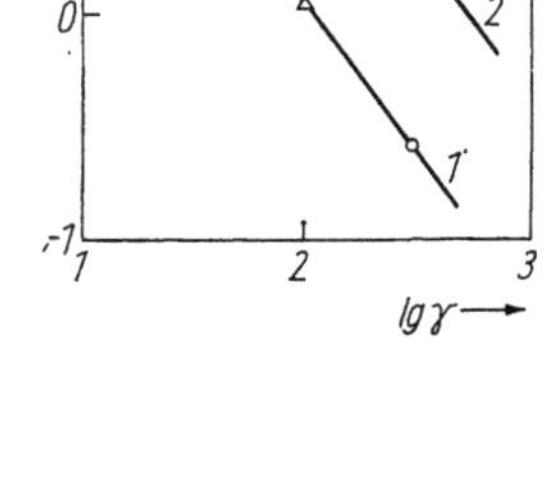

Fig. 5. Dependence of time to fracture τ of metals on the magnitude of stacking fault energy γ [47]; (1) $\sigma/E = 8.1 \times 10^{-5}$, (2) $\sigma/E = 5 \times 10^{-5}$, $D_v = 10^{-11}$ cm^2 s^{-1}; □ Pt, △ Pd, ◇ Ni, ▽ Ag, ○ Al

Table 5

Comparison of quantity ε_0 and stacking fault energy γ

metal	$10^2\, \varepsilon_0$	$10^2\, \tau\dot{\varepsilon}$	γ (mJ/m²)
silver	2	2	25
nickel	4.5	4.5	40
copper	6	—	70
palladium	—	16.5	90
platinum	—	16.0	120
aluminium	32	33	250

The values cited in columns 2 and 3 coincide, and the parameter ε_0 increases with increase in the stacking fault energy of the metals.

The data given in the table permit to plot ε_0 versus stacking fault energy γ (Fig. 6). The plot shows a linear relation between plasticity at equilibrium creep stage and stacking fault energy.

Comparison of a sufficiently large body of experimental creep test data for materials with f.c.c. lattice leads to the important conclusion that the plasticity of metals at the equilibrium creep stage is proportional to the magnitude of the stacking fault energy in them. This relation between the magnitude of plasticity at equilibrium creep stage and stacking fault energy explains not only the constancy of the product $\dot{\varepsilon}\tau$ for a particular material but also the experimentally observed differences in the plasticity of the pure f.c.c. metals studied.

It is highly essential that all these metals possess an approximately equal and sufficiently high plasticity under active extension conditions both at room temperature and in the region of high temperatures. This allows one to speak of the existence of the phenomenon of embrittlement of materials under creep, when plastic deformation is controlled by the climb of edge dislocations. It is a simple matter to see that the degree of embrittlement increases with decreasing stacking fault energy γ. A decrease of the stacking fault energy from 250 to 25 mJ/m², by approximately a factor of 10, leads to a decrease in plasticity from 33 to 2 units, i.e. by about an order of magnitude.

Under these deformation conditions the fracture occurs due to void formation processes, as a result of vacancy coagulation [64 to 66]. At present there exist a great deal of investigations concerned with the void formation under creep conditions at temperatures of materials of about $0.5T_{\mathrm{melt}}$. We mention here only some of them [67 to 75]. All of them confirm the earlier concepts about the mechanism of fracture under creep as resulting from the diffusion growth of voids due to excess vacancies that arise by the action of thermal energy and plastic deformation. It should, however,

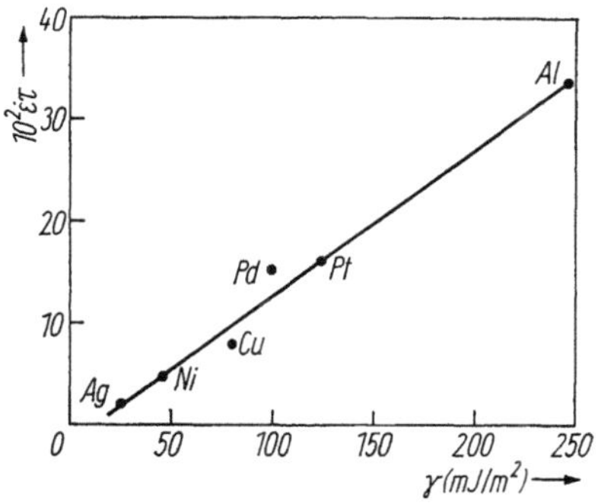

Fig. 6. Dependence of $\tau\dot{\varepsilon}$ on γ [58]

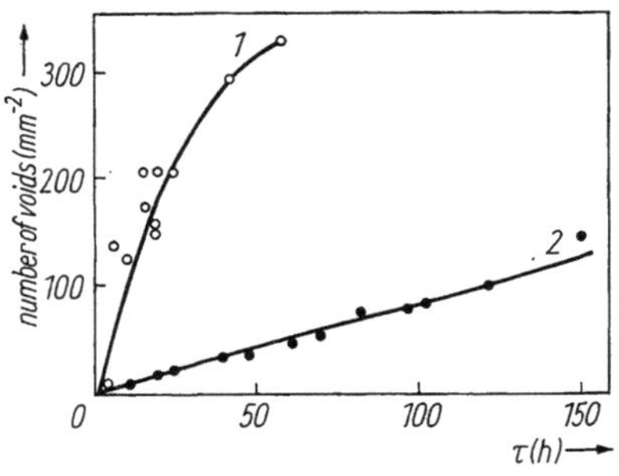

Fig. 7. Creep duration dependence of void density for silver (1) tested at 773 K and for copper (2) tested at 848 K [76]

be noted that the void nucleation mechanism has not yet been ascertained. Some authors propose that nucleation centres in a non-distorted material exist in advance.

Quite naturally, it appears necessary to study the void formation in metals with different stacking fault energy. Relevant investigations have been carried out in [76]. The materials investigated were silver, copper, and aluminium, with stacking fault energies of 25, 70, and 250 mJ/m^2, respectively. All the metals had a purity as high as 99.99%. The experiments were carried out at the same homoglogous temperatures $T/T_{\text{melt}} = 0.6$ and at 10^{-7} s^{-1} strain rate at equilibrium creep stage.

Fig. 7 presents void density versus creep duration curves for silver and copper. It is seen that the void nucleation rate for silver is much higher than that for copper. The technique employed for silver and copper did not permit the authors to detect the void formation in aluminium. Therefore, the density of the material was measured after different creep test durations. Fig. 8 presents curves depicting the variation of the density of aluminium and copper with increase in test duration. Throughout the experiment, no density variations were found to occur in aluminium, whereas the density in copper decreased appreciably, by about 4%, due to void formation. Consequently, it may be said that the phenomenon of swelling of materials exists also under normal creep conditions, similar to the phenomenon of increase in volume which occurs in metals irradiated by high-energy particles.

The data obtained thus evidence that the void nucleation during creep is largely affected by the stacking fault energy. These results are plotted in Fig. 9. The plot illustrates the void formation process during plastic deformation as a function of stacking fault energy under creep conditions at temperatures of about $0.6T_{\text{melt}}$, when the flow of material is governed by the climb of dislocations due to the dislocation diffusion of point defects such as vacancies or interstitial atoms.

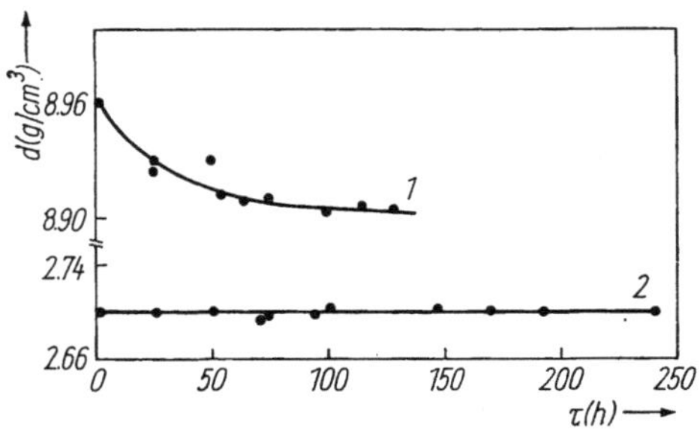

Fig. 8

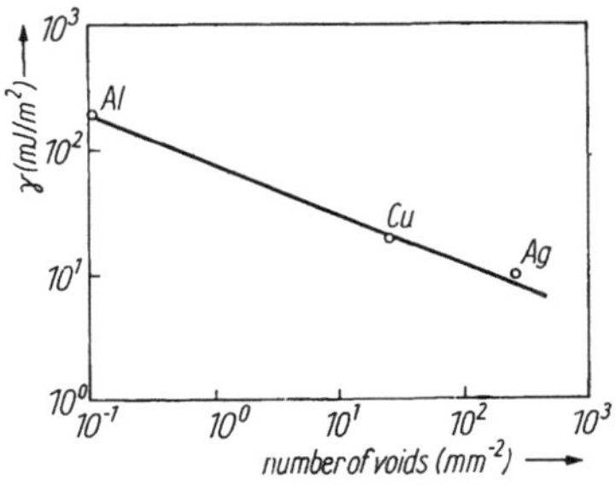

Fig. 9

Fig. 8. Density variation in creep tests for copper (1) at 848 K and for aluminium (2) at 583 K [76]

Fig. 9. Stacking fault energy γ dependence of density of voids arising during creep

The most typical case, under normal test conditions, is the formation of voids at grain boundaries. However, when the density of defects becomes very high for some or another reason, the void formation occurs not only at grain boundaries but also within them. This situation takes place, for example, in the case of evaporating zinc from brass or irradiating metals and alloys by high-energy particles, when the density of point defects becomes high. This may be illustrated with reference to the example of irradiation of nickel by Ni^+ ions at 877 K to doses of the order of 25 dpa or irradiation of niobium by 8×10^{19} neutrons/cm^2 at 533 K [77, 78]. For normal creep conditions, just as in the case of radiative irradiation, the strain rate thus consists of two parts:

$$\dot{\varepsilon} = c\dot{\varepsilon}_0 + p\dot{s} ,$$

where $\dot{\varepsilon}_0$ is the rate of creep and $\dot{s}$ the rate of swelling. But the quantity $\dot{s}$ is much smaller compared to that in the case of radiation-initiated creep.

6. Effect of Relative Density of Screw Dislocations on Creep of Metals and Alloys

It has been noted in the foregoing that in their nuclear structure screw dislocations differ significantly from edge dislocations. A pure screw dislocation is not capable of climbing. A splitting screw dislocation can climb, but the rate of screw dislocation diffusion is a factor of five lower than that of edge dislocation diffusion (see Table 1), and therefore the climbing power of an extended dislocation is extremely low. It is of interest to study the effect of screw dislocations on the behaviour of materials subject to load in the temperature range where the deformation is controlled by the climb of edge segments.

In [79, 80] the effect of the relative length of screw dislocations on the mechanical properties of armco iron at room temperature and at higher temperatures was studied. The dislocation structure was characterized by the ratio L_{hkl}/L, where L_{hkl} is the length of all the dislocations aligned with the projection in $\langle hkl \rangle$ direction, and L is the length of all the dislocations measured. By hot and cold rolling and subsequent annealing at 953 K (680 °C), specimens of armco iron with different ratios of dislocations oriented along $\langle 111 \rangle$ directions were obtained.

The point is that as the ratio L_{111}/L decreases the relative length of low-mobility screw dislocations diminishes, which leads to an incraese in the plasticity of the material. In [80] specimens were investigated at high temperatures under creep conditions when the plastic deformation is controlled by the edge dislocation climbing process. The deformation mode employed to produce specimens with a given dislocation structure and the quantity L_{111}/L specifying the relative length of screw dislocations are presented in Table 6.

Table 6

specimen	deformation mode	L_{111}/L
A	hot rolling	0.8
B	cold rolling 60%	0.5
C	cold rolling 70%	0.5
D	cold rolling 80%	0.1

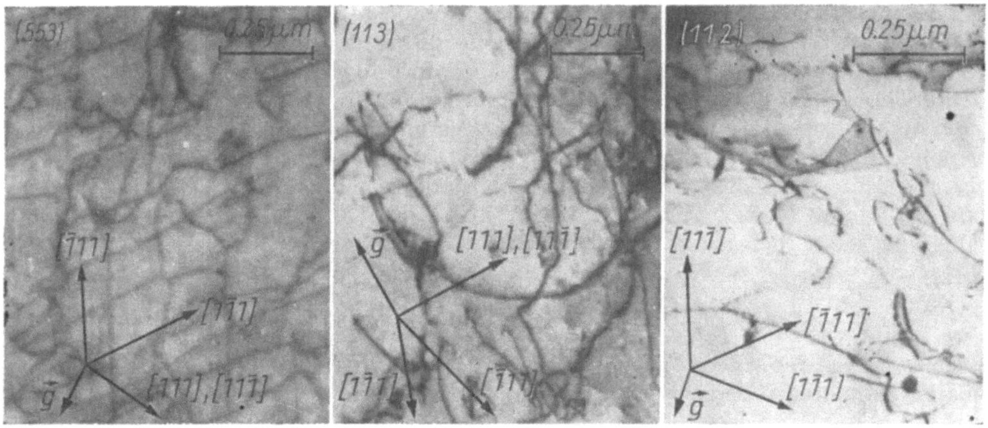

a b c

Fig. 10. Dislocation structure of armco iron a) after hot rolling followed by annealing at 953 K b) after cold rolling by 60% and annealing at 953 K, c) after cold rolling by 90% and annealing at 953 K [79]

Fig. 10 shows the dislocation structure of iron after hot and cold rolling. Upon hot rolling (Fig. 10a), nearly all the dislocations within the grains lie along the ⟨111⟩ directions. In iron cold-rolled by 60%, dislocations oriented along other directions may be found (Fig. 10b). When the degree of preliminary cold deformation is high, dislocations aligned exactly with the ⟨111⟩ directions (Fig. 10c) are difficult to find.

The main measurements were made at a temperature of 873 K and at stresses from about 255 to 115.5 MPa, which corresponds to the variation of the creep rate within four orders of magnitude from 10^{-4} to 10^{-8} s^{-1}. The creep activation energy was determined with the temperature being varied between 873 and 987 K.

Fig. 11 depicts the variation of lg $\dot{\varepsilon}$ as a function of stress σ at 873 K. It is seen that at high stresses specimens of iron with different values of L_{111}/L at a stress of $\approx$ 150 MP will have approximately the same steady-state creep rate.

At stresses smaller than 49 MPa the creep rates of specimens with the largest and smallest values of $\dot{\varepsilon}$ differ by about a factor of 10.

As is seen from Fig. 12, when $\sigma = 33.9$ MPa samples with $L_{111}/L \approx 0.1$ deform at a rate of 1.16×10^{-7} s^{-1} and fracture after 315 h. With increase in the relative rate of screw dislocations up to $L_{111}/L \approx 0.8$ the steady-creep rate decreases to $\dot{\varepsilon} \approx 1.2 \times 10^{-8}$ s^{-1}, and the time to fracture exceeds 1000 h.

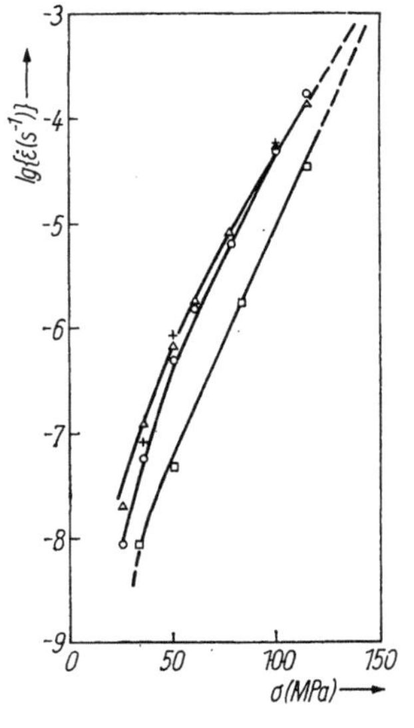

Fig. 11. lg $\dot{\varepsilon} = f(\sigma)$ dependence for armco iron with different values of $L_{111}/L = 0.1$ (+), 0.5 (○, △), 0.8 (□) [80]

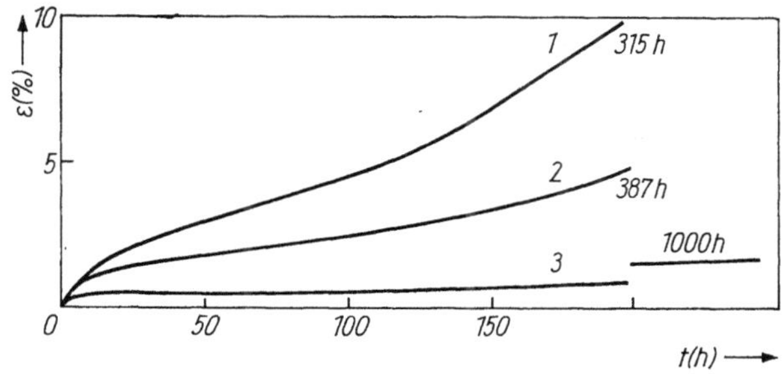

Fig. 12. Creep curves for armco iron specimens with different ratios $L_{111}/L = 0.1$ (1), 0.5 (2), (3) 0.8 [80]; $\dot{\varepsilon} = \approx 4 \times 10^{-2}$ (1), 2×10^{-2} (2) 10^{-3}%/h (3)

The conclusion may be drawn that the effect of screw dislocations on the behaviour of armco iron during creep manifests itself, chiefly, at low strain rates, when diffusion processes begin to play a significant part in the creep process. This phenomenon boils down into the decrease of the steady-state creep rate and the increase of the time to fracture with increasing relative length of screw dislocations.

The activation energy of creep in the 873 to 973 K range was estimated for two sorts of specimens, A and B. It has turned out that the activation energy for specimens A amounts to approximately 445 kJ/mol and for specimens B 315 kJ/mol. Thus, the activation energy of motion of screw dislocations is considerably higher than that of edge dislocations, which is equal to 273 kJ/mol [81].

For the time being, it would be premature to speak of a mechanism of plastic deformation during creep with screw dislocations prevailing in the structure. Further systematic research should apparently be conducted in this direction. It may only be stated that the dislocation structure (produced in materials) with the determined ratio of screw dislocations, L_{111}/L, has turned out to be sufficiently stable during further plastic deformation under creep conditions.

7. Conclusions

1. Dislocations and grain boundaries are sites of high diffusion mobility and are good point defect sinks. The ratio of the dislocation diffusion activation energy to the material volume activation energy Q_d/Q_v, varies from 0.3 to 0.8 and depends on the properties of the dislocations. The rate of diffusion decreases with decreasing Burgers vector and with increasing degree of dislocation splitting, i.e. with decreasing stacking fault energy γ. The rate of extended screw dislocation diffusion is approximately a factor of five lower than that of edge dislocation diffusion. In the case of grain boundary diffusion the energy ratio Q_g/Q_v varies between 0.2 and 0.7.

2. The dislocation diffusion processes determine the rate of creep under the deformation conditions when the governing mechanism is the climb of edge dislocations.

3. The creep rate and the time to fracture depend strongly on the magnitude of the stacking fault energy. When the stacking fault energy decreases by a factor of 10 the creep rate decreases by approximately two orders of magnitude, and the time to fracture by one order of magnitude.

4. The fracture under these deformation conditions arises from the formation of voids. A decrease in stacking fault energy is conductive to a more intensive void formation, which is accompanied by a decrease in the density of the material during creep and by the development of embrittlement.

5. An increase of the relative length of screw dislocations and of the quantity L_{111}/L under the conditions of creep which is controlled by the climb of edge dislocations is accompanied by a decrease of the creep rate and by an increase of the time to fracture.

References

[1] V. A. PAVLOV, Fizicheskie osnovy kholodnoi deformatsii OTsK metallov, Izd. Nauka, Moskva 1976.
[2] M. A. BOLSHANINA and V. A. PAVLOV, Zh. eksper. teor. Fiz. 7, 518 (1937).
[3] D. TURNBULL and R. E. HOFFMANN, Acta metall. 2, 419 (1954).
[4] G. LOVE and F. G. SHEWON, Acta metall. 11, 899 (1963).
[5] R. F. CANON and J. P. STARK, J. appl. Phys. 40, 4360, 4366 (1969).
[6] W. R. UPTHAGROVE and M. J. SINOFT, Trans. ASM 50, 1031 (1958).
[7] M. WUTTIG and H. K. BIRNBAUM, Phys. Rev. 247, 495 (1966).
[8] T. E. VOLIN, K. H. LIE, and R. W. BALLUFFI, Rep. No. 1336, Materials Science Centre, Cornell Univ., Ithaca (N.Y.) 1970.
[9] C. BEKER, M. WUTTIG, and H. K. BIRNBAUM, Trans. Japan. Inst. Metals, Suppl. 9, (1968).
[10] E. C. OREN and O. T. BAUER, Acta metall. 15, 773 (1967).
[11] R. W. BALLUFFI, phys. stat. sol. 42, 11 (1970).
[12] J. MIMKES, Phys. Rev. 9, 5320 (1974).
[13] J. G. E. M. BACKUS, H. BAKKER, and H. MEHRER, phys. stat. sol. (b) 64, 151 (1974).
[14] T. E. VILIN, K. H. LIE, and R. W. BALLUFFI, Acta metall. 19, 263 (1971).
[15] M. A. KRISHTAL and M. A. VYBOISCHIK, Fiz. Metallov i Metallovedenie 38, 356 (1974).
[16] YU. G. BORODKO, A. F. VYATKIN, and A. YU. RASULOV, Fiz. tverd. Tela 22, 3707 (1980).
[17] S. A. GOLOVIN, L. A. USTINOV, and D. M. LEVIN, Fiz. Metallov i Metallovedenie 40, 151 (1980).
[18] K. M. MILLER, K. W. JUGLE, and A. G. CROKER, Acta metall. 29, 1519 (1981).
[19] N. I. NOSKOVA, V. A. PAVLOV, and S. A. NEMNONOV, Fiz. Metallov i Metallovedenie 20, 920 (1965).
[20] J. WEERTMAN, Trans. AIME 233, 2069 (1965).
[21] D. GUPTA, J. appl. Phys. 43, 4455 (1973).
[22] A. GANGULEE and F. M. d'HAUVLE, Scripta metall. 7, 1027 (1973).
[23] S. M. KLOTSMAN, YA. RABARSKII, V. K. TALINSKII, and A. N. TIMOFEEV, Fiz. Metallov i Metallovedenie 28, 1025 (1969).
[24] A. HOSHINO and T. ARAKI, Trans. Nat. Res. Inst. Metals 13, 99 (1971).
[25] S. V. ZEMSKII, V. I. GREGORKIN, and L. N. MOSKALEVA, Khim. i Metallurgiya 13, 86 (1970).
[26] A. PAUL and R. P. AGARWAI, Metallurg. Trans. 2, 2691 (1971).
[27] V. G. BRIK, L. N. LARIKOV, and V. M. FILCHENKO, Ukr. fiz. Zh. 20, 397 (1975).
[28] R. A. PAUL, M. S. NAIK, and K. N. G. KAIMAL, J. nuclear Mater. (Amsterdam) 58, 205 (1975).
[29] N. N. DE RECE and C. A. PAMPILO, Scripta metall. 9, 1355 (1975).
[30] T. RASSO and C. SABATINE, Scripta metall. 6, 51 (1972).
[31] G. MARTIN, Acta metall. 23, 697 (1975).
[32] G. SCHWITZGEBEN, Acta metall. 20, 1297 (1972).
[33] G. BARREAU, G. BRNEL, and P. LACOMBE, C.R. Acad. Sci. (France), Sér. C 270, 516 (1970).
[34] J. C. M. HWANG and R. W. BALLUFFI, J. appl. Phys. 50, 1359 (1979).
[35] S. M. KLOTSMAN, A. N. TIMOFEEV, and I. SH. TRAKHTENBERG, Fiz. Metallov i Metallovedenie 23, 257 (1967).
[36] D. GUPTA and K. K. KIM, J. appl. Phys. 51, 120 (1980).
[37] A. N. ALESHIN, B. S. BOKSHTEIN, and L. S. SHVINDERMAN, Fiz. tverd. Tela 19, 3511 (1977).
[38] S. M. KLOTSMAN, Doctor theses, Sverdlovsk 1970.
[39] V. A. PAVLOV, N. I. NOSKOVA, and R. I. KUZNETSOV, Fiz. Metallov i Metallovedenie 24, 947 (1976).
[40] J. FRIEDEL, Dislokatsii, Izd. Mir, Moskva 1967 (p. 644).
[41] A. SEEGER, Mekhanizm skolzheniya i uprochneniya v kubicheskikh plotnoupakovannykh metallakh, Izd. in. Lit., Moskva 1960 (pp. 179 to 268).
[42] D. MAK-LIN, Tochechnye defekty i mekhanicheskiye svoistva metallov i splavov pri vysokikh temperaturakh, Vakansii i tochechnye defekty, Metallurgizdat, Moskva 1961 (pp. 197 to 248)
[43] C. R. BARRET and O. D. SHERBY, Trans. AIME 233, 1116 (1965).
[44] R. M. BONSTEEL and O. D. SHERBY, Acta metall. 14, 385 (1966).
[45] H. OIKAWA and S. KARESHIMA, Scripta metall. 5, 909 (1972).

[46] V. A. Pavlov and V. I. Shalaev, Legirovanie i svoistva zharoprochnykh materialov, Izd. Nauka, Moskva 1971 (p. 61 to 68).
[47] V. I. Shalaev, I. B. Tkachenko, and V. A. Pavlov, Fiz. Metallov i Metallovedenie **27**, 331 (1969).
[48] A. Orlova, Kovove Mater. 7, 242 (1969).
[49] G. Ya. Kozyrskii, N. P. Okrainets, and V. G. Pischak, Fiz. Metallov i Metallovedenie **34**, 607 (1972).
[50] A. I. Dekhtyar and V. A. Kononenko, Fiz. Metallov i Metallovedenie **46**, 612 (1978).
[51] R. R. Vandervort, Trans. AIME **242**, 345 (1968).
[52] R. R. Vandervort, Trans. AIME **245**, 2269 (1969).
[53] K. R. Wheeler, E. R. Gilbert, and F. L. Duran, Acta metall. **19**, 21 (1971).
[54] S. L. Robinson and O. D. Serby, Acta metall. **17**, 109 (1969).
[55] W. V. Green, Trans. AIME **233**, 1818 (1965).
[56] R. I. Kuznetsov, V. A. Pavlov, and B. T. Shmatov, Fiz. Metallov i Metallovedenie **24**, 345 (1967).
[57] V. I. Shalaev, R. I. Kuznetsov, and V. A. Pavlov, Fiz. Metallov i Metallovedenie **25**, 755 (1968).
[58] V. I. Shalaev, V. A. Pavlov, and I. B. Tkachenko, Fiz. Metallov i Metallovedenie **28**, 183 (1969).
[59] F. S. Mankman and N. I. Grant, Proc. Amer. Soc. Testing Mater. **56**, 593 (1956).
[60] S. N. Zhurkov and T. P. Sanfirova, Zh. eksper. teor. Fiz. **28**, 1719 (1958).
[61] N. Dzh. Grant, in: Atomnyi mekhanizm razrusheniya, Metallurgizdat, Moskva 1963 (pp. 25 to 29).
[62] V. S. Ivanova, Zavodsk. Labor. **2**, 212 (1955).
[63] E. S. Makhlin, see [61] (pp. 212 to 216).
[64] P. W. Davis and B. Wilshire, Rep. Symp. Iron and Steel Inst. and Inst. of Metals, May 3rd and 4th, 1961, London and Bradford (pp. 334 to 343).
[65] I. A. Oding, V. S. Ivanova, V. V. Burdukskii, and V. N. Geminov, Teoriya polzuchesti i dlitelnoi prochnosti metallov i splavov, Metallurgizdat, Moskva 1959.
[66] V. M. Rozenberg, Polzuchesti metallov, Izd. Metallurgiya, 1967.
[67] N. G. Weedman and T. Gladman, Metal Sci. J. **44**, 64 (1980).
[68] V. Sklandicka, I. Sax, and J. Cudec, Kovove Mater. **17**, 459 (1979).
[69] E. L. Svenson and G. L. Dunlop, Canad. metall. Quart. **18**, 39 (1979).
[70] J. B. Cane and C. I. Middleton, Metal Sci. J. **15**, 297 (1981).
[71] I. W. Chen and A. S. Argou, Acta metall. **29**, 3321 (1981).
[72] L. E. Svenson and G. L. Danpol, 5th Internat. Conf. Adv. Fract. Res. Proper., Froct Cones, 1981, Vol. 4, Oxford (Ca.) 1981 (p. 1627).
[73] V. Vitek, Metal Sci. J. **14**, 403 (1980).
[74] I. I. Minz, M. M. Steinberg, and N. G. Shulgin, Fiz. Metallov i Metallovedenie **50**, 832 (1980).
[75] A. Needleman and J. R. Rice, Acta metall. **28**, 1315 (1980).
[76] V. A. Pavlov, V. V. Melnikova, and N. L. Pechorkina, Fiz. Metallov i Metallovedenie **21**, 309 (1966).
[77] C. W. Chan and R. W. Buttri, Radiat. Eff. **56**, 219 (1981).
[78] C. P. Chan and C. W. Chen, Crystal Lattice Defects **7**, 91 (1977).
[79] N. L. Pechorkina, V. P. Ketova, and V. A. Pavlov, Fiz. Metallov i Metallovedenie **42**, 609 (1976).
[80] V. A. Pavlov, N. L. Pechorkina, V. I. Shalaev, and L. N. Lebedeva, Fiz. Metallov i Metallovedenie **44**, 1320 (1976).
[81] K. L. Murty, M. Gold, and A. L. Rouf, J. appl. Phys. **41**, 4917 (1970).

(Received February 13, 1984)

phys. stat. sol. (a) 85, 29 (1984)

Subject classification: 1.4; 10.1; 21; 21.1

Rijksuniversitair Centrum Antwerpen[1])

The Nature of Some Planar Defects in 2H Martensite of Cu–Al Alloys as Determined by HREM

By

F. C. Lovey[2]), G. van Tendeloo, and S. Amelinckx

The nature of some planar defects in 2H martensite of Cu–Al alloys is investigated by means of high resolution electron microscopy. It is found that the planar defects are parallel either to the {001} or {121} planes. They constitute the boundaries between translation variants related by a displacement vector $\boldsymbol{R} = \pm\boldsymbol{a}/3 + \boldsymbol{c}/2$. The defects parallel to the basal plane are more specifically described as single faults involving one violation of the hexagonal stacking rule. The possible relation of the planar defects with the growth process of the martensite is discussed.

Die Art einiger planarer Defekte in 2H-Martensit von Cu-Al-Legierungen wird mittels hochauflösender Elektronenmikroskopie untersucht. Es wird gefunden, daß die Planardefekte entweder parallel zur {001}- oder {121}-Ebene liegen. Sie bilden die Grenzen zwischen Translationsvarianten, die durch einen Verschiebungsvektor $\boldsymbol{R} = \pm\boldsymbol{a}/3 + \boldsymbol{c}/2$ verknüpft sind. Die parallel zur Basisebene liegenden Defekte werden spezifischer als einzelne Stapelfehler beschrieben, die die Verletzung der hexagonalen Stapelregel einschließen. Die mögliche Verknüpfung der Planardefekte mit dem Wachstumsprozeß von Martensit wird diskutiert.

1. Introduction

A 2H or γ' martensite can be obtained in Cu–Al alloys, in a composition range between ≈25 and ≈28 at% Al, by quenching from the β (DO_3) phase [1, 2].

The crystal structure was first determined by Kurdyumov et al. [1] from X-ray diffraction experiments. They found that the structure is basically h.c.p., having a superstructure with orthorhombic symmetry as shown in Fig. 1a. This result was later confirmed by serveral other authors from electron diffraction observations [2, 3].

The γ' martensite has been found to be finely internally twinned [2, 4], the twinning plane being of the {201} or {121} type. The matrix and twins of the martensite plates are further subdivided by planar defects causing striations in the electron microscopy images. Trace analysis indicated that these defects are parallel either to the twin plane {202}, to {122}, to the basal plane (001), or to the normal γ' twinning planes {201} or {121}. They were attributed to twins of higher order or to stacking faults [2].

In order to obtain more detailed information about the nature of these defects a high resolution electron microscopy study has been carried out. The results which are presented and discussed here show that the 2H martensite plates are subdivided into translation variants whose boundaries can account for most of the observed striations.

---

[1]) Groenenborgerlaan 171, B-2020 Antwerp, Belgium.

[2]) On leave from Centro Atómico Bariloche, 8400-S.C. de Bariloche, Argentine; this co-author is indebted to the Consejo Nacional de Investigaciones Cientificas y Tecnicas of Republica Argentina for financial support.

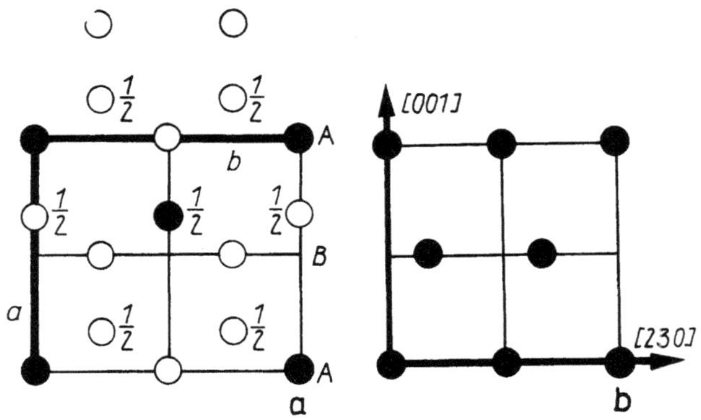

Fig. 1. a) The orthorhombic unit cell of 2H martensite projected on the (001) plane. The lattice parameters are $a = 0.451$, $b = 0.520$, and $c = 0.422$ nm. Open and full circles represent Cu and Al, respectively. b) Projection of the 2H structure along the [3̄10] zone axes

2. Experimental Procedure

A Cu–27 at% Al alloy was prepared by melting in an argon arc furnace. The obtained button was homogenized by annealing at 900 °C for 24 h in an evacuated quartz tube. The martensitic transformation was produced by quenching in ice water. Thin foils suitable for electron microscopy were prepared by jet polishing with a mixture of 50% water, 25% phosphoric acid, and 25% methanol followed by a final polishing in a saturated solution of CrO_3 and phosphoric acid at 1.5 V.

The specimens were investigated in a Jeol 200 CX high resolution electron microscope.

3. Observations and Results

Most of the striations are found to be caused by planar defects observed edge-on along the [21̄0] or [210] zone axis. Fig. 2 shows a bright field image in which one of the two twinned variants is observed along a direction close to the [21̄0] zone axis. Planar defects are parallel to the basal plane of the 2H structure. Some of these defects associated with the basal plane are intercepted by other defects on (121) and (1̄2̄1) planes.

A high resolution image taken in a thin region of the foil containing planar defects parallel to the basal plane is shown in Fig. 3. The defocus (≈ -140 nm) was chosen in order to obtain images exhibiting white dots arranged as the atom column projection along the [21̄0] direction (see Fig. 1b). An enlarged image of the upper planar defect in Fig. 3 is reproduced in Fig. 4.

In order to establish the exact relationship between the white dots in the experimental images and the real atom positions, a full computer image simulation was performed. Using the real space method [5] and the model shown in Fig. 1b high resolution images were simulated for a range of crystal thicknesses up to 30 nm and defocus values ranging from -45 to -165 nm. It was found that for any thickness there is always a range of defocus values (of approximately 10 nm width) in which the atom columns are imaged as white dots. This range is slightly shifted from one thickness to another. Nevertheless, all ranges are in the vicinity of (or include) the defocus value of -145 nm. One of these simulated images is shown as an inset in Fig. 4.

Useful high resolution information about the {121} type defects (see Fig. 2 for a low magnification) is represented in Fig. 5 for a somewhat thicker area. Two planar defects parallel with the basal plane are linked by another defect having its trace along (121).

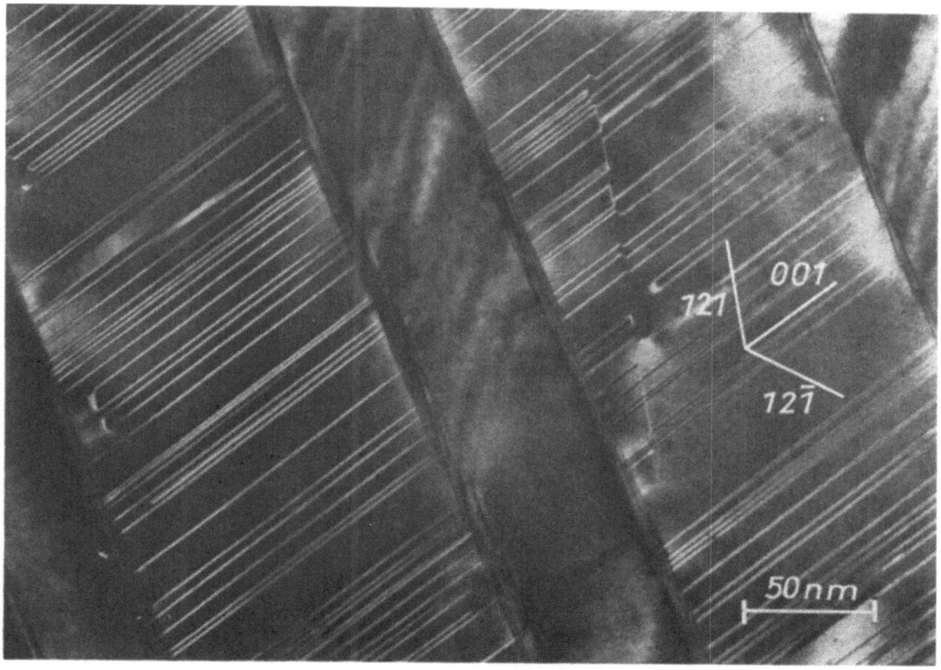

Fig. 2. Microstructure of 2H martensite plates along ≈ [210] direction. Bright field image

Since ordering is not revealed in this projection the specimen was tilted in order to investigate the possible relation between the APB's and the planar defects described above. It was confirmed, by means of conventional dark field imaging using some superlattice reflections, that the planar defects under study are not APB's, in agreement with the results of Swann and Warlimont [2]. Thus the ordering can be ignored for the analysis of the planar defects and the 2H martensite can be considered as having a simple hexagonal close-packed structure. The stacking sequence can be directly read from the images (see e.g. Fig. 3 to 5). The 2H martensite contains a number of basal plane stacking faults, i.e. interfaces between different stacking variants of the hexagonal structure such as ABAB ..., ACAC ..., BCBC ..., etc. They have been indicated in the photographs.

The defect associated with the {121} type of plane (see Fig. 5) constitutes the boundary between two differently stacked hexagonal variants, i.e. BCBC ... on the left and BABA ... on the right; it is therefore also a translation interface. In this case the B-plane continues unchanged through the interface, while A changes into C.

Other striations mentioned by Swann and Warlimont [2] parallel with the traces of the {202}, {122}, and {201} planes are not associated with the defects under consideration; they are related to differently oriented martensite variants probably having a twin relationship. We will restrict ourselves to the analysis of the different stacking variants within a single martensite orientation variant.

4. Structural Considerations

From a geometrical point of view the martensitic transformation in Cu–Al alloys can be described as a Bain-type transformation of the $\{110\}_{DO_3}$ planes into close-packed planes followed by a reshuffling of the planes to obtain the required stacking sequence.

For any product phase showing a symmetry relationship with the parent phase one can deduce the number of translation variants as well as the number of orientation variants from space group symmetry considerations [6]. In the present context we are only interested in the different translation variants and we will address the problem how to deduce the different stacking variants as well as antiphase related variants from a given 2H martensite (Fig. 1).

Ignoring the ordering for a moment, three stacking positions A, B, and C are possible the respective planes are depicted in Fig. 6. Denoting by $T_{a/3}$ and $T_{-a/3}$ the operators

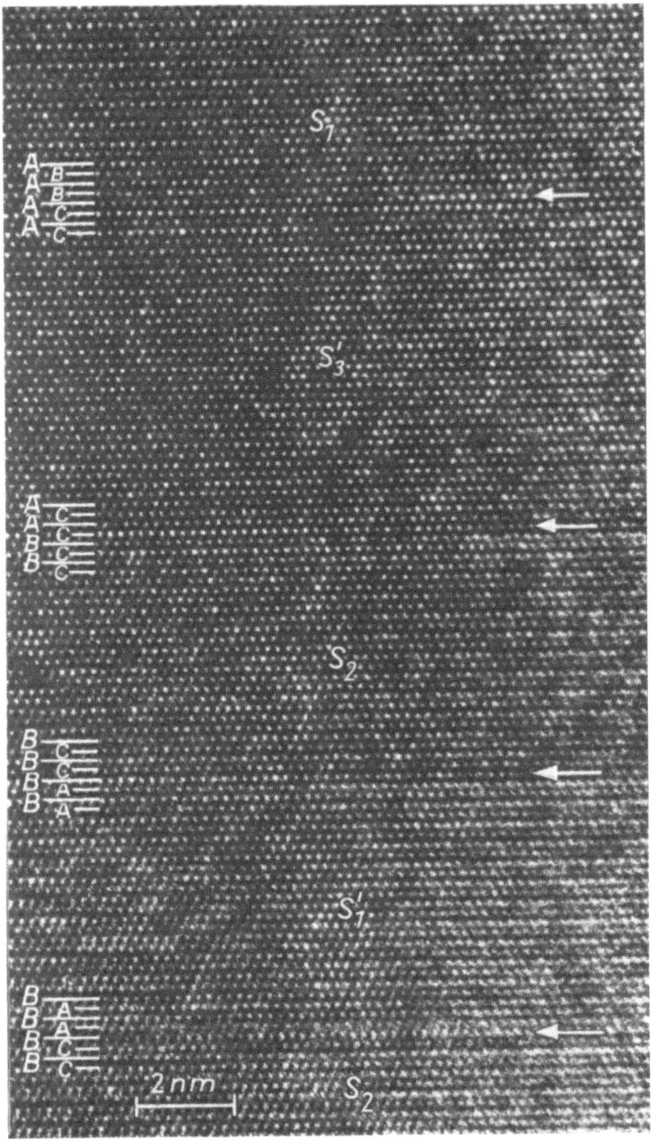

Fig. 3. High resolution image along the [2̄10] direction exhibiting some planar defects parallel to the basal plane indicated by arrows. Five different translation variants are present

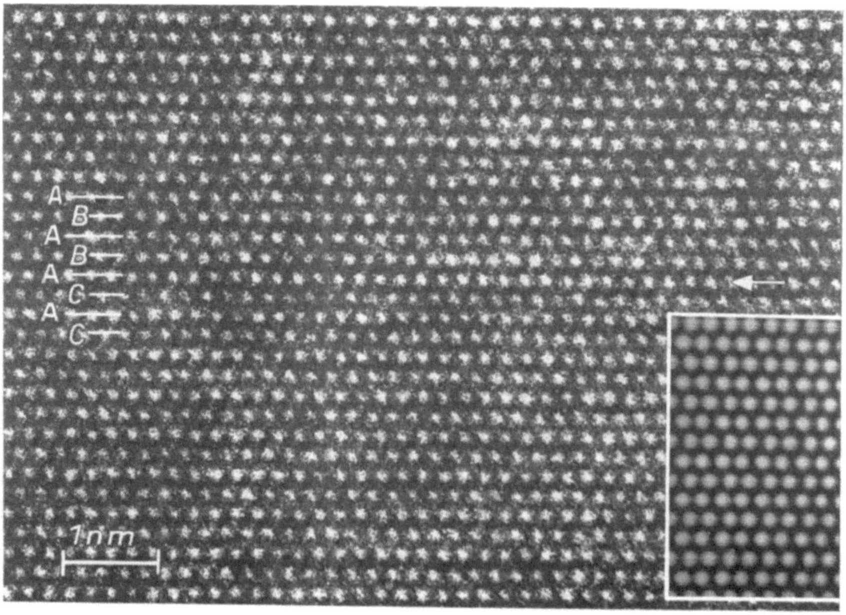

Fig. 4. Enlarged image of the upper planar defect observed in Fig. 3. The inserted computer simulated image corresponds to a thickness of 30 nm and a defocus value of −142.5 nm

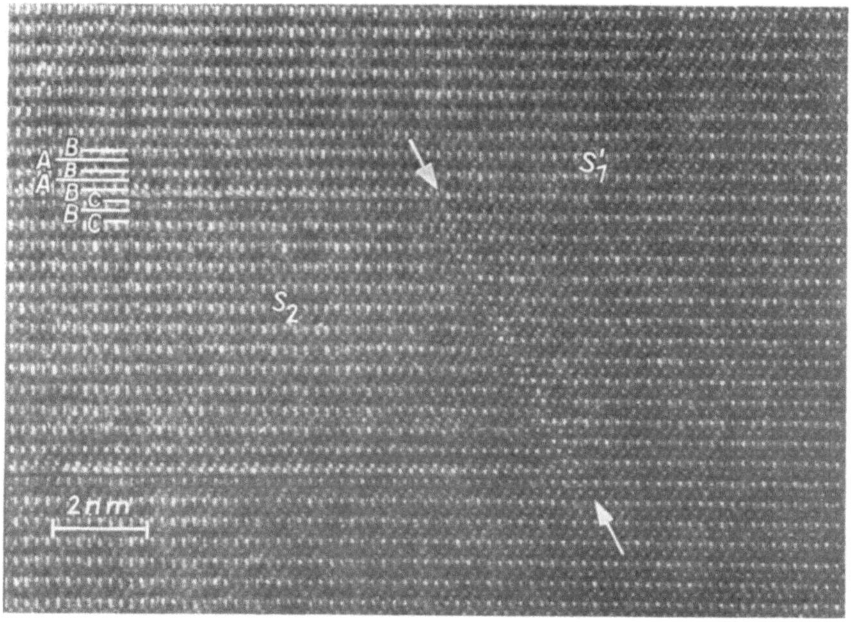

Fig. 5. High resolution image along the [2̄10] direction showing two defects in basal planes linked by a planar defect parallel to the (121) plane. See arrows

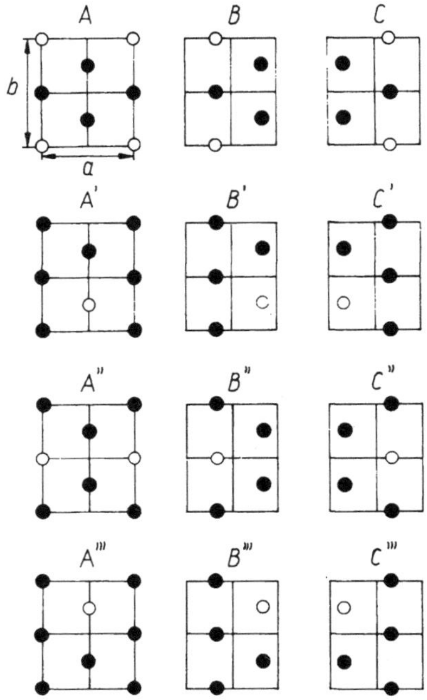

Fig. 6. Orthorhombic superstructure. Atomic arrangements of the different (001) planes. Full circles correspond to copper atoms and open circles to aluminum atoms

associated with the respective translations $\boldsymbol{R}_1 = \boldsymbol{a}/3$ and $\boldsymbol{R}_2 = -\boldsymbol{a}/3 = 2\boldsymbol{a}/3$ we can consider the group

$$\mathrm{T_S} = \{E,\ T_{a/3},\ T_{-a/3}\} \tag{1}$$

which, starting from an initial variant $S_1 \equiv$ ABAB ... will generate the stacking variants $S_2 \equiv$ BCBC ... and $S_3 \equiv$ CACA ... by a pure shear mechanism.

If we furthermore want to describe identical stacking variants but displaced over $\boldsymbol{R}_3 = \boldsymbol{c}/2$ (i.e. insertion of one extra plane) such as e.g.

$$\underbrace{\mathrm{AB}}\ \underbrace{\mathrm{AB}}\ \mathrm{C}\ \underbrace{\mathrm{AB}}\ \underbrace{\mathrm{AB}}$$

we will have to introduce the group of operators

$$\mathrm{T_c} = \{E,\ T_{c/2}\} \tag{2}$$

which will allow to transform an initial variant $S_1 \equiv$ ABAB ... into $S_1' \equiv$ BABA ...

The complete set of stacking variants starting from S_1 can be obtained by applying the set of operators included in the stacking variant generating group SVGG which is the direct product of $\mathrm{T_S}$ and $\mathrm{T_c}$,

$$\mathrm{SVGG} = \mathrm{T_S} \otimes \mathrm{T_c}\,. \tag{3}$$

It contains six elements given by

$$\begin{aligned} &S_1 \equiv \mathrm{ABAB}\ \dots, && S_1' \equiv \mathrm{BABA}\ \dots, \\ &S_2 \equiv \mathrm{BCBC}\ \dots, && S_2' \equiv \mathrm{CBCB}\ \dots, \\ &S_3 \equiv \mathrm{CACA}\ \dots, && S_3' \equiv \mathrm{ACAC}\ \dots\,. \end{aligned} \tag{4}$$

The relation between these variants can be briefly expressed as

$$\begin{aligned} &T_{c/2} S_i = S_i'\,, \\ &T_{\pm a/3} S_i = S_j\,, \\ &T_{c/2} T_{\pm a/3} S_i = S_j'\,. \end{aligned} \tag{5}$$

Taking into account the orthorhombic superstructure each stacking variant in (4) will split into a number of antiphase related translation variants. Their number equals the volume of the ordered primitive cell divided by the volume of the disordered primitive cell or

$$n = \det |M|\,,$$

where M is the transformation matrix between both primitive cells [6]. In the present case

$$M = \begin{pmatrix} 2 & 1 & 0 \\ 0 & 2 & 0 \\ 0 & 0 & 1 \end{pmatrix}$$

so that $n = \det |M| = 4$.

The four translation variants are interrelated by the symmetry translation of the h.c.p. structure which were lost on ordering. The set of four translation operators can be chosen to form a group

$$\mathrm{T_{APB}} = \{E, T_R, T_R^2, T_R^3\} \tag{6}$$

which correspond with translations of 0, $\boldsymbol{R} = \frac{1}{4}[2\boldsymbol{a} + \boldsymbol{b}]$, $2\boldsymbol{R}$ and $3\boldsymbol{R} \equiv -\boldsymbol{R}$, respectively. These four translation variants for one plane are depicted in Fig. 6 as projected along [001].

In the ordered phase each Al atom is surrounded by six Cu. So the initial variant can be taken, for example, as AB″AB″ ..., then by applying the elements of the group $\mathrm{T_{APB}}$ given by (6) the other antiphase related variants will be A′B‴A′B‴ ... , A″BA″B ..., and A‴B′A‴B′ ...

The complete set of all possible translation variants is obtained by applying the operators belonging to the direct product

$$\mathrm{SVGG'} = \mathrm{T_{APB}} \otimes \mathrm{T_S} \otimes \mathrm{T_c} \,. \tag{7}$$

This group contains 24 elements and determines the relationship between the 24 translation variants of 2H Cu–Al martensite.

5. Discussion of the Results

From the experimental observations it was concluded that the defects under discussion are certainly no antiphase boundaries and we can limit ourselves to the six possible different stacking variants S_1, S_2, S_3 and S_1', S_2', S_3' given by (4).

Using this notation the sequence observed in Fig. 3 and 5 (from top to bottom) can be written as $S_1S_3'S_2S_1'S_2$ and $S_1'S_2$, respectively. Comparing with (5) it is immediately clear that all pairs of adjacent variants are related by a combined translation operation $T_{c/2}T_{\pm a/3}$. No pairs of translation variants related by a single shear $T_{\pm a/3}$ nor by a translation $T_{c/2}$ have been found after observations in several martensite plates.

The translation variants are separated by boundaries parallel with the basal plane, i.e. stacking faults, and boundaries parallel with the {121} planes.

Stacking faults in close-packed structures have been earlier described by Amelinckx [7] in terms of violations with respect to the hexagonal sequence. For example, the boundary between variants related by a combined translation $T_{c/2}T_{\pm a/3}$ (extraction of one plane followed by a shear of the close-packed planes) can be described as a single fault involving only one violation of the hexagonal stacking rule,

ABAB$\underline{\mathrm{AB}}$CBCB ...

On the other hand, the vicinity of a boundary between two variants related by a single shear $T_{\pm a/3}$ consists of a double fault containing two violations of the stacking rule,

ABABABCACACA ...

Finally a triple fault containing three violations of the stacking rule is found in the vicinity of the boundary between two subvariants related by a translation $T_{c/2}$,

ABABABCABABAB ...

In this case one extra plane (C) is inserted into the AB sequence.

So far we have restricted ourselves to defects parallel with the basal plane and we have shown that they consist of a single violation in the hexagonal stacking rule. They constitute the main part of the boundaries between two variants and are related by a combined translation $T_{c/2}T_{\pm a/3}$. The other type of interfaces parallel with the $\{121\}$ type of plane which separate the same pair of translation variants, can also be analysed. In this type of boundary, for example, an B-plane crosses the interface unchanged while A changes into C as shown in Fig. 5. Since the A- and C-planes are shifted one with respect to the other by a pure shear translation $T_{a/3}$ the close-packed structure cannot be retained when they join along the boundary. This mismatch can be reduced if atom displacements take place around the defect. The sense and magnitude of the atom shifts will be very complex and they can involve several atom planes in the vicinity of the interface as can be judged from the width of the boundary observed in the high resolution image of Fig. 5. This kind of defect can still be classified under the translation defects since there is a well-defined translation vector $\boldsymbol{R} = \boldsymbol{a}/3 + \boldsymbol{c}/2$ relating the two crystal parts in either side of the interface; the boundary region itself is smeared out over several atom planes.

The presence of translation variants in a martensite plate is probably related to the growth process of the martensite. Two possible mechanisms can be considered: (i) The translation variants (or stacking faults) appear during the martensite growth by the introduction of layers in wrong positions, following a mechanism similar to that proposed by Andrade et al. [8] for the stacking fault formation in 18R Cu–Zn–Al martensite on the basis of Ahlers' model [9]. (ii) The shape deformation associated with a growing martensite plate acts on the surrounding b.c.c. phase producing anticipated nucleations (of the same orientation variant) just ahead of the advancing front. Different nucleated regions will be in general out of phase with respect to one another and translation variants are generated. This mechanism has been recently proposed [10] to explain the presence of domains with different stacking sequence in 18R martensite in Cu–Al alloys.

Two possible conditions can determine the only (or preferential) occurrence of a pair of adjacent variants related by a translation $T_{c/2}T_{\pm a/3}$. First, it has been suggested by Sato et al. [3] that the 2H type of stacking is the most stable one for the martensitic phase around the alloy composition under study. From this point of view it can be assumed that the boundary between variants related by a translation $T_{c/2}T_{\pm a/3}$ will have the lower energy since it involves only one infraction against the hexagonal stacking. Second, another reason could be related to the habit plane requirement during the transformation. It has been shown by Lovey et al. [11], using the phenomenological theory of Wechsler et al. [12], that a continuous set of differently oriented low-distorted interfaces between the DO_3 matrix and one variant of martensite can be obtained by gradually changing the stacking sequence from 18R to 2H. The situation, however, appears more complex in the present case since the macroscopic habit plane is determined by the ratio of the thicknesses of the two twin orientations [13] rather than from the interface of the individual plates of 2H with the DO_3 matrix. The selection of the variants related by a translation $T_{c/2}T_{\pm a/3}$ would result from the combined conditions for a low-distorted interface of the individual plates with the matrix (determined by the density of faults with respect to the hexagonal sequence) and the twin thickness ratio in order to obtain a macroscopic, undistorted habit plane.

6. Conclusions

The planar defects in the 2H martensite parallel with the basal plane and the {121} type of planes were found to be the boundaries between two translation variants related by a displacement vector $\boldsymbol{R} = \pm \boldsymbol{a}/3 + \boldsymbol{c}/2$.

The boundaries parallel with the basal plane contain a single infraction against the hexagonal stacking rule, i.e. ... ABABCBCB ...; they constitute single faults.

The boundaries parallel with the {121} type of planes involve extra atom displacements but they can still be described as translation boundaries.

The fact that only adjacent variants related by a translation $\boldsymbol{R} = \pm \boldsymbol{a}/3 + \boldsymbol{c}/2$ occur is probably due to a combination of the lower energy of this type of defect and the habit plane requirement during the transformation.

References

[1] G. V. Kurdyumov, V. Miretskii, and T. Stelletskaya, Soviet Phys. – J. tech. Phys. 8 1959 (1938).
[2] P. R. Swann and H. Warlimont, Acta metall. **11**, 511 (1963).
[3] H. Sato, R. S. Toth, and G. Honjo, Acta metall. **15**, 1381 (1967).
[4] H. Warlimont and L. Delaey, Progr. Mater. Sci. **18**, 1 (1974).
[5] D. van Dyck, J. Microscopy **119**, 141 (1980).
D. van Dyck and W. Coene, submitted to Ultramicroscopy (1984).
W. Coene and D. van Dyck, submitted to Ultramicroscopy (1984).
[6] G. van Tendeloo and S. Amelinckx, Acta cryst. **A30**, 431 (1974).
[7] S. Amelinckx, Dislocation in Solids, Vo. 2, Ed. F. Nabarro, North-Holland Publ., Co., Amsterdam 1979 (p. 87).
[8] M. Andrade, M. Chandrasekaran, and L. Delaey, Acta metall., to be published (1984).
[9] M. Ahlers, Z. Metallk. **65**, 636 (1974).
[10] F. C. Lovey, G. van Tendeloo, J. van Landuyt, L. Delaey, and S. Amelinckx, submitted to Acta metall. (1984).
[11] F. C. Lovey, M. Chandrasekaran, and M. Ahlers, Z. Metallk. **72**, 43 (1981).
[12] M. S. Wechsler, D. S. Lieberman, and T. A. Read, Trans. AIME **197**, 1503 (1953).
[13] J. K. Mackenzie and J. S. Bowles, Acta metall. **5**, 137 (1957).

(Received June 5, 1984)

phys. stat. sol. (a) 85, 39 (1984)

Subject classification: 2; 22.5

Department of Physics, Malaviya Regional Engineering College, Jaipur[1])

Partial Higher-Order Structure Factors in the Long-Wavelength Limit for Molten Cuprous Chloride

By

R. V. SHARMA

A method is developed for the calculation of the partial triplet and higher-order structure factors for molten halides in the long-wavelength limit. These functions are calculated for the molten cuprous chloride. The magnitudes of these functions are larger than those obtained for liquid metals by at least one or two orders which indicates that the multiparticle correlation is stronger in electrolytes.

Es wird eine Methode der Berechnung des partiellen Triplet-Strukturfaktors und der Strukturfaktoren höherer Ordnung von geschmolzenen Halogeniden im Grenzfall langer Wellen entwickelt. Diese Funktionen werden für geschmolzenes Kupferchlorid berechnet. Die Werte dieser Funktionen sind um mindestens ein oder zwei Größenordnungen höher als die für flüssige Metalle und zeigen, daß Vielteilchenkorrelation in Elektrolyten stärker ist.

1. Introduction

Ballentine and Lakshmi [1] have recently made a formalism which enables them to introduce the n-th-order structure factor for a single component system defined as the ensemble average of a product of n Fourier components of the atomic density. More recently Sharma and Sharma [2] have generalized Ballentine and Lakshmi's results in the case of liquid metals.

Parrinello and Tosi [3] have independently introduced a function related to the third-order partial structure factor, in order to discuss the dependence of the partial structure factors in a liquid binary alloy on concentration and density in the long-wavelength limit.

In fact, over the past few years there has been a renewed interest in the properties of the higher-order correlation functions and their Fourier transforms which although not amenable to direct experimental determination, play a very important role, not only in the development of liquid structure theories, but also in the evaluation of macroscopic properties of a liquid [4]. Furthermore, the long-wavelength limits of the higher-order structure factors, in all their wave vectors, are related to derivations of the number density with respect to the fugacity (at constant volume and temperature), a result originally due to Yvon [5].

More recently, Gray and Silbert [6] have generalized the results to multicomponent systems. They used the method of functional differentiation which has the advantage amongst others, of spelling out the relationship between the n-th-order structure functions and the corresponding n-th-order correlation functions in a proper statistical mechanical framework.

Here we generalize the ideas of Ballentine and Lakshmi to the higher-order partial structure functions for molten halides in the long-wavelength limits. In order to calculate the partial higher-order structure functions, we take the compressibility from

[1]) Jaipur 302017, India.

the dense gas concepts of molten halides [7]. The calculations are made for the partial higher-order structure factors for molten cuprous chloride.

2. Theoretical Framework

The equilibrium structure of a molten salt [8] may be described in terms of the three partial radial (pair) correlation functions: $g_{++}(r)$ or $g_{11}(r)$, $g_{--}(r)$ or $g_{22}(r)$, and $g_{+-}(r)$ or $g_{12}(r)$. Hansen and McDonald [8] discussed the problem in terms of only two correlation functions as

$$g_1(r) = \tfrac{1}{2}\,[g_{11}(r) + g_{22}(r)]\,, \tag{1}$$

$$g_2(r) = g_{12}(r)\,; \tag{2}$$

corresponding to $g_1(r)$ and $g_2(r)$ we have two total correlation functions, $h_1(r)$ and $h_2(r)$, two direct correlation functions $C_1(r)$ and $C_2(r)$, and two structure factors:

$$\begin{aligned} S_1(k) &= 1 + \tfrac{1}{2}\, nh_1(k) \\ &= \tfrac{1}{2}\,[S_{11}(k) + S_{22}(k)] \end{aligned} \tag{3}$$

(for details see Appendix) and

$$S_2(k) = S_{12}(k)\,. \tag{4}$$

The charge neutrality [9] condition requires that

$$\lim_{k\to 0} S_1(k) = \lim_{k\to 0} S_2(k) \tag{5}$$

or

$$\lim_{k\to 0} \tfrac{1}{2}\,[S_{11}(k) + S_{22}(k)] = \lim_{k\to 0} S_{12}(k)\,. \tag{6}$$

Fluctuations in density are described in k-space by the structure factor $S(k)$, given by

$$S(k) = S_1(k) + S_2(k)\,. \tag{7}$$

The long-wavelength limits are determined by

$$\begin{aligned} \lim_{k\to 0} S(k) &= \lim_{k\to 0}\,[S_1(k) + S_2(k)] \\ &= nk_{\mathrm{B}}T\chi_T\,. \end{aligned} \tag{8}$$

Now from (5), (6), and (8) we get

$$S_{11}^{(2)}(0) + S_{22}^{(2)}(0) = nk_{\mathrm{B}}T\chi_T \tag{9}$$

and

$$S_{12}^{(2)}(0) = \tfrac{1}{2}\, nk_{\mathrm{B}}T\chi_T\,. \tag{10}$$

The superscript (2) simply corresponds to the order of the correlation function. $S_{11}(k)$, $S_{22}(k)$, and $S_{12}(k)$ are the structure factors corresponding to $g_{11}(r)$, $g_{22}(r)$, and $g_{12}(r)$ and $S_{11}^{(2)}(0)$, $S_{22}^{(2)}(0)$, and $S_{12}^{(2)}(0)$ are the values of $S_{11}(k)$, $S_{22}(k)$, and $S_{12}(k)$ in the long-wavelength limits, respectively.

In order to evaluate $S_{11}^{(2)}(0)$, $S_{22}^{(2)}(0)$, and $S_{12}^{(2)}(0)$ we have adopted a simple procedure. In a recent publication Sharma and Sharma [10] have derived an expression for the structure factor in the square well system, in which the square well (SW) potential has been treated as a perturbation on the hard-sphere potential for liquid metals. Further the calculation of the partial static structure factors of molten cuprous chloride has been reported [11] in which the expression given by Powles [12] for a molecular liquid has been utilized. According to his argument the cuprous chloride could be treated as a molecular liquid with the chlorine atom as the molecular centre. The object of this paper is to calculate the partial higher-order structure factors of molten cuprous chloride using SW potential. The potential parameters of SW λ and ε which represent breadth and depth, respectively, are taken from the earlier calculations [13] where these parameters were determined from the chlorine–chlorine structure factor [$S_{\text{Cl-Cl}}(k)$] by comparing its first peak with the experimental data. The expression of $S_{\text{Cl-Cl}}(k)$ is given by

$$S_{\text{Cl-Cl}}(k) = \frac{1}{1 + C(k)}. \tag{11}$$

The $C(k)$ in the SW potential is given by

$$\begin{aligned} C(k) = &-\frac{24\eta}{(k\sigma)^6}\Big\{\alpha(k\sigma)^3 [\sin k\sigma - k\sigma \cos k\sigma] + \\ &+ \beta(k\sigma)^2 [2k\sigma \sin k\sigma - (k^2\sigma^2 - 2) \cos k\sigma - 2] + \\ &+ \gamma[(4k^3\sigma^3 - 24k\sigma) \sin k\sigma - (k^4\sigma^4 - 12k^2\sigma^2 + 24) \cos k\sigma + 24] - \\ &- \frac{\varepsilon}{k_B T}(k\sigma)^3 [\sin \lambda k\sigma - \lambda k\sigma \cos \lambda k\sigma + k\sigma \cos k\sigma - \sin k\sigma]\Big\}, \end{aligned} \tag{12}$$

where α, β, and γ are given in [10]. In the long-wavelength limits $S_{\text{Cl-Cl}}(0)$ or $S_{22}^{(2)}(0)$ is reduced to

$$S_{22}^{(2)}(0) = \left[\frac{1 + 4\eta + 4\eta^2 - 4\eta^3 + \eta^4}{(1 - \eta)^4} - \frac{8\eta\varepsilon}{k_B T}(\lambda^3 - 1)\right]^{-1}. \tag{13}$$

For the compressibility of the system we use the dense gas concepts [7] which yielded results of compressibility and critical constants, etc. in reasonable agreement with the experimental data for molten alkali halides. According to these concepts the equation of state may be written as

$$\frac{PV}{Nk_B T} = \frac{P_h V}{Nk_B T} - A\left(\frac{V_m}{V}\right)^{1/3} \frac{k_B T_m}{k_B T}, \tag{14}$$

where V is the molar volume and T the temperature (K). P is the total pressure, P_h corresponds to the pressure of a hard-sphere system which is given by the Carnahan and Starling (CS) equation of state [14].

Differentiating (14) and using the CS equation of state we get the equation for the compressibility as

$$nk_B T\chi_T = \left[\frac{1 + 4\eta + 4\eta^2 - 4\eta^3 + \eta^4}{(1 - \eta)^4} - \frac{B\eta^{1/3}}{T}\right]^{-1}. \tag{15}$$

It yields, using (10),

$$S_{12}^{(2)}(0) = \frac{1}{2}\left[\frac{1 + 4\eta + 4\eta^2 - 4\eta^3 + \eta^4}{(1 - \eta)^4} - \frac{B\eta^{1/3}}{T}\right]^{-1}, \tag{16}$$

where the constant $B = (4/3)\, AT_m/\eta_m^{1/3}$ and A is given in [7]. The packing fraction η is related to the hard-sphere diameter σ by $\eta = \pi n\sigma^3/6$ where n is the number density of the molecules. The subscript m denotes the melting point. Combining (9), (13), and (15) we get

$$\begin{aligned} S_{11}^{(2)}(0) = &\left[\frac{1 + 4\eta + 4\eta^2 - 4\eta^3 + \eta^4}{(1 - \eta)^4} - \frac{B}{T}\eta^{1/3}\right]^{-1} - \\ &- \left[\frac{1 + 4\eta + 4\eta^2 - 4\eta^3 + \eta^4}{(1 - \eta)^4} - \frac{8\eta\varepsilon(\lambda^3 - 1)}{k_B T}\right]^{-1}. \end{aligned} \tag{17}$$

The expression [1] which is used to evaluate the third-order and higher-order partial structure factors in the long-wavelength limit is

$$S^{(n)}(\boldsymbol{k}_1, \dots, \boldsymbol{k}_{n-1}, 0) = nk_BT \frac{\partial}{\partial P} \varrho^{(n-1)}(\boldsymbol{k}_1, \dots, \boldsymbol{k}_{n-1}) + \\ + S^{(2)}(0)\, S^{(n-1)}(\boldsymbol{k}_1 \dots \boldsymbol{k}_{n-1}) . \tag{18}$$

The derivatives of $S^{(2)}(0)$ and $S^{(3)}(0, 0, 0)$ with respect to pressure are governed by [1]

$$nk_BT \frac{\partial f(\eta)}{\partial P} = \eta S^{(2)}(0) \frac{\partial f(\eta)}{\partial \eta} . \tag{19}$$

3. Results and Discussion

The calculated values of $S_{11}^{(2)}(0)$, $S_{22}^{(2)}(0)$, $S_{12}^{(2)}(0)$, $S_{11}^{(3)}(0, 0, 0)$, $S_{22}^{(3)}(0, 0, 0)$, $S_{12}^{(3)}(0, 0, 0)$, $S_{11}^{(4)}(0, 0, 0, 0)$, $S_{22}^{(4)}(0, 0, 0, 0)$, and $S_{12}^{(4)}(0, 0, 0, 0)$ for molten cuprous chloride at the

Table 1

Partial higher order structure factors in the along wavelength limit for molten cuprous chloride at T_m

	$S_{11}(0)$	$S_{22}(0)$	$S_{12}(0)$
$S^{(2)}(0)$	-2.594×10^{-2}	6.866×10^{-2}	2.136×10^{-2}
$S^{(3)}(0, 0, 0)$	-8.846×10^{-3}	-39.919×10^{-3}	-2.567×10^{-3}
$S^{(4)}(0, 0, 0, 0)$	-74.411×10^{-4}	604.943×10^{-4}	7.633×10^{-4}

melting point 703 K are given in Table 1. The parameters used for the calculation are $\eta = 0.45$, $\varepsilon/k_B = 650$ K, and $\lambda = 1.95$. To the author's knowledge there are no experimental data available to judge the approximations used in this paper. However, the results are larger than those obtained for a monatomic liquid [2] $^2(H_g)$ at least by one or two orders of magnitude, which indicates that the multiparticle correlations are greater in molten electrolytes. It may be attributed due to the strong long-range electrostatic interaction and such a result is not unexpected. We anticipate that the partial structure factors for molten halides may play an important role in understanding the physical properties of the system and the data become available in near future. Once such calculations in the long-wavelength limit are verified, it will provide a way to investigate the wave-vector dependence of these higher-order partial structure factors.

Appendix

Relation with the partial static structure functions

$$\begin{aligned} S_1(k) &= 1 + \tfrac{1}{2}\, nh_1(k) \\ &= 1 + \tfrac{1}{2}\, n \int_V e^{i\boldsymbol{k}\cdot\boldsymbol{r}} [g_1(r) - 1]\, d\boldsymbol{r} \\ &= 1 + \tfrac{1}{2}\, n \int_V e^{i\boldsymbol{k}\cdot\boldsymbol{r}} \{\tfrac{1}{2} [g_{11}(r) + g_{22}(r)] - 1\}\, d\boldsymbol{r} \\ &= \tfrac{1}{2} [(1 + \tfrac{1}{2}\, n \int_V e^{i\boldsymbol{k}\cdot\boldsymbol{r}} \{g_{11}(r) - 1\}\, d\boldsymbol{r}) + (1 + \tfrac{1}{2}\, n \int_V e^{i\boldsymbol{k}\cdot\boldsymbol{r}} \{g_{22}(r) - 1\}\, d\boldsymbol{r})] \\ &= \tfrac{1}{2} [\{1 + \tfrac{1}{2}\, nh_{11}(k)\} + \{1 + \tfrac{1}{2}\, nh_{22}(k)\}] \\ &= \tfrac{1}{2} [S_{11}(k) + S_{22}(k)] . \end{aligned}$$

References

[1] L. E. Ballentine and A. Lakshmi, Canad. J. Phys. **53**, 372 (1975).
[2] R. V. Sharma and K. C. Sharma, Chem. Phys. Letters **44**, 253 (1976).
[3] M. Parrinello and M. P. Tosi, Nuovo Cimento **25B**, 242 (1975).
[4] P. A. Egelstaff, Ann. Rev. Phys. Chem. **24**, 159 (1973) and references therein.
[5] J. Yvon, Correlations and Entropy in Classical Statistical Mechanics, Pergamon Press, Oxford 1969.
[6] P. Gray and M. Silbert, Phys. Letters A **56**, 192 (1976).
[7] R. V. Sharma and K. C. Sharma, Indian J. Phys. **49**, 477 (1975); **52B**, 340 (1978).
[8] J. P. Hansen and I. R. McDonald, Phys. Rev. A **11**, 2111 (1975).
[9] F. H. Stillinger and R. Lovett, J. chem. Phys. **49**, 1991 (1975).
[10] R. V. Sharma and K. C. Sharma, Physica (Utrecht) **89A**, 213 (1977).
[11] R. V. Sharma and K. C. Sharma, phys. stat. sol. (b) **78**, K93 (1976).
[12] J. G. Powles, J. Phys. C **8**, 895 (1975).
[13] R. V. Sharma and P. S. Mahesh, Indian J. pure appl. Phys. **16**, 784 (1978).
[14] N. F. Carnahan and K. E. Starling, J. chem. Phys. **51**, 635 (1969).

(Received March 16, 1984)

phys. stat. sol. (a) 85, 45 (1984)

Subject classification: 1.5 and 13.3; 21.7

Central Institute of Electron Physics, Academy of Sciences of the GDR, Berlin[1]) *(a) and Institute of Physics, Academy of Sciences of the Ukrainian SSR, Kiev (b)*

Quantitative Interference Electron Microscopy for Spherical Objects

By

H. BERGER (a), YU. A. KULYUPIN (b), S. A. NEPIJKO (b), and N. N. SEDOV (b)

The use of interference electron microscopy for measuring the mean inner potential of spherical objects is investigated theoretically and experiments are carried out to determine the mean inner potential of Bi spheres. The experiments clearly show the decrease of the mean inner potential near the surface of the Bi spheres.

Die Anwendung der Interferenz-Elektronenmikroskopie zur Bestimmung des mittleren inneren Potentials von kugelförmigen Proben wird theoretisch und experimentell untersucht. Bei den an Bi-Kügelchen durchgeführten Experimenten kann die Veränderung des mittleren inneren Potentials an der Oberfläche der Bi-Kügelchen nachgewiesen werden.

1. Introduction

The interference electron microscopy [1, 2] is a unique method for investigating the potential distribution with high spatial resolution. It was applied to the determination of the mean inner potential of thin films [3] and the electric field on the surface of an object [4], and at p–n junctions [5]. If the potential distribution for an object is known, one can calculate the shape of the interference fringes and compare it with the experimental one (the direct problem), and vice versa from the experimentally determined fringe shape one can calculate the potential distribution for an object (the inverse problem). This paper considers methods of solving the direct and inverse problems of the quantitative interference electron microscopy for a spherical object. It will be shown that this method can be used to investigate the potential distribution near the boundary of a small particle.

2. Experimental Scheme and Condition of Interference

We consider the scheme of Möllenstedt and Düker [2] modified by Drahoš and Delong [6] used in our experiments (Fig. 1). In the coordinate system (x, y, z) the axis z is directed parallel to the electron beam. The filament cathode (image K) and the filament B of the electrostatic Fresnel biprism are oriented parallel to the axis $\boldsymbol{x}$. The object is a sphere with radius R. It is characterized by its projection onto the object plane O. L is the objective lens. The plane $z = 0$ is the image plane of the object. The biprism forms two imaginary images of the cathode at the points A $(0, -l, D)$ and C $(0, l, D)$, whose distance $2l$ depends linearly on the positive potential V_B of the biprism and is about tens of microns. The resulting interference pattern without an object is a system of parallel and nearly equidistant fringes oriented along the axis $\boldsymbol{x}$.

[1]) Hausvogteiplatz 5—7, DDR-1086 Berlin, GDR.

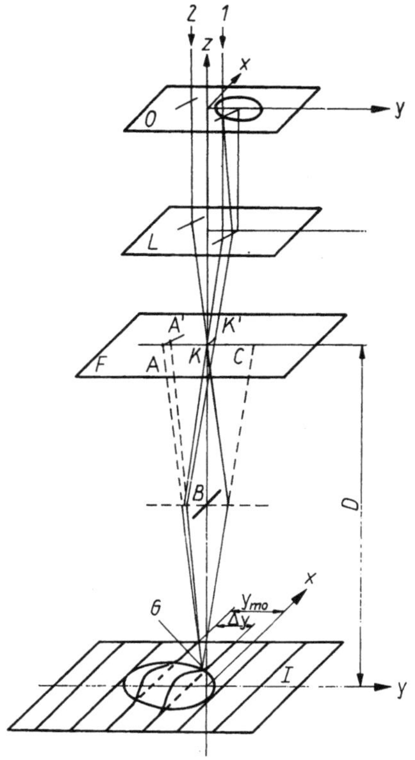

Fig. 1. Scheme of the interference electron microscope. O object plane, L objective lense, F focus plane of the objective lense, K image of the cathode, B electrostatic Fresnel biprism, I image plane

The condition for the interference fringe maximum to be at the point $G(x, y_m)$ is

$$\mathrm{AG} - \mathrm{CG} = \frac{(\mathrm{AG})^2 - (\mathrm{AC})^2}{2D} = \frac{2y_{m0}l}{D} = m\lambda \,, \tag{1}$$

λ is the wavelength including the relativistic correction and m an integer.

The separation between the interference fringes is then given by $\Delta y = D\lambda/2l$. If a beam passes an object characterized by a potential $U(x, y, z)$, then the phases of the two coherent beams get an additional difference $\Delta\varphi$. Thus the interference fringe at the point $G(x, y_{0m})$ is shifted to a point given by the condition

$$\frac{2y_m l}{D} = m\lambda + \frac{\Delta\varphi}{2\pi}\lambda$$

or

$$\frac{y_m - y_{m0}}{\Delta y} - \frac{\Delta\varphi}{2\pi} = 0 \,. \tag{2}$$

This is the basic equation of the interference electron microscopy.

3. Electron Interference on Spherical Particles

For the direct problem an analytical expression has to be derived for the phase difference appearing in (2). The increase in phase of an electron wave travelling along the length increment dz is given by $d\varphi = k\, dz = 2\pi/h\ (2\, em_e V)^{1/2}\, dz$. Hence the expression for the electron wave phase will be

$$\varphi = \frac{2\pi}{h} (2em_e)^{1/2} \int_s V^{1/2}\, dz \,, \tag{3}$$

where m_e is the relativistic electron mass, e the electron charge, h the Planck constant, V the potential, and s the electron path. The potential is the sum of the acceleration voltage V_0 and the potential of the object U. The phase difference between the electron beam probing the object and the coherent reference electron beam is

$$\Delta\varphi = \frac{2\pi}{h} (2em_e)^{1/2} \left(\int_s (V_0 + U)^{1/2}\, dz - \int_s V_0^{1/2}\, dz\right) .$$

In the electron microscope always $U \ll V_0$. Therefore, $(V_0 + U)^{1/2} \approx V_0^{1/2}\,(1 + U/2V_0)$ and it follows

$$\Delta\varphi = \frac{\pi}{h}\left(\frac{2m_e e}{V_0}\right)^{1/2} \int\limits_s U\,\mathrm{d}z\,. \tag{4}$$

The shift of the interference fringe at the point $G(x, y_{m0})$ is then given by (2) and (4),

$$\frac{y_m - y_{m0}}{\Delta y} = \frac{1}{h}\left(\frac{em_e}{2V_0}\right)^{1/2} \int\limits_s U\,\mathrm{d}z\,. \tag{5}$$

For a spherical object $U = U(\varrho)$. Therefore, we replace z by $(\varrho^2 - r^2)^{1/2}$ (Fig. 2) and (5) will be

$$\frac{y_m - y_{m0}}{\Delta y} = g(r) = \frac{1}{h}\left(\frac{2em_e}{V_0}\right)^{1/2} \int\limits_r^{R\to\infty} \frac{U(\varrho)\,\varrho\,\mathrm{d}\varrho}{(\varrho^2 - r^2)^{1/2}}\,. \tag{6}$$

A simple example is the usually considered step-like form of the mean inner potential. For a spherical particle $U(\varrho) = U_0$ for $\varrho \leqq R$ and 0 for $\varrho > R$, where R is the radius of the particle. In this case (6) has the simple form

$$\frac{y_m - y_{m0}}{\Delta y} = \frac{1}{h}\left(\frac{2me}{V_0}\right)^{1/2} U_0(R^2 - r^2)^{1/2} \tag{7}$$

for $R \geqq r$ and $y_m - y_{m0}/\Delta y = 0$ for $r > R$.

From the knowledge of the mean inner potential U_0 one can plot the interferogram and compare it with the experimental one. However, this procedure showed clearly [7] that the assumption of a step function for the mean inner potential is only a rough approximation. This is also clear from the physical point of view. Therefore the inverse problem is of more practical interest, in which the real potential shape of a particle is determined from the interference fringe bend obtained in the experiment.

Upon replacing the variables $t = r^2$ and $\tau = \varrho^2$ and introducing the notations $G(t) = g(r^2)$, $V(\tau) = (2m_e e/V_0)^{1/2}\,U(\varrho)/2h$ (6) takes the form of the well-known Abel integral equation

$$G(t) = \int\limits_t^{\infty} \frac{V(\tau)\,\mathrm{d}\tau}{(\tau - t)^{1/2}}\,. \tag{8}$$

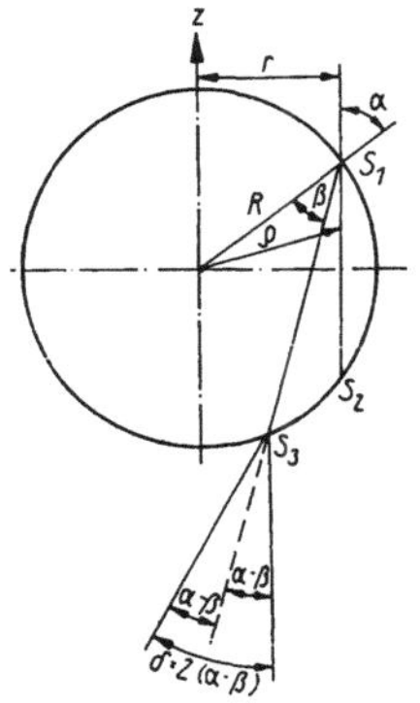

Fig. 2. Refraction of electrons at spherical particles

Solution of (8) for the function $V(\tau)$ is of the form

$$V(\tau) = -\frac{1}{\pi}\frac{d}{d\tau}\int_{\tau}^{\infty}\frac{G(t)\,dt}{(t-\tau)^{1/2}}. \qquad (9)$$

Returning to the previous notations, we obtain

$$U(\varrho) = \frac{h}{\pi}\left(\frac{2V_0}{em_e}\right)^{1/2}\frac{d}{\varrho\,d\varrho}\int_{\varrho}^{\infty}\frac{g(r)\,r\,dr}{(r^2-\varrho^2)^{1/2}}. \qquad (10)$$

Thus using the experimental function $g(r)$ one can obtain the dependence of the potential on the distance from the centre of the spherical particle, by numerically calculating the integral.

Till now we always assumed that the electrons passing an additional potential suffer only a shift of their phases. The change of the direction of these electrons was neglected.

To estimate this influence we consider the case of a spherical particle. We assume for the mean inner potential $U = U_0$ for $\varrho \leqq R$ and $U = 0$ for $\varrho > R$. The electron beam enters the particle with the impact parameter r at the point S_1 and leaves it at the point S_3 with an angle δ to the initial direction of the beam (Fig. 2). The refraction angle $\delta = 2(\alpha - \beta)$ is given by the Snellius law $\sin\alpha = n \sin\beta$, where n is the refractive index of the particle. Since $n = (1 + U_0/V_0)^{1/2}$ [8] and the mean inner potential U_0 of crystals is of the order of 10 eV and the acceleration voltage of the order of 10^5 eV, the difference between the refraction coefficients of the object and vacuum $n - n_0$ is of the order $U_0/V_0 \approx 10^{-4}$. Therefore the refraction angles are also very small. The maximum δ-value results for $\alpha = 90°$, and is equal to $\delta_{max} = 1.4 \times 10^{-2}$ rad, if we assume a mean inner potential of 20 eV and $N_0 = 10^5$ V. From the small refraction angle two important consequences follow:

(i) Since in the case of exact focussing the planes of the object O and of the image I (Fig. 1) are conjugate, the beams transmitted from any point of the object at an arbitrary, small angle and arriving at the corresponding point on its image will have equal phase, in spite of the fact that because of refraction the cathode image K and position of one of the imaginary sources A shift into K′ and A′, respectively. Due to small refraction angles the beams travel near the optical axis of the microscope reducing to minimum distortion of the aberration phase at the lens edge. When the planes O and I are defocused, they are no longer conjugate. In this case it is necessary to apply a correction for refraction.

(ii) A trajectory bend of the particle can be neglected. Indeed, the bend of the interference fringes is determined by the internal potential and the object thickness. However, due to the smallness of δ, the apparent path difference $\Delta_1 = S_1S_2(n-1) = 2R(n-1)\cos\alpha$ differs from the real one $\Delta_2 = S_1S_3(n-1) = 2R(n-1)\cos(\alpha - \delta/2) \approx \Delta_1 \cos(\delta/2) + 2R(n-1)\sin\alpha\sin(\delta/2) \approx \Delta_1 + 2R(n-1)\,r\delta/2R$ by $\Delta_2 - \Delta_1 = (n-1)\,r\delta \ll \lambda$. This enables us to replace the path S_1S_3 by S_1S_2 and thereby to simplify the problem.

4. Experimental Results

The experiments were carried out in an electron microscope Tesla BS 613 with an equipment for electron beam interferences. In the electron microscope an evaporator for in situ preparation of thin films is mounted. The details of the experimental arran-

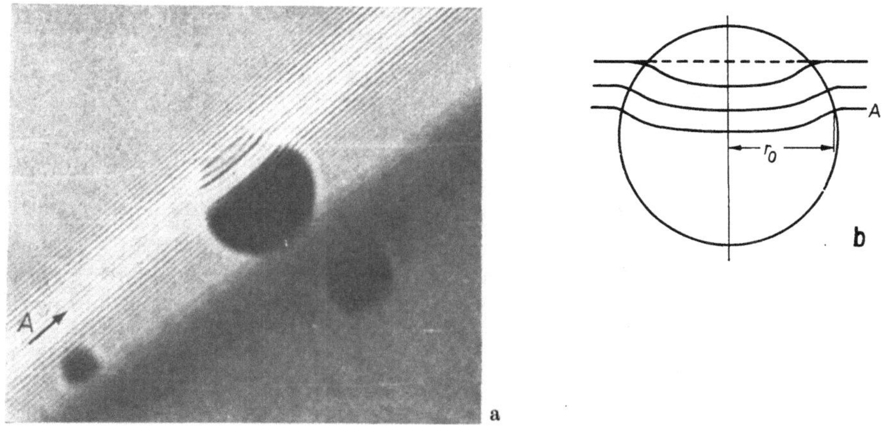

Fig. 3. a) Interference fringes bent by a spherical Bi particle, $R = 20$ nm, b) Selected lines of minimum intensity for the same interferogram

gements are described in [7]. In the same paper first results on the inner potential of Bi were published. The results showed clearly that there is a difference between the measured and the calculated interference lines for a mean inner potential $U = U_0$ for $\varrho \leqq R$ and $U = 0$ for $\varrho > R$ on the surface of the spheres (Fig. 4 in [7]). In order to investigate the mean inner potential near the surface we repeated the experiments on a Bi island film which was evaporated at 140 °C. Subsequently the island film was warmed up to 320 °C. At this temperature the Bi is liquid and the islands are nearly ideal spheres. Fig. 3a shows the interferogram of a Bi sphere with a radius $R = 20$ nm and Fig. 3b the intensity minima on the interferogram measured with a coordinate meter. From line A the function $g(r) = (y_m - y_{m0})/\Delta y$ defined in (6) was subsequently calculated. The results are shown in Fig. 4. For this selected interference line the mean inner potential was calculated as a function of the radius ϱ by (10).

For the numerical calculation of the potential the points for integration on the interference line were chosen in such a way that the radius vector satisfies the con-

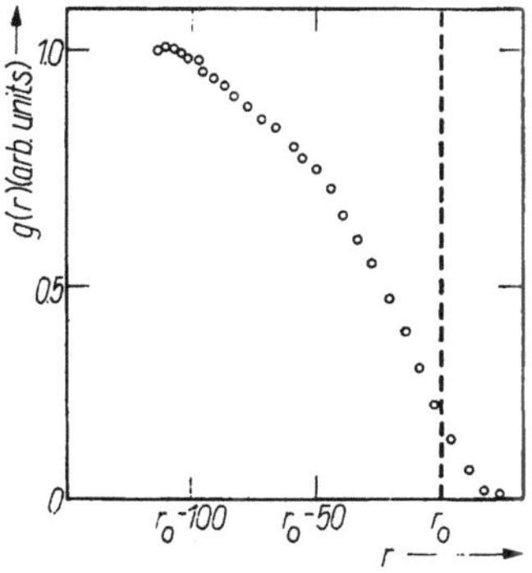

Fig. 4

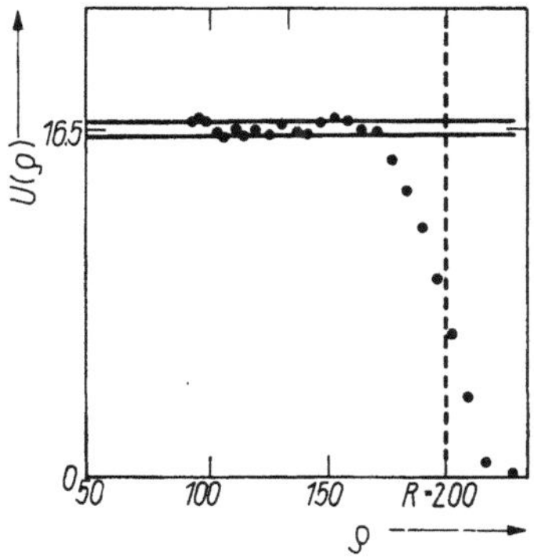

Fig. 5

Fig. 4. Function $g(r)$ calculated for line A in Fig. 3

Fig. 5. Mean inner potentia as a function of radius of a Bi sphere, calculated for line A in Fig. 3

dition $r_{i+1} - r_i = \text{const.}$ If the number of points i exceeds 28, the result is independent of i. The curve for the mean inner potential calculated in this way is represented in Fig. 5. The value of 16.5 eV for the interior of the spheres corresponds to the values in [6]. At a distance of approximately 2.5 nm from the surface U begins decreasing. At the surface the potential is only one half of its value in the volume. It is also clear that the potential has a noticeable value up to 2 nm outside the spheres. The range of $\approx$ 5 nm within which the mean inner potential decreases from $U_0 = 16.5$ eV to zero seems very large. This result shows clearly that for an understanding of the details of the diffraction of electrons in the first atomic layers at the surface the assumption of a constant mean inner potential is a very rough approximation. In any case, the mean inner potential at the surface should be smaller than in the volume of a solid.

References

[1] L. MARTON, Phys. Rev. **85**, 1057 (1952).
G. F. MISSIROLI, G. POZZI, and U. VALDRÈ, J. Phys. E **14**, 649 (1981).
[2] G. MÖLLENSTEDT and H. DÜKER, Naturwissenschaften **42**, 41 (1955).
G. MÖLLENSTEDT and H. DÜKER, Z. Phys. **145**, 377 (1956).
[3] G. MÖLLENSTEDT and H. KELLER, Z. Phys. **148**, 34 (1957).
[4] YU. A. KULYUPIN, S. A. NEPIJKO, N. N. SEDOV, and V. G. SHAMONYA, Optik **52**, 102 (1978/79).
[5] P. G. MERLI, G. F. MISSIROLI, and G. POZZI, J. Microsc. **21**, 11 (1974)
[6] V. DRAHOŠ and A. DELONG, Čes. Čas. Fys. **13**, 278 (1963); **15**, 476 (1965).
[7] H. BERGER, YU. A. KULYUPIN, S. A. NEPIJKO, I. A. OBUCHOV, and V. G. SHAMONYA, Z. Phys. **B37**, 23 (1980).
[8] H. BETHE, Ann. Phys. (Leipzig) **87**, 55 (1928).

(Received May 29, 1984)

phys. stat. sol. (a) 85, 51 (1984)

Subject classification: 1.4

Ural Forest Technological Institute (a)
and Institute of Metallurgy,
Academy of Sciences of the USSR, Ural Scientific Centre, Sverdlovsk[1]) *(b)*

Analysis of the Ground State of a Solid Solution with Interactions up to *k*-th Coordination Spheres of the Crystal Lattice

Selection of the Subdivision of the Lattice into Sublattices

By

B. A. MEN (a) and M. L. LEVITAN (b)

A selection rule for the divisions of the lattice into sublattices corresponding to the ground state superstructures of the solid solution is proposed. This rule permits to conclude that the selection does not depend on the quantity and the sort of the solution components, but depends only on the number of the coordination spheres in which the interaction takes into account. A basis of the rule is given for the binary solutions only. The rule has a hypothetic nature if the number of components in the solution is more than two. The rule is examined numerically for magnetic alloys. A special point-number diagram technique is used to choose the sublattices and superstructures.

Предлагается правило отбора разбиений решетки на подрешетки, соответствующих сверхструктурам основного состояния твердого раствора. Формулировка правила такова, что отбор не зависит от числа и индивидуальности компонент раствора, а только от числа координационных сфер решетки, взаимодействие в которых учитывается. Обоснование правила дается в работе лишь для бинарных растворов. Правило носит гипотетический хатактер, если число компонент в растворе больше двух. Для магнитных сплавов оно прошло численную проверку. При отборе подрешеток и сверхструктур используется специальный аппарат точечно-числовых диаграмм.

1. Introduction

Recently there have been a number of works on the ground state superstructure determination in alloys [1 to 3] and magnetics [4] under the condition that the interaction between atoms is taken into account up to k-th coordination spheres of the lattice. The method offered in [1 to 4] is based on the formation of a system of inequalities. Then the vertices of the polyhedron described by these inequalities are enumerated. However, if we have more than two components in solution (in alloy) or if $k > 3$, then the analysis of such a system of inequalities is difficult. The approach proposed below is connected with the study of the lattice subdivision into sublattices.

The physical non-equivalence of lattice sites determines the division of the lattice of a solution into sublattices. The statistical description of the phase diagram of the ordering of the alloy [5], the study of the crystallographical properties of the chemical compounds, and, in this paper, the enumeration of the possible superstructures of the ground state are based on the lattice division. The present paper involves the basis

[1]) S. Kovalevskaya 20, Sverdlovsk K-66, USSR.

and the numerical consideration of the following hypothesis: the subdivision of the lattice into sublattices for superstructures, on which the minimum of the internal energy of the solution E is realized, does not depend on the quantity of its components. It only depends on the number of coordination spheres (k) taken into account. If this hypothesis is right, then the well-studied superstructures of binary alloys can be used a priori for the consideration of a three-component systems (it is known, for example, that the superstructure of the alloy Cu_2MnAl is connected with the same b.c.c. lattice division as Fe_3Al), and for the study of magnetic alloys and solutions with defects. Moreover, an essential information about the potential and the interaction radius for a great number of solutions with known superstructures may be obtained from this hypothesis.

Some selection rules are necessary because the number of variants of lattice subdivisions into sublattices is infinite. The symmetrical selection rules based on the Landau concept and the Lifshits conditions [6 to 8] are not connected with the potential and the interaction radius. That is why a considerable part of the superstructures corresponding to the minimum of the internal energy E is excluded. A new selection rule for the lattice divisions which depends on the number of the coordination spheres taken into account, is proposed. This rule is founded for binary systems and numerically examined for some systems with three states on the lattice sites.

2. Indication

1. R is a set of lattice sites, r_k the distance from a fixed site to the sites which belong to the k-th coordination sphere, z_k the k-th coordination number.

2. S is a set of states on a site. If the solution contains the l sorts of magnetic atoms A_i with two spin projections and the m sorts of non-magnetic atoms B_j, then $S = \{A_1\uparrow, A_1\downarrow, A_2\uparrow, A_2\downarrow, \ldots, A_l\uparrow, A_l\downarrow, B_1, B_2, \ldots, B_m\}$; $s = |S| = (2l + m)$ is the number of states on the site.

3. The microstructure (or filling) is some function $f\colon R \xrightarrow{f} S$. It is possible to determine the following macrocharacteristics (statistical parameters) for the given microstructure:

$$p_{A_i}, p_{A_i\uparrow}, p_{A_i\downarrow}, p_{B_j}, p^{(k)}_{A_iA_{\bar{i}}}, p^{(k)}_{B_jB_{\bar{j}}}, p^{(k)}_{A_iB_j}, p^{(k)}_{A_i\uparrow A_{\bar{i}}\uparrow}, p^{(k)}_{A_i\uparrow A_{\bar{i}}\downarrow}, p^{(k)}_{A_i\downarrow A_{\bar{i}}\downarrow}, p^{(k)}_{A_i\uparrow B_j}, p^{(k)}_{A_i\downarrow B_j}, \tag{1}$$

where, for example, $p_{A_i\uparrow}$ is the probability of the filling of the site by the state $A_i\uparrow$, $p^{(k)}_{A_iB_j}$ is the probability of the filling of the pair of sites at distance r_k by the states A_i and B_j. If we consider only the case $k \leqq k_0$, then in the list (1) there are $d = 3l + m + k_0(\frac{5}{2} l^2 + \frac{1}{2} m^2 + 3lm + \frac{3}{2} l + \frac{1}{2} m)$ parameters connected by $l + 1 + k_0(\frac{1}{2} l^2 + lm + \frac{3}{2} l + m)$ equations.

4. Let the d-dimensional point be a permissible one if its coordinates correspond to the macrocharacteristics (1) of some filling. It is easy to show that the set M of permissible points is convex and it is possible to suggest that M is a polyhedron (the polyhedron of the solution). If only a part of the parameters (1) is included into the expression for the internal energy of the solution, then the projection of the polyhedron M on the corresponding coordination subspace forms the polyhedron $\overline{M}$.

5. We conclude that the non-intersecting subsets $R_1, R_2, \ldots, R_t$ of the set R are sublattices if $R_i\tau = R_j$ (where τ is any translation of the lattice) and $\cup R_i = R$. The enumeration of the sublattices (or of the lattice divisions) is an independent problem. G is the group of transformations of the sublattices under the action of the translations. $|G| = t$ for s.c., b.c.c., f.c.c. lattices. If Φ_0 is the group of the lattice translations then the subdivision into sublattices determines the homomorphism of the group Φ_0

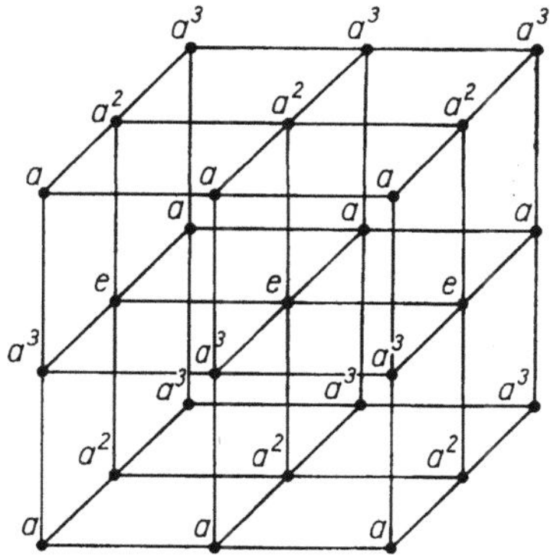

Fig. 1. The four sublattices as identified by the elements of the group G

on G. On the other hand, if the homomorphism of Φ_0 on some t-order group is given, it is possible to obtain the corresponding lattice division into t-sublattices. Let τ_1, τ_2, τ_3 be the base of the translation group Φ_0. Thus, for the problem of the division it is sufficient to give the abstract Abelian group G and to define arbitrarily the homomorphous images $\bar{\tau}_1$, $\bar{\tau}_2$, $\bar{\tau}_3$ in it. The division of the s.c. lattice into four sublattices for the variant $G = \langle a^4 \rangle$, $(\overline{100}) = e$, $(\overline{010}) = a$, $(\overline{001}) = a^2$ is given in Fig. 1. (The sublattices are identified by the elements of the group G. The origin point (the lattice site) is fixed. After that for any site of the lattice we determine the translation from the origin to this site and find its homomorphous image in the group G.)

6. Let then $\Phi_0^{(k)}$ the be set of translations at the distance r_k. We retain only one translation from every pair of translations τ, τ^{-1} entering $\Phi_0^{(k)}$, for example, τ. We obtain the set $\tilde{\Phi}_0^{(k)}$ of the order $z_k/2$. For the given lattice division (G, $\bar{\tau}_1$, $\bar{\tau}_2$, $\bar{\tau}_3$) we introduce the following characteristics: q_k ($k = 1, 2, \ldots, k_0$) is the number of translations from $\Phi_0^{(k)}$, the unit element of group G being a homomorphous image of them, i.e. of the translations permitting only sites within the sublattices.

7. Finally, the microstructure f corresponds to the lattice division into sublattices (G, $\bar{\tau}_1$, $\bar{\tau}_2$, $\bar{\tau}_3$) if the sites of every sublattice are filled by atoms of one sort (taking into account the spin projection for the magnetic atoms).

3. Principles of Analysis

If the magnetic field equals zero, then the internal energy of the magnetic alloy is expressed as follows:

$$E = \sum_i \lambda_i p_{A_i} + \sum_j \lambda_j p_{B_j} + \sum_{\substack{i \le \bar{i} \\ k \le k_0}} I_{i\bar{i}}^{(k)} p_{A_iA_{\bar{i}}}^{(k)} + \sum_{\substack{i,j \\ k \le k_0}} I_{ij}^{(k)} p_{A_iB_j}^{(k)} + \sum_{\substack{j \le \bar{j} \\ k \le k_0}} I_{j\bar{j}}^{(k)} p_{B_jB_{\bar{j}}}^{(k)} + \\ + \sum_{i \le \bar{i},\, k \le k_0} \bar{I}_{i\bar{i}}^{(k)} (p_{A_i\uparrow A_{\bar{i}}\uparrow}^{(k)} + p_{A_i\downarrow A_{\bar{i}}\downarrow}^{(k)} - p_{A_i\uparrow A_{\bar{i}}\downarrow}^{(k)} - p_{A_i\downarrow A_{\bar{i}}\uparrow}^{(k)}) \tag{2}$$

(I, $\bar{I}$ are crystallographic and magnetic energy parameters; λ is the Lagrangian multiplier corresponding to the condition of constant concentration). The polyhedron M has the dimension $\nu = l + m - 1 + k_0(l + m)(l + m - 1)/2 + k_0 l(l + 1)/2$. The independent statistical variables from the list (1) which are involved in (2) are

$$p_{A_i},\ p_{B_j} (j \neq m),\ p_{A_iB_j}^{(k)},$$

$$Q_{i\bar{i}}^{(k)} = \frac{1}{2} (p_{A_i\uparrow A_{\bar{i}}\downarrow}^{(k)} + p_{A_i\downarrow A_{\bar{i}}\uparrow}^{(k)}),\ p_{A_iA_{\bar{i}}}^{(k)} \begin{pmatrix} i < \bar{i} \le l \\ \text{or} \\ i \le \bar{i} < l \end{pmatrix},\quad p_{B_jB_{\bar{j}}}^{(k)} \begin{pmatrix} j < \bar{j} \le m \\ \text{or} \\ j \le \bar{j} < m \end{pmatrix}. \tag{3}$$

In particular, for a non-magnetic binary alloy ($l = 0, m = 2$), $\nu = 1 + k_0$, the basic parameters are p_{B_1}, $p^{(k)}_{B_1B_2}$, or $p^{(k)}_{B_1B_1}$ and for the binary alloy with one sort of magnetic atoms ($l = 1, m = 1$), $\nu = 1 + 2k_0$ the basic parameters are p_A, $p^{(k)}_{AB}$, $Q^{(k)}_{AA} = p^{(k)}_{A\uparrow A\downarrow}$.

Finding the microstructure with minimum internal energy is equivalent to the minimization of the linear function (2) on the polyhedron $\overline{M}$. According to the known theorem of linear programming [9] the minimum corresponds to one of the vertices of the polyhedron. Thus,

problem 1 is to enumerate the vertices of polyhedron $\overline{M}$;

problem 2 is to find the microstructures represented by every verte, the macrocharacteristics of which correspond to the coordinates of this vertex.

In [1 to 4] the following method of solution of the problem 1 "from above" proposed.

1. A system of linear inequalities connecting the basic parameters (3) is formed. The question of the completeness of the system is not discussed, consequently, the system determines some polyhedron $M' \supseteq \overline{M}$.
2. The vertices of M' are enumerated.
3. A vertex is permissible if there is a suitable microstructure for it.

The drawbacks of this method are:

a) a large volume of calculations even for the simplest case $l = 0$, $m = 2$, $k_0 = 2$;

b) if it is not possible to choose a microstructure, the situation is not clear: whether we have an incomplete system of inequalities or the investigation should be continued.

In the present paper a method of solution of the problem "from below" is proposed.

1. The lattice of the alloy is divided into sublattices. The microstructures corresponding to these divisions [10] are found. Some finite set of permissible points N is obtained.
2. A convex envelope for the N points is formed. It is a polyhedron $M'' \subseteq \overline{M}$. Its vertices are enumerated.
3. If with the increase of the number of sublattices the polyhedron M'' does not grow, then the procedure is stopped.

The choice of the maximum of the sublattice quantity is a difficult problem, that is why the selection rule is necessary. The essential advantages of this approach are:

a) there is no necessity to solve the problem 2;

b) if the hypothesis of independence is right, and the selection rule works for a system with $s > 2$ also, the volume of calculations even for the case of a high dimension is comparatively small.

4. The Diagram Lemma

The most difficult moment of the procedure is the choice of the extreme points of the set N forming the vertices of the polyhedron M''. The choice is easily done on the plane visually ($\nu = 2$); at $\nu = 3$ the visual choice is difficult, and at $\nu = 4$ is impossible. The bounds of visuality are extended by the using of point-number diagrams. The selection rule is formulated in terms of point-number diagrams.

The element of the $(p + 1)$-dimensional diagram $(y_1, y_2, \ldots, y_p)\ Y$ is the p-dimensional point $(y_1, y_2, \ldots, y_p)$ in which the number Y is located. The convex combination of several elements is determined as usual: $(\sum \alpha_i y_1^{(i)}, \sum \alpha_i y_2^{(i)}, \ldots, \sum \alpha_i y_p^{(i)})\ \sum \alpha_i Y^{(i)}$ $(\sum \alpha_i = 1, \alpha_i \geqq 0)$.

Now let us determine the rule of the element crossing-out on the diagram. Let $\overline{N}$ be a set of point-numbers forming the λ-diagram.

The rule: The element $(y_1, y_2, \dots, y_p)\, Y$ must be crossed out if the element $(y_1, y_2, \dots, y_p)\, \overline{Y}$ can be obtained by the convex combination of other elements, where $\overline{Y} \leqq Y$.

The retained elements of the diagram after crossing-out are basic ones.

Let the set of points $\mathrm{N} = \{x_1^{(i)}, x_2^{(i)}, \dots, x_w^{(i)}\}$ meet the requirements:

1. $x_j^{(i)} \geqq 0$;
2. $(0, 0, \dots, 0) \in \mathrm{N}$ (let us suppose that this is the first point of the set, i.e. $i = 1$);
3. $x_1^{(i)} > 0$ for $i \neq 1$.

Then we form the λ-diagram, the element $(x_2^{(i)}/x_1^{(i)}, x_3^{(i)}/x_1^{(i)}, \dots, x_w^{(i)}/x_1^{(i)})\; 1/x_1^{(i)}$ corresponding to every point, except the first one. It is obvious that the correspondence between the points $\mathrm{N}/(0, 0, \dots, 0)$ and the elements of the diagram is reciprocally unambiguous.

The lemma: There is one and only one basic element of the λ-diagram for any extremal point of the set N $\big($except $(0, 0, \dots, 0)\big)$ and vice versa.

4.1 Proof

Let us prove that the crossed-out elements correspond to the internal points of the set N and vice versa. Let $\{x_1^{(i)}, x_2^{(i)}, \dots, x_w^{(i)}\}$ be an internal point of N, then it is a convex combination of the other points,

$$(x_1^{(i)}, x_2^{(i)}, \dots, x_w^{(i)}) = \sum \alpha_j (x_1^{(j)}, x_2^{(j)}, \dots, x_w^{(j)})\,; \quad \alpha_i = 0\,, \quad \alpha_j \geqq 0\,, \quad \sum \alpha_j = 1\,. \tag{4}$$

But then

$$\left(\frac{x_2^{(i)}}{x_1^{(i)}}, \frac{x_3^{(i)}}{x_1^{(i)}}, \dots, \frac{x_w^{(i)}}{x_1^{(i)}}\right) \frac{1}{x_1^{(i)}} = \sum \beta_j \left[\left(\frac{x_2^{(j)}}{x_1^{(j)}}, \frac{x_3^{(j)}}{x_1^{(j)}}, \dots, \frac{x_w^{(j)}}{x_1^{(j)}}\right) \frac{1}{x_1^{(j)}}\right], \tag{5}$$

where the factors

$$\beta_j = \frac{\alpha_j x_1^{(j)}}{\sum \alpha_j x_1^{(j)}} \tag{6}$$

satisfy all conditions of a convex combination, β_i being equal to zero. According to the rule the element $(x_2^{(i)}/x_1^{(i)}, x_3^{(i)}/x_1^{(i)}, \dots, x_w^{(i)}/x_1^{(i)})\; 1/x_1^{(i)}$ is crossed out. On the other hand, if the element $(x_2^{(i)}/x_1^{(i)}, x_3^{(i)}/x_1^{(i)}, \dots, x_w^{(i)}/x_1^{(i)})\; 1/x_1^{(i)}$ is crossed out, then for some $X \leqq 1/x_1^{(i)}$

$$\left(\frac{x_2^{(i)}}{x_1^{(i)}}, \frac{x_3^{(i)}}{x_1^{(i)}}, \dots, \frac{x_w^{(i)}}{x_1^{(i)}}\right) X = \sum \beta_j \left[\left(\frac{x_2^{(j)}}{x_1^{(j)}}, \frac{x_3^{(j)}}{x_1^{(j)}}, \dots, \frac{x_w^{(j)}}{x_1^{(j)}}\right) \frac{1}{x_1^{(j)}}\right], \quad \begin{array}{l} \beta_i = 0\,, \\ \sum \beta_j = 1\,. \end{array} \tag{7}$$

Now

$$\left(\frac{1}{X}, \frac{x_2^{(i)}}{x_1^{(i)} X}, \dots, \frac{x_w^{(i)}}{x_1^{(i)} X}\right) = \sum \alpha_j (x_1^{(j)}, x_2^{(j)}, \dots, x_w^{(j)})\,, \tag{8}$$

where the factors

$$\alpha_j = \frac{\beta_j / x_1^{(j)}}{\sum \beta_j / x_1^{(j)}} \tag{9}$$

satisfy to the all conditions of the convex combination. So, the point $C(1/X,\, x_2^{(i)}/x_1^{(i)} X,$

... $x_w^{(i)}/x_1^{(i)}X$) is an internal point of the set N. In its turn, the point

$$(x_1^{(i)}, x_2^{(i)}, \dots, x_w^{(i)}) = X x_1^{(i)} \left(\frac{1}{X}, \frac{x_2^{(i)}}{x_1^{(i)} X}, \dots, \frac{x_w^{(i)}}{x_1^{(i)} X} \right) + (1 - X x_1^{(i)}) \, (0, 0, \dots, 0) \quad (10)$$

located between the point C and the origin will be internal also.

5. The Selection Rule

Let us connect the element $(q_1, q_2, \dots, q_{k_0})\, t$ of the $(k + 1)$-dimensional point-number λ-diagram (the determination of q_k see Section 2, item 6); t is the quantity of sublattices) with every division of the solution lattice into sublattices.

The rule: the basic elements of the λ-diagram correspond to the optimal divisions of the lattice (for the given interaction radius r_{k_0}).

The basic elements of the two-dimensional diagram ($k_0 = 1$) for s.c., b.c.c., and f.c.c. lattices are given in Fig. 2, the basic elements for the three-dimensional diagram ($k_0 = 2$) for s.c., b.c.c. and f.c.c. lattices are given in Fig. 3. Let us describe the division of the lattice corresponding to the basic elements of the diagrams. The base of the translation groups are: for the s.c. lattice $\tau_1 = (100)$, $\tau_2 = (010)$, $\tau_3 = (001)$; for the b.c.c. lattice $\tau_1 = (-\frac{1}{2}\,\frac{1}{2}\,\frac{1}{2})$, $\tau_2 = (\frac{1}{2}\,-\frac{1}{2}\,\frac{1}{2})$, $\tau_3 = (\frac{1}{2}\,\frac{1}{2}\,-\frac{1}{2})$; for the f.c.c. lattice $\tau_1 = (0\,\frac{1}{2}\,\frac{1}{2})$, $\tau_2 = (\frac{1}{2}\,0\,\frac{1}{2})$, $\tau_3 = (\frac{1}{2}\,\frac{1}{2}\,0)$.

The result of the description is given in Table 1. At present the four-dimensional selection diagrams are formed for all cubic lattices. The corresponding results and the optimal substructures for binary non-magnetic alloys will be discussed in a separate paper. Let us notice that with taking into account the interaction in two coordination spheres of the f.c.c. lattice the maximum number of sublattices must be equal to six (see Table 1; 19 to 26), corresponding to [1, 2]. Therefore, the division of the f.c.c. lattice into eight sublattices which was used in [5] (p. 32) to describe the phase diagram

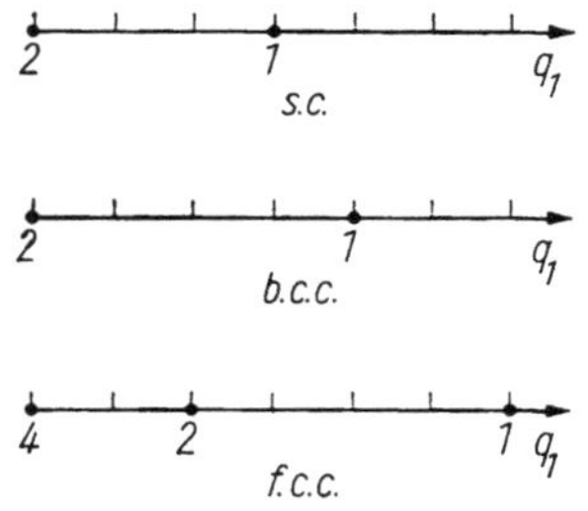

Fig. 2. Basic elements of the two-dimensional diagrams

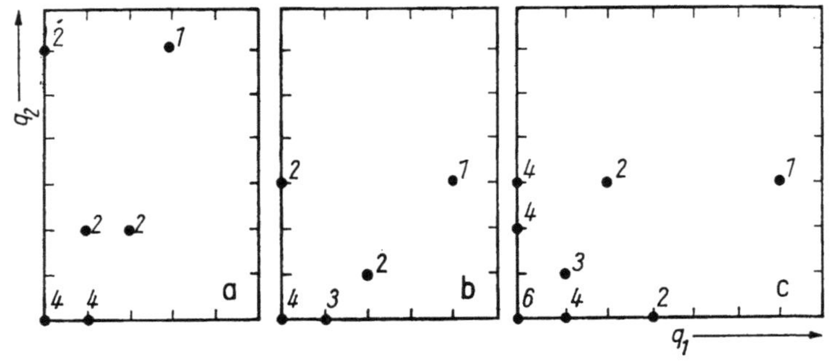

Fig. 3. Basic elements of the three-dimensional diagrams a) for the s.c. lattice, b) for the b.c.c. lattice, c) for the f.c.c. lattice

Table 1

The divisions of the lattice corresponding to the basic elements of diagrams

n.n.	Fig.	lattice	element	G	$\bar{\tau}_1$	$\bar{\tau}_2$	$\bar{\tau}_3$	note
1	2	s.c.	(0) 2	$\langle a\rangle^2$	a	a	a	NaCl
2	2	s.c.	(3) 1	$\langle e\rangle$	e	e	e	
3	2	b.c.c.	(0) 2	$\langle a^2\rangle$	a	a	a	CsCl
4	2	b.c.c.	(4) 1	$\langle e\rangle$	e	e	e	
5	2	f.c.c.	(0) 4	$\langle a^4\rangle$	a	a^2	a^3	Al_3Ti
$\bar{5}$				$\langle a^2\rangle\langle b^2\rangle$	a	b	ab	Cu_3Au
6	2	f.c.c.	(2) 2	$\langle a^2\rangle$	e	a	a	CuAuI
7	2	f.c.c.	(6) 1	$\langle e\rangle$	e	e	e	
8	3a	s.c.	(00) 4	$\langle a^2\rangle\langle b^2\rangle$	a	b	ab	
9	3a	s.c.	(10) 4	$\langle a^4\rangle$	e	a	a^2	
$\bar{9}$				$\langle a^2\rangle\langle b^2\rangle$	e	a	b	
10	3a	s.c.	(06) 2	$\langle a^2\rangle$	a	a	a	coincide with 1
11	3a	s.c.	(12) 2	$\langle a^2\rangle$	e	a	a	
12	3a	s.c.	(22) 2	$\langle a^2\rangle$	e	e	a	
13	3a	s.c.	(36) 1	$\langle e\rangle$	e	e	e	coincide with 2
14	3b	b.c.c.	(00) 4	$\langle a^4\rangle$	a	a	a	Fe_3Al, NaTl
15	3b	b.c.c.	(10) 3	$\langle a^3\rangle$	e	a	a	
16	3b	b.c.c.	(03) 2	$\langle a^2\rangle$	a	a	a	coincide with 3
17	3b	b.c.c.	(21) 2	$\langle a^2\rangle$	e	e	a	
18	3b	b.c.c.	(43) 1	$\langle e\rangle$	e	e	e	coincide with 4
19	3c	f.c.c.	(00) 6	$\langle a^6\rangle$	a	a^2	a^4	
20	3c	f.c.c.	(10) 4	$\langle a^4\rangle$	e	a	a^2	
$\overline{20}$				$\langle a^2\rangle\langle b^2\rangle$	e	a	b	
21	3c	f.c.c.	(02) 4	$\langle a^4\rangle$	a	a^2	a^3	5
22	3c	f.c.c.	(03) 4	$\langle a^2\rangle\langle b^2\rangle$	a	b	ab	coincide with $\bar{5}$
23	3c	f.c.c.	(11) 3	$\langle a^3\rangle$	e	a	a^2	
24	3c	f.c.c.	(30) 2	$\langle a^2\rangle$	e	e	a	CuPt
25	3c	f.c.c.	(23) 2	$\langle a^2\rangle$	e	a	a	coincide with 6
26	3c	f.c.c.	(63) 1	$\langle e\rangle$	e	e	e	coincide with 7

in Cu–Pt alloys is inconsistent. Such a division appears first when the third coordination sphere is taken into account.

6. Basis of the Selection Rule for Binary Systems ($s = 2$)

Consider a non-magnetic binary alloy ($l = 0$, $m = 2$, $\nu = 1 + k_0$) with independent statistical parameters p_{B_1}, $p^{(k)}_{B_1B_1}z_k/2$ (instead of the parameters $p^{(k)}_{B_1B_1}$ from Section 3). We determine the microstructure of the $(B_1)_1(B_2)_{t-1}$ composition for every division of the alloy lattice into sublattices as follows: we fill one of the sublattices by atoms of the sort B_1 and the other by those of the sort B_2. Then we calculate the coordinates of the permissible point $(1/t, q_1/t, q_2/t, \dots, q_{k_0}/t)$ ($p_{B_1} = 1/t$, $p^{(k)}_{B_1B_1} = 2q_k/z_kt$).

The set of permissible points N is obtained by listing all different divisions, and the minimum of the internal energy can be achieved only at extremal points of the set (at the vertices of the convex envelope of M''). The point (0, 0, ..., 0) which corresponds to the filling of all sites of the lattice by atoms of sort B_2 is added to the set N. It is easy to show that all conditions of the diagram lemma 4 are satisfied, hence, it is possible to find the basic elements of the λ-diagram instead of the extremal points of the set N. According to the rule of replacement 4,

$$\left(\frac{1}{t}, \frac{q_1}{t}, \frac{q_2}{t}, \dots, \frac{q_{k_0}}{t}\right) \to (q_1, q_2, \dots, q_{k_0})\, t\,. \tag{11}$$

Thus, the selection rule for a binary alloy is a search for the optimal superstructures at their cut-down set when the atoms of one sort fill not more than one sublattice. Nevertheless, the numerical examination made for all cubic lattices at $t \leqq 8$, $k_0 \leqq 3$ shows that the minimization of the internal energy at the full set of fillings does not give any new optimal divisions of the lattice as compared with the search at the cut-down set.

7. Numerical Examination of the Selection Rule for Binary Alloys with one Magnetic Atom Sort ($s = 3$)

Consider a binary alloy with one magnetic atom sort ($l = 1$, $m = 1$, $\nu = 1 + 2k_0$). We take into account the interaction of nearest neighbours only ($k_0 = 1$). The independent statistical parameters are p_A, $p^{(1)}_{AB}$, $p^{(1)}_{A\uparrow A\downarrow}$ (see Section 3). The point (0, 0, 0) corresponds to the filling of all the sites of the lattice by atoms of sort B. The conditions of the diagram lemma are satisfied for all other permissible points ($p_A > 0$).

Let us realize the procedure described in Section 3 and compare the geometry of the optimal superstructures with the divisions obtained as a result of the selection rule for $k_0 = 1$ (Fig. 2 and Table 1, 1 to 7).

s.c. lattice: Let us consider at first the s.c. lattice. There are three different ways of divisions into two sublattices; 3–3; 9–4; 5–5; 13–6; 8–7; 24–8. If $t \leqq 4$, then 840 coordinated microstructures (including the geometrically and physically equivalent ones) are possible for such 16 divisions. All possible divisions of the s.c. lattice into sublattices in terms of (G, $\bar{\tau}_1$, $\bar{\tau}_2$, $\bar{\tau}_3$) (see Section 2) are given in Table 2 ($t \leqq 4$).

Note that if $k_0 = 1$, then the divisions 1 and 4 are basic ones according to the rule of crossing-out (see Table 1 and Fig. 2). Then we fill the lattice sites by atoms A↑, A↓,

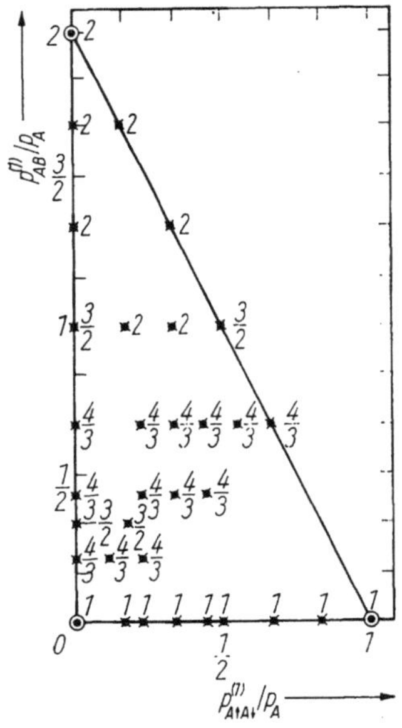

Fig. 4. The point-number diagram with $(p^{(1)}_{A\uparrow A\downarrow}/p_A,\ p^{(1)}_{AB}/p_A)$ $1/p_A$ elements for the s.c. lattice

Table 2

The divisions of the s.c. lattice into one, two, three, and four sublattices

n.n.	t	G			
1	1	⟨e⟩	e	e	e
2	2	⟨a²⟩	e	e	a
3	2	⟨a²⟩	e	a	a
4	2	⟨a²⟩	a	a	a
5	3	⟨a³⟩	e	e	a
6	3	⟨a³⟩	e	a	a
7	3	⟨a³⟩	a	a	a
8	4	⟨a⁴⟩	e	e	a
9	4	⟨a⁴⟩	e	a	a
10	4	⟨a⁴⟩	e	a	a²
11	4	⟨a⁴⟩	a	a	a
12	4	⟨a⁴⟩	a	a	a²
13	4	⟨a⁴⟩	a	a²	a²
14	4	⟨a²⟩⟨b²⟩	e	a	b
15	4	⟨a²⟩⟨b²⟩	a	a	b
16	4	⟨a²⟩⟨b²⟩	a	b	ab

B for every division given in Table 2. The sites belonging to one and the same sublattice must be filled by the same atoms. Now, generally speaking, $1/p_A \neq t$. Hence, the discussion of this case is more difficult than that of $s = 2$. Let us build the point-number diagram with $(p^{(1)}_{A\uparrow A\downarrow}/p_A, p^{(1)}_{AB}/p_A)$ $1/p_A$ elements and determine the basic elements of this diagram. The point-number diagram for the s.c. lattice with these elements is represented in Fig. 4. When the point-number diagram is formed we have 33 elements.[2]) After further crossing-out only three basic elements are retained. Along with point (0, 0, 0) we get four extremal points of the set N:

1. "pure" ferromagnetic A↑ (Table 1; 2);
2. "pure" metal B (Table 1; 2);
3. "pure" antiferromagnetic A↑A↓ (Table 1; 1);
4. alloy AB (Table 1; 1).

b.c.c. lattice: There are two different ways of divisions into two sublattices; 2–3; 7–4; 5–5; 10–6; 7–7; 20–8 (it is interesting that we have for the f.c.c. lattice as many divisions as for the b.c.c. lattice). If $t \leq 4$, then there are 13 divisions, 669 coordinated structures, 29 elements on the diagram, three basic elements and four extremal points of the set N:

1. "pure" ferromagnetic A↑ (Table 1; 4);
2. "pure" metal B (Table 1; 4);
3. "pure" antiferromagnetic A↑A↓ (Table 1; 3);
4. alloy AB (Table 1; 3).

f.c.c. lattice: In this case we have seven basic elements on the diagram, eight extremal points of the set N:

1. "pure" ferromagnetic A↑ (Table 1; 7);
2. "pure" metal B (Table 1; 7);

---

[2]) If several numbers correspond to the same point of the diagram, then the minimal number should be retained according to the rule of crossing-out.

3. "pure" antiferromagnetic A↑A↓: a) superstructure with two sublattices (Table 1; 7), b) superstructure with four sublattices (Table 1; 5), the sublattices e, a are filled by atoms A↑, and a^2, a^3 by A↓;

4. alloy AB: a) superstructure with two sublattices (Table 1; 6), b) superstructure with four sublattices (Table 1; 5), the sublattices e, a are filled by atoms B, and a^2, a^3 by A (only A↑ or only A↓);

5. alloy AB_3: superstructure with four sublattices (Table 1; 5, $\bar{5}$), one of which is filled by atoms of sort A, and three other ones by B.

6. alloy A_3B: superstructure with four sublattices (Table 1; 5, $\bar{5}$), one of which is filled by atoms of sort B, and three other ones by A;

7. magnetic superstructure A↑A↓B_2 (Table 1; 5, $\bar{5}$), two sublattices are filled by atoms of sort B, one by A↑, the other by A↓;

8. magnetic superstructure $(A\uparrow)_2A\downarrow B$ (Table 1; 5, $\bar{5}$), two sublattices are filled by atoms A↑, one by A↓, the other by B.

It should be noted that the microstructures 1–4 in case of the s.c. lattice correspond to the optimal divisions 1 and 4 (Table 2). Analogous statements are correct for b.c.c. and f.c.c. lattices also. It is the numerical confirmation of the selection rule.

The calculations show that the selection rule for the case of $l = 1$, $m = 1$, $k_0 = 1$ is correct. The difficulties occurring in more complicated cases are connected with the high dimension of the diagrams. Let us notice that if the rule is correct a priori, the volume of the necessary calculations decreases considerably.

References

[1] D. van Dyck, R. de Ridder, and S. Amelinckx, phys. stat. sol. (a) 59, 513 (1980).
[2] R. de Ridder, D. van Dyck, and S. Amelinckx, phys. stat. sol. (a) 61, 231 (1980).
[3] D. van Dyck and R. de Ridder, phys. stat. sol. (a) 67, 603 (1981).
[4] J. Kanamori, Progr. theor. Phys. (Kyoto) 35, 16 (1966).
[5] L. N. Larikov, V. V. Geitchenko, and V. M. Faltchenko, Diffuzionny protsessy v uporyadotchenny splavach, Izd. Naukova dumka, Kiev 1975.
[6] L. D. Landau and E. M. Lifshits, Statisticheskaya Fizika, Vol. 1, Izd. Nauka, Moskva 1976.
[7] E. M. Lifshits, Zh. eksper. theor. Fiz. 11, 255 (1941).
[8] E. M. Lifshits, Zh. eksper. theor. Fiz. 11, 269 (1941).
[9] F. I. Karnelevich and L. E. Sadovskii, Elementy lineinoi algebry i lineinogo programmirovaniya, Izd. Nauka, Moskva 1967.
[10] B. A. Men, Izvestiya vuzov, Fizika 1, 50 (1983).

(Received January 10, 1984)

phys. stat. sol. (a) 85, 61 (1984)

Subject classification: 1.4; 22.1.2

Department of Materials Science, State University of New York[1])

A Dark-Field TEM Method for Crystal Structure Determination of Cu-Rich Phases on Twins in Silicon

By

D. M. VANDERWALKER

A dark-field TEM method is developed to identify the crystal structures of second phases which is based on the fact that the difference in structure factor between matrix and precipitate is proportional to the intensity difference. It can distinguish between phases whose structure factors are equal in magnitude but opposite in sign. The method is applied to Cu silicide phases on twin boundaries in Si. Three phases are observed, a cloud-like phase, a small spherical phase embedded in the cloud, and a layer which forms on the top and bottom edges of the twin. The cloud and the layer are found to be CuSi and the small spherical phase a transition phase of the Si lattice with Cu on octahedral sites. Weak-beam lattice images show the cloud-like and spherical phases are coherent and that vein-like voids can emanate from the precipitate interface.

Um die Kristallstrukturen der zweiten Phase zu identifizieren, wird eine TEM-Dunkelfeldmethode entwickelt, die darauf beruht, daß die Differenz der Strukturfaktoren zwischen Matrix und Präzipitat proportional zur Intensitätsdifferenz ist. Es kann zwischen Phasen unterschieden werden, deren Strukturfaktoren in der Größe gleich, jedoch im Vorzeichen unterschiedlich sind. Die Methode wird auf Cu-Silizidphasen auf Zwillingsgrenzen in Si angewendet. Drei Phasen werden beobachtet, eine wolkenähnliche Phase, eine kleine kugelförmige Phase, die in der Wolke eingebettet ist, und eine Schicht, die sich an den oberen und unteren Kanten des Zwillings ausbildet. Es wird gefunden, daß die Wolke und die Schicht CuSi ist und die kleine kugelförmige Phase eine Übergangsphase des Si-Gitters mit Cu auf Oktaederplätzen darstellt. Schwachstrahlgitterdiagramme zeigen, daß die wolkenähnliche und die kugelförmige Phase kohärent sind und daß aderförmige Hohlräume von der Präzipitatgrenzfläche ausgehen können.

1. Introduction

The crystal structures of second phases in a solid can in most cases be determined from X-ray diffraction experiments [1]. However, only the magnitude of the structure factor be found, and phases whose structure factors are equal in magnitude but opposite in sign are indistinguishable. In the study of transformations involving metal silicides, it is important to differentiate between the structures having structure factors of opposite sign. The transition phase of the Si sublattice with metal atoms on octahedral sites has a structure factor which is opposite in sign to the phase of the Si sublattice with metal atoms on tetrahedral sites.

The crystal structures of the phases in the Cu–Si system identified using X-ray diffraction are listed in Pearson [2, 3]. In 1955, Dash examined the decoration of hexagonal dislocation loops in Si with Cu precipitates using an infrared image tube [4]. It is not possible to use X-ray diffraction methods to identify the crystal structure of the particles on the dislocations. Thus, this paper introduces a new method to identify crystal structures using dark-field transmission electron microscopy and applies it to study the formation of Cu-rich phases on twin boundaries in Si.

---

[1]) Stony Brook, New York 11794, USA.

2. Experimental Procedure

A layer of Cu was evaporated on a cast Si specimen before annealing for six hours at 500 °C in a He furnace and quenching in air. The specimens were thinned for transmission electron microscopy (TEM) in two steps, first mechanically then chemically with a 6/7 nitric acid, 1/7 hydrofluoric acid solution. All of the transmission electron microscopy (TEM) was performed on the JEM 200CX at Stony Brook except the high resolution lattice images which where taken on the Phillips 400 ST at IBM. The high resolution lattice images were taken using "weak-beam" diffracting conditions. Weak-beam conditions [5] are set by tilting the beam such that $\boldsymbol{g}$ is on the optic axis and $w = s\xi_g > 5$, where s is the deviation parameter and ξ_g the extinction distance. In the weak-beam lattice images $\boldsymbol{g}$ was placed on the optic axis and a large aperture positioned to include a portion of the $2\boldsymbol{g}$ and 0 spots.

3. Results and Discussion

3.1 The use of dark-field precipitate reflection images to identify the crystal structure of second phases

A method has been developed to identify the crystal structure of small precipitates which involves comparing intensity ratios to structure factor ratios. It is the only method available to distinguish between two structures whose structure factors are equal in magnitude but opposite in sign.

When precipitates are coherent, the main contribution to the image is structure factor contrast [6]. The change in intensity due to the presence of a particle is thought of as an effective change in thickness. The intensity change is proportional to the difference in the structure factor. The following relations from Hirsch et al. [6] are used in this analysis:

$$\xi_g = \frac{\pi V_c \cos\theta}{\lambda F_g}, \tag{1}$$

$$\Delta I = \pi \Delta t \left[\frac{1}{\xi_g^P} - \frac{1}{\xi_g^M}\right] \sin\left(\frac{2\pi t}{\xi_g}\right), \tag{2}$$

$$\Delta I = -\pi \Delta t \left\{\frac{\lambda}{\pi V_c \cos\theta} (F_P - F_M)\right\} \sin\left(\frac{2\pi t}{\xi_g}\right), \tag{3}$$

$$\Delta F = F_P - F_M, \tag{4}$$

where ξ_g^P is the extinction distance in the precipitate, ξ_g^M is the extinction distance in the matrix, Δt is the particle thickness, λ the electron wavelength, F_g the structure factor, and V_c is the volume of the unit cell.

Dark-field precipitate reflection images are taken with a known *hkl*, and the intensity of each phase present measured relative to the background. Since the change in intensity is proportional to $\Delta F = F_P - F_M$, ΔF values can be calculated and the ratio of the ΔF's for two phases compared to the measured intensity ratios.

3.2 Application of the method to the Cu–Si system

The method developed to identify the crystal structures of small second phases was applied to the precipitation of Cu rich phases on twin boundaries in Si. There are three phases present, a cloud-like phase (Fig. 1 and 2), a small phase which is surrounded by the cloud (Fig. 1 and 2), and a layer of second phase which forms along the top

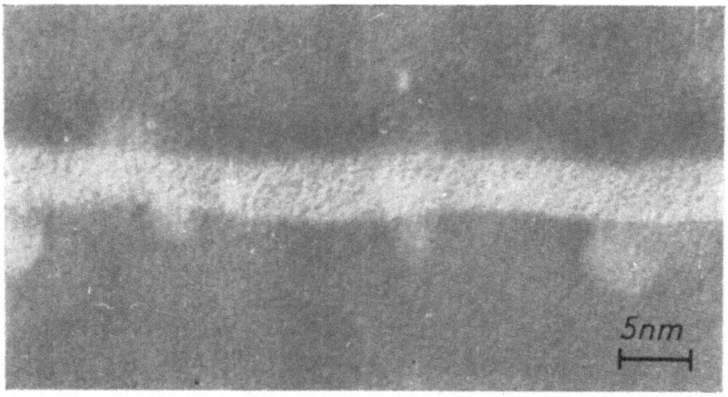

Fig. 1. Weak-beam lattice image of precipitation on a crack. Both phases are coherent

and bottom edges of the twin boundary (Fig. 3). In Fig. 4, the small phase is periodically spaced within the layer on the edge of the boundary. The dark-field precipitate reflection image in Fig. 5a shows the layer of second phase to be in strong contrast and the region of twin boundary in the center to produce no contrast. Fig. 5 b is a dark-field precipitate reflection image of the same boundary tilted to a different orientation. The cloud appears grey and the small phase white. In this experiment, the intensities of the phases were measured with respect to background by taking profiles across the phases in Fig. 5 with a microdensitometer. The ratio of the intensities of the white phase to the layer in Fig. 5a and the ratio of the intensities of the white phase to the grey phase in Fig. 5b were both found to be about 2.6. The phases can be identified by finding the two whose ratio of structure factor differences is 2.6.

The assumption that Δt is the same for each phase was used in this analysis because there do not appear to be any intensity variations within the precipitate when its thickness changes.

The structure factors were calculated from the positions of the atoms in a unit cell (x, y, z) and the atomic scattering factor f_i [1],

$$F_{hkl} = \sum_i f_i \, e^{2\pi i(hx+ky+lz)} .$$

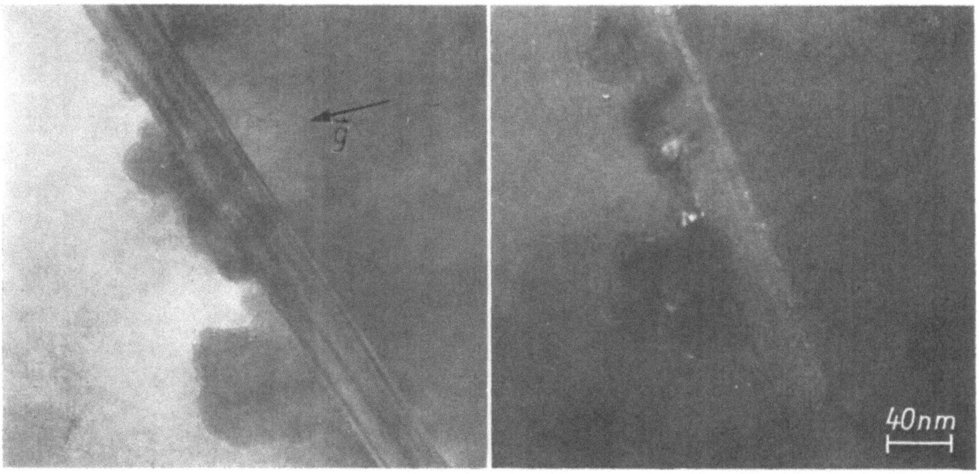

a b

Fig. 2. Bright field and weak-beam images of precipitates on twin boundary (B2). Two phases are visible, a cloud-like phase and a smaller phase which is within the cloud

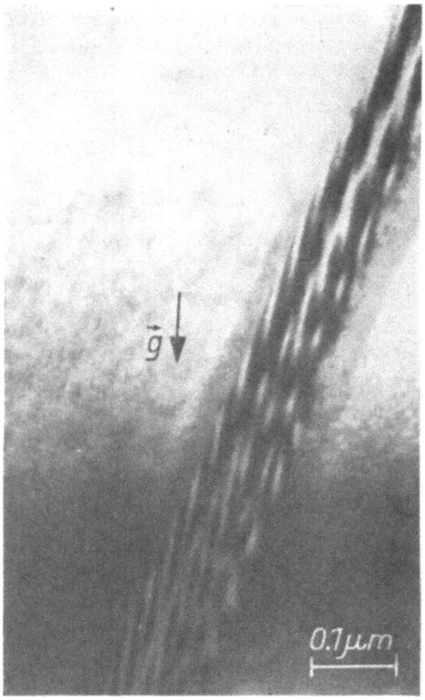

Fig. 3. A strip on each side of the twin (B2) has been replaced by a layer of second phase

Table 1 lists the structure factor values for the crystal structures of Cu–Si phases listed in Pearson [2, 3] when *hkl* = 020. The reflection *hkl* = 020 was chosen because it is a forbidden reflection in Si. The phases considered are CuSi (A2), CuSi (B_3) [7], CuSi (A3), $Cu_{15}Si_4$ ($D8_6$), Cu_5Si (A13), Cu_3Si (A_2), and two $CuSi_2$ transition phases. The transition phases are the Si sublattice with Cu atoms on either the tetrahedral or octahedral sites. Their structure factors are equal in magnitude but opposite in sign. The ratio of the change in structure factor can be taken from this chart. The closest value to 2.6 is attained when the cloud-like phase and the layer are both CuSi and the small spherical phase is the transition phase of the Si sublattice with Cu on octahedral sites.

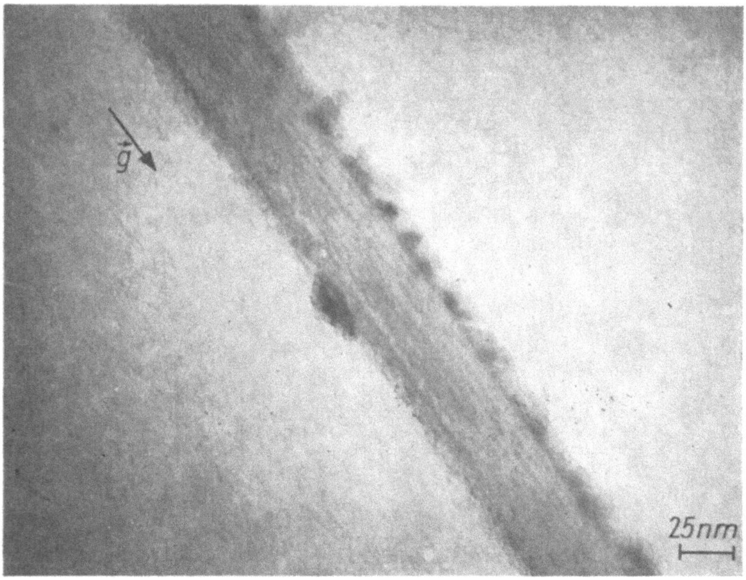

Fig. 4. Twin boundary (B2) shows that most of the precipitates are periodically spaced within the layer region on the side of the boundary

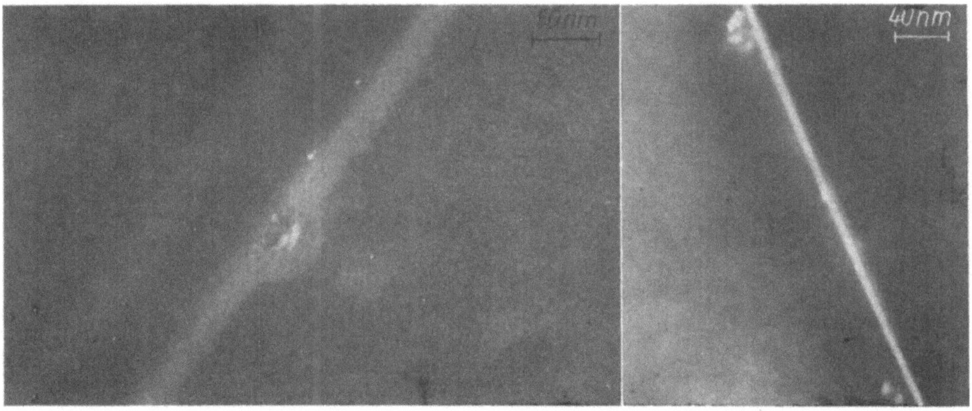

Fig. 5. Dark field precipitate reflection images of the same twin boundary (B2) taken at two different orientations. $g = 020$. a) The layers on each side of the twin are in contrast while the middle section is out of contrast. b) The cloud-like phase appears grey and the small phase white

Table 1

A list of Cu–Si second phases and their structure factors for the $hkl = 020$ reflection

	F_P	F_M	ΔF
CuSi (A3) disordered	− 6.25	0	− 6.25
CuSi (A2)	+12.5	0	12.5
CuSi (B_3, dc)	+ 2	0	2
$Cu_{15}Si_4$ ($D8_6$)	0	0	0
Cu_5Si (A13)	0	0	0
$CuSi_2$ (Si sublattice with Cu on tetrahedral sites)	−26	0	−26
$CuSi_2$ (Si sublattice with Cu on octahedral sites)	+26	0	+26
Cu_3Si (A_2) disordered	+12.75	0	+12.75

3.3 The structure of the twin boundary

The twin boundary was characterized by finding the boundary plane, angle and axis of rotation. By tilting the crystal so the boundary plane is parallel to the electron beam, the boundary plane was found to be $(\bar{1}20)$. The misorientation angle was measured by taking diffraction patterns one each side of the boundary and measuring the magnitude of the shift in Kikuchi lines. The angle corresponding to the shift is 3.7°. This method is similar to that used by Young et al. [8] to measure misorientation angles in low-angle grain boundaries. The rotation axis, found by taking the cross product between the beam direction and the shift vector, is $a_2 = (12\bar{1})$. A second boundary (B1) which contains the Cu rich phases has a misorientation angle of 2.5° and an axis of rotation $a_1 = (\bar{6}\bar{1}1)$.

Dislocations appear in the twin boundary shown in Fig. 6. Since the second phases are only observed on the top and bottom edges of the twin boundary, the stress fields of the defect must promote the transformation by either reducing the strain

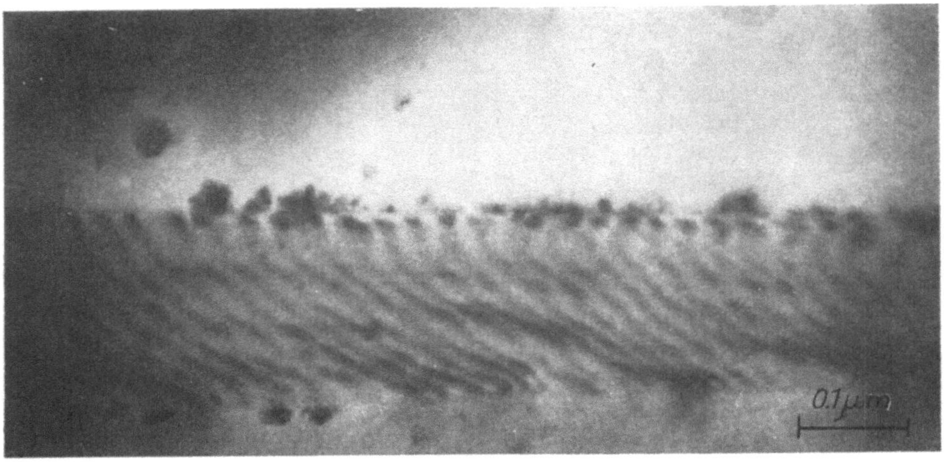

Fig. 6. Dislocation array in twin boundary (B2). Precipitation occurs at the top and bottom edges of the twin

energy or acting as a fast diffusion path enabling a critical concentration of solute to be accumulated [9]. The dislocations terminate at the edges of the twin plate. Since there are changes in the pressure and stress field near the end of a terminating dislocation [10, 11], the region near the termination point is a preferred nucleation site.

The most unusual observation is the lines of contrast which start from the interface of either the small spherical phase or the cloud-like phase and penetrate the crystal in branches (see Fig. 7). The branches cannot be dislocations because dislocations cannot end in the middle of a crystal. The lines may be voids whose formation is associated with the emission of vacancies, which has been suggested to occur in order to relieve strain associated with the transformation [12].

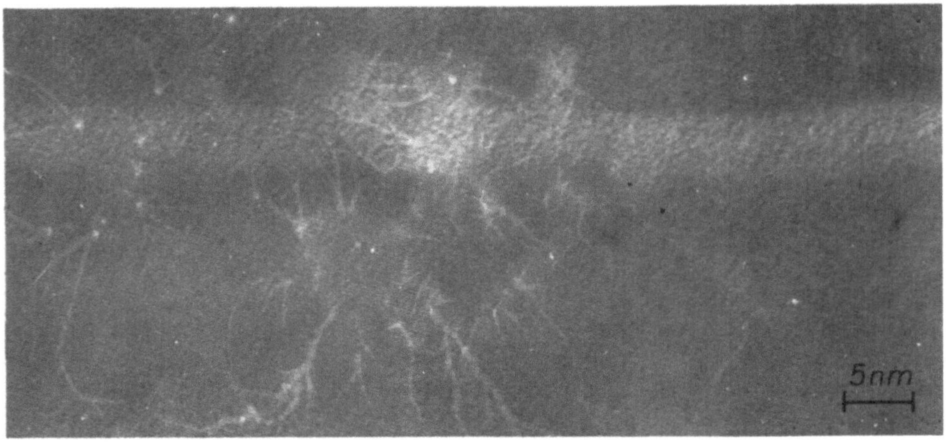

Fig. 7. Weak-beam lattice image which shows vein-like voids emanating from the precipitate interfaces

4. Summary and Conclusions

(i) A dark field TEM method was developed to identify the crystal structures of second phases based on the concept that the change in intensity between precipitate and matrix is proportional to the structure factor difference. It can distinguish between phases whose structure factors are equal in magnitude but opposite in sign.

(ii) The method was applied to Cu-rich phases on twin boundaries in Si. Three phases were observed, a cloud-like phase, a small phase embedded in the cloud, and a layer at the top and bottom edges of the twin. The cloud and layer were found to be CuSi and the small phase a transition phase of Si sublattice with Cu on octahedral sites.

(iii) Dislocations were found to be present in the twin boundary. The region near where the dislocations terminate at the edge of the twin plate is a preferred nucleation site because the gradients in the dislocation's pressure and stress field can assist the transformation.

(iv) Weak-beam lattice images show that the cloud and small phases are coherent and that vein-like voids emanate from the precipitate's interface.

Acknowledgements

The author would like to thank Dr. K. N. Tu, Dr. D. A. Smith, Dr. C. R. M. Grovenor, and Dr. P. E. Batson of IBM, Dr. J. B. Warren of Brookhaven National Laboratory, and Prof. A. H. King and Prof. R. Reeder of SUNY Stony Brook for their assistance.

References

[1] B. D. Cullity, Elements of X-Ray Diffraction, Addison-Wesley Publ. Co., Reading (MA) 1978.
[2] W. B. Pearson, Internat. Ser. Monographs Metal Physics and Physical Metallurgy, Pergamon Press, New York 1958.
[3] W. B. Pearson, A Handbook of Lattice Spacings and Structures of Metals and Alloys, Vol. 2, Pergamon Press, New York 1967.
[4] W. C. Dash, J. appl. Phys. 27, 1193 (1956).
[5] D. J. H. Cockayne, J. Microscopie 98, 116 (1973).
[6] P. B. Hirsch, A. Howie, R. B. Nicholson, D. W. Pashley, and M. J. Whelan, Electron Microscopy of Thin Crystals, Krieger Publ. Co., New York 1977.
[7] G. Das, J. appl. Phys. **44**, 4459 (1973).
[8] C. T. Young, J. H. Steele, Jr., and J. L. Lytten, Metallurg. Trans. **4**, 2081 (1973).
[9] D. M. Vanderwalker, Proc. Ann. Electron Microscopy Soc. Amer., Vol. 42, Ed. G. W. Bailey, 1984 (p. 346).
[10] S. J. Shaibani and P. M. Hazzledine, Phil. Mag. **44**, 657 (1981).
[11] D. M. Vanderwalker and J. B. Vander Sande, Proc. Internat. Conf. Solid–Solid Phase Transformations, New York 1982 (p. 667).
[12] R. Bullough and R. C. Newman, Rep. Progr. Phys. **33**, 101 (1976).

(Received June 19, 1984)

phys. stat. sol. (a) 85, 69 (1984)

Subject classification: 1.3 and 1.5; 22

Department of Electrical Engineering, University of Nebraska, Lincoln[1]) *(a) and NASA Lewis Research Center, Cleveland*[2]) *(b)*

Structural and Chemical Analysis of a Silicon Nitride Film on GaAs by Null Ellipsometry

By

S. A. ALTEROVITZ[3]) (a), G. H. BU-ABBUD (a), J. A. WOOLLAM (a), and D. C. LIU (b)

Multiple-wavelength-angle-of-incidence null ellipsometry, together with the effective medium approximation, are used to evaluate both the composition and the geometrical structure of an insulating film made of silicon nitride and silicon oxide on a GaAs substrate. The results are found to be in good agreement with AES measurements on the same film.

L'ellipsometric nulle, à ondes et angles d'incidence multiples, avec l'approximation du milieu effectif, sont utilisées pour evaluer la composition et la structure geometrique d'une couche dielectrique mince constituée de nitride du silicon et d'oxyde du silicon sur GaAs. Les resultats cont affirmés par les measurements en AES faits sur le même echantillon.

1. Introduction

An accurate, non-destructive testing of thin films on substrates is one of the most important applications of ellipsometry [1, 2]. The ultimate goal of the test is the chemical composition of the film as a function of depth. However, this type of result can be obtained in practice only for rather simple structures, and only where the optical properties of the materials included in the film can be estimated ahead of time. This is not a major restriction in one of the most important applications, namely, the study of insulating thin films on top of semiconducting (or metal) substrates. In these cases, the films are deposited or grown in such a way that the major chemical components of the film are known, and only their distribution is being examined. Under these conditions, ellipsometry as a non-destructive method, is superior to AES depth profiling by ion milling, which is destructive and lacks the depth resolution of ellipsometry.

The experimental arrangements used in ellipsometry can be divided into two general categories-spectroscopic, and fixed wavelength. In scanning spectroscopic ellipsometry [1, 3 to 6], measurements are made to obtain two ellipsometric parameters ψ and Δ (or a combination of the sin and cos of these parameters) at a large number of wavelengths. Usually, this technique is used at a fixed angle of incidence. However, the spectroscopic ellipsometry technique is not trivial to implement. First, the hardware of a completely working system is not commercially available today. In addition, a sophisticated and hardware dependent on-line computer is essential for driving the system and acquiring and analyzing data. Thus the number of such systems in operation is small.

---

[1]) W 194 Nebraska Hall, Lincoln, Nebraska 68588, USA.

[2]) Cleveland, Ohio 44135, USA.

[3]) Now at NASA Lewis Research Center, Cleveland, Ohio 44135, USA.

The other experimental category includes all the systems where single wavelength measurements are considered to be a distinct aspect of the experiment. The largest example in this category is null ellipsometry [2]. More generally this class includes any ellipsometer that needs one or more manual operations to change the wavelength. Thus, commercial "automatic" ellipsometers are included in this second class. The most common type, the null ellipsometer, is now a standard tool in many research and development laboratories. It is simple to use, commercially available in various forms, and (in many cases) can be used at multiple angles of incidence. All ellipsometers in this class are limited to a small number of wavelengths, because of the manual operation mentioned above, and (for null ellipsometry) the need for a strong light source, e.g., a laser line or a strong gas emission line. Null ellipsometry has never been used (to our knowledge) to get information on both structure and composition of a thin film. Yet, many more null or semiautomated ellipsometers are in use than automated systems.

In this paper, we show that null or fixed wavelength ellipsometry can be used in a similar way to the scanning automatic technique. The material studied is a silicon nitride film on a GaAs substrate. It is known that silicon nitride films tend to be non-stoichiometric Si_3N_4, i.e., they contain other components, unless special precautions are taken. For this reason, we have tested several films made under typical industrial laboratory conditions. Layers with extra oxide or a-Si were expected. However, we had no a priori knowledge of the existence of these layers, nor their position or chemical composition. Thus, the ellipsometric measurement has been put to a difficult test in order to show the problems encountered in growing these films. The results are given in terms of a model assuming several layers, each with its particular chemical composition.

2. Experimental Method and Results

The experimental method is essentially based on the multiple wavelength–angle of incidence ellipsometric technique [7, 8]. However, the data analysis has been changed in order to obtain composition rather than optical constants. The method includes two parts: (i) experimental, and (ii) data analysis. In the first part, the ellipsometric parameters ψ and Δ are measured at multiple angles of incidence (MAI) φ_i, and at several wavelengths λ_j. In the second part, we apply an inversion process that includes several steps: (I) Modeling the composition and structure of the sample under examination; (II) finding a first estimate of the free parameters required to fit the experimental results to the model of steps I; and (III) using a minimization process to find the values of the free parameters that give the best fit to the experimental ψ and Δ.

In the present case, silicon nitride has been sputtered on GaAs. Therefore, step I includes the assumption that the films contain one or more of the following: Si_3N_4, SiO_2, amorphous silicon (a-Si), voids and SiO_xN_y. The structure includes one or more layers, each one parallel to the substrate and homogeneous in composition. The optical parameters of the components except SiO_xN_y were taken from the literature [9 to 11]. Since no optical data are available for SiO_xN_y, we modeled this as being as Si_3N_4 plus SiO_2 physical mixture. The optical parameters of the GaAs substrate were measured and corrected for overlayers [7]. The result is in good agreement with published [12, 13] results on GaAs. In step II we combined the dielectric constant of the m-th component in the n-th layer $\varepsilon_{m,n}$ to obtain the effective dielectric constant of the layer (ε_n) using the effective medium approximation (EMA) i.e. [14, 15]

$$\sum_m f_{m,n} \frac{\varepsilon_{m,n} - (\varepsilon_n)}{\varepsilon_{m,n} + 2(\varepsilon_n)} = 0\,, \qquad \sum_m f_{m,n} = 1\,, \tag{1}$$

where $f_{m,n}$ is the volume fraction of the m-th material in the n-th layer. The expected ψ' and Δ' were calculated using an assumed set of thicknesses d_i ($i = 1, ..., n$) of the layers, using standard equations [2]. In step III, we use a Marquardt algorithm to minimize the error function F [16, 17], or ϱ, where

$$F = \sum_{ij} [(\psi - \psi')^2 + (\Delta - \Delta')^2] \tag{2}$$

and

$$\varrho = \frac{F}{(N - P)^{1/2}}\,. \tag{3}$$

Here $\psi = \psi(d_i, \lambda_j)$ and $\Delta = \Delta(\varphi_i, \lambda_j)$ are the experimental values, N is the total number of experimental points, and P is the number of free parameters. In this algorithm the values of the free parameters in the problem, i.e., the volume fractions $f_{m,n}$ and the thicknesses of the layers, are changed in finite steps in such a way that F is made smaller. After a finite number of steps, a local minimum of F is found. The value of the free parameters at this minimum is the final answer.

The actual experimental set-up has been described earlier [8] and will not be repeated here. We will only mention that a commercial null ellipsometer was used, with a mercury lamp and an argon laser as light sources. Sample preparation was also described in an earlier paper [18].

Measurements were taken in the visible for up to seven wavelengths, and, at several angles of incididence for each wavelength. A summary of the results obtained from the least square fits are given in Tables 1 and 2, and in Fig. 1. No real minimum was

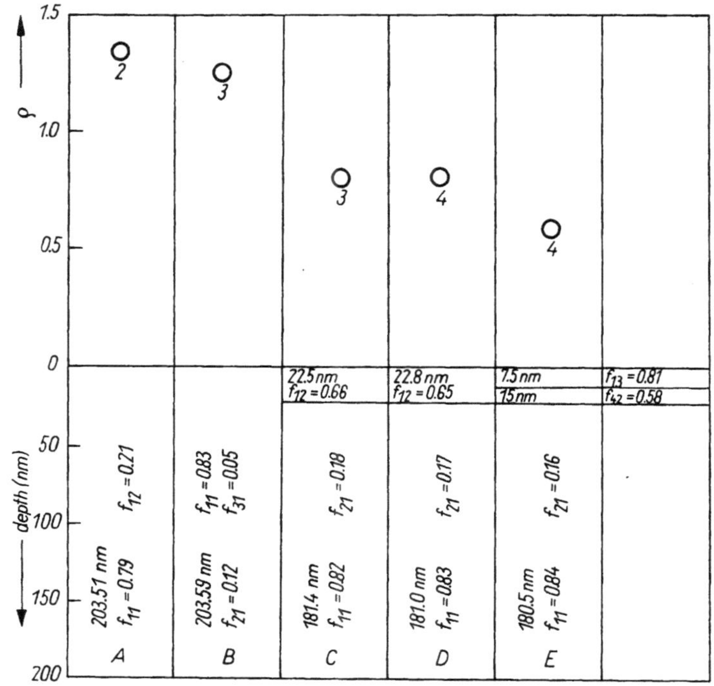

Fig. 1. Bar diagram of the five assumed model configurations shown in Table 1. The upper part shows ϱ (defined by (3)), and the numbers on the graph indicate the number of free parameters. The lower part is a schematic of the five structures

Table 1

Results of least squares fits for ≈ 204 nm film

No. of free parameters	d_1 (nm)	f_{21}	f_{31}	f_{41}	d_2 (nm)	f_{22}	f_{32}	d_3 (nm)	f_{32}	ϱ	for Fig. 1
2	203.51	0.210	—	—	—	—	—	—	—	1.33	A
3	203.59	0.172	—	0.050	—	—	—	—	—	1.25	B
3	206.73	0.161	0.05	—	—	—	—	—	—	9.4	—
3	181.40	0.177	—	—	22.5*)	0.345	—	—	—	0.78	C
4	182.12	0.197	0.01	—	22.5*)	0.350	—	—	—	1.20	—
4	181.08	0.175	—	—	22.81	0.348	—	—	—	0.77	D
4	180.50	0.163	—	—	15.0*)	0.420	—	7.5*)	0.192	0.62	E
4	180.4	0.162	—	—	10.0*)	0.474	—	12.5*)	0.249	0.62	—

*) Fixed.

Data taken at seven wavelengths in the range 404.7 to 546.1 nm, ϱ is defined by (3).

Table 2

Results of the least squares fits for a ≈ 64 nm film

No. of free parameters	d_1 (nm)	f_{21}	f_{31}	f_{41}	d_2 (nm)	f_{22}	ϱ
2	63.92	0.307	—	—	—	—	0.76
3	63.91	0.129	—	0.100	—	—	0.70
3	63.92	0.496	−0.105	—	—	—	0.83
3	54.0*)	0.283	—	—	12.8	0.62	0.65
3	59.0*)	0.289	—	—	9.5	0.99	0.63

*) Fixed.

Data taken at five wavelengths in the range of 435.8 to 514.5 nm.

found when assuming the presence of just one material ($m = 1$) in either a one ($n = 1$) or a two ($n = 2$) layer model. We therefore started with $m = 2$ and $n = 1$. A marked decrease in ϱ was observed when three unknown parameters were assumed. However, reliable results were not obtained for more than three parameters (for the ≈ 64 nm film), or four parameters for the ≈ 204 nm film), because of high correlations [8]. Adding a physically unmeaningful parameter will not decrease ϱ. Adding a parameter which is correlated to others will increase the dependence of the final results on the first estimates. It will also increase the 95% confidence limits, and will give very high correlation coefficients [8]. We have not shown the 95% confidence limits in the tables, because they are exceedingly small, of order 0.1 nm for thicknesses, and 0.005 for volume fractions. (These errors do not reflect systematic errors, which are probably much higher.)

We therefore used $m \leqq 3$ in our analysis. That is we assumed the presence of either voids *or* a-Si, in addition to Si_3N_4 and SiO_2. (In the tables we use the following designations: $m = 1$ for Si_3N_4, $m = 2$ for SiO_2, $m = 3$ for a-Si, and $m = 4$ for voids. For each layer $f_{1,n} = 1 - \sum_{m=2}^{4} f_{m,n}$.) A relatively large decrease in ϱ was observed when we added a second layer ($n = 2$). In order to avoid an unnecessary increase in the number of parameters, we used a "slicing" method for analyzing composition versus depth. In this approach, the thickness of all layers except one was assumed to be constant. Several examples are shown in Table 2. The improvement in the ψ and Δ fits to the experiment are shown in Fig. 2 and 3. We see that the "slicing" method works well

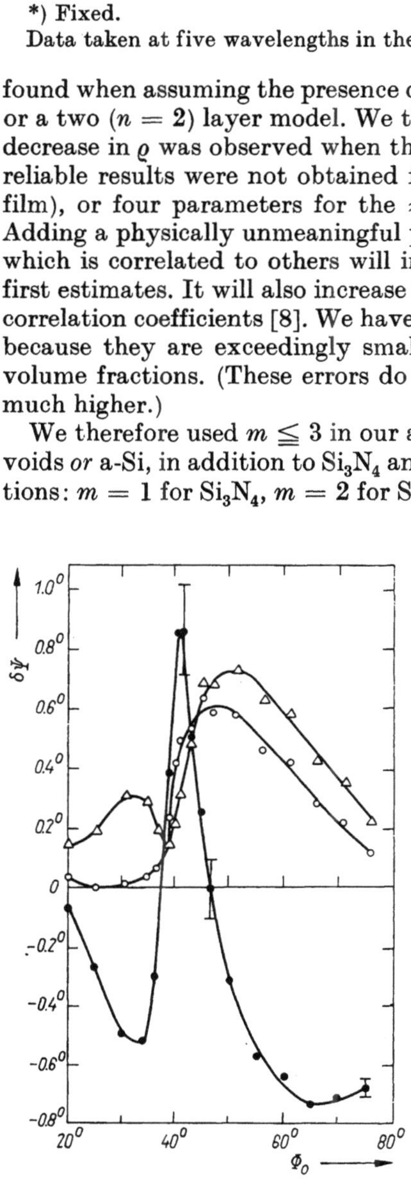

Fig. 2. Difference between the experimental and the calculated ellipsometric angle ψ vs. the angle of incidence φ_0, at $\lambda = 496.5$ nm. The calculated functions were obtained using the following values: 2 parameters $d_1 = 203.5$ nm, $f_{21} = 0.210$ (●); 3 parameters $d_1 = 181.6$ nm, $f_{21} = 0.177$, $d_2 = 22.5$ nm, $f_{22} = 0.365$ (○); 4 parameters $d_1 = 180.5$ nm, $f_{21} = 0.163$, $d_2 = 15.0$ nm, $f_{22} = 0.420$, $d_3 = 7.5$ nm, $f_{32} = 0.192$ (△) (A, C, and E in Table 1 respectively)

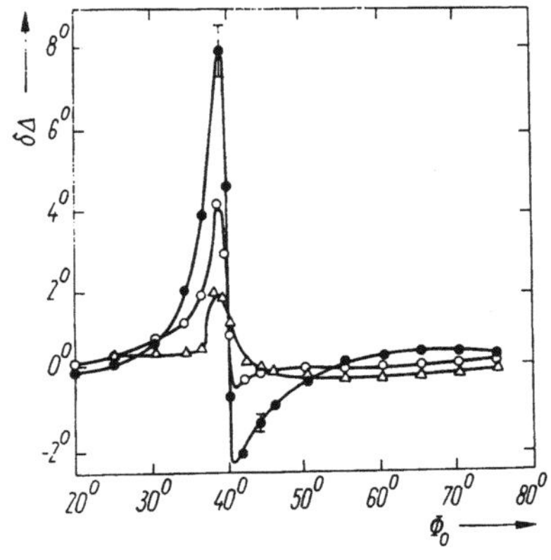

Fig. 3. Difference between the experimental and the calculated ellipsometric angle Δ vs. the angle of incidence, at $\lambda = 496.5$ nm. The calculated functions were obtained using the same values of the parameters given for Fig. 2

as an effective use of the free parameters to minimize ϱ. The reason for this is that composition gradients are probably present in the SiO_2 (or Si_3N_4). We tried other possible "slices" for the film in addition to those given in Table 2, but all gave higher values of ϱ. For example, we tried using four parameters to describe the following structures: a) an oxide rich interface layer $\leqq 5$ nm thick adjacent to the substrate, on top of which was a thick layer, and finally a top oxide rich layer, b) a thick layer, followed by *a* ≈ 20 nm thick layer, and then finally (on top) a very thin layer, c) a thick layer and two ≈ 20 nm thick top layers. In all these cases, the only two components considered were SiO_2 and Si_3N_4. We also tried to introduce a-Si into the top layer using the a-Si fraction as a fourth parameter in the 22.5 nm "slice".

A summary of the results for the ≈ 204 nm film is a follows:

1. There are no signs of any measurable oxide rich layer just on top of the substrate. This fact suggests that steps taken to preserve the substrate from oxidation prior to deposition were successful.

2. There are no measurable sign of a-Si, suggesting that all the sputtered silicon was either oxidized or nitrided.

3. The SiO_2 volume fraction, which is probably below 0.16 in most of the body of the film, increases markedly in the top ≈ 25 nm. In the uppermost layer, probably less that 10 nm thick, we have indications that the SiO_2 fraction decreases.

4. The top ≈ 25 nm has a concentration gradient. The absence of abrupt concentration changes is the reason that the "slicing" method makes best use of the free parameters for this film.

For the ≈ 64 nm we could reliably use only three parameters, probably due to the restricted spectral range and the resulting correlations. In summary: (i) There is no sign of a-Si in the film. (ii) There are voids in the film. However, a correlation between the void fraction and the SiO_2 fraction did not permit an accurate evaluation of these quantities. (iii) There is a concentration gradient in the SiO_2 plus void fractions within the film. An almost constant fraction (0.03) of SiO_2 (assuming no voids) is found in most of the film, except the uppermost ≈ 10 nm. An AES measurement of this film is shown in Fig. 4. We use the ellipsometric result to calibrate the thickness versus AES sputter time. AES finds an oxide rich top 10 nm layer, and fairly constant oxide and nitride concentrations in the remainder of the film. These results are in excellent agreement with our ellipsometric data on this film. The results on both films indicate a constant oxidation during the plasma enhanced sputtering, and an additional oxidation of the uppermost layer, probably after deposition. A similar result was reported for LPCVD Si_3N_4 films by spectroscopic ellipsometry [6].

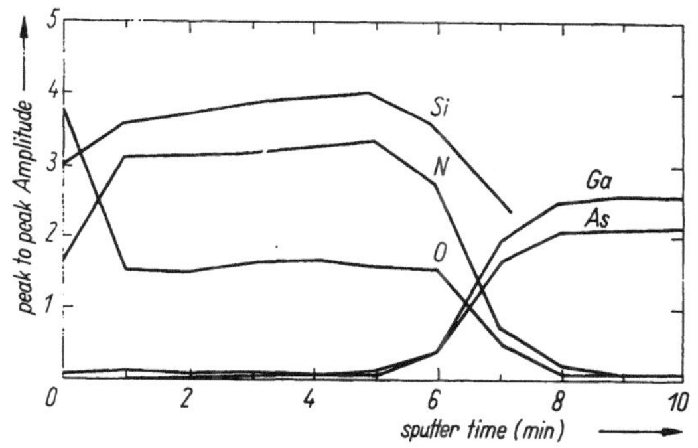

Fig. 4. Auger electron spectroscopy (AES) depth profile for the ≈ 64 nm silicon nitride–silicon oxide sputtered film on a GaAs substrate

3. Conclusion

In summary, we have shown that using null elipsometry operated at multiple wavelengths and multiple angles of incidence it is possible to analyze nondestructively and semiquantitatively the structure and chemical composition of a silicon nitride film on a GaAs substrate. The results have been analyzed in terms of their importance to the film composition and preparation procedure. Only a small number (up to 4) of parameters could be reliably estimated. However, even in the case of spectroscopic ellipsometry [6] a large number of parameters (6 for example) became correlated with data not completely reliable. (Thickness and composition of one layer became dependent on angle of incidence, for example.) We found that the most effective way of analyzing data when there are only a small number of parameters is to use the "slicing" method.

Acknowledgements

We wish to thank Dr. Y. Chung of Universal Energy Systems and Dr. D. Langer of the Avionics Laboratory, Wright Patterson Air Force Base, for support in preparing samples. This research was supported by Grant NAG3-154 from NASA Lewis Research Center, and Contract F33601-81-MJ869 from the Avionics Laboratory of Wright Patterson Air Force Base.

References

[1] D. E. Aspnes, Optical Properties of Solids: New Developments, Ed. B. O. Seraphin, North Holland Publ. Co. 1976 (p. 799).
[2] R. M. A. Attam and N. M. Bashara, Ellipsometry and Polarized Light, Chap. 4., North Holland Publ. Co., 1977.
[3] P. S. Hauge, Surface Sci. **96**, 108 (1980).
[4] D. E. Aspnes, J. Vacuum Sci. Technol. **18**, 289 (1981).
[5] J. B. Theeten and D. E. Aspnes, Ann. Rev. mater. Sci. **11**, 97 (1981).
[6] F. H. P. M. Habraken, A. E. T. Kuiper, A. V. Vostrum, Y. Tamminga, and J. B. Theeten, J. appl. Phys. **53**, 404 (1982).
[7] G. H. Bu-Abbud, S. A. Alterovitz, N. M. Bashara, and J. A. Woollam, J. Vacuum Sci. Technol. **A1**, 619 (1983).
[8] S. A. Alterovitz, G. H. Bu-Abbud, J. A. Woollam, and D. C. Liu, J. appl. Phys. **54**, 1559 (1983).
[9] I. H. Malitson, J. Opt. Soc. Amer. **55**, 1205 (1965).
[10] H. R. Phillip, J. Electrochem. Soc. **120**, 295 (1973).
[11] D. E. Aspnes, J. B. Theeten, and F. Hottier, Phys. Rev. B **20**, 3292 (1979).
[12] D. E. Aspnes and A. A. Studna, Phys. Rev. B **27**, 985 (1983).

[13] D. E. ASPNES, G. P. SCHWARTZ, G. J. GAULTIERI, A. A. STUDNA, and B. SCHWARTZ, J. Electrochem. Soc. **128**, 590 (1981).
[14] C. GRANDQUIST and O. HUNDERI, Phys. Rev. B **18**, 2897 (1978).
[15] D. E. ASPNES, Thin Solid Films **89**, 249 (1982).
[16] G. H. BU-ABBUD and N. M. BASHARA, Appl. Optics **20**, 3020 (1981).
[17] D. L. MARQUARDT, J. Soc. Indust. appl. Math. **2**, 431 (1963).
[18] S. A. ALTEROVITZ, G. H. BU-ABBUD, J. A. WOOLLAM, D. C. LIU, Y. CHUNG, and D. LANGER, Appl. Phys. Commun. **1**, 163 (1982).

(Received June 13, 1984)

phys. stat. sol. (a) 85, 77 (1984)

Subject classification: 1.1; 22.1.2

Department of Physics, Faculty of Technology, Gottwaldov[1])

Coherence Properties of Radiation Diffracted by an Elastically Bent Crystal

By

Z. JANÁČEK

The properties of X-rays scattered by an elastically bent crystal are examined theoretically under various coherence conditions. The degree of coherence of the Laue diffracted beam in the case of an incoherent incident beam is studied more thoroughly. Using the approximation of large deformations it is shown that the coherence depends both, on the magnitude of the deformation and on the sign of the diffraction vector. This effect may be interpreted as the influence of crystal anisotropy.

Die Eigenschaften von den an einem homogen gebogenen Idealkristall abgebeugten Röntgenstrahlen werden theoretisch untersucht. Es werden verschiedene Kohärenzbedingungen berücksichtigt. Der Fall der Laue-Beugung und unkohärenten einfallenden Welle wird gründlicher behandelt. In der Näherung großer Deformationen wird gezeigt, daß der Kohärenzbereich von der Deformationsgröße sowie von der Richtung des Diffraktionsvektors abhängt. Dies kann als Effekt der Anisotropie des Kristalls interpretiert werden.

1. Introduction

Recently a new attention has been paid to the study of the dynamical X-ray diffraction using the formalism of mutual coherence functions (MCF) [1]. It has been shown that the coherence approach enables us to compute easily the diffracted intensity and its angular distribution.

Kuběna [2] studied diffraction from mosaic crystals within the kinematical theory of diffraction. He showed that the relation between the coherence width (length) and the characteristic dimension of the imperfection plays the main role in the diffraction from imperfect crystals.

Diffraction from perfect crystals has been studied by Bezirganyan using both kinematical [3] and Darwin dynamical [4] theories. It was shown that the influence of coherence effects depends on the ratio of the coherence width (length) and the extinction length. Decreasing the time coherence leads to the reduction of the visibility of the interference effects and to modifications of the diffraction profile in the kinematical approximation. The visibility of the moiré fringes in X-ray interferometers depends directly on the coherence length of the incidet radiation [5].

The growing interest in using bent crystals as a part of the diffraction equipments for clasical X-rays [6], synchrotron [7] and neutron [8] radiations requires to investigate more deeply the properties of the radiation diffracted by bent crystals.

In this paper we study the coherence properties of such radiation. The formula for the MCF of the wave field emerging from the bent crystal will be analyzed under various coherence conditions. In the case of spatially incoherent primary beam the absolute value of the normalized coherence function is calculated in the approximation of large deformations.

---

[1]) POB 72, nám. Rudé armády 275, 76272 Gottwaldov, Czechoslovakia.

2. Theory

Let us suppose that a quasimonochromatic beam with the mutual coherence function $\Gamma_e(x, x'; t, t')$ falls on the entrance crystal surface. The MCF of the radiation emitted from the exit surface is given by the relation [9]

$$\Gamma_j(y, y'; t, t') = \int dx \int dx' \, \Gamma_e\left[x, x'; t, t' + \left(\Delta - \frac{\delta_0}{c}\right)(x - x') - \right.$$
$$\left. - \left(\Delta - \frac{\delta_j}{c}\right)(y - y')\right] G_j(x, y; \nu) \, G_j^*(x', y'; \nu) \,, \tag{1}$$

where $G_j(x, y; \nu)$ is the Green function of the crystal. In deriving (1) we assumed that the dependence of the Green function on the frequency ν is very weak in the spectral range of one characteristic X-ray line.

Formula (1) cannot be easily analyzed in the general case of an arbitrary wave falling on the crystal entrance surface. For a better understanding of the diffraction process we investigate only the limiting cases.

2.1 Point source on the entrance surface

The point source of X-rays placed at one point of the entrance surface is apparentely the simplest case for analyzing (1). This situation can be experimentally arranged with an incident beam having a divergency much greater than the reflection curve of the crystal. The width of the beam should be much smaller than the width of the base of the Borrmann triangle.

Under these assumptions the MCF of the incident beam may be written in the form

$$\Gamma_e(x, x'; t, t') = \delta(x - x_0) \, \delta(x' - x_0) \, \gamma(t - t') \,, \tag{2}$$

where $\gamma(t - t')$ is the Fourier transform of the spectral density $g(\nu)$ of the incident radiation.

Putting (2) into (1) we obtain

$$\Gamma_j^p(y, y'; t, t') = G_j(x_0, y; \nu) \, G_j^*(x_0, y; \nu) \, \gamma\left[t - t' - \left(\Delta - \frac{\delta_j}{c}\right)(y - y')\right] . \tag{3}$$

The scattered radiation is spatially coherent as can be seen from (3). The time coherence is determined by the time coherence of the incident radiation. It is due to the approximation used in which the Green function was supposed to depend weakly on the frequency of radiation.

2.2 Spatially coherent primary beam

Let us suppose that the crystal is irradiated by a "bundle" of plane waves with various frequencies from the range of one characteristic X-ray line. In the experimental arrangement we can use the beam diffracted from the first crystal in the double crystal diffractometer with the second crystal being bent. The divergency of this beam should be much smaller than the width of the reflection curve. The spatial width should be greater than that of the Borrmann triangle. The MCF of such a beam can be expressed by

$$\Gamma_e(x, x'; t, t') = \int d\nu \, g(\nu) \exp\left\{2\pi i \left[\nu(t - t') - \left(\frac{\nu}{c}\delta_0 + \eta_x\right)(x - x')\right]\right\} , \tag{4}$$

where η_x characterizes the deviation of the incident beam from the kinematical Bragg position.

Putting (4) into (1) we obtain

$$\Gamma_j^c(y, y'; t, t') = \int d\nu\, g(\nu) \exp\left\{2\pi i\nu\left[t - t' + \left(\Delta - \frac{\delta_j}{c}\right)(y - y')\right]\right\} \times$$
$$\times\, R_j(\nu\Delta + \eta_x; y)\, R_j^*(\nu\Delta + \eta_x; y')\,, \tag{5}$$

where we have introduced the local reflectivity of the bent crystal by the relation

$$R_j(q; y) = \int dx\, G_j(x, y; \nu) \exp(-2\pi i q x)\,. \tag{6}$$

The intensity of the scattered radiation is then the convolution of the spectral density and the square of the local reflectivity,

$$I_j^c(y) = \int d\nu\, g(\nu)\, |R_j(\nu\Delta + \eta_x; y)|^2\,. \tag{7}$$

In the case of a monochromatic incident beam the scattered radiation is completely coherent. Formula (5) has then the form

$$\Gamma_j^c(y, y'; t, t') = V_j(y)\, V_j^*(y') \exp[2\pi i\nu(t - t')]\,, \tag{8}$$

where

$$V_j(y) = R_j(\nu\Delta + \eta_x; y) \exp\left[2\pi i\left(\Delta - \frac{\delta_j}{c}\right) y\right].$$

2.3 Spatially incoherent primary beam

The MCF of the spatially incoherent incident beam can be written in the form

$$\Gamma_e(x, x'; t, t') = \delta(x - x')\, \gamma(t - t')\,. \tag{9}$$

For the MCF of the scattered radiation we can derive, using (9) and (1),

$$\Gamma_j^i(y, y'; t, t') = \gamma\left[t - t' + \left(\Delta - \frac{\delta_j}{c}\right)(y - y')\right] \int dx\, G_j(x, y; \nu)\, G_j^*(x, y'; \nu)\,. \tag{10}$$

The scattered radiation is partially coherent even when the incident beam is time coherent (monochromatic). In order to analyze the coherence properties thoroughly we have to introduce the actual form of the Green function of the elastically bent crystal. In the Laue case of diffraction it has the form [9]

$$G_h(x, y; \nu) = \alpha\gamma_0(x)\, {}_1F_1(y - x) \exp\left(-\frac{\mu T}{2}\sin\omega\right) \exp\left[-\frac{\mu}{2}(y - x)\cos\omega\right] \times$$
$$\times \exp[i\Phi(x, y)]\, \theta(y - x - T_0)\, \theta(T_h - y + x)\,. \tag{11}$$

${}_1F_1(y - x)$ is the confluent hypergeometrical function, ω the angle between the normal to the diffraction planes and the inner normal to the crystal entrance surface.

$\Phi(x, y)$ is a quadratic form with coefficients depending on the experimental arrangement and deformation, $\theta(x)$ is the usual step function.

Generally the radiation scattered by the bent crystal is not homogeneous, i.e. the MCF depends on the coordinates of two points y, y' rather than on their difference $\varrho = y' - y$. In such a case the MCF should be calculated locally in the vicinity of each point at the exit surface.

Putting (11) into (10) we can find that the absolute value of the MCF does not depend on the position of the observation point y,

$$|\Gamma_h^i(y, y+\varrho; t, t')| = \left|\gamma\left[t - t' - \left(\Delta - \frac{\delta_h}{c}\right)\varrho\right]\right| \exp\left[-\mu\left(T \sin\omega + \right.\right.$$

$$\left.\left. + \frac{\varrho}{2}\cos\omega\right)\right] \left|\int dv\, {}_1F_1(v)\, {}_1F_1^*\,(v+\varrho) \exp\left[(ic_3\varrho - \mu\cos\omega)\,v\right]\right| \times$$

$$\times\ \theta(v - T_0)\,\theta(T_h - v)\,\theta(v + \varrho - T_0)\,\theta(T_h - v - \varrho)\,. \tag{12}$$

The calculation of the convolution in (12) cannot be carried out analytically in the general case. However, if we restrict ourselves to the approximation of large deformations [10], the approximated value of the integral in (12) can be found supposing the confluent hypergeometrical function to be constant over the integration path.

If we define the normalized MCF by the formula

$$T_h(\varrho) = \frac{|\Gamma_h(y, y+\varrho; 0, 0)|}{|\Gamma_h(y, y; 0, 0)|}, \tag{13}$$

we obtain under these assumptions

$$T_h^2(\varrho) = \frac{1}{\cosh(ML) - 1}\,\frac{\cosh[M(L-\varrho)] - \cos[c_3\varrho(L-\varrho)]}{1 + (c_3/M)^2\varrho^2}. \tag{14}$$

Moreover, we supposed, for the sake of simplicity, the incident radiation to be monochromatic. Deriving (14) we introduce parameters $M = \mu\cos\omega$, $L = T_h - T_0$, and the new variable $\varrho = y' - y$.

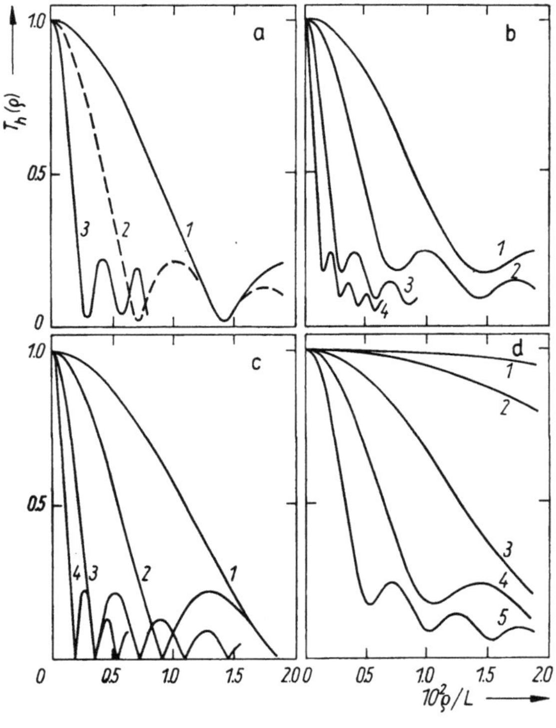

Fig. 1. The normalized mutual coherence functions of the Laue diffracted beam. Silicon single crystal, thickness 0.4 mm. Deformation parameter $\varepsilon = 0.2 \times 10^{-4}$ (1), 0.4×10^{-4} (2), 1.0×10^{-4} (3), 2.0×10^{-4} (4), 4.0×10^{-4} (5). a) Asymmetrical diffraction ($\bar{1}11$), MoKα radiation, b) asymmetrical diffraction ($\bar{1}11$), CuKα radiation, c) symmetrical diffraction ($2\bar{2}0$), MoKα radiation, d) asymmetrical diffraction ($1\bar{1}\bar{1}$), CuKα radiation

The parameter c_3 depends on the deformation and the diffraction conditions. The actual form can be derived using results of our previous papers [9, 10],

$$c_3 = \frac{\pi \gamma_0 \gamma_h}{\sin 2\vartheta} \frac{A}{R}$$

and

$$A = \frac{2}{d \sin 2\vartheta} \left[\cos \omega \left(\delta_0 \delta_h - \frac{S_{13}}{S_{33}} \gamma_0 \gamma_h + 2 \sin^2 \omega\right) - 2\gamma_0 \gamma_h \sin \omega \frac{S_{53}}{S_{33}}\right], \tag{15}$$

where R is the radius of curvature of the crystal, d the interplanar spacing of the diffracting planes, S_{ij} are the values of the compliances with respect to the given crystallographic orientation.

The normalized MCFs are plotted in Fig. 1 a and b for various deformations and for two wavelengths. They show a very sharp maximum near $\varrho = 0$. With increasing deformation parameter $\varepsilon = T/2R$ the first minima move to lower ϱ. Their position ϱ_m can be easily estimated. In the range of large deformations the values of the parameter c_3 become high so that the first minimum occurs for $\varrho \ll L$. From (14) and (15) it follows

$$\varrho_m \doteq |A\varepsilon|^{-1} . \tag{16}$$

One can see from (16) that the position of the first minimum, which corresponds to the coherence width, does not depend on the crystal thickness. This result is in contradiction with that for the perfect crystal [1]. It is probably due to the approximation used in deriving the formula for the normalized MCF (14).

For symmetrical transmission we can simplify (14) to the form

$$T_h^{\text{sym}}(\varrho) = \frac{|\sin (c_3 L \varrho/2)|}{c_3 L \varrho/2} . \tag{17}$$

The corresponding curves are plotted in Fig. 1c. The curves for both wavelengths are identical. The position of the first minimum (zero) is determined again by (16).

3. Discussion

The numerical calculations showed that the normalized MCF has a sharp maximum near $\varrho = 0$ (Fig. 1a). We have defined the width (ϱ_m) of the NMCF as the distance of the first minimum of the NMCF from the zeroth maximum. The radiation diffracted from a bent crystal under the above coherence conditions is thus partially coherent in those pairs of points whose distance is smaller than ϱ_m. In the approximation given it depends neither on the crystal thickness nor on absorption. With increasing deformation the width of the NMCF decreases, i.e. the spatial coherence of the diffracted beam decreases, too. In the limiting case of very large deformations (kinematical limit), the width of the NMCF tends to zero. The diffracted beam becomes spatially incoherent.

Reversing the sign of the diffraction vector (asymmetrical diffraction) causes substantial changes in the NMCF widths. In Fig. 1b and d the NMCFs are plotted against the distance ϱ of two points on the exit surface. The ratio of the widths for diffractions $(\bar{1}11)$ and $(1\bar{1}\bar{1})$ can be estimated from (16) to be $\varrho_m(\bar{1}11)/\varrho_m(1\bar{1}\bar{1}) = 0.135$. For symmetrical diffraction no difference between the MCFs occurs for the (hkl) and $(\bar{h}\bar{k}\bar{l})$ diffractions. Taking into account the results of our previous paper [11] we can estimate the effect of reversing the sign of the diffraction vector on the width of the reflection curves. For our special case it is $\Delta\theta(\bar{1}11)/\Delta\theta(1\bar{1}\bar{1}) = 1.7$. This result is in qualitative agreement with that for the NMCF width. The narrower the reflection curve, the smaller is the divergency of the diffracted beam. Hence, the coherence width of such a beam should be larger.

We can conclude that the effect of reversing the diffraction vector on the NMCF is due to the anisotropy of the crystal. This statement is supported by two facts:

(i) If we suppose a crystal to be an isotropic material, then in (15) $S_{53} = 0$ and $S_{13}/S_{33} = -\nu$ (Poisson ratio). The absolute value of the parameter A in (15) does not depend on the sign of the diffraction vector. From (16) it follows that the NMCF width behaves the same way.

(ii) In perfect nondeformed crystal the MCF is not influenced by the reversion of the diffraction vector as can be easily derived using the results of [1].

However, it will be necessary to explain why the reflection curves are sensitive to the reversing of the diffraction vector even in the isotropic model of a crystal medium. It will be unevitable to derive more exact formula than that from [11].

4. Conclusion

Studying the coherence properties of radiation diffracted by an elastically bent crystal we have analyzed three special cases:

(i) A very narrow divergent beam incident on the entrance crystal surface (point source of radiation on the surface). The scattered beam is completely coherent irrespective of a deformation (3).

(ii) A wide plane wave (completely coherent radiation) incident on a crystal. The scattered radiation is completely coherent, too (8).

(iii) Spatially incoherent radiation incident on a crystal. We examined this case more closely. For the sake of simplicity we supposed the primary beam to be monochromatic. Using the explicit form of the Green function we derived expression (14) for the normalized mutual coherence function.

Because of difficulties with computing integrals comprising the confluenct hypergeometrical function we restricted ourselves to the approximation of large deformations. The radiation diffracted from the bent crystal was partially coherent. The degree of coherence depended on the magnitude of deformation.

References

[1] V. HOLÝ, phys. stat. sol. (b) **101**, 575 (1980).
[2] J. KUBĚNA, Czech. J. Phys. **B20**, 106 (1970).
[3] P. A. BEZIRGANYAN, Zh. tekh. Fiz. **40**, 377 (1970).
[4] P. A. BEZIRGANYAN and A. G. GASPARYAN, Izv. Akad. Nauk Arm. SSR, Fizika **6**, 106 (1971).
[5] E. V. SHULAKOV and V. V. ARISTOV, Acta cryst. **A35**, 201 (1979).
[6] D. W. BEREMAN, Phys. Rev. B **19**, 560 (1979).
[7] K. KOHRA and M. ANDO, Nuclear Instrum. and Methods **177**, 117 (1980).
[8] G. ALBERTINI, A. BOEUF, B. KLAR, S. LAGOMARSINO, S. MAZKEDIAN, S. MELONE, P. PULITI, and F. RUSTICHELLI, phys. stat. sol. (a) **44**, 127 (1977).
[9] Z. JANÁČEK, Ph. D. Thesis, Purkyně University, Brno 1984.
[10] Z. JANÁČEK, J. KUBĚNA, and V. HOLÝ, phys. stat. sol. (a) **50**, 285 (1978).
[11] Z. JANÁČEK, J. KUBĚNA, and V. HOLÝ, Czech. J. Phys. **B31**, 1397 (1981).

(Received May 28, 1984)

phys. stat. sol. (a) 85, 83 (1984)

Subject classification: 1.4; 10.2; 22.4.2

Institute of Physics, Polish Academy of Sciences, Warsaw[1]) *(a),*
Arrhenius Laboratory, Department of Inorganic Chemistry, University of Stockholm (b),
and Institut für Festkörperphysik und Elektronenmikroskopie
der Akademie der Wissenschaften der DDR, Halle (Saale) (c)

Defect Structure of ZnSe Crystals Investigated by Electron Microscopy

By

E. MIZERA (a), M. SUNDBERG (b), and P. WERNER (c)

The defects in the zinc selenide crystals with zinc blende structure grown from the melt by the Bridgman method under argon pressure are examined by electron microscopy. The microscopical observations are performed using high voltage, high resolution, and conventional microscopes. Results of investigations of twins, stacking faults, dislocations, precipitates, and pores in ZnSe crystals are presented and discussed.

Mittels Elektronenmikroskopie werden die Defekte in Zinkselenidkristallen mit Zinkblendestruktur, die mit der Bridgmanmethode unter Argondruck aus der Schmelze gezogen werden, untersucht. Die mikroskopischen Beobachtungen werden mittels Hochspannungs-, Hochauflösungs- und konventioneller Mikroskopie durchgeführt. Ergebnisse von Untersuchungen an Zwillingen, Stapelfehlern, Versetzungen, Präzipitaten und Poren in ZnSe-Kristallen werden mitgeteilt und diskutiert.

1. Introduction

ZnSe is one of II–VI semiconductor compounds with a zincblende structure which are used in optoelectronics, especially for luminescence diodes. The physical properties of this material and its applications depend not only on dopands or other impurities but also on lattice defects caused by crystal growth or manufacturing processes. It is rather difficult to obtain large ZnSe single crystals with a high perfection. As a consequence of the nearly identical environments of atoms in the hexagonal and cubic close-packed structure, the "real crystals" we meet in practice contain faulted arrangement of the layers. The stacking faults and the twins are one of the principal defects in ZnSe structure.

The stacking faults and the twins in ZnSe crystals have been analysed by TEM and discussed by Cutter et al. [1]. Shiojiri et al. [2] have presented HRTEM results of the observation of twins in ZnSe crystals which had been prepared by solid–solid reaction. Dislocations in deformed ZnSe single crystals have been studied by weak-beam electron microscopy and discussed by Aristov et al. [3]. Irradiation damage of ZnSe crystals in a high voltage electron microscope have been presented by Yoshiie et al. [4].

Taking into consideration the importance of information on the defects in crystals of the II–VI compounds we have examined the real defect structure of ZnSe by means of transmission electron microscopy.

2. Experimental

The examined ZnSe crystals with resistivity 10^6 to 10^9 Ω cm were grown from the stoichiometric melt by the Bridgman method under argon pressure. Cylindrical boules

[1]) Al. Lotnikow 32/46, 02-668 Warsaw, Poland.

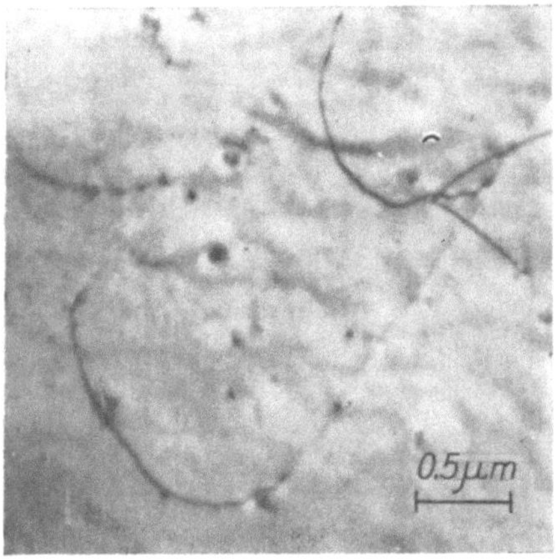

Fig. 1. Mobile dislocations in the [110]-oriented ZnSe foil

of 9 mm diameter with the growth direction close to ⟨110⟩ were produced. For TEM examination the specimens were sectioned into slices parallel to {110} planes perpendicular to the growth direction. After grinding and mechanical polishing, slices of 2.3 and 3 mm diameter were chemically polished down in HCl + CrO_3 solution [3]. A part of the specimens were thinned by the ion-milling procedure. In the latter case the specimens 20 to 30 μm thick, with chemically polished faces, were thinned down using 3 to 6 keV Ar^+ ions. In both cases the crystal surfaces were very rough and covered with thin amorphous layers. For HRTEM observation some crystal fragments were crushed. ⟨110⟩-oriented, crushed thin crystal fragments protruding over the holes in the carbon film on a supporting copper net were also examined. ({110} is the ZnSe principal cleavage plane.)

Microscopy was performed using Halle-JEOL 1000, Warsaw-JEM 200A TEM's, Halle-JEM 100C (C_s = 1.4 mm) HRTEM, and Stockholm-JEM 200CX (C_s = 1.2 mm, TEM equipped with a high resolution stage and a LaB_6 electron source. In HRTEM, the images were recorded under axial illumination. A large objective aperture was sometimes also used. In these experimental conditions the HRTEM image quality was destroyed due to the preparation and contamination by the amorphous layers covering the foils.

It was found that the specimens deteriorate easily in all experimental TEM conditions (HV, HR, and CTEM). Deterioration times lasted typically several minutes in HRTEM. It should be also emphasized that the dislocations are very mobile under the action of electron-beam irradiation. The dislocations in the thin parts of the foil moved, especially at the early stage of the examination when considerable temperature gradients appear in the specimens. The dislocations were mobile even in thicker regions of the foils. As an example, Fig. 1 shows the slip traces of mobile dislocations with a total Burgers vector $(a/2)$ ⟨110⟩ in the thicker part of the [110]-oriented ZnSe foil prepared by chemical polishing.

3. Results and Discussion

Light microscopy inspection of ZnSe crystals showed that the main growth defects were pores (negative crystals). The density of the pores varied from one technological process to another from very low to high density. Pores have approximately the same form. Their one flat well-formed plane has an equivalent orientation in the matrix. SEM-EDAX and microprobe JXA-50A X-ray analyses of the chemical composition of regions containing pores do not show any change in the Se and Zn concentrations in

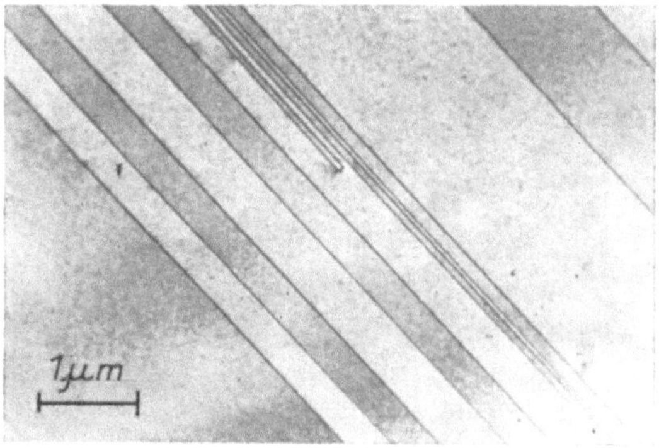

Fig. 2

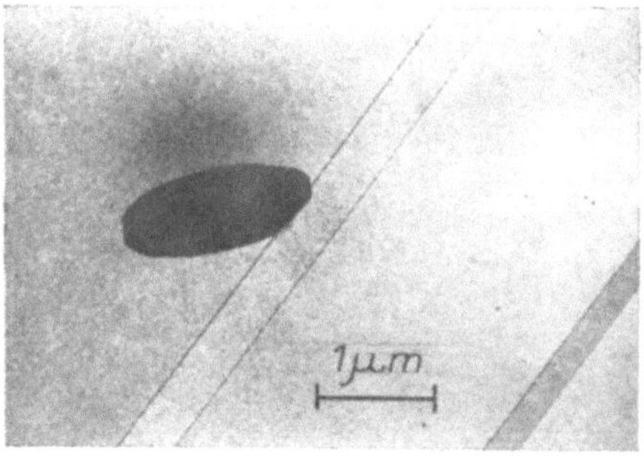

Fig. 3

Fig. 2. Group of narrow twins in ZnSe crystal

Fig. 3. Inclusion in ZnSe crystal

comparison with the matrix. This may mean that the pores were not accompanied by precipitates, or they were not accompanied by precipitates containing elements heavier than oxygen.

TEM inspection showed that the dominant defects in the zinc selenide were narrow twins extending completely across the crystals parallel to the growth axis. These twins were often found close together in small groups which were separated by larger regions of untwinned material. This observation is consistent with investigations of vapour-grown ZnSe crystals [1]. Twinning in the ZnSe zincblende structure has a ⟨111⟩ twin axis and composition surface parallel to {111}. In the investigated crystals only one of the {111} planes takes part in twin formation. In Fig. 2 a group of narrow twins in ZnSe is shown.

Fig. 3 show an inclusion in a ZnSe crystal. We also meet much smaller precipitates decorating the twin boundaries, especially twin-boundary jogs (Fig. 4). Occasionally we meet precipitates in the thicker region of the foils. Since their density was very low, we have not yet identified their crystallographic and chemical nature.

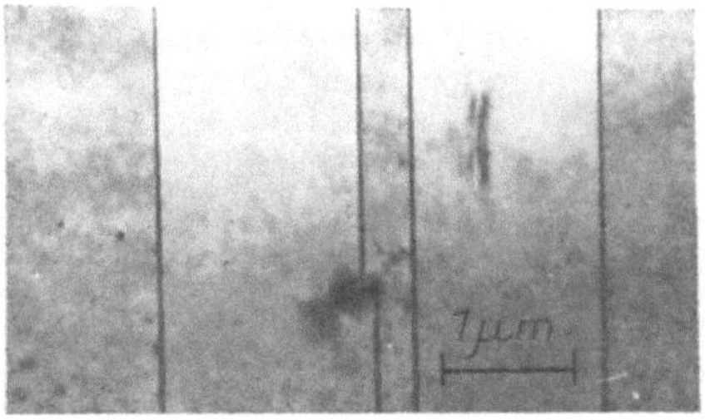

Fig. 4

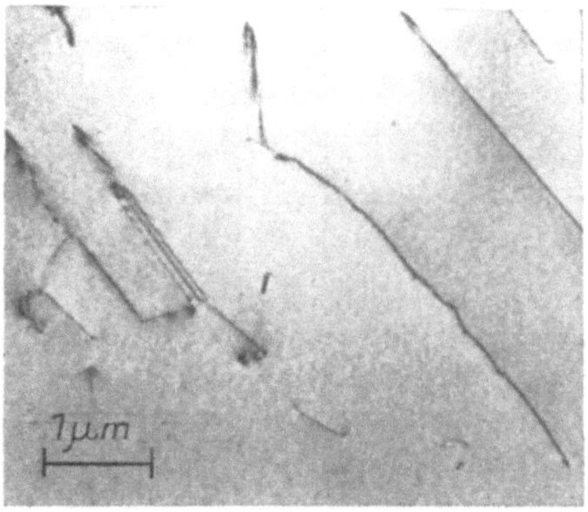

Fig. 5

Fig. 4. Precipitate decorating the twin-boundary jog in ZnSe crystal

Fig. 5. Dislocations in the as-grown ZnSe crystal

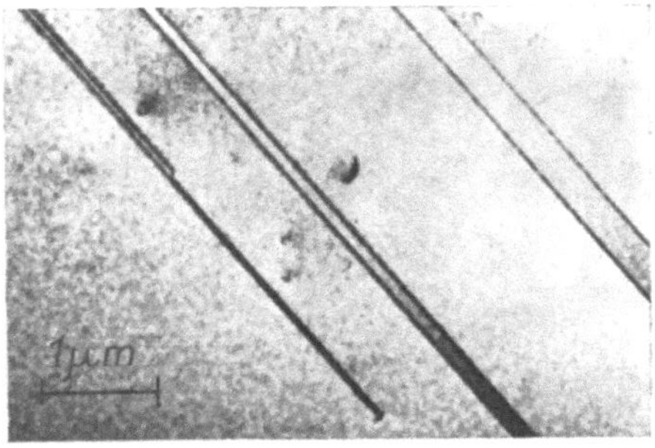

Fig. 6. Twin terminating in the partial dislocations accompanied by a stacking fault in ZnSe crystal

The incidence of dislocation in the as-grown crystals of zinc selenide was very low. The dislocations were observed only in larger regions of untwinned material. In some regions we have observed a higher dislocation density (an example is shown in Fig. 5), but it was not representative for the investigated material. In general we meet individual dislocations separated by large distances.

Dipoles were formed, but no dislocation loops we observed in the investigated material. The dislocations in the as-grown crystals of zinc selenide have a total Burgers vector of $(a/2)$ ⟨110⟩ type.

Stacking faults were not frequent, nevertheless they often accompanied twins which are terminating in the partial dislocations. Fig. 6 shows a twin accompanied by a stacking fault. Those configurations of stacking faults and twins may form during growth or cooling.

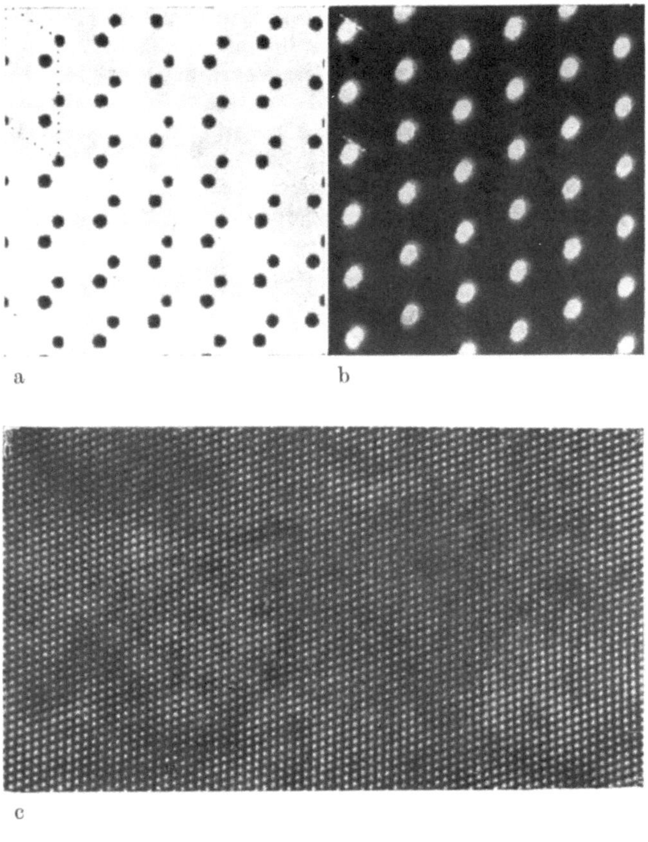

Fig. 7. [011] zincblende ZnSe: a) projected crystal potential, N beam = 139; b) 115 beam calculation, R = 4.2 nm^{-1}, Δ = 5 nm, thickness 3 nm, the objective lens defocus −63 nm; (Calculation was performed by the multi-slice method as programmed by J. A. Skarnulis with some algorithms by P. Fejes, adapted for VAX and modified by L. Kihlborg.) c) lattice image of the ZnSe crystal in the [011] projection

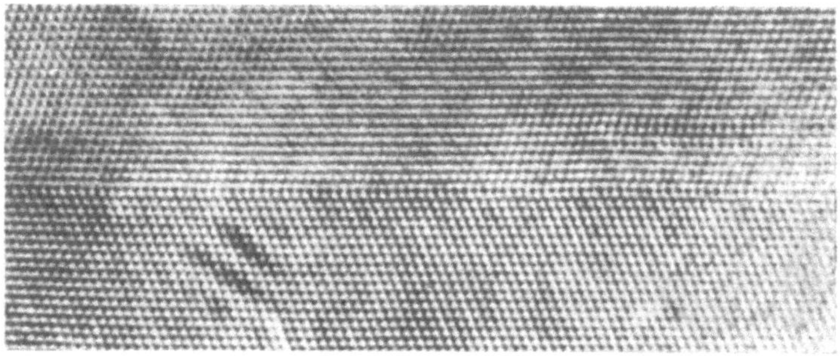

Fig. 8. The lattice image of the twin boundary in ZnSe crystal

Taking into account the evidence that the stacking fault may easily form from twins one can support the assumptions made in [7]. Movement of twin boundaries, formation or disappearance of stacking faults are realized by the slip of a partial dislocation with a glide set.

The ZnSe zincblende structure is best imaged by high resolution TEM with a ⟨110⟩ direction parallel to the electron beam. The projected crystal potential distribution in this orientation is shown in Fig. 7a. Zinc and selenium atoms are aligned in columns separated by 0.142 nm according to the spacing of the (004) plane for ZnSe with the lattice parameter $a = (0.56686 \pm 0.00001)$ nm determined by X-ray diffraction. Fig. 7b shows an example of the computed image. In Fig. 7c the experimental lattice image of ZnSe is shown as projected along the [011] direction. The (111) and (200) planes with the distances of 0.327 and 0.283 nm, respectively, are visible. The Zn and

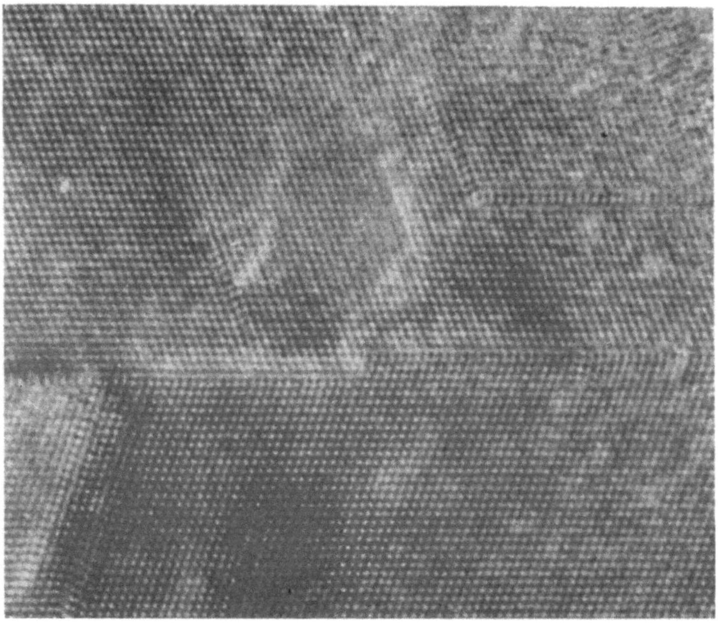

Fig. 9. The lattice image of the twin-boundary region in ZnSe crystal. Twin-boundary jog and number of defects are distinguished

Se are not completely separated in the images, the etch image spot is asymmetrical, reflecting the projected potential distribution.

Fig. 8 shows a lattice image of the twin boundary in ZnSe crystal, where the stacking sequence of the atoms along the [111] direction is clearly changed.

The atomic positions on the defect cannot be determined unambiguously from the analogy of the image contrast in the perfect crystal. The image calculations on a proposed atomic model of the boundary are necessary. We have not yet done them. Nevertheless, taking into account twin-boundary movement [7], formation of stacking fault from twin by movement of partial dislocation, creation of a twin by movement of partial dislocations on successive (111) planes (in CdTe; F. A. Ponce) we assume a simpler twin-boundary structure than discussed by Shiojiri et al. [2]. The twin structure results from the introduction of single stacking faults during growth.

In Fig. 9 the twin-boundary region is shown. The twin-boundary jog and numerous defects are clearly distinguished. It is surprisingly small distortion of planes around the jog region.

Among the defects recorded were: dissociated dislocations, intrinsic and extrinsic stacking faults. Defect characterization of the faults was possible from the position of the image spots, as had been found in Si [5] and other semiconductors such as CdTe [6].

Acknowledgements

The authors are grateful to Dr. M. Demianiuk providing the ZnSe crystals. We would like to thank Prof. J. Auleytner, Prof. L. Kihlborg, and Prof. J. Heydenreich for their encouragement and for the use of the microscopes and other facilities at their Institutes. We also thank Dr. Dr. G. Jasiołek, T. Niklewski, and L. Göthe for performing the X-ray measurements. One of us (E.M.) is grateful to Swedish Institute for scholarship.

References

[1] J. R. CUTTER, G. J. RUSSELL, and J. WOODS, J. Crystal Growth **32**, 179 (1976).
[2] M. SHIOJIRI, C. KAITO, S. SEKIMOTO, and N. NAKAMURA, Phil. Mag. A **46**, 495 (1982).
[3] V. V. ARISTOV, A. V. ZARETSKII, Y. A. OSIPYAN, V. F. PETRENKO, G. K. STRUKOVA, and I. I. KHODOS, phys. stat. sol. (a) **75**, 101 (1983).
[4] T. YOSHIIE, H. IWANAGA, N. SHIBATA, K. SUZUKI, M. ICHIHARA, and S. TAKEUCHI, Phil. Mag. A **47**, 315 (1983).
[5] P. REZ and O. L. KRIVANEK, Proc. 9th Internat. Cong. on Electron Microscopy, Vol. 1, Toronto 1978 (p. 288).
[6] A. OLSEN and J. C. H. SPENCE, Phil. Mag. A **43**, 945 (1981).
[7] S. A. OMELCHENKO, S. I. BREDIKHIN, P. A. BERLOV, M. F. BULANIY, S. Z. SHMURAK, and A. Y. YAKUNIN, Fiz. tverd. Tela **24**, 2803 (1982).

(Received June 28, 1984)

phys. stat. sol. (a) 85, 89 (1984)

Subject classification: 12.1; 21.1.1

Fracture and Deformation Division, Center for Materials Science, National Bureau of Standards, Boulder[1])

Monocrystal–Polycrystal Elastic Constants of a Stainless Steel

By

H. M. LEDBETTER

For a face-centered-cubic steel, new measurements of the monocrystal Voigt elastic-stiffness constants C_{11}, C_{12}, C_{44} are given. The monocrystal steel, Fe–19Cr–10Ni, corresponds closely to the well-known AISI-304 austenitic stainless steel. Considering seven theories for the monocrystal–polycrystal elastic constants, it is found that the Hershey-Kröner-Eshelby theory agrees best with measurements. It predicts the shear modulus within 2% of observation, where the Voigt-Reuss first-order bounds differ by 49%. Ten sets of Fe–Cr–Ni C_{ij} results are reviewed with the finding that both Zener's elastic anisotropy and the C_{12}/C_{11} ratio are constant within 5%.

Für kubisch-flächenzentrierten Stahl werden neue Meßergebnisse der einkristallinen Voigtschen Elastizitätskonstanten: C_{11}, C_{12} und C_{44} mitgeteilt. Der einkristalline Stahl Fe–19Cr–10Ni entspricht gut dem bekannten rostfreien Austenitstahl AISI 304. Durch Berücksichtigung von sieben Theorien für die einkristallinen–polykristallinen Elastizitätskonstanten wird gefunden, daß die Hershey-Kröner-Eshelby-Theorie am besten mit den Messungen übereinstimmt. Sie sagt den Schermodul innerhalb von 2% zu den Beobachtungen voraus, während die Voigt-Reuss-Sätze erster Ordnung um 49% abweichen. Es werden zehn Sätze von C_{ij}-Ergebnissen in Fe–Cr–Ni diskutiert und es wird gefunden, daß sowohl die Zenersche elastische Anisotropie als auch das C_{12}/C_{11}-Verhältnis innerhalb 5% konstant sind.

1. Introduction

The relationship between monocrystal physical properties (represented by tensors) and polycrystal physical properties (scalars) remains one of the most intriguing unsolved problems in crystal physics. Tensor 'averages' have been proposed for many physical properties, including dielectric constant, diffusivity, elastic constants, electrical conductivity, magnetic permeability, magnetorestriction, piezoelectric constants, and thermal conductivity.

In terms of averaging, elastic constants have received more attention than most other physical properties. Voigt [1] was first to suggest how the fourth-rank elastic-stiffness tensor, C_{ijkl}, could be averaged to obtain the two independent quasi-isotropic elastic constants: the bulk modulus, B, and the shear modulus, G. Forty years later, Reuss [2] showed that Voigt's solution is not unique. Landau and Lifshitz [3] point out that "there is therefore no general relation between the moduli of elasticity of a polycrystal and those of a single crystal of the same substance". Many ingenious approximate averaging methods have been developed, too many to enumerate here. Most of these are either described or referenced in reviews by Hearmon [4], Beran [5], Ledbetter [6], Kröner and Koch [7], and Watt et al. [8].

Until a generally accepted averaging method arises, interest remains high in testing various models against careful measurements. For both ambient and low temperatures, Ledbetter [9, 10] recently did this for copper. He considered eight models: Voigt, Hashin-Shtrikman, Hershey-Kröner-Eshelby, Hill (arithmetic), Hill (geometric),

[1]) Boulder, Colorado 80303, USA.

Aleksandrov, Ledbetter-Naimon, and Reuss. Among these, the Hershey-Kröner-Eshelby model worked best, predicting the quasi-isotropic polycrystal shear modulus within 1%.

The study reported here tests the various monocrystal-polycrystal models for the elastic constants of an f.c.c. Fe–Cr–Ni alloy. This alloy possesses technological interest because it approximates the chemical composition of AISI-304 stainless steel, probably the best known and most widely used stainless-steel alloy. This study represents the first test of elastic-constant averaging models for a nondilute ternary alloy. While the alloy exhibits macroscopic paramagnetism at ambient temperature, all constituents, Fe, Cr, Ni, exhibit ferromagnetism, antiferromagnetism, or both. Thus, local interatomic interactions may be nonparamagnetic. Because the alloy is concentrated, contains magnetic species, and possesses considerable technological interest, we undertook the present study.

2. Experiment

2.1 Monocrystal

An alloy of nominal weight-percent composition Fe–19Cr–10Ni was vacuum melted and vacuum cast from high-purity metals. Monocrystals were grown by a Bridgman technique, homogenized at 1350 °C for 20 h, cut to shape and size with a spark cutter

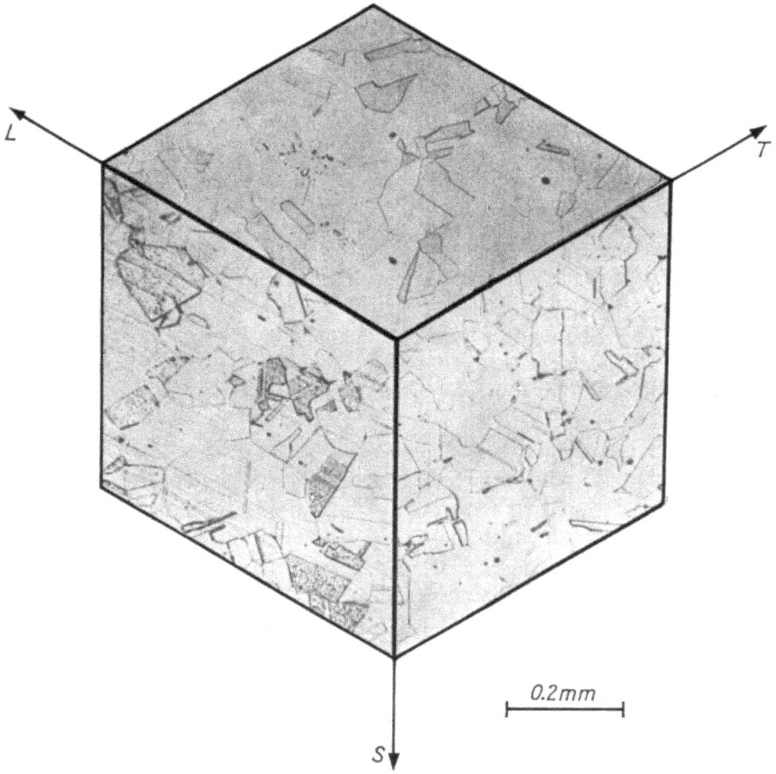

Fig. 1. Microstructure of a typical polycrystalline AISI 304 LN stainless-steel alloy in its wrought-and-annealed condition. LN denotes low carbon content and high nitrogen content. The specimen came from a commercial plate; L denotes rolling direction, T transverse in-plate direction, and S through-plate-thickness direction

and annealed at 1100 °C for 20 h. Back-reflection Laue X-ray diffraction was used to establish the crystal orientation, which was 2.5° from a [110] axis. (In subsequent measurement and analysis, we ignored this small departure from a [110] axis.) Specimen dimensions were $1.5 \times 1.5 \times 0.7\ cm^3$. Mass density measured by Archimedes's method was $7.937\ g/cm^3$.

2.2 *Polycrystal*

For these, we used the twenty AISI-304-type stainless-steel alloys studied previously by Ledbetter et al. [11] in their as-received wrought-and-annealed conditions. Their average weight-percent chemical composition was 18.6 Cr, 9.3 Ni, 1.3 Mn, with other trace elements. Their average Archimedes method mass density was $7.880\ g/cm^3$. Their average grain size measured 88 μm, ranging from 68 to 105. Fig. 1 shows a typical microstructure.

2.3 *Measurements*

We measured sound velocities, longitudinal and transverse, by a pulse-echo-overlap method described previously [11]. For the polycrystals, the percent-standard error was 0.3% for v_l and 0.6% for v_t. This includes all possible error sources: chemical composition, thermal-mechanical treatment, measurement method, and so forth. For the monocrystals, the percent-standard error was approximately 0.1%.

3. Results

Table 1 gives, for both monocrystal and polycrystal, the measured sound velocities

Table 1
Sound-velocity and elastic-constant results
(ϱ in g/cm^3; v in cm/μs; C, B, and E, in $10^{11}\ N/m^2$)

	monocrystal	polycrystal
ϱ	7.937	7.880
v_l	0.6121	0.5759
v_{t1}	0.3988	0.3134
v_{t2}	0.2052	—
C_L	2.974	2.614
$(C_{11} - C_{12})/2$	0.334	0.774
C_{44}	1.262	—
C_{11}	2.046	2.614
C_{12}	1.377	1.066
B	1.600	1.582
E	—	1.996
ν	—	0.290

together with the derived elastic-stiffness constants calculated from

$$C = \varrho v^2 , \tag{1}$$

where C denotes elastic stiffness, ϱ mass density, and v sound velocity. For the polycrystal,

$$C_{11} = C_l = \varrho v_l^2 \tag{2}$$

and

$$C_{44} = G = \varrho v_t^2 , \tag{3}$$

where C_l denotes longitudinal modulus and G shear modulus. For a monocrystal, along a [110] direction,

$$C_L = \varrho v_l^2 = \frac{C_{11} + C_{12} + 2C_{44}}{2}. \tag{4}$$

$$C_{t1} = \varrho v_{t1}^2 = C_{44}, \tag{5}$$

and

$$C_{t2} = \varrho v_{t2}^2 = \frac{C_{11} - C_{12}}{2}. \tag{6}$$

From (4) to (6) one obtains the usual three independent Voigt elastic stiffnesses: C_{11}, C_{12}, and C_{44}. For the monocrystal, the bulk modulus is

$$B = \frac{C_{11} + 2C_{12}}{3} \tag{7}$$

while for the polycrystal it is

$$B = C_l - \tfrac{4}{3} G. \tag{8}$$

Based on seven averaging methods, Table 2 gives predicted quasi-isotropic elastic

Table 2

Polycrystalline elastic constants predicted by averaging monocrystal elastic constants

E (10^{11} N/m^2)	G (10^{11} N/m^2)	ν	method
2.255	0.891	0.265	Voigt
2.025	0.785	0.289	Hashin-Shtrikman, upper
1.850	0.708	0.307	Hashin-Shtrikman, lower
1.938	0.746	0.298	Hashin-Shtrikman, average
1.960	0.756	0.296	Hershey-Kröner-Eshelby
1.934	0.745	0.299	Hill, arithmetic
1.928	0.742	0.299	Aleksandrov
1.901	0.730	0.302	Hill, geometric
1.596	0.598	0.334	Reuss

constants. For this table, E denotes Young modulus, G shear modulus, and ν Poisson ratio. All averaging methods predict the same bulk modulus, $B = 1.600 \times 10^{11}$ N/m^2, following (7). For details of the mathematical relationships associated with the various methods, see Ledbetter [6] or references to the original studies given in [9].

4. Discussion

Results in Table 2 show clearly that, among the seven averaging methods considered, one method agrees best with observation: the Hershey-Kröner-Eshelby method. It predicts the shear modulus within 2%, the Young modulus within 2%, the bulk modulus within 1%, and the Poisson ratio within 2%. Part of the discrepancy may arise from the slightly different chemical compositions of the monocrystal (19Cr–10Ni) and the average polycrystal (18.6Cr–9.3Ni–1.3Mn, plus trace elements). Note also that observation lies much closer to the Hashin-Shtrikman upper bound than to the lower. Especially, this occurs for Poisson's ratio. If real, this suggests that the polycrystal is not perfectly disordered, that it contains texture, characteristic grain shape, or characteristic grain-boundary geometry. But, present results permit no conclusions concerning these matters.

For cubic-symmetry monocrystals where (7) gives the bulk modulus, the averaging problem amounts to predicting another elastic constant. Most models predict the shear modulus. The Hershey-Kröner-Eshelby method does this by considering an anisotropic nonhomogeneity embedded in an isotropic matrix. Eshelby [12] gave a relatively simple discussion of this inclusion problem, which leads to a cubic equation for the shear modulus G:

$$G^3 + \alpha G^2 + \beta G + \gamma = 0\,, \tag{9}$$

where

$$\alpha = \frac{5C_{11} + 4C_{12}}{8}, \tag{10}$$

$$\beta = -\frac{C_{44}(7C_{11} - 4C_{12})}{8}, \tag{11}$$

and

$$\gamma = -\frac{C_{44}(C_{11} - C_{12})\,(C_{11} + 2C_{12})}{8}, \tag{12}$$

where in (12) we give γ correctly as twice that reported in [12].

An interesting feature of (9) is that it contains the bulk modulus. In the H-K-E-model, the cubic-symmetry shear moduli are not sufficient for predicting G. Other models — Voigt, Reuss, Hill, Aleksandrov, for example — do not require B to predict G. The question arises: why does the resistance to shear deformation depend on the resistance to dilation ? In isotropic media, we know since Stokes [13] that G and B are the two independent elastic constants representing extreme types of deformation: shape change without volume change and volume change without shape change, respectively. We can factor any general deformation into dilation and shear components. Thus, is this $G(B)$ dependence real or an artifact of the model ? Will the final model be of the form $G(C_{44}, C', B)$ or simply $G(C_{44}, C')$?

Recently, Gairola and Kröner [14] derived a simple formula for calculating all the bounds for and the self-consistent (infinite bound) value of the shear modulus of a macroscopically isotropic cubic-symmetry polycrystalline aggregate. The numerical results agree exactly with the Hershey-Kröner-Eshelby method. The approach differs from the spherical-inclusion problem described above, but it leads to the same results. Using a unified method [15, 16], they derive the result (in their notation):

$$G = \frac{\alpha_2 G_0^2 + \beta_2 G_0 + \gamma}{G_0^2 + \alpha_1 G_0 + \beta_1}, \tag{13}$$

where

$$\alpha_1 = \frac{45\varkappa + 24\mu + 36\nu}{40}, \tag{14}$$

$$\alpha_2 = \frac{2\nu + 3\mu}{5}, \tag{15}$$

$$\beta_1 = \frac{3\varkappa(2\mu + 3\nu)}{20}, \tag{16}$$

$$\beta_2 = \frac{3(6\varkappa\nu + 9\varkappa\mu + 20\mu\nu)}{40}, \tag{17}$$

$$\gamma = \frac{3\varkappa\mu\nu}{4}, \tag{18}$$

$$\varkappa = \frac{C_{11} + 2C_{12}}{3}, \tag{19}$$

$$\nu = \frac{C_{11} - C_{12}}{2}, \tag{20}$$

$$\mu = C_{44}. \tag{21}$$

One can easily verify that putting $G_0 = \infty$ and $G_0 = 0$ leads to the Voigt and Reuss bounds, G_V and G_R, the so-called first-order upper and lower bounds, $G^{(1)}$ and $G_{(1)}$. If instead one puts $G_0 = \mu = C_{44}$ and $G_0 = \nu = (C_{11} - C_{12})/2$ (μ and ν are the zeroeth-order bounds $G^{(0)}$ and $G_{(0)}$), then one obtains the Hashin-Shtrikman bounds, $G^{(2)}$ and $G_{(2)}$. This self-consistent renormalization scheme yields either the odd-order or the even-order bounds. Either set leads to $G^{(\infty)} = G_{(\infty)} = G$ (self-consistent) in exact agreement with (9). Some authors call the third-order bounds, $G^{(3)}$ and $G_{(3)}$, the Kröner bounds.

Finally, we collect and discuss briefly all previously reported monocrystal stainless-steel elastic constants. Table 3 contains ten sets of reported elastic constants, converted

Table 3

Monocrystal elastic constants of face-centered-cubic Fe–Cr–Ni alloys. Units of elastic constants are 10^{11} N/m², except A, which is dimensionless

alloy (Cr–Ni)	C_{11}	C_{12}	C_{44}	B	C'	A	source
12–12	2.109	1.403	1.225	1.638	0.353	3.47	Bradfield [17]
12–18	2.332	1.627	1.226	1.862	0.353	3.48	Salmutter, Strangler [18]
18–12	1.912	1.179	1.386	1.423	0.367	3.78	Kikuchi [19]
18–12	2.159	1.446	1.289	1.684	0.357	3.62	Bradfield [17]
18–14	1.98	1.25	1.22	1.49	0.365	3.34	Mangalick, Fiore [20]
18–19	1.91	1.19	1.24	1.43	0.360	3.44	Mangalick, Fiore [20]
19–10	2.05	1.38	1.26	1.60	0.334	3.78	present
19–10	2.07	1.32	1.23	1.57	0.375	3.28	Kupperman [21]
19–14	2.05	1.33	1.27	1.57	0.360	3.58	Kupperman [21]
19–19	2.04	1.33	1.26	1.57	0.355	3.55	Kupperman [21]
Fe	2.51	1.57	1.12	1.89	0.470	2.38	Johnson [22]
Fe	2.760	1.735	1.363	2.08	0.513	2.66	Diesburg [23]
Co	2.398	1.634	1.334	1.89	0.382	3.49	Leamy, Warlimont [24]
Ni	2.516	1.544	1.220	1.87	0.251	2.51	Alers, Neighbours, Sato [25]

where necessary to the C_{ij}: C_{11}, C_{12}, and C_{44}. Table 3 contains also B and C' together with the Zener anisotropy

$$A = \frac{C_{44}}{C'}. \tag{22}$$

The table shows also results for f.c.c. Fe, Co, and Ni. The Ni results were measured. The Co results were obtained by extrapolation from measurements on Ni–Co alloys.

Diesburg's result [23] for Fe was also obtained by extrapolation, using Fe–Ni alloys. Johnson's [22] result for Fe came from theory based on fitting a two-body central-force interatomic potential to Fe's second-order b.c.c. elastic constants. For the ten Fe–Cr–Ni alloys in Table 3, the Zener anisotropy shows remarkable near-constancy: $A = 3.53 \pm 0.17$, a standard error of 5%. Also nearly constant, we find the ratio $C_{12}/C_{11} = 0.652 \pm 0.025$, a standard error of 4%. These empirical ratios provide an opportunity for calculating the three monocrystal elastic constants from the two quasi-isotropic polycrystal elastic constants. Ledbetter [26] used this approach to estimate the monocrystal elastic constants of 316-type stainless steel. In that study, based on only four Fe–Cr–Ni alloys, he used $A = 3.51$ and $C_{12}/C_{11} = 0.635$.

5. Conclusions

For an austenitic stainless-steel alloy, we considered the monocrystal–polycrystal elastic-constant relationship. Among seven models, the Hershey-Kröner-Eshelby model works best.

Acknowledgements

This study was supported by the U.S. DoE Office of Fusion Energy. Mr. J. T. Ward assisted with measurements. Mr. M. W. Austin contributed in many miscellaneous ways. Dr. N. F. Fiori of Cabot Corporation loaned the monocrystals. Dr. D. T. Read contributed the photomicrograph.

References

[1] W. Voigt, Ann. Phys. (Leipzig) **38**, 573 (1889).
[2] A. Reuss, Z. angew. Math. Mech. **9**, 49 (1929).
[3] L. D. Landau and E. M. Lifshitz, Theory of Elasticity, Pergamon Press, London/Oxford 1959 (p. 40).
[4] R. F. S. Hearmon, in: Physics of the Solid State, Academic Press, Inc., New York/London 1969 (p. 401).
[5] M. J. Beran, phys. stat. sol. (a) **6**, 365 (1971).
[6] H. M. Ledbetter, J. appl. Phys. **44**, 1451 (1973).
[7] E. Kröner and H. Koch, Solid Mech. Arch. **1**, 183 (1976).
[8] J. P. Watt, G. F. Davies, and R. J. O'Connell, Rev. Geophys. Space Phys. **14**, 541 (1976).
[9] H. M. Ledbetter, J. Phys. D **13**, 1879 (1980).
[10] H. M. Ledbetter, phys. stat. sol. (a) **66**, 477 (1981).
[11] H. M. Ledbetter, N. F. Frederick, and M. W. Austin, J. appl. Phys. **51**, 305 (1980).
[12] J. D. Eshelby, in: Progress in Solid Mechanics, Vol. 2, North-Holland Publ. Co., Amsterdam 1961 (p. 130).
[13] G. G. Stokes, in: Mathematics and Physics Papers, Vol. 1, Cambridge University Press, 1880 (p. 75).
[14] B. K. D. Gairola and E. Kröner, Internat. J. Engng. Sci. **19**, 865 (1981).
[15] E. Kröner, J. Phys. F **8**, 2261 (1978).
[16] B. K. D. Gairola, in: Proc. Continuum Models of Discrete Systems, Waterloo University Press, Waterloo (Canada) 1980 (p. 387).
[17] G. Bradfield, J. Iron Steel Inst. **202**, 616 (1964).
[18] K. Salmutter and F. Stangler, Z. Metallk. **51**, 1 (1960).
[19] M. Kikuchi, Trans. Japan Inst. Met. **12**, 417 (1971).
[20] M. C. Mangalick and N. F. Fiore, Trans. MS AIME **242**, 363 (1968).
[21] D. S. Kupperman, private communication.
D. S. Kupperman and K. J. Reimann, IEEE Trans. Sonics Ultrasonics **27**, 7 (1980).

[22] R. A. JOHNSON, Phys. Rev. **145**, 423 (1966).
[23] D. DIESBURG, Ph. D. Thesis, Iowa State University, 1971.
[24] H. J. LEAMY and H. WARLIMONT, phys. stat. sol. **37**, 523 (1970).
[25] G. A. ALERS, J. R. NEIGHBOURS, and H. SATO, J. Phys. Chem. Solids **13**, 40 (1960).
[26] H. M. LEDBETTER, Brit. J. Non-Destructive Testing **34**, 286 (1981).

(Received June 14, 1984)

phys. stat. sol. (a) 85, 97 (1984)

Subject classification: 4 and 20.1; 22.5; 22.5.2

Sektion Chemie (a) and Sektion Physik (b) der Friedrich-Schiller-Universität Jena[1])

Structure and Molar Refraction and Its Wavelength Dependence at Different Alkali and Ammonium Halides

By

W. BURCKHARDT (a), P. BUSSEMER (b), and U. DIETRICH (b)

The wavelength dependence of the molar refraction of AB-compounds, crystallizing in the NaCl- or CsCl-structure type, is analysed by use of a two-term Sellmeier dispersion formula. In the range of the validity of a single-oscillator model a clear distinction between series with Na Clor Cs Clstructure is possible on the base of the influence of the parameters λ_1, the resonance wavelength of the electronic absorption, and n_{eff}, the effective number of electrons involved in the electron transition, on the molar refraction and its slope. The Sellmeier model is compared with the dispersion model of Wemple.

Die Wellenlängenabhängigkeit der Molrefraktion von AB-Verbindungen, die im NaCl- oder im CsCl-Strukturtyp kristallisieren, wird unter Verwendung einer zwei-gliedrigen Sellmeier-Dispersionsformel analysiert. Im Gültigkeitsbereich eines Ein-Oszillator-Modellsist auf der Grundlage des Einflusses der Parameter λ_1, Resonanzwellenlänge der Elektronenabsorption, und n_{eff}, effektive Zahl der am Elektronenübergang beteiligten Elektronen, auf die Molrefraktion und deren Anstieg eine eindeutige Unterscheidung zwischen Reihen mit NaCl- bzw. CsCl-Struktur möglich. Dem Sellmeier-Modell wird das Dispersionsmodell von Wemple gegenübergestellt.

1. Introduction

In the literature different models are represented for the description of the wavelength dependence of the refractive index. Besides dispersion formulae, which exhibit pure interpolation character and whose parameters have no physical significance, models are used, in which the dispersion parameters are related to structural quantities. Wemple and Di Domenico [2 to 4] use for the characterization of the refractive index dispersion at wavelength on the upper side of the absorption edge a single-oscillator model of the form $n^2 - 1 = E_d E_0/(E_0^2 - E^2)$. The dispersion energy is empirically related to structure parameters like coordination number and bond length. It is the aim of the present paper to show the influence of the structure on refractive index and dispersion in the case of alkali and ammonium halides, which crystallize in the NaCl or CsCl structure, on the base of a dispersion model of the Sellmeier-type. In previous papers [5, 6] and by Hirota and Izumitani [7] it has been shown that the parameters entering the dispersion formula are closely related to the position of the absorption bands and their intensities. In [1] it could be demonstrated that the structure of optical media has a decisive influence for the combination of different materials of a two-lens achromate.

[1]) Max-Wien-Platz 1, DDR-6900 Jena, GDR.

2. Dispersion Calculations

For a number of alkali halides the wavelength dependence of the refractive index was mathematically modelled according a procedure described by Stavroudis and Sutton [8] with a two-term dispersion formula of the Sellmeier-type

$$n^2 - 1 = \frac{A_1\lambda^2}{\lambda^2 - \lambda_1^2} + \frac{A_2\lambda^2}{\lambda^2 - \lambda_2^2}. \tag{1}$$

Refractive indices from the literature have been used. The first term of the dispersion formula (1) considers the contribution of the optical interband transitions, the second term the influence of oscillations of atoms or ions in the structure groups on the refractive index.

In Table 1 the parameters A_i and λ_i for the different alkali halides obtained by dispersion calculations are summarized.

Table 1
Dispersion parameters of different alkali halides

substance	A_1	λ_1 (μm)	A_2	λ_2 (μm)	ref.
LiF	0.92733	0.07196	7.23983	33.05	[9]*)
NaF	0.74701	0.07659	4.02895	44.08	
NaCl	1.32957	0.11851	2.20520	49.52	[10]
KCl	1.17269	0.11806	1.13332	48.35	[11]
KBr	1.36092	0.13281	1.09515	60.04	[12, 13]
KI	1.64879	0.15818	1.59408	82.62	[14, 15]
CsBr	1.78274	0.13494	0.80553	65.45	[16, 17]
CsI	2.03095	0.16428	1.06279	97.26	[18]

*) Values of the refractive index.

3. Results and Discussion

The investigations demonstrate that the wavelength dependence of the refractive index can be described in an extensive range by a single-oscillator model. The region of validity of a single-oscillator model results from the representation $(n^2 - 1)^{-1} = f(\lambda^{-2})$. As shown in Fig. 1 for three different alkali halides linearity is given in

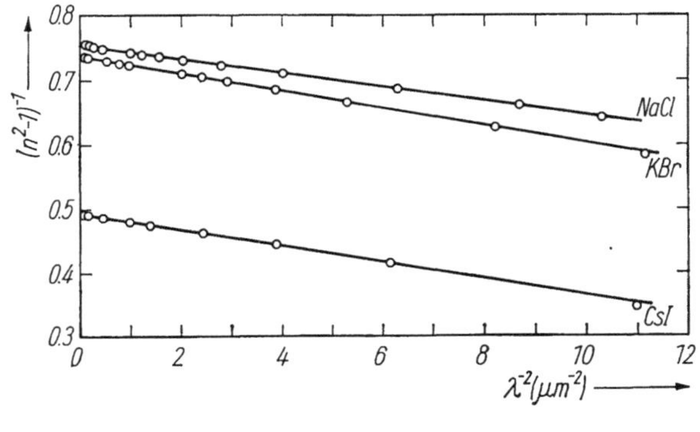

Fig. 1. Region of validity of the one-oscillator model for different alkali halides

the wavelength range from about 1 μm up to the beginning slope of the edge of the electronic absorption.

To compare refractive index and dispersion of different materials it is necessary to normalize on a constant number of particles. The molar refraction is given in the single-oscillator model by

$$R_M = \overline{V} \frac{A_1 \lambda^2}{\lambda^2 - \lambda_1^2}, \tag{2}$$

$\overline{V}$ is the molar volume, and the slope of the molar refraction curve at any wavelength by

$$-\frac{dR_M}{d\lambda} = 2\lambda \overline{V} \frac{A_1 \lambda_1^2}{(\lambda^2 - \lambda_1^2)^2}. \tag{3}$$

By series expansion in the parameter $\lambda_1/\lambda \ll 1$ and substitution of

$$A_1 = \frac{e^2}{\pi c^2 m_1} N_1 f_1 \lambda_1^2, \tag{4}$$

where N_1 is the electron density, f_1 the oscillator strength, and m_1 the electron mass, from (2) and (3) follows

$$R_M = \frac{e^2 N_L}{\pi c^2 m_1} n_{\text{eff}} \lambda_1^2 \left[1 + \left(\frac{\lambda_1}{\lambda}\right)^2 + \dots\right] \tag{5}$$

and

$$-\frac{dR_M}{d\lambda} = \frac{2e^2 N_L}{\pi c^2 m_1} n_{\text{eff}} \lambda_1 \left(\frac{\lambda_1}{\lambda}\right)^3 \left[1 + \left(\frac{\lambda_1}{\lambda}\right)^2 + \dots\right]^2 \tag{6}$$

with

$$n_{\text{eff}} = n_1 f_1 \quad \text{and} \quad n_1 = \frac{N_1 \overline{V}}{N_L}.$$

The number of electrons per average atom taking part in the transition is given by n_1. Different authors [2, 3, 18] assume that n_1 is equal to the number of the valence electrons, that means for instance $n_1 = \text{const} = 4$ for all alkali halides. According to (4) in combination with $n_1 = N_1 \overline{V}/N_L$ it is only possible to calculate $n_{\text{eff}} = n_1 f_1$. Each assumption for n_1 is immediately transferred on f_1. For this reason in the previous paper [1] n_{eff} is only used. A comparison of the molar refraction R_{M_j} and $-dR_{M_j}/d\lambda_j$ of the different alkali halides at wavelengths λ_j which are determined by $\lambda_{1j}/\lambda_j =$ = const results from (5) and (6) in the relations

$$R_{M_j} = C_1 n_{\text{eff}\,j} \lambda_{1j}^2; \qquad C_1 = \text{const}, \tag{7}$$

$$-\frac{dR_{M_j}}{d\lambda_j} = C_2 n_{\text{eff}\,j} \lambda_{1j}; \qquad C_2 = \text{const}. \tag{8}$$

Fig. 2 and 3 illustrate the influence of the quantities n_{eff} and λ_1^2 on the molar refraction at wavelengths determined by $\lambda_{1j}/\lambda_j = 0.25$. In the representation different alkali and ammonium halides, for which n_{eff} and λ_j^2 were calculated according to values of Wemple and Di Domenico [2] and Wemple [3], are incorporated. In Fig. 4 and 5 the

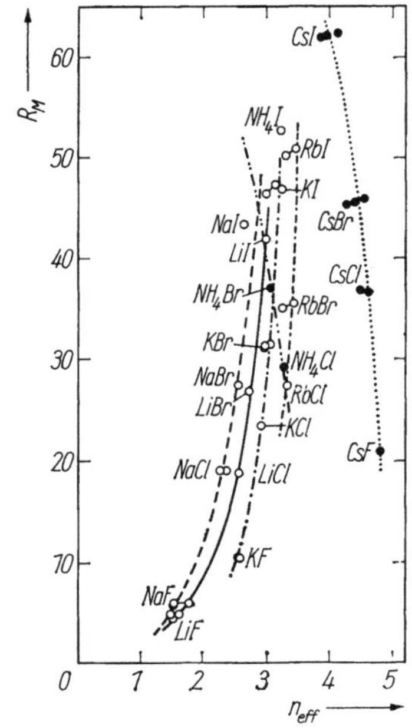

Fig. 2

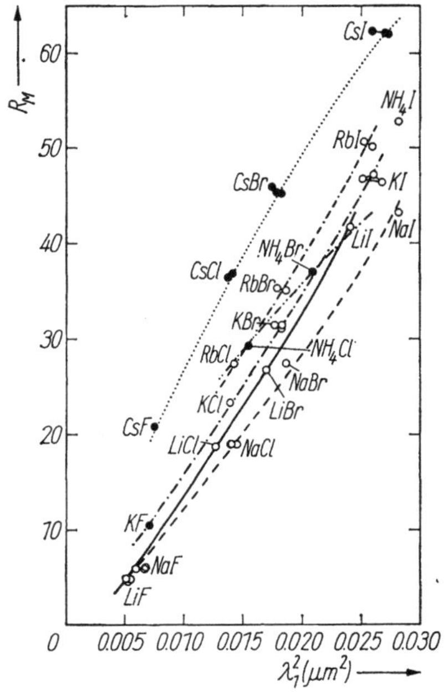

Fig. 3

Fig. 2. Dependence of the molar refraction of different alkali and ammonium halides on n_{eff} at $\lambda_1/\lambda = \text{const} = 0.25$

Fig. 3. Influence of λ_1^2 on R_M of alkali and ammonium halides at $\lambda_1/\lambda = \text{const} = 0.25$

influence of n_{eff} and λ_j on $-dR_{Mj}/d\lambda_j$ at $\lambda_{1j}/\lambda_j = 0.25$ is demonstrated. It is a characteristic feature, that homologous series with NaCl structure are arranged in Fig. 2 and 4 on curves with steadily increasing slope, while alkali and ammonium halides with CsCl structure are ordered on curves with steadily decreasing slope. The influence of the structure on R_M and $-dR_M/d\lambda$ in Fig. 3 and 5 emerges even more clearly. For homologous series with rocksalt-structure R_M and $-dR_M/d\lambda$ increase strongly with increasing values of n_{eff}, whereas for series with CsCl structure the molar refraction and the slope of the molar refraction curve decrease with increasing values of n_{eff}.

In the series of the ammonium halides NH_4Cl and NH_4Br crystallizing in the CsCl structure are arranged on curves, which are analogous to those of the homologous series of the caesium halides. NH_4I crystallizing in the rocksalt structure diverges strongly from these curves.

The oscillator strength f_1 is connected with the resonance wavelength λ_1 by the relation

$$f_1 = \frac{8\pi^2 m_1 c}{3h} \frac{1}{\lambda_1} \hat{r}^2 , \tag{9}$$

where $\hat{r}$ is the matrix element of the transition.

Because one cannot calculate f_1 independently without special assumption for n_1, in Fig. 6 $n_{eff} = n_1 f_1$ is represented as a function of $1/\lambda_1$ according to (9). The difference between NaCl and CsCl structures emerges also clearly in this representation. While in homologous series with rocksalt structure a change of n_{eff} to higher wavelength is

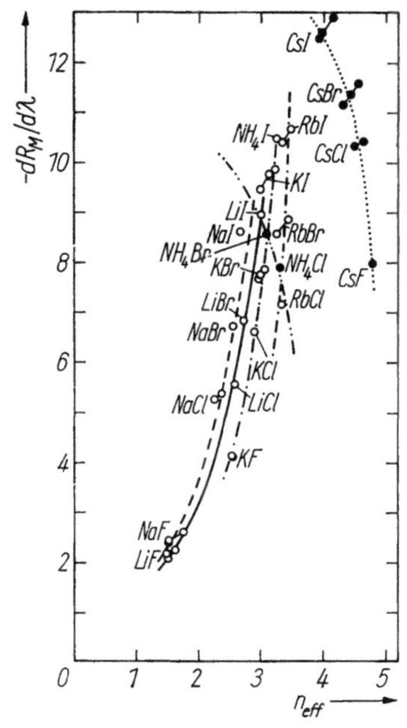

Fig. 4

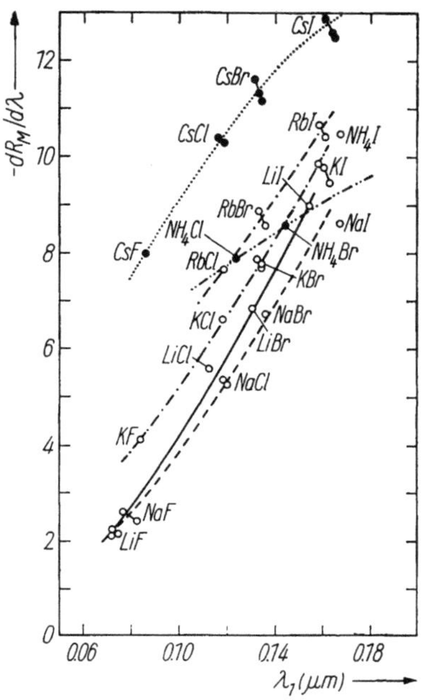

Fig. 5

Fig. 4. $-dR_M/d\lambda$ as function of n_{eff} at $\lambda_1/\lambda = \text{const} = 0.25$

Fig. 5. $-dR_M/d\lambda$ as function of λ_1 at $\lambda_1/\lambda = \text{const} = 0.25$

connected with an increase of the number of electrons taking part in the transition effectively, in homologous series with CsCl structure n_{eff} is decreasing with increasing values of the resonance wavelength λ_1.

The investigations illustrate in the case of isotropic solids crystallizing in two typical structures of AB-compounds the influence of the structure on the refractive index and its dispersion. The use of a dispersion model of the Sellmeier-type does neither require assumptions for several quantities entering the dispersion parameters λ_i and A_i nor the introduction of other parameters which combine structural quantities together.

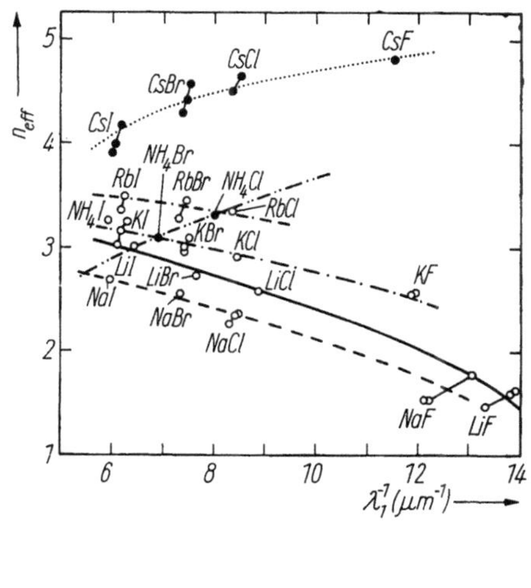

Fig. 6. Dependence of n_{eff} on $1/\lambda_1$ for different alkali and ammonium halides

4. Comparison with the Single-Oscillator Model of Wemple and Di Domenico

Wemple and Di Domenico [2, 3] describe the wavelength dependence of the refractive index of different classes of materials by use of a single-oscillator model of the form

$$n^2 - 1 = \frac{E_d E_0}{E_0^2 - E^2}, \tag{10}$$

where E_d is the dispersion energy and E_0 the oscillator energy.

For the dispersion energy E_d is stated the empirical relation

$$E_d = f n_e Z_a N_a d^3 , \tag{11}$$

where n_e is the number of valence electrons per anion, Z_a the formal chemical valence of the anion ($Z_a = 1$ for halides), N_a the volume density of the anions, d the bond length, and f an empirical factor. A comparison of the different materials at energies, which are determined by $E/E_0 = \text{const}$, see (7), results in the relation

$$R_M = C_1 \overline{V} \frac{E_d}{E_0} ; \qquad C_1 = \text{const} \tag{12}$$

for the molar refraction and for the slope of the molar refraction curve at any energy in the connection

$$\frac{dR_M}{dE} = C_2 \overline{V} \frac{E_d}{E_0^2} ; \qquad C_2 = \text{const} . \tag{13}$$

The product $n_e N_a$ in (11) is equal to $N_1 = n_1 N_L / \overline{V}$. Z_a is equal to 1 for all alkali halides. Substitution of (11) into (12) and (13) by use of $E_0 = hc/\lambda_0$ yields

$$R_M = C_1 \frac{N_L}{hc} n_1 f d^3 \lambda_0 \tag{14}$$

and

$$\frac{dR_M}{dE} = C_2 \frac{2N_L}{h^2 c^2} n_1 f_1 d^3 \lambda_0^2 . \tag{15}$$

Even with the assumption of special values for n_1, as made by Wemple [3] with $n_1 = 4$ for all alkali halides, R_M and dR_M/dE are functions of three independent quantities. A comparison of (14) and (7) by inclusion of (9) results in

$$f = \frac{8\pi e^2}{3} \frac{\hat{r}^2}{d^3} . \tag{16}$$

With the assumption that the quotient $\hat{r}^2/d^3$ does not change for a special structure type (15) and (16) would strongly simplify. In fact the factor f varies strongly in the case of the alkali halides. Fig. 7 illustrates the dependence of the molar refraction on f, which was calculated with the assumption $n_1 = 4$. For homologous series with rocksalt structure the molar refraction increases exponentially with increasing values of f, in homologous series with CsCl structure the molar refraction decreases exponentially with increasing f. An exception are the series of the rubidium halides, in which the molar refraction is nearly independent of f.

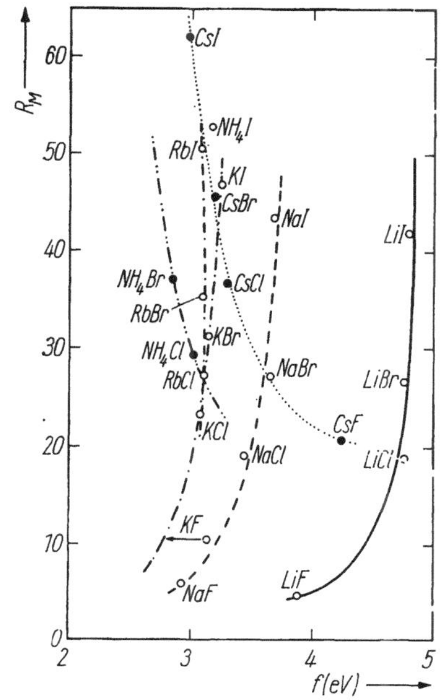

Fig. 7. Molar refraction R_M of alkali and ammonium halides as function of the empirical parameter f

With exception of this homologous series a distinction between both structure types is possible like in Fig. 2 to 5. But Fig. 7 simultaneously demonstrates that in spite of the coupling of the dispersion energy with the coordination number, the chemical valency and the bond length an essential influence of the structure on the molar refraction is arranged by the quantity f, introduced by Wemple [3] and taken to be constant for different classes of materials.

For an illustration of the influence of the structure on the molar refraction and its wavelength dependence it is not necessary to define this quantity. Without introduction of this additional parameter the Sellmeier model yields the same statements.

References

[1] W. Burckhardt, Exper. Tech. Phys., to be published.
[2] S. H. Wemple and M. Di Domenico, Phys. Rev. B 3, 1338 (1971).
[3] S. H. Wemple, J. chem. Phys. 67, 2151 (1977).
[4] S. H. Wemple, Appl. Optics 18, 31 (1979).
[5] W. Burckhardt, J. non-crystall. Solids 50, 173 (1982).
[6] W. Burckhardt, Wiss. Z. FSU Jena, math.-naturwiss. R. 32, 205 (1983).
[7] S. Hirota and T. Izumitani, J. non-crystall. Solids 29, 109 (1978).
[8] O. N. Stavroudis and L. E. Sutton, J. Opt. Soc. Amer. 51, 368 (1961).
[9] T. Radhakrishnan, Proc. Indian Acad. Sci. (A) 27, 165 (1948).
[10] G. N. Ramachandran, Proc. Indian Acad. Sci. (A) 25, 481 (1947).
[11] A. Feldman, D. Horowitz, R. M. Waxler, and M. J. Dodge, NBS Technical Note 993 (1979).
[12] R. J. Spindler and W. S. Rodney, J. Res. Nat. Bur. Standards 49, 253 (1952).
[13] R. E. Stephens, E. K. Plyler, W. S. Rodnay, and R. J. Spindler, J. Opt. Soc. Amer. 43, 110 (1953).
[14] K. Korth, Z. Phys. 84, 617 (1933).
[15] H. Harting, S.-B. Dtsch. Akad. Wiss. Berlin, Kl. Math. allg. Naturwiss. Nr. 4 (1948).
[16] W. S. Rodney and R. J. Spindler, J. Res. Nat. Bur. Standards 51, 123 (1953).
[17] W. S. Rodney and R. J. Spindler, J. Opt. Soc. Amer. 42, 431 (1952).
[18] W. S. Rodney, J. Opt. Soc. Amer. 45, 987 (1955).
[19] G. Lucovsky, Phys. Rev. B 15, 5762 (1977).

(Received June 12, 1984)

phys. stat. sol. (a) 85, 105 (1984)

Subject classification: 1.2; 10.1; 21.1

Department of Structural Properties of Materials, Technical University of Denmark and Institute of Metal Physics, Academy of Sciences of the USSR, Ural Scientific Centre, Sverdlovsk[1])

Kinetics of Vacancy Mechanism of the Solid–Liquid Transition in F.C.C. Metals

By

V. G. Chudinov and V. I. Protasov

Results are given on simulation of the phase transition solid–liquid in copper by the molecular dynamics technique (MDT), pressure and volume being constant. A direct evidence for the possibility of vacancy mechanism of melting predicted by Frenkel is obtained. In the presence of the free surface (p = const) when the temperature of the medium approaches the melting point vacancies are formed at the surface monolayer ($\approx$ 800 K). At higher temperatures an intense diffusion of defects takes place. The appearance of vacancies is associated with "premelting". Real melting starts (T_m) with an inward diffusion of the "surface" vacancies, simultaneously the whole volume being characterised by generation of Frenkel pairs ($\approx$ 1050 K). For V = const T_m increases appreciably ($\approx$ 1710 K) due to thermal pressure. Mechanism in this case is similar to that for p = const. There is an instability of the lattice along the ⟨100⟩ direction that results in generation of Frenkel pairs at the melting point. The mechanism is associated with simultaneous perpendicular movement of the two neighbour atoms along the above directions. The lifetime of an unstable pair is about $\approx 10^{-13}$ s. Then it annihilates or falls into stable defects. An intense diffusion by a vacancy and interstitial mechanism is observed ($\approx 5 \times 10^{-5}$ cm²/s).

Es werden Ergebnisse der Simulation des Phasenübergangs fest–flüssig in Kupfer mittels der dynamischen Molekulartechnik (MDT) unter konstantem Druck und Volumen angegeben. Es werden direkte Hinweise für die von Frenkel vorhergesagte Möglichkeit eines Vakanzenmechanismus des Schmelzprozesses erhalten. Beim Vorliegen einer freien Oberfläche (p = const) werden, wenn die Temperatur des Mediums den Schmelzpunkt ($\approx$ 800 K) erreicht, an der Oberflächenmonoschicht Vakanzen gebildet. Bei höheren Temperaturen findet eine intensive Diffusion der Defekte statt. Das Auftreten der Vakanzen ist mit dem „Vorschmelzprozeß" verknüpft. Reales Schmelzen tritt ein mit einer nach innen gerichteten Diffusion der „Oberflächen"-Vakanzen, wobei das gesamte Volumen durch die Generation von Frenkelpaaren ($\approx$ 1050 K) charakterisiert ist. Für V = const nimmt T_m beträchtlich ($\approx$ 1710K) durch den thermischen Druck zu. Der Schmelzmechanismus ist in diesem Fall mit dem für p = const ähnlich. Die Generation von Frenkelpaaren am Schmelzpunkt führt zur Instabilität des Gitters in Richtung ⟨100⟩. Der Mechanismus ist verknüpft mit gleichzeitiger senkrechter Bewegung der beiden benachbarten Atome in den oben angegebenen Richtungen. Die Lebensdauer eines instabilen Paares ist etwa $\approx 10^{-13}$ s. Dann annihiliert es oder fällt in stabile Defekte. Eine intensive Diffusion über einen Vakanzen- und Zwischengittermechanismus wird beobachtet ($\approx 5 \times 10^{-5}$ cm²/s).

1. Introduction

Melting (i.e. the solid–liquid transition) is likely to be the only phenomenon in solid state physics the microscopic mechanism of which has not been understood completely up to now. Main disagreements between the total phenomenological theory and the experimental data at T_m are noteworthy [1]:

---

[1]) ul. S. Kovalevskoi 18, 620219 Sverdlovsk GSP-170, USSR.

1. According to the theory during the phase transition the linear expansion coefficient and the compressibility should tend to infinity, and the shear moduli to zero. This is observed for no crystal in the solid state.

2. At certain volume and pressure values all media should have a critical point characterized by no differences between liquid and solid. Such a point has been found for no medium even at very high pressure.

3. The theory predicts that in the absence of centres for liquid phase formation a solid may be overheated. Small overheating (usually by tenths of a degree) has been repeatedly observed experimentally. However, usually it turns out that it is connected with other phenomena (i.e. internal stress [2], finite rate of movement of the phase interface [3], etc.).

The difficulties (including those mentioned above) in the classical theory of phase transitions have stimulated an alternative approach to the melting problem. According to it melting in a solid is assoicated with the formation of some defects, viz. interstitials [4], vacancies [5], or dislocations [6, 7]. According to the first hypothesis the critical point must exists at certain pressure and temperature values (for example, for Ar: $p_{cr} \approx 2400$ at and $T_{cr} \approx 100$ K). But this turned out to be a failure, because experimentally there is no medium with such a point. The vacancy mechanism of melting also has not gained wide recognition, since no model of avalanche-like growth of vacancies at T_m has been proposed, which is the necessary condition for this mechanism [8]. In recent years the dislocation model of melting has seemed to be improved greatly both experimentally [9] and theoretically [6, 7]. The most convincing results have been obtained by MDT [10 to 12]. In these papers f.c.c. structures are simulated using potentials of Lenard-Jones form. In [15] a shift of the transverse [111] mode towards larger wavelengths and a slip of neighboured (111) planes was observed. The slip of the (111) planes along three possible $\langle 112 \rangle$ directions was also observed in [11]. This leads, finally, to the formation of partial Shockley dislocations. However, one cannot rely upon complete identification of dislocations at high temperatures. Besides, the potentials of only one type do not allow one to make a unique conclusion about the dislocation mechanism even in simple compounds with f.c.c. structures. It is to be noted that a universally recognized model of melting is absent today.

The present paper attempts to develop further the view that melting is associated with an instability of the crystal lattice to form some defects.

Melting with a harder branch (as compared to [11]) of attraction has been investigated by MDT at constant pressure and volume. In every temperature interval we have been searching for both point (vacancy and interstitial) defects and spatial ones: various types of three-dimensional formations and dislocations.

2. Calculation Technique

The calculations have been carried out by the program MOLDYN on a crystallite consisting of 1554 copper atoms at $p = \text{const}$ and 2136 atoms at $V = \text{const}$. The form of the crystallite was close to a spherical one. During the solid–liquid phase transition physical processes proceed by MDT time scales slowly hence, while developing the program, main attention has been paid to maximum speed of operation [13]. The program guarantees calculation by a computer with the speed 10^6 operations/s of a time step in a crystallite consisting of 1000 atoms for 4 to 5 s. The value of a time step is treated as the geometric mean of two values,

$$\delta t = \frac{\delta t_1 \, \delta t_2}{2(\delta t_1 + \delta t_2)} \, .$$

The value of δt_1 is determined from

$$\delta t_1 = K_1 \frac{R_{\min}}{V_{\max}},$$

where $R_{\min}$ is the minimum distance between atoms, $V_{\max}$ the maximum velocity of an atom, K_1 the fraction of $R_{\min}$ covered by an atom per time step.

The value of δt_2 is determined proceeding from the frequency criterion

$$\delta t_2 = 2\pi K_2 \sqrt{\frac{m}{\varphi''(R_{\min})}},$$

where K_2 is the fraction of oscillations of an atom at $R_{\min}$ in the equivalent parabolic potential, m is the atomic mass. Constants K_1 and K_2 are chosen experimentally. As shown by the calculations performed with different interaction potentials within a wide range of atom energy (from 0 to 10 keV) $K_1 = 0.1$ and $K_2 = 0.05$ (for the second-order Eller method), which guarantees the absence of divergence up to (5 to 10) $\times$ 10^3 steps.

Besides, to test the accuracy of the numerical solution of an equation (with the above method of choosing a step) a comparison has been made with an analytical solution for the case of two atoms. After 1000 steps the difference in atom coordinates amounted to 5% compared to the analytical solution. Therefore, one may expect that an error (caused by an inaccuracy of the numerical methods) will not appreciably affect the sequence of diffusion acts ($\approx$ 100 steps) which finally results in melting. The total number of these acts (per unit volume and time) is not associated with the accuracy in solving the equations of motion and will be defined by statistic laws.

A special block of the program was responsible for searching point defects at arbitrary temperature. This is a complicated problem since near T_m the amplitudes of atomic oscillations amounted to $\approx$ 10% of the value of the lattice constant. With this in view the whole crystallite has been divided into Wigner-Seitz cells. It has been assumed that a cell with one atom is a common site of the crystalline lattice, an empty cell is a vacancy: a chell with two atoms an interstitial, and a cell with three atoms a diinterstitial, etc. A combination of a vacancy and an interstitial at neighbouring sites of the lattice has been treated as an unstable Frenkel pair.

Spatial defects have been identified by typical decoration pictures of point defects at the boundary. A more detailed analysis has been carried out using the graphical layer-by-layer output lists of atomic coordinates.

With this program one can also observe the trajectories of every atom and determine its energy at any instant.

While simulating melting at constant pressure the surface of the crystallite remains free, while at constant volume 1088 external atoms are fixed.

The energy has been introduced into a crystallite by two ways: 1. All the atoms have equal velocities, but are randomly oriented in space. After some time an equilibrium was settled in the system.[2]) 2. At every time step the rate of every atom was multiplied by a value greater than unity. In the first case the calculations have been carried out within time intervals of $\approx 5 \times 10^{-12}$ s, while in the second the rate of heating did not exceed 10^{-4} K per a time step. In all cases the equilibrium Maxwell distribution was established within $\approx 10^{-12}$ s. During the calculations the values of kinetic,

[2]) The system was supposed to be in equilibrium when (within statistical error) Maxwell velocity distribution and equilibrium between the kinetic and potential energies far from T_m were established.

potential, and total energy were controlled both in the crystallite and in its central part. The values of the lattice constant, radial density function, the distribution by velocities were also controlled.

A spherically symmetric pair interaction potential was used in the calculations [14]. It satisfactorily describes formation and migration energies of vacancies and interstitials, the elastic constants, dispersion relations, etc. for copper. At $0 < r < 0.203$ nm a Born-Mayer potential was used, at $0.203 \leqq r \leqq 0.2866$ nm a Morse potential, and at $0.2866 < r < 0.4034$ nm the polynomial of third power

$$\varphi(r) = 0.045 \exp\left[-\frac{13.154\,(r - 0.2551)}{0.2551}\right] (\text{eV}) ;$$

$$\varphi(r) = 0.19\,\{\exp[-2 \cdot 1.3588\,(r - 0.2866)] - 2 \exp[-1.3588\,(r - 0.2866)]\}\ (\text{eV}) :$$

$$\varphi(r) = -240r^3 + 248r^2 - 83.1r + 8.9\ (\text{eV}) .$$

At $r = 0.4034$ nm the derivatives were equal to zero. A time step varied from 2×10^{-14} to 0.8×10^{-14} s with temperature. This helped to avoid the divergence in the equation of motion and to keep the total energy in the system constant.

The calculations were carried out with "Burrough-7800" in the North European University Computer Centre (NEUCC).

3. Calculation Results

Fig. 1a and b show the temperature (T) of the medium and the number (N) of different defects in equilibrium as a function of the total energy (E) given to a crystallite with free surfaces (thermal equation of state). Within experimental error a linear dependence was observed up to 800 K. In this case the kinetic and potential energies were equal. Diffusion was absent. However, when the crystallite attained high energy at the surface monolayer the formation of unstable Frenkel pairs began, the lifetime being $\approx 10^{-13}$ s and the number constant for any instant (Fig. 2a). A large amount of these pairs annihilate at this temperature. Sometimes, however, some atoms move towards the "surface" steps where they occupy the "right" sites of the crystal lattice. Since formation of these pairs needs some energy (≈ 0.5 eV per atom), the caloric curve deviates from the linear dependence, while the heat capacity begins increasing. At $E \approx 1900$ K and $T \approx 900$ K the pairs decompose and intensity diffusion begins over the

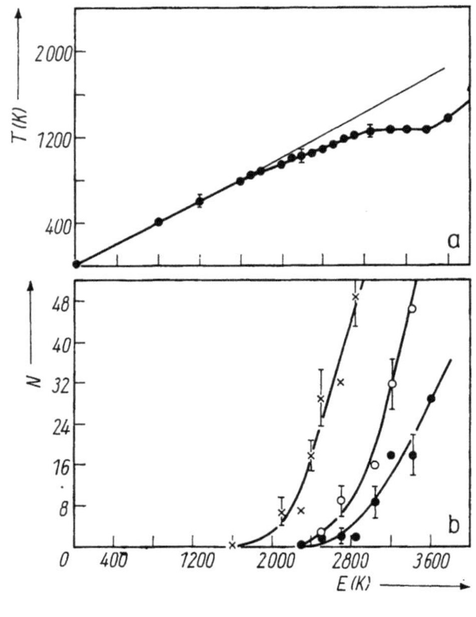

Fig. 1. a) Caloric equation for a crystallite with free surface. b) The number of point defects as a function of the total crystal energy per atom. × "surface" vacancies, ● "inside" vacancies, ○ unstable Frenkel pairs

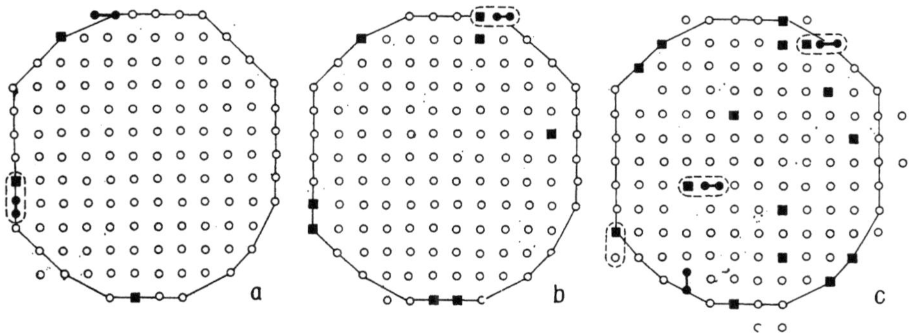

Fig. 2. The structure of the central crystalline planes ((100) plane) at different temperatures: a) $T = 1800$, b) 2300, c) 3000 K; ○ atom near the lattice sites, ■ vacancy, ● dumb-bell)

surface (at $T = 1000$ K the surface diffusion coefficient is 8×10^{-5} cm²/s). Inside the crystallite all atoms keep their sites. This phenomenon is likely to be connected with "pre-melting" and is systematically observed experimentally at $p =$ const [15].

Real melting starts at $E \approx 2300$ K and $T \approx 1050$ K when the "surface" vacancies begin penetrating into the crystal (Fig. 2b). Simultaneously, within the crystal Frenkel pairs begin to be generated (Fig. 2c), a part of which decomposes in the course of time into stable defects. In principle, the kinetics of the two processes is similar. In the latter case, if two neighbouring atoms with rather high energy have impulses in the direction close to ⟨100⟩, but perpendicular to each other they, moving simultaneously in the above directions, form at nearest sites a vacancy and a dumb-bell. In the first case an atom moving in the same way occupies simply a vacant site. Unstable Frenkel pairs, as a rule, annihilate after $\approx 10^{-13}$ s. Sometimes (i.e. at T_m) however, it turns out that an "alien" atom from one of the neighbour sites succeeds in occupying a vacant site and hence, two stable defects are formed. Fig. 3 shows the trajectories and positions of atoms in a part of the (100) pane. Preferential movement of atoms in the ⟨100⟩ direction, formation and decomposition of an unstable Frenkel pair is well seen in this figure. Then stable defects diffuse easily: interstitials by a "looping" mechanism and vacancies by the above mechanism of jumps. On collision of vacancies and dumb-bells the reverse process is possible.

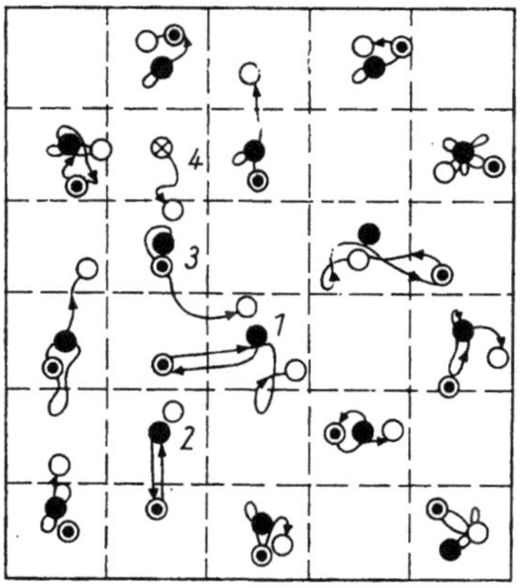

Fig. 3. Kinetics of nucleation and decomposition of an unstable Frenkel pair. Arrangement of atoms at ● $\tau = 1.6 \times 10^{-13}$, ⊙ 8×10^{-13}, ○ 1.6×10^{-12} s; ⊗ atom from another plane, ——— trajectories of atomic movement, – – – boundaries of imaginary "cells". (100) plane, $T = 1100$ K. At 8×10^{-13} s atoms 1 and 2, 3 and 4 formed the unstable Frenkel pairs, respectively, in the figure and perpendicular planes. The former decomposed at $\tau = 1.6 \times 10^{-12}$ s, while in the latter atom 3 was replaced by atom 4. Atom 3 in its turn formed with atom 1 a new dumb-bell. Finally two stable defects were formed

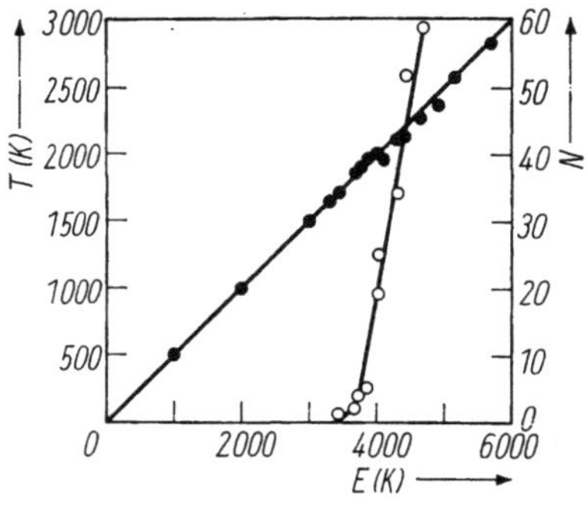

Fig. 4. Caloric equation for a crystallite with constant volume (● caloric relations, ○ number of Frenkel pairs)

Owing to their lower migration energy interstitials reach the free surface more readily and become equivalent sites of the lattice. As a result the bulk of the crystallite is continually enriched by vacancies.

Beginning with $E \approx 2800$ and $T \approx 1280$ K the temperature of the crystallite does not change up to $E \approx 3400$ K, while the number of defects continually increases up to a concentration of $\approx 4\%$. (It is to be noted that in the transition region from 1050 to 1280 K the time for defect formation is likely to be commensurable with the heating rate of the crystallite. Hence, their equilibrium concentration was not obtained in our calculations). In computer calculations the temperature interval of the transition region is accounted for by limitations in time only, while in real situations it should be absent. At $E \approx 3400$ K the temperature begins increasing once more. This implies that melting is finished and there occurs the transition to the liquid state.

At constant volume melting occurs also by the vacancy mechanism according to the scheme described above. However, it starts at higher temperature because of thermal pressure. At 1710 K unstable Frenkel pairs begin forming within the whole volume, part of them transform further into common vacancies and interstitials. In this case (unlike at constant p) in agreement with the Clausius-Clapeyron law no typical plateau is observed on the caloric dependence at T_m (Fig. 4). The kinetic and potential energies are equal at any temperature within experimental error. The lifetime of defects at T_m is less than 10^{-13} s.

It should be noted that in the interval of $\approx 10^{-12}$ s flat clusters of five to six vacancies appeared. The structure of these clusters was similar to that of Frank dislocation loops with Burgers vector $a/3\langle 111 \rangle$, but it was not formed completely. At the edge of these loops

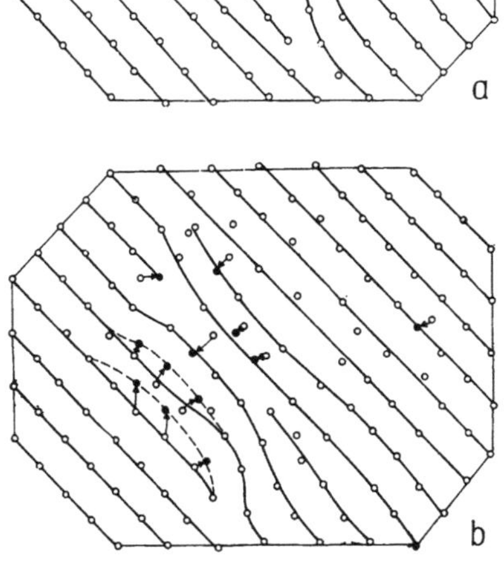

Fig. 5. The structure of a vacancy cluster ($\tau = 3 \times 10^{-12}$ s, $T = 1240$ K). a) $a_0/2$ atomic layer parallel to the (100) plane, b) nearest-neighbour layer

dislocation dipoles were observed, the planes of which made with the plane of the loop an angle of 90°. Their lifetime did not exceed 2×10^{-13} s. If they dissociate they may be identified as Shockley dislocations by the technique described in [11]. Fig. 5a and b show two central atomic layers parallel to the (100) plane which crossed one of these clusters. Within 10^{-11} s clusters of interstitials were not observed.

4. Discussion

The solid–liquid transition has the same character for all crystalline solids in spite of the different natures of interatomic interactions, types of crystal structure, etc. Hence, despite of the fact that our calculations have to a great extent model character, the qualitative picture should be the same.

The results of the present paper should be treated as experimental ones, though they are obtained with a computer.

The observed mechanism of melting in the presence of a free surface was predicted in general by Frenkel about 40 years ago [16]. It confirms the point of view that the reason of the mechanism lies in the instability of the crystal lattice to defect formation. The proper melting occurs by a vacancy mechanism, while the final liquid structure is associated with dislocations characterized by short lifetimes. In principle, the observed transition shows the features of both vacancy and dislocation mechanisms of melting.

Now let us dwell upon the mechanism of defect nucleation. Analysing numerous data one can claim that the kinetics of defect generation is of general nature, i.e. loss of stability of the f.c.c. lattice along the ⟨100⟩ direction. It concerns melting at constant volume and pressure. The presence of a free surface makes the process of vacancy formation easy and leads to pre-melting due to the formation of surface vacancies. Indeed, at constant volume no anomalies are seen on the heat capacity versus temperature curves near T_m, while at constant pressure they are seen practically for every medium.

We have not found any differences in temperatures of dissolving of surface vacancies and formation of Frenkel pairs inside the crystallite. According to some general thermodynamic considerations [17] the former is believed tor pecede the latter. This fact, however, has not been established reliably because of the small sizes of the crystallites and the limitations of our experiment. Nevertheless, it is obvious that the difference in temperature is not appreciable.

Probably it is this melting mechanism that results in no anomalies in the dispersion dependences of aluminium [9] along the ⟨111⟩ directions near T_m in f.c.c. metals. These anomalies must appear in aluminium assuming that neighbouring (111) planes shift along the ⟨112⟩ direction.

It is likely that in compounds with other structures and interaction potentials melting could occur due to an instability in any direction. For example, while simulating melting in the f.c.c. structure [18] using hard sphere potentials the transition was observed with increasing the volume by 1.55 as compared to that of the closest packing. On the other hand, it was shown in [19] that to cross the atom lens in the (110) plane along the ⟨110⟩ direction freely volume should be increased by a factor of 1.54. Indeed, the coincidence is surprising and seems hardly to be accidental.

Our results support the concept according to which during the phase transitions oscillating modes are frozen in along any direction [20]. Apparently, in this case the compressibility along this direction tends to infinity, and the elastic moduli to zero. A similar situation is likely to take place in our experiment along the ⟨100⟩ direction at T_m. It leads finally to Frenkel pairs and vacancy diffusion.

In conclusion, it should be noted that the origin of such an instability in a solid has not been clearly understood up to now and requires further investigation.

Acknowledgements

The authors would like to thank Dr. R. M. J. Cotterill and other collaborators of the Department of Structural Properties of Materials for moral support, interesting discussions and kind help during the investigations without which the present paper would not have appeared at all.

References

[1] A. UBELLODE, Plavlenie i kristallicheskaya struktura, Izd. Mir, Moskva 1969.
[2] M. KÄSS and S. MAGNN, J. Kristallogr. Miner. **116**, 354 (1961).
[3] N. G. AINSLIE, J. D. MACKENZIE, and D. TURNBULL, J. phys. Chem. **65**, 1718 (1961).
[4] J. E. LENARD-JONES and A. F. DEVONSHIRE, Proc. Roy. Soc. **A170**, 464 (1939).
[5] J. FRENKEL, Kineticheskaya teoriya zhidkosti, Idz. Akad. Nauk SSSR, Moskva 1955.
[6] K. MIZUSHIMA, J. Phys. Soc. Japan **15**, 70 (1960).
[7] D. KUHLMANN-WILSDORF, Phys. Rev. **140**, A1599 (1965).
[8] V. I. VLADIMIROV, Fiz. tverd. Tela **10**, 2639 (1968).
[9] R. M. J. COTTERILL and J. K. KRISTENSEN, Phil. Mag. **36**, 453 (1977).
[10] E. J. JENSEN, W. DAMGAIRD KRISTENSEN, and R. M. J. COTTERILL, Phil. Mag. **27**, 623 (1973).
[11] R. M. J. COTTERILL, W. DAMGARD KRISTENSEN, and E. J. JENSEN, Phil. Mag. **30**, 245 (1974).
[12] J. M. DICKEY and A. PASKIN, Phys. Letters **A30**, 209 (1969).
[13] V. I. PROTASOV and V. G. CHUDINOV, Modelirovanie na EVM defektov v kristallakh, Idz. FTI, Leningrad 1980 (p. 105).
[14] YU. M. PLISHKIN and I. E. PODCHINYONOV, Fiz. Metallov i Metallovedenie **32**, 254 (1971).
[15] G. BORELING, Solid State Phys. **15**, 1 (1963).
[16] YA. I. FRENKEL, Vvedenie v teoriyu metallov, Izd. Nauka, Leningrad 1947.
[17] L. D. LANDAU and E. N. LIVSHITZ, Statisticheskaya fiz., Izd. Nauka, Moskva 1964.
[18] T. E. FABER, An Introduction to the Theory of Metals, Cambridge University Press 1972 (p. 94).
[19] R. M. J. COTTERILL, Proc. Nato Advanced Study Institute Physics of Superionic Conductors and Electrode Material, Odense University, 4 to 22 August, 1980 (p. 1).
[20] K. MATSUNO, J. Phys. Soc. Japan **35**, 364 (1973).

(Received January 31, 1984)

phys. stat. sol. (a) 85, 113 (1984)

Subject classification: 1.4; 10.1

Instituto de Física, Universidad Nacional Autónoma de México[1])

On the Identification of the Nature of Stacking Faults in H.C.P. Materials

By

R. PÉREZ

In f.c.c. materials there are two kinds of stacking faults: intrinsic and extrinsic. However, hexagonal structures have three kinds of stacking faults. Previous methods of identification of the nature of stacking faults in f.c.c. materials are based on strong anomalous absorption being present. However in hexagonal structures this is not always the case. The image contrast characteristics of stacking faults in h.c.p. cobalt are explored under strong beam conditions. Some of the more commonly appearing systematic sets are used as the imaging diffraction Conditions. A method of identification of the three different stacking faults in h.c.p. structures is suggested. The method is based on differences in image contrast for the three types of faults obtained in wedge crystals under strong beam conditions. A theoretical explanation of these differences in terms of the two-beam approximation of the dynamical theory is also provided.

Im f.c.c.-Material existieren zwei Arten von Stapelfehlern: intrinsische und extrinsische. Hexagonale Strukturen zeigen jedoch drei Arten von Stapelfehlern. Frühere Identifizierungsmethoden der Stapelfehler in f.c.c.-Materialien basieren auf einer auftretenden starken anomalen Absorption. Jedoch in hexagonalen Strukturen ist dies nicht immer der Fall. Es werden die Bildkontrastcharakteristiken von Stapelfehlern in h.d.p.-Kobalt unter Starkstrahlbedingungen untersucht. Dabei werden einige der üblicherweise auftetenden systematischen Sätze benutzt. Eine Identifizierungsmethode für die drei unterschiedlichen Stapelfehler in h.d.p.-Strukturen wird vorgeschlagen. Die Methode beruht auf Unterschieden im Bildkontrast für die drei Arten von Fehlern, die in keilförmigen Kristallen unter Starkstrahlbedingungen erhalten werden. Eine theoretische Erklärung dieser Unterschiede wird mit der Zweistrahl-Näherung der dynamischen Theorie angegeben.

1. Introduction

In the h.c.p. structures the close-packed planes are the (002) planes. These basal planes are also the most frequently occurring slip planes in h.c.p. materials. In these types of structures there are three kinds of stacking faults. Two of them are intrinsic I_1 and I_2 and the other is extrinsic E [1]. The intrinsic fault I_1 can be formed by the removal of a basal plane and a shear, as is illustrated in Fig. 1a. An estimate of the stacking fault energy can be obtained using a scheme presented by Hirth and Lothe [1]. In this scheme the number of pairs of second nearest neighbor planes which are not in the proper stacking sequence are counted. Thus, as is indicated above, for the intrinsic fault I_1, there is one pair of second nearest neighbor planes in the wrong stacking sequence. The fault I_2 can be formed directly by shear as is illustrated in Fig. 1b. As can be seen from this diagram, there are two violations of the second nearest neighbor sequence and therefore the relative stacking fault energy is about twice that for the I_1 case. The extrinsic fault E is formed by inserting an extra plane in the perfect crystal stacking sequence and is shown in Fig. 1c. There are in this case,

[1]) Apartado Postal 877, Ensenada B.C.N. Mexico.

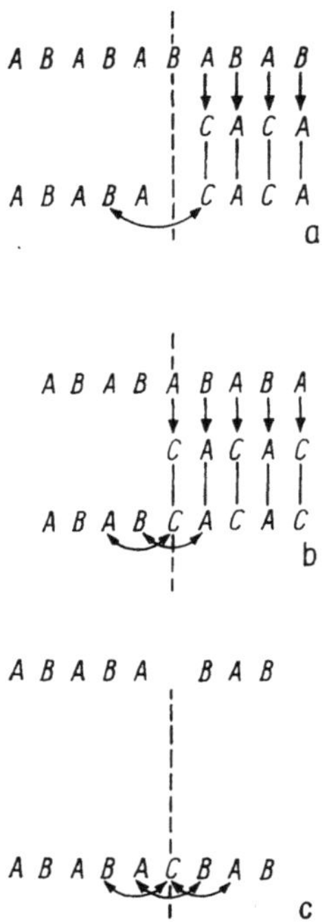

Fig. 1. Arrangements of close-packed planes for the three types of stacking faults in h.c.p. structures. a) intrinsic I_1, b) intrinsic I_2, and c) extrinsic E

three violations of second nearest neighbor sequence and therefore the stacking fault energy of this fault is about three times the energy of an I_1 and one and a half that of an I_2 fault [1]. A problem of considerable interest to the materials scientist is the development of an experimental method for the determination of the nature of stacking faults. In the past, the effects of systematic reflections in f.c.c. materials which contain stacking faults have been extensively studied [2 to 4]. As a result of these types of investigation methods which permit a determination of the nature of the two possible types of stacking faults found in f.c.c. structures have been developed [5 to 7]. It is important to mention that Blank et al. [8] have developed a method to identify the intrinsic nature of stacking faults (I_1 or I_2) in h.c.p. materials. In their approach, two dark field images are required, one for l even and one for l odd for the same value of $h - k$. If the nature of the outer fringes is the same for l even and for l odd the fault is of type I_2. If they are different the fault is of type I_1. However, stacking faults in h.c.p. structures are of three different types. A method to differentiate between these three different types of faults have not been presented in the literature. The work presented in this paper has been carried out to investigate this question. The approach adopted has been the study of image contrast characteristics of stacking faults when some of the more commonly appearing systematic sets of reflections in h.c.p. materials are excited ((100), (002), and $(1\bar{1}1)$).

2. Displacement Vectors and Phase Angles of Stacking Faults in H.C.P. Structures

The displacement vectors $\boldsymbol{R}$ corresponding to the three different types of stacking faults in h.c.p. materials can be obtained using a representation given by Berghezan et al. [9], which is shown in Fig. 2. The displacement vectors using this representation are given by

$$\begin{aligned} I_1&: \boldsymbol{R} = \overline{A\sigma} + \overline{\sigma S} = \overline{AS}\,, \\ I_2&: \boldsymbol{R} = \overline{A\sigma}\,, \\ E&: \boldsymbol{R} = \overline{S\sigma}\,. \end{aligned} \tag{1}$$

On the other hand, some of the more commonly occurring sets of systematic reflections in h.c.p. structures are the (100), (002), and $(1\bar{1}1)$ sets. One of the most important

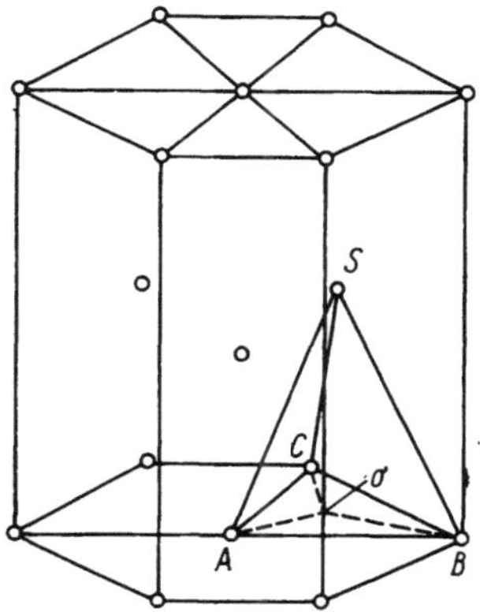

Fig. 2. Displacement vectors $\boldsymbol{R}$ in the hexagonal close-packed lattice: $\boldsymbol{R} = \overline{AS}, \overline{A\sigma}$, and $\overline{\sigma S}$

parameters influencing the role that systematic reflections play on stacking fault image contrast is the phase angle of the fault, $\alpha = 2\pi \boldsymbol{g} \cdot \boldsymbol{R}$. The stacking fault phase angle α obtained for each of the above-mentioned types of stacking faults (I_1, I_2, and E) using the h.c.p. low-order systematic sets of reflections ((100), (002), and (1$\bar{1}$1)) are given in Table 1. Table 1 shows, for example, that using the commonly occurring (100) set of systematic reflections, the intrinsic stacking faults I_1 and I_2 have the same phase angle and therefore the image contrast exhibited by these two types of faults would be identical under these diffraction conditions. On the other hand, the extrinsic stacking fault produces a phase angle $\alpha = 0$ and therefore under these circumstances this type of fault would be invisible.

Table 1

Stacking fault phase angle α obtained using the h.c.p. low-order reflections (100), (002), and (1$\bar{1}$1)

	I_1	I_2	E
(1 0 0)	$\pm 2\pi/3$	$\pm 2\pi/3$	0
(0 0 2)	$\pm 2\pi$	0	$\pm 2\pi$
(1 $\bar{1}$ 1)	$\pm \pi/3$	$\pm 2\pi/3$	$\pm \pi$

When the (002) set of systematic reflections is excited, Table 1 shows that all three types of stacking faults produce the same phase angle α ($\pm 2\pi$ or 0) and therefore under these diffraction conditions all stacking faults in h.c.p. materials will be invisible.

The most interesting systematic set from the point of view of imaging stacking faults is the (1$\bar{1}$1) set of reflections. Table 1 shows that for this set each one of the stacking faults (I_1, I_2, and E) gives rise to a different phase angle. Thus, for example, the extrinsic fault E produces a value of $\alpha = \pm\pi$ and the intrinsic stacking faults I_1 and I_2 give rise to different phase angles α, $\pm\pi/3$, and $\pm 2\pi/3$ respectively. This result suggests the possibility that stacking fault images obtained using the (1$\bar{1}$1) set can be used to distinguish between the three different types of stacking faults in h.c.p. materials.

3. Image Contrast Characteristics of Stacking Faults when the ($\bar{1}11$) Systematic Set is Excited

In order to investigate the image contrast characteristics of the three types of stacking faults in h.c.p. materials when the ($1\bar{1}1$) systematic set is excited, theoretical calculations have been carried out under strong beam diffraction conditions (i.e. the ($1\bar{1}1$) reflection at its Bragg condition) and for both bright and dark field images. The material used for this investigation was the h.c.p. phase of cobalt. The bright field results obtained from these two-beam calculations for $\alpha = \pi/3$, $2\pi/3$, and π stacking faults are shown in Fig. 3 a, b, and c. The profiles were all obtained for a crystal thickness equal to 171 nm ($\approx$ 5.5 extinction distances) and an accelerating voltage of 150 kV. These figures clearly show that there are significant differences between the profiles corresponding to the three types of stacking faults suggesting a possible method of fault identification. However, this approach has the disadvantage that the thickness of the specimen in which the fault lies must be known. Also, subsidiary peaks are strongly dependent on the values of the absorption coefficients used in the theoretical calculations [10], inaccuracies in these coefficients could result in incorrect conclusions being drawn. An alternative method for fault identification in h.c.p. materials might be to carry out observations in a wedge crystal specimen rather than one of constant thickness. In order to explore this possibility a two-dimensional display of theoretical stakking fault images based on the two-beam approximation of the dynamical theory has been used. Fig. 4 a, b, and c show the result obtained for bright field images of $\alpha = \pi/3$, $2\pi/3$, and π faults, respectively. These calculations have been obtained assuming that the fault lies in a wedge crystal and the excited reflection ($1\bar{1}1$) satisfies its Bragg condition. It can clearly be seen from these figures that the overall image contrast of the three types of stacking faults (I_1, I_2, and E) are quite different. Thus, for example, the extrinsic E, in Fig. 4c displays parallel fringes which do not change appreciably as the crystal thickness is increased. The image of a $2\pi/3$ (intrinsic I_2) fault shown in Fig. 4b indicate that the fringes in this case are similar to the image of a π fault except that each of the fringes is strongly modulated as the thickness is increased. The $\pi/3$ (intrinsic I_1) fault shown in Fig. 4a, on the other hand, can be seen to be quite different from either of the images shown in Fig. 4 c, b. These differences can be seen to be due to a modulation of the fringes with increasing thickness which gives rise to regions of

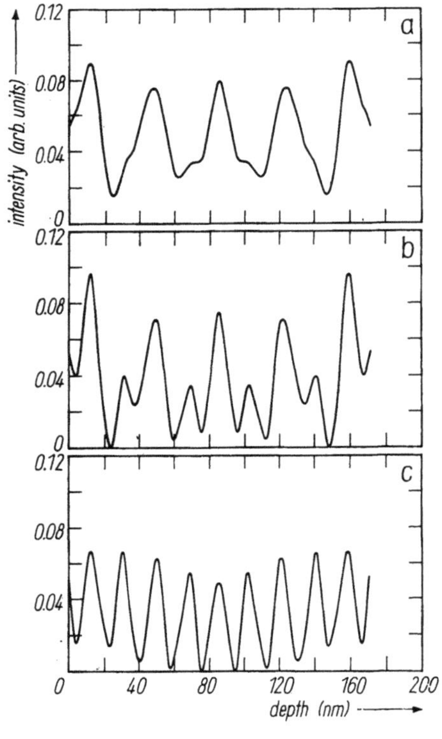

Fig. 3. Stacking fault profiles obtained when the ($1\bar{1}1$) reflection satisfies its Bragg condition and for a crystal thickness of 171 nm. (a) $\alpha = \pi/3$, (b) $\alpha = 2\pi/3$, and (c) $\alpha = \pi$

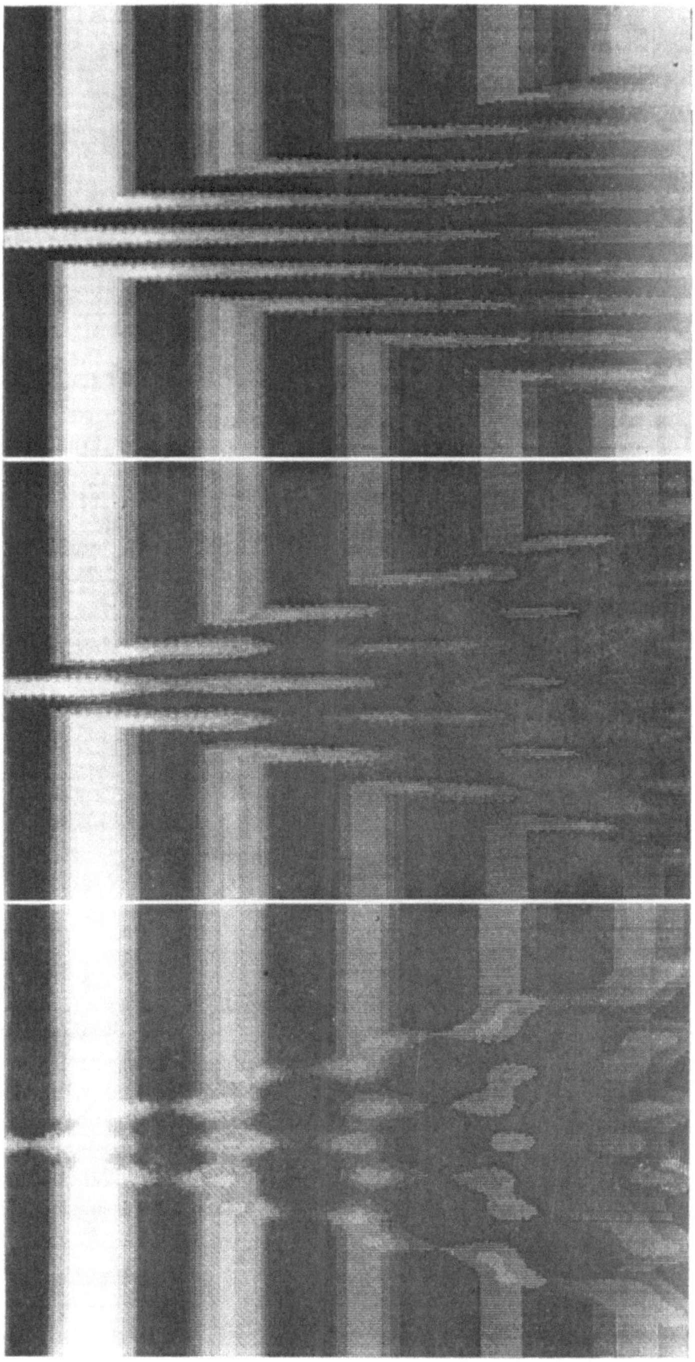

a b c

Fig. 4. Two-dimensional displays of theoretical stacking faults obtained when the $(1\bar{1}1)$ reflection satisfies its Bragg condition in a wedge crystal. a) $\alpha = \pi/3$, b) $\alpha = 2\pi/3$, and c) $\alpha = \pi$

high contrast at thickness of about $n\xi_0$ and regions of low contrast at thicknesses of about $(n + 1/2)\ \xi_0$, (ξ_0 is the extinction distance). It is important to note that similar results to these were obtained in the dark field theoretical calculations.

4. Analysis of the Theoretical Results in Terms of the Two-Beam Approximation

The behavior of the theoretical images of the three types of stacking faults in h.c.p. materials presented in Fig. 4 a, b, and c has shown that the overall image contrast is different under these diffraction conditions. The differences of these faults as a function of the crystal thickness z can be understood by considering an expression for the bright field intensity $I_0(z)$, which was obtained by Whelan and Hirsch [11]. This expression is based on the two-beam approximation when the lowest-order reflection g satisfies its Bragg condition. Also absorption is not taken into account. It is important to note, as can be seen from Section 3, that the difference in image contrast for the three types of stacking faults are more pronounced in thin regions of the specimen. Under these circumstances, absorption can be neglected and therefore an analysis in terms of the two-beam approximation of the dynamical theory without absorption might give a qualitative description of image contrast.

The transmitted beam intensity at the Bragg condition of the first-order reflection can be written as [11]

$$I_0(t, z) = \cos^2\left(\frac{\alpha}{2}\right)\cos^2\left(\frac{\pi t}{\xi_0}\right) + \sin^2\left(\frac{\alpha}{2}\right)\cos^2\left(\frac{2\pi z}{\xi_0}\right), \tag{2}$$

where ξ_0 is the extinction distance, z the crystal thickness, and t the distance measured from the centre of the crystal to the fault in a direction perpendicular to the surface.

Setting $t = 0$ or $t = n\xi_0/2$ where n is an integer, (2) can be written for the three types of stacking faults as

$$\pi/3 \text{ case:} \quad I_0\left(0 \text{ or } \frac{n\xi_0}{2}, z\right) = \frac{3}{4}\cos^2\left(\frac{\pi z}{\xi_0}\right) + \frac{1}{4}, \tag{3}$$

$$2\pi/3 \text{ case:} \quad I_0\left(0 \text{ or } \frac{n\xi_0}{2}, z\right) = \frac{1}{4}\cos^2\left(\frac{\pi z}{\xi_0}\right) + \frac{3}{4}, \tag{4}$$

$$\pi \text{ case:} \quad I_0\left(0 \text{ or } \frac{n\xi_0}{2}, z\right) = 1\,. \tag{5}$$

Thus, according to (5), the centre of a π-fault (and other points for which $t = (n\xi_0)/2$) should have the same intensity, independent of the crystal thickness. However, the $2\pi/3$-fault, (4) should show intensity modulations (as a function of crystal thickness) between the values 3/4 to 1 (assuming 1 to be the incident intensity). In the $\pi/3$ case, (3), the same intensity modulations in the stacking fault fringes are present but in this case this modulation is much stronger, between 1/4 to 1. It is, therefore, clear that the behavior of the stacking fault fringes as the crystal thickness is increased for the three different types of stacking faults (I_1, I_2, and E) can be explained by using the two-beam approximation of the dynamical theory.

5. Conclusions

In an f.c.c. material the basis for determining the nature of stacking faults depends on strong anomalous absorption being present [5 to 7].

In h.c.p. cobalt when the commonly (100) set of systematic reflections is excited and $\Delta\theta_{100} = 0.0$ (i.e. the reflection (100) satisfies its Bragg condition), anomalous absorption is almost absent [12]. The stacking faults obtained under these diffraction conditions are therefore symmetric with respect to the middle of the foil both in the bright and dark field cases. Furthermore, the phase angle obtained for both types of intrinsic (I_1 and I_2) is the same and therefore the same image contrast is obtained.

Clearly, the identification of the nature of these defects under these particular circumstances presents serious difficulties. On the other hand, when the (002) systematic set of reflections is excited the phase angles α obtained for the different types of stakking faults are $\pm 2\pi$ or 0 and therefore they are invisible under these diffraction conditions and nothing can be learned about their nature.

The results presented in this paper leads to the obvious conclusion that when the identification of the nature of stacking faults in h.c.p. cobalt is required the best systematic set of reflections to which an electron microscopist might resort is the $(1\bar{1}1)$. It is important to point out that this method would need wedge crystals but only one micrograph would be required, the bright or dark field image of the stacking fault obtained under strong beam conditions, also it is worthwhile to mention that these results are more general and can be applied to different h.c.p. materials. Also the same imaging properties presented in this work for the $(1\bar{1}1)$ set are found in other systematic sets of reflections which are in the same class, as defined by Jones [13]. Therefore, there is a number of other systematic sets of reflections commonly found in h.c.p. materials which can be used in the identification procedure.

Acknowledgements

The author would like to thank Prof. S. S. Sheinin for useful discussions and suggestions and also for providing the facilities to carry out this investigation. The author is also grateful to Drs. M. J. Yacamán, A. Gómez, D. Romeu, P. Shabes, and V. Castaño for the reading of the manuscript.

References

[1] J. P. Hirth and J. Lothe, Theory of Dislocations, McGraw-Hill Publ. Co., New York 1968.
[2] C. J. Humphreys, A. Howie, and G. R. Booker, Phil. Mag. **15**, 507 (1967).
[3] S. S. Sheinin and K. Z. Botros, phys. stat. sol. (a) **12**, 549 (1972).
[4] K. Z. Botros and S. S. Sheinin, phys. stat. sol. (a) **31**, 625 (1975).
[5] P. B. Hirsch, A. Howie, R. B. Nicholson, D. W. Pashley, and M. J. Whelan, Electron Microscopy of Thin Crystals. Robert E. Krieger Publ. Co., New York 1977.
[6] J. Van Landuyt, R. Gevers, and S. Amelinckx, phys. stat. sol. **18**, 167 (1966).
[7] A. Art, R. Gevers, and S. Amelinckx, phys. stat. sol. **3**, 697 (1963).
[8] H. Blank, P. Delavignette, R. Gevers, and S. Amelinckx, phys. stat. sol. **7**, 747 (1964).
[9] A. Berghezan, A. Fourdeux, and S. Amelinckx, Acta metall. **9**, 464 (1961).
[10] S. S. Sheinin, K. Z. Botros, and C. D. Cann, phys. stat. sol. (a) **3**, 537 (1970).
[11] M. J. Whelan and P. B. Hirsch, Phil. Mag. **2**, 1121 (1957).
[12] R. Pérez, S. S. Sheinin, and K. Z. Botros, phys. stat. sol. (a) **83**, 445 (1984).
[13] I. P. Jones, phys. stat. sol. (a) **47**, 163 (1978).

(Received March 8, 1984)

phys. stat. sol. (a) 85, 121 (1984)

Subject classification: 10.1 and 12.1; 21

Institute of Materials Science, University of Tsukuba, Ibaraki[1])

Dislocation Pinning at Low Temperatures in Nb

By

N. KURAMOCHI, H. MIZUBAYASHI, and S. OKUDA

The movement of hydrogen atoms in high-purity Nb below 100 K is studied through changes in dislocation motion where dislocation pinnings due to hydrogen atoms are observed as an increase in the elastic modulus and a decrease in the internal friction. Dislocations in specimens containing a few at. ppm hydrogen are pinned by hydrogen atoms at low temperatures. The dislocations are cleared from the pinnings by three methods: dislocation breakaway by vibrating at high amplitude, quenching to liquid helium from room temperature, and deformation at liquid helium temperature. During warmup after these treatments, four stages of dislocation pinning are observed. Among these stages, the first one probably starting below 6 K is attributable to dislocation rearrangement. The second stage which starts at ≈ 22 K is transient. The last two stages at higher temperatures at 36 to 50 K and 60 to 80 K, respectively, coincided with the recovery stages observed in resistivity measurements after quenching. Since a hydrogen free specimen shows no pinning, these pinnings are apparently caused by hydrogen atoms. These pinnings are studied for various specimens and discussed in detail. Further, in a heavily deformed specimen, a large α-peak (≈ 40 K for 600 Hz) is observed and its activation volume suggests that the peak is caused by kink diffusion along screw dislocations.

Anhand der Änderungen der Versetzungsbewegung wird die Wasserstoffatomwanderung in hochreinem Nb unterhalb 100 K untersucht, wobei Versetzungsverankerung infolge der Wasserstoffatome als Anstieg des Elastizitätsmoduls und Verringerung der inneren Reibung beobachtet wird. Versetzungen in Proben, die wenige At. ppm Wasserstoff enthalten, werden durch Wasserstoffatome bei niedrigen Temperaturen verankert. Die Versetzungen werden von den Verankerungen mittels dreier Methoden befreit: Versetzungsabreißen durch Schwingungen hoher Amplitude, Abschrecken auf die Temperatur des flüssigen Heliums von Zimmertemperatur aus und Verformung bei Heliumtemperatur. Während des Erwärmens nach diesen Behandlungen werden vier Stufen der Versetzungsverankerung beobachtet. Von diesen Stufen ist die erste, die wahrscheinlich unterhalb 6 K einsetzt, einer Versetzungsumordnung zuzuordnen. Die zweite Stufe, die bei ≈ 22 K beginnt, ist nichtstationär. Die letzten beiden Stufen bei höheren Temperaturen, 36 bis 50 K bzw. 60 bis 80 K, stimmen mit den Erholungsstufen überein, die nach Abschrecken bei Widerstandsmessungen beobachtet werden. Da eine wasserstoffreie Probe keine Verankerung zeigt, werden diese Verankerungen anscheinend durch Wasserstoffatome verursacht. Diese Verankerungen werden für verschiedene Proben untersucht und ausführlich diskutiert. Ferner wird in einer stark deformierten Probe ein großes α-Maximum (≈ 40 K bei 600 Hz) beobachtet, dessen Aktivierungsvolumen daraufhin weist, daß das Maximum durch Kinkendiffusion in Richtung der Schraubenversetzungen verursacht wird.

1. Introduction

The behavior of hydrogen (H) at low temperatures in Va metals has been studied extensively because of interests in diffusion or precipitation problems. Since the mobility of H below 100 K in the α-phase is very low, conventional techniques cannot be applied in the study of diffusion at low temperatures. Migration of H atoms has been usually studied by measurements of recovery in electrical resistivity after quenching to liquid helium temperature [1]. There is another method to investigate the

[1]) Sakura-mura, Ibaraki 305, Japan.

behavior of H at low temperatures. That is an observation of dislocation pinnings due to H [2], where dislocations are used as a probe to detect the motion of H.

Dislocations in b.c.c. metals are known to be very sensitive to H and three kinds of, dislocation motions are considered to occur below 100 K [2 to 5]. In the present work, dislocation pinnings which suppress these motions were studied after various treatments in high-purity Nb.

2. Experimental Procedure

2.1 Specimens

Specimens used were made from Nb polycrystalline wire of 0.5 mm in diameter purchased from Materials Research Corporation. The wires were purified in UHV as follows: They were decarburized at 1900 to 2100 K in an oxygen atmosphere of (1.2 to 7.0) $\times$ 10^{-3} Pa and annealed at 2200 to 2500 K for 30 min to 10 h in vacuum of 7 $\times$ 10^{-8} to 2 $\times$ 10^{-7} Pa. After the annealing, two different treatments were made. In one case, an oxide film was formed on the specimen surface at 500 K by introducing oxygen gas into the UHV furnace to avoid re-absorption of H (oxide coated specimen). The residual H concentration after this treatment was estimated to be below 10 at.ppm from the difference in resistivity at liquid helium temperature (LHeT) between quenched and slowly cooled states [1]. Then, some specimens were deformed at RT in air by bending (below 10%) or by rolling (40 to 50%) before mounting to the measuring bed. In another case, a thin Pd film was formed on the specimen surface at RT by evaporation in the UHV furnace without exposure to air, then this specimen was deformed to 7.0% by bending and annealed at 420 K for 8 days in air in order to remove the residual H in the specimen [6, 7]. This specimen will be called the Pd coated specimen. The residual resistivity ratios were 1600 to 2500 for the oxide coated specimens.

2.2 Measurements

An apparatus used to measure internal friction was described in [8]. The specimen was fixed into a holder and the holder was mounted to the measuring bed in a vacuum chamber. In most cases, the specimen was cooled from RT to LHeT at the rate of $\approx$ 1 K/min (slow cooldown without vibration (off-cooldown)). Then, internal friction (IF) and the resonant period (P) which corresponds to elastic modulus were measured during warmup. Typical warmup rate was $\approx$ 0.5 K/min and the maximum strain amplitude of the specimen (ε_{max}) in measuring runs was $\approx 1 \times 10^{-6}$ when not specified otherwise. The atmospheric pressure was kept below 10^{-3} Pa during the measurements. Dislocations in the oxide coated specimen after the slow off-cooldown are pinned down strongly by H atoms. For the observation of the motion of H atoms at low temperatures through the dislocation motion, this strong pinning must be broken. Three different methods were used to break the pinning. The first method was high-temperature breakaway treatment (HTB) [2,5]. Typical procedure was as follows: After the slow cooldown to LHeT, the specimen was vibrated at a high ε_{max} of $\approx 2 \times 10^{-5}$ during warmup to 80 to 100 K, then cooled down to LHeT again while having kept the high amplitude of vibration (on-cooldown with high ε_{max}). During this thermal cycling, dislocations were broken away from some pinners and hydrogen concentration along dislocation lines decreased. After the HTB, ε_{max} was reduced to $\approx 1 \times 10^{-6}$ and P and IF were measured as a function of temperature during warmup. The second method was quenching [2]: The specimen in its holder was dropped into liquid helium (LHe) from RT. After the quenching, the holder was mounted on the measuring bed in LHe, then the LHe in the chamber was pumped

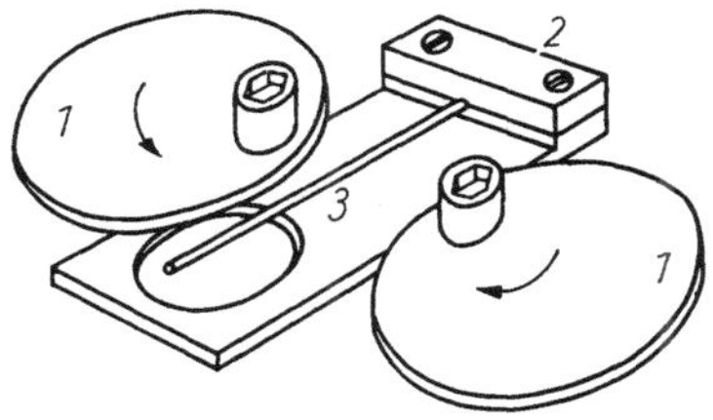

Fig. 1. The bender used for the in situ deformation in LHe: 1 rotors, 2 holder, and 3 specimen. The specimen can be bent by turning the rotors

out and measurements of IF and P were started. The third method was deformation at LHeT: The deformation was performed in LHe using a special tool set at the bottom in the chamber as shown in Fig. 1. The two rotors could be rotated using a driver installed from outside of the chamber. The specimen fixed in the holder was deformed by bending toward either side by the rotors. After the specimen had been deformed by a few %, the specimen was set in the measuring bed. Then, the same procedure was followed as that after the quenching.

3. Experimental Results

Fig. 2 shows P and IF versus temperature in an annealed specimen (oxide coated). Curves 1 and 1′ were obtained during warmup with $\varepsilon_{max} = 9.5 \times 10^{-7}$ immediately after the slow off-cooldown from RT. As can be seen, there is no relaxation peak in IF. This indicates that dislocations are pinned down strongly after the slow off-cooldown. Next, during vibration with $\varepsilon_{max} = 2.3 \times 10^{-5}$, the specimen was warmed up to 75 K and cooled down to LHeT (HTB). Hereafter, this HTB will be indicated as HTB (2.3×10^{-5}, 75). After this HTB (2.3×10^{-5}, 75), ε_{max} was reduced to 9.5×10^{-7} again and curves 2 and 2′ were obtained during warmup. P at 6 K is seen to have increased suggesting that dislocation motion below 6 K increased after the HTB. In curve 2′, there appears a distinctive IF peak at $\approx$ 25 K. This is the α_1-peak which is considered to be caused by the kink diffusion along screw dislocations. On the other hand, the α_2-peak which is thought to be attributed to the kink pair formation on non-screw dislocations and takes place at $\approx$ 60 K [5] was only observed partly because the dislocation pinning began at 36 K as seen in curves 2 and 2′ (compare with Fig. 3). From the temperature region of the recovery and the fact that no pinning was observed in the Pd coated specimen as shown in Fig. 6, these pinners are considered to be H atoms. The H pinners are released from dislocations when the dislocations are in high-amplitude vibration, because the effective trapping potential of dislocations becomes shallower. The released H atoms may be trapped by impurity atoms around the dislocations. During warmup, the H atoms detrap from the impurities above critical temperatures and pin down dislocations again during the warmup measurements with $\varepsilon_{max} = 9.5 \times 10^{-7}$. Curves 2 and 2′ also show that P and IF almost recovered at $\approx$ 60 K. Fig. 3 shows the results on the oxide coated specimen deformed 7.9% by bending at RT which were observed for measurements with various ε_{max}'s after HTB. Curves 1 and 1′ were obtained in measurement with $\varepsilon_{max} = 1.6 \times \times 10^{-6}$ after the slow off-colldown from RT. Small α_1- and α_2-peaks are barely seen at $\approx$ 25 and $\approx$ 60 K, respectively. Anomalous spikes seen in IF curves as in curve 1′ are considered to be the IF-spikes sometimes observed in Va metals [9]. Curves 2 to 6 and 2′ to 6′ were obtained with various ε_{max}'s after the same HTB (2.4×10^{-5}, 100). As can be seen in the figure, the temperature where the dislocation pinning starts increased from 44 to 53 K as ε_{max} increased (curves 2 to 6). With respect to the internal friction peaks, the α_1- and α_2-peaks became larger after HTB and could be distinguished clearly (curves 2′ to 6′). The peak temperatures of the α_1-peak after HTB are

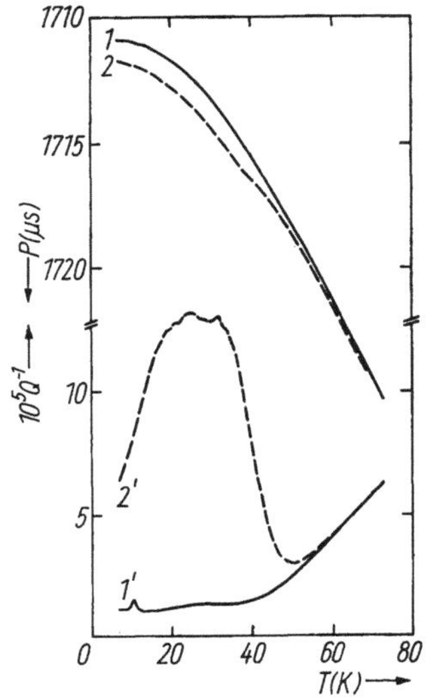

Fig. 2

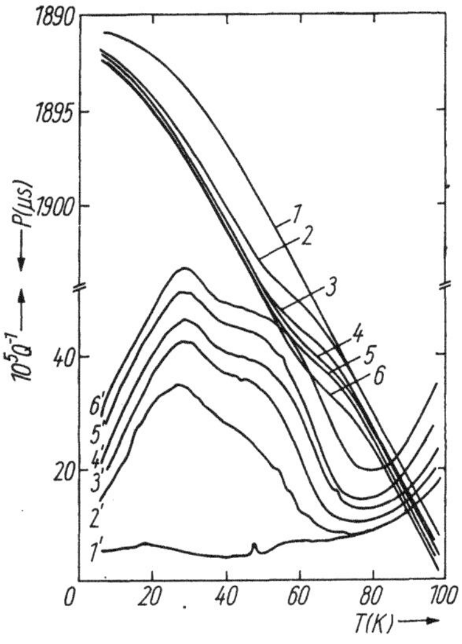

Fig. 3

Fig. 2. An example of dislocation pinning observed in an annealed specimen (oxide coated Nb) after high-temperature breakaway treatment (HTB). Curves 1 and 1′ (———) were obtained in measurements during warmup after the slow off-cooldown from RT. Curves 2 and 2′ (— — —) were obtained after HTB (2.3×10^{-5}, 75). Curve 2 shows that the dislocation pinning starts from $\approx$ 36 K. These measurements were carried out with $\varepsilon_{max} = 9.5 \times 10^{-7}$

Fig. 3. Effect of HTB on deformed, oxide coated Nb (7.9% bent at RT) for various amplitudes ε_{max}. Curves 1 and 1′ were obtained with $\varepsilon_{max} = 1.6 \times 10^{-6}$ during warmup after the slow off-cooldown from RT. All other curves were obtained with various ε_{max}'s during warmups after HTB's (2.4×10^{-5}, 100). ε_{max}'s were as follows: (2, 2′) 1.6×10^{-6}, (3, 3′) 2.4×10^{-6}, (4, 4′) 3.2×10^{-6}, (5, 5′) 4.8×10^{-6}, and (6, 6′) 6.4×10^{-6}

higher than that observed after the slow off-cooldown. This is probably due to the increase in the free length of dislocations after HTB. On the other hand, the α_2-peak after HTB was affected by the dislocation pinning and its complete shape could not be observed. Results of isochronal annealing performed after HTB are shown in Fig. 4. This was obtained after measurements shown in Fig. 3. It is seen that there is only one large pinning stage starting at $\approx$ 44 K. Fig. 5 shows the results on a oxide coated specimen deformed 41% by rolling at RT. The specimen was slowly off-cooled down to LHeT. Curves 1 and 1′ were obtained with $\varepsilon_{max} = 1.7 \times 10^{-5}$ after HTB (1.7×10^{-5}, 95). After that, the repeated measurements of P and IF with $\varepsilon_{max} = 1.7 \times 10^{-5}$ below 95 K gave the same curves as the curves 1 and 1′ but not shown here, i.e. they were reproducible. After HTB (1.7×10^{-5}, 95), the measurements with the lower amplitude ε_{max} of 9.0×10^{-7} during warmups to increasingly higher temperatures were repeatedly made. The results are shown by the curves 2 to 4 and 2′ to 4′. It is seen that a pronounced dislocation pinning due to H atoms started from $\approx$ 80 K. This temperature changed from 60 to 80 K in various rolled specimens after HTB (not shown here). The results of the final run (curves 4 and 4′) were the same as the results obtained in the warmup run with the same ε_{max} made immediately after off-cooldown from RT (not shown here), i.e. an initial state of pinning was resumed after warmup to 100 K. The results in Fig. 6 were obtained on the Pd coated specimen. The specimen was

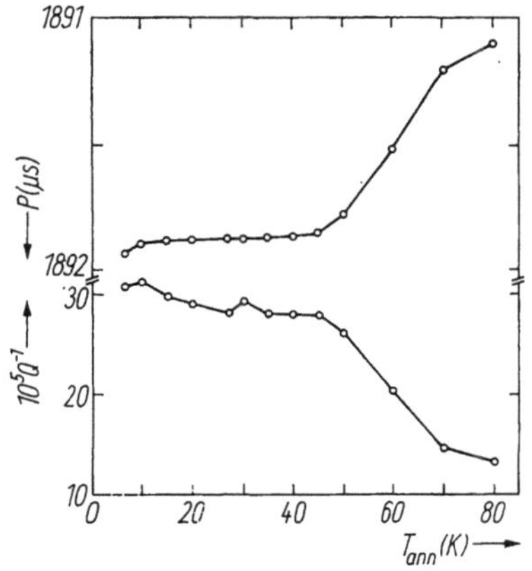

Fig. 4. Changes in resonant period and internal friction in a slightly deformed specimen (the same specimen used in the experiment shown in Fig. 3) observed in isochronal annealing after HTB (2.4×10^{-5}, 100). Changes are plotted against the annealing temperature (T_{ann}). Holding time was 2 min for each annealing and measurements were made at 6.2 K with $\varepsilon_{max} = 1.6 \times 10^{-6}$

off-cooled down slowly from RT and curves 1 and 1′ were obtained during warmup measurements with ε_{max} of 1.1×10^{-6}. There is no other peak up to RT but not shown here. This suggests that the residual H was completely removed in this specimen [7]. Curves 2 and 2′ were obtained with the same ε_{max}, 1.1×10^{-6} after HTB (1.7×10^{-5}, 100). As seen in the figure, this HTB had little effect on P and IF and no

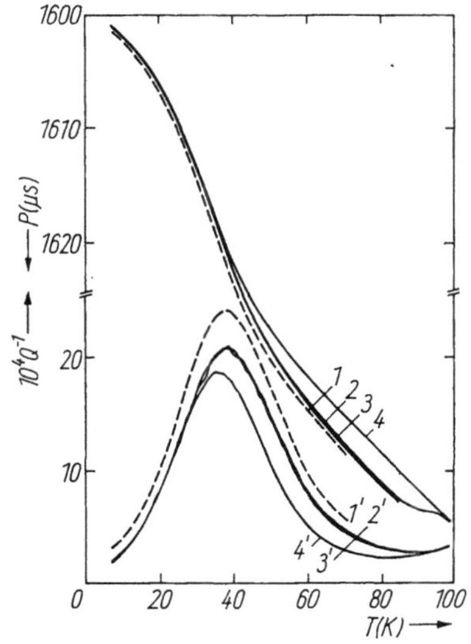

Fig. 5

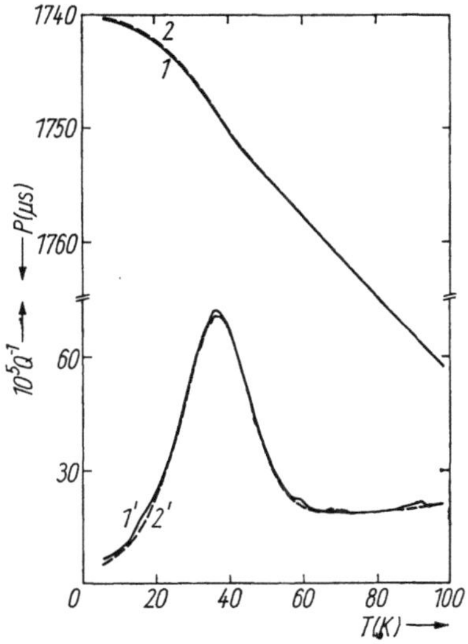

Fig. 6

Fig. 5. Changes in resonant period and internal friction in a rolled specimen (41% at RT, oxide coated Nb). Curves 1 and 1′ (– – –) were measured with $\varepsilon_{max} = 1.7 \times 10^{-5}$ during warmup after HTB (1.7×10^{-5}, 95). After on-cooldown with $\varepsilon_{max} = 1.7 \times 10^{-5}$ from 70 K following (1) and (1′), Curves 2 and 2′ (———) were obtained during warmup with $\varepsilon_{max} = 9.0 \times 10^{-7}$. Then, curves 3, 3′, 4, and 4′ (———) were successively obtained also during warmups with $\varepsilon_{max} = 9.0 \times 10^{-7}$ after the on-cooldown with $\varepsilon_{max} = 9.0 \times 10^{-7}$ from 85 and 100 K, respectively

Fig. 6. Changes in resonant period and internal friction observed in a Pd coated Nb before and after HTB. The specimen was deformed 7.0% and annealed at 420 K in the air to remove the residual hydrogen. Curves 1 and 1′ (———) were obtained during warmup after the off-cooldown from RT. Curves 2 and 2′ (– – –) were obtained during warmup after HTB (1.7×10^{-5}, 100). Both measurements were carried out with $\varepsilon_{max} = 1.1 \times 10^{-6}$

pinning was observed. Therefore the pinning observed after HTB in oxide coated specimens is concluded to be caused by the residual H atoms.

Fig. 7 shows pinning processes observed after HTB in the annealed (curves 1), deformed (by bending) (curves 2, 3, and 5), and rolled (curve 4) specimens. Curves 1 to 4 were obtained using oxide coated specimens and curve 5 was obtained in the Pd coated specimen. In this figure, ΔP which is the difference in P's between before and after HTB was plotted to show the pinning clearly. The Pd coated specimen shows no pinning.

With regard to the pinning in the oxide coated specimens, the temperature where a main stage starts seems to increase from $\approx$ 35 to $\approx$ 80 K as the amount of deformation increases. The rolled specimen has another minor and transient pinning stage at around 25 K. Furthermore, the rolled specimen is characterized by very small ΔP at 6 K after HTB. Fig. 8 shows the results of isochronal annealing in a rolled specimen (52% rolled at RT) after HTB (2.0×10^{-5},100). The transient behavior of the 25 K stage is clearly seen. It is worth noting that the amount of recovery in a stage at $\approx$ 80 K (curve 4 in Fig. 7) is very small when measured at 6 K. This means that the dislocation motion suppressed in this stage is mostly the dislocation motion observed in the temperature range of the α-peak.

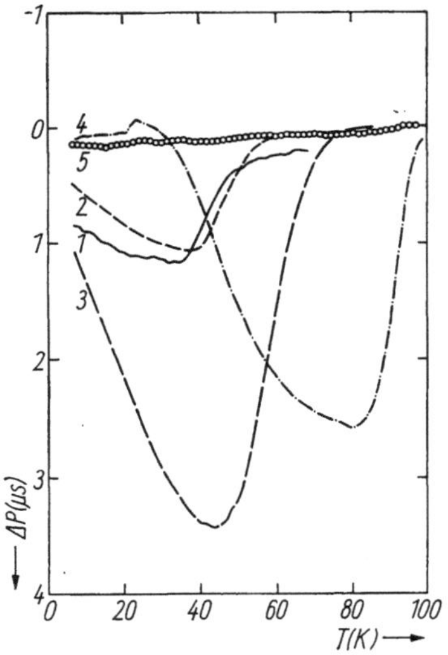

Fig. 7

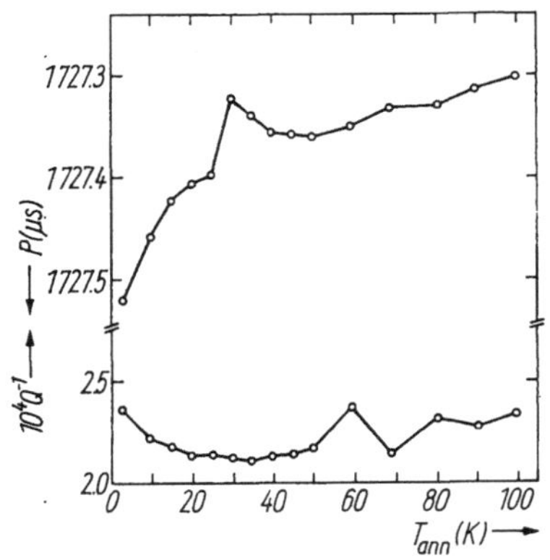

Fig. 8

Fig. 7. Dislocation pinnings observed after HTB for various specimens. ΔP represents the difference in P's between before and after HTB, e.g. curve 1 in the figure shows (curve 2 — curve 1) in Fig. 2. Curves 1 to 4 were obtained in oxide coated specimens and curve 5 was obtained in a Pd coated specimen. Specimen states, HTB conditions and measuring ε_{max}'s were as follows: (1) (———) as-annealed, (2.3×10^{-5}, 75), 9.5×10^{-7}; (2) (— — —) 4.6% bent at RT, (1.0×10^{-5}, 100), 6.6×10^{-7}; (3) (— — —) 7.9% bent at RT, (2.4×10^{-5}, 100), 1.6×10^{-6}; (4) (— · —) 41% rolled at RT, (1.7×10^{-5}, 95), 9.0×10^{-7}; (5) (○) 7.0% bent at RT, (1.7×10^{-5}, 100), 1.1×10^{-6}

Fig. 8. Results of isochronal annealing in a rolled oxide coated specimen (52% rolled at RT) after HTB (2.0×10^{-5}, 100). Holding time was 2 min for each annealing and measurements were made at 6.2 K with $\varepsilon_{max} = 1.4 \times 10^{-6}$

Fig. 9 shows the results of quenching experiments. Here ΔP shows the difference in P's obtained during warmups between immediately after quench and after 100 K annealing in oxide coated specimens with various treatments. During quenching from RT, some H atoms are trapped by impurities in the matrix before they can reach at dislocations and less dislocation pinnings would result. During warmup after the quench, the impurity trapped H atoms are freed from impurities depending on their binding energies and pin down dislocations. Curve 1 shows the result for the annealed specimen with $\varepsilon_{max} = 1.0 \times 10^{-6}$, curve 2, for the specimen deformed 5.9% at RT by bending and charged electrolytically with H of 130 at.ppm ($\varepsilon_{max} = 1.3 \times 10^{-6}$). Curves 3a to 3c were obtained for the 46% rolled specimen with similar ε_{max}. They show the results of successive measurements immediately after the quench, after warmups to 19 and 52 K. A transient pinning at around 25 K and a main pinning from ≈ 35 K are seen in curves 2 and 3, but they are not very clear in curve 1.

Fig. 10 shows the result of LHeT deformation. The annealed oxide coated specimen was deformed ≈ 3% in LHe by bending. Measurements during warmups were carried out with $\varepsilon_{max} = 1.4 \times 10^{-6}$. In this case, the P's after annealing at 80 K were taken as standard for ΔP. Curves a to d were obtained in a successive way, immediately after the deformation (a), after on-cooldown from 19 K (b), 32 K (c), and 41 K (d), respectively. The transient pinning at around 20 K and a main pinning from ≈ 35 K can be seen. Furthermore, as is seen in comparing curves a with b, ΔP recovers even below 20 K. Such a recovery below 20 K can also noticed in curve 3a in Fig. 9.

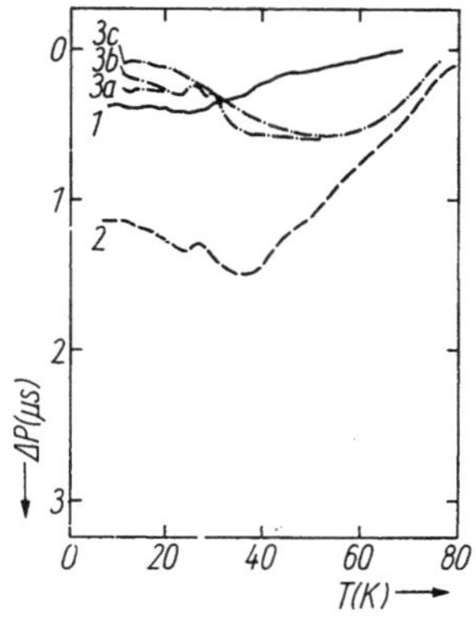

Fig. 9

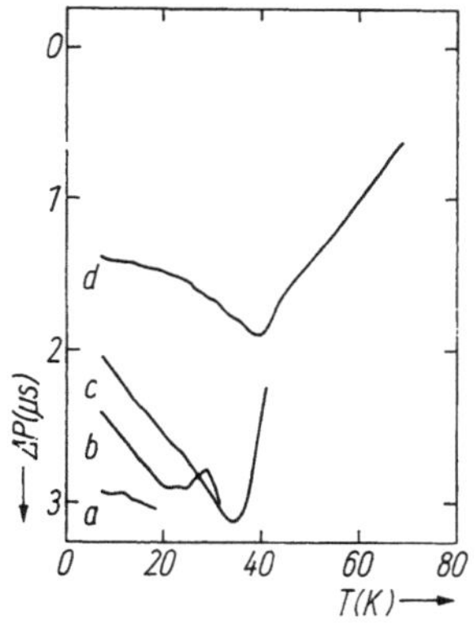

Fig. 10

Fig. 9. Dislocation pinnings observed in oxide coated specimens after quenching. ΔP represents the difference in P's obtained during warmups between immediately after quench and after 100 K annealing. Specimen states and ε_{max}'s were as follows: (1) (———) as-annealed, 1.0×10^{-6}; (2) (– – –) 5.9% bent at RT + hydrogen charged (130 at.ppm), 1.3×10^{-6}; (3) (— · —) 46% rolled at RT, (1 to 2) $\times 10^{-6}$; (3a) immediately after quenching, (3b) and (3c) successively after on-cooldown from 19 and 52 K, respectively

Fig. 10. Dislocation pinnings observed in a oxide coated specimen immediately after deformation at 4.2 K. ΔP represents the difference in P's obtained during warmups between after deformation and after annealing at 80 K. ε_{max} was 1.4×10^{-6}. Curve a was obtained immediately after the deformation. Curves b, c, and d were obtained successively after the on-cooldown from 19, 32, and 41 K, respectively

4. Discussion

4.1 Dislocation pinning

It is considered that there are three kinds of dislocation motions below 100 K in Nb which cause relaxation peaks in the temperature ranges shown in Table 1 [5]. In Nb specimens which contain a few at.ppm of H, these motions are hindered due to dislocation pinnings by H atoms after an ordinary cooling to LHeT. The number of these pinnings can be lessened by the treatments such as HTB, quenching, and deformation at LHeT. During warmups after these treatments, the H atoms which have been set free from dislocations and probably trapped by impurities in the matrix, are detrap-

Table 1

Three kinds of dislocation motions below 100 K which cause relaxation peaks in the temperature ranges shown

	below 6 K	15 to 30 K	60 to 70 K
non-screw dislocation	kink diffusion		kink pair formation
screw dislocation		kink diffusion	

Table 2

Dislocation pinning stages observed in Nb after various treatments

oxide coated Nb	HTB	annealed bent
		rolled
	quenching	annealed bent
		rolled
	LHeT deformation	bent
Pd coated Nb	HTB	bent
hydrogen charged Nb	resistivity measurements after quenching	Hanada [11]
		Engelhard [13]
		Gotoh et al. [14]

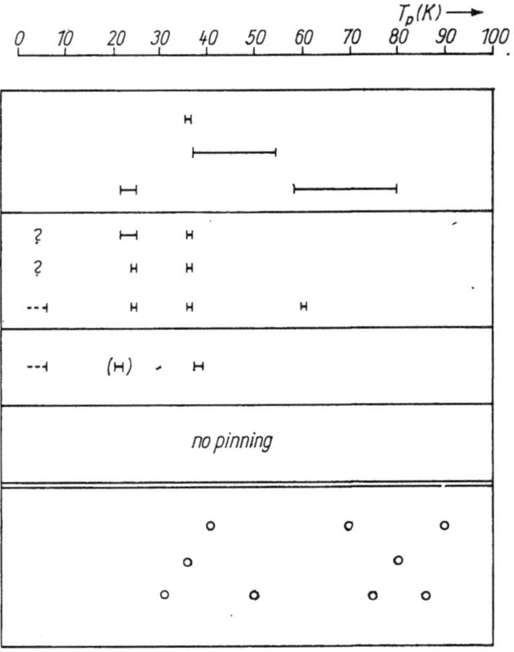

The temperatures indicated (T_p) are those from which apparent pinnings were observed to start. T_p's indicated are taken from all the results some of which were not shown in the figures shown here. The 22 K stage was not observed after LHeT deformation in some specimens.

ped at certain temperatures and come back to dislocations because the binding energy with dislocation is larger than with impurities [3, 10 to 12]. These pinning stages observed after various treatments are summarized in Table 2, where the temperatures (T_p) at which the decrease in P became apparent during warmup measurements are shown. The result that no pinning was observed in the Pd coated specimen demonstrated that the pinning agents are H atoms. In Table 2, the recovery temperatures of electrical resistivity after quenching for specimens doped with H below a few hundred at.ppm are also given [11, 13, and 14] for comparison. The present results show that there are four pinning stages below 100 K, i.e. the first stage started from the lowest temperature below 20 K (probably below 6 K) and for the other three T_p's are at around 22, 36 to 50, and 60 to 80 K, respectively. In the following, the origin of these stages will be discussed in turn.

4.1.1 Below 20 K

This stage was observed clearly only after deformation at LHeT and after quenching in rolled specimens (not known but possibly also in annealed and bent specimens). Here, P decreased continuously during the first warmup below 20 K. Since it was observed only after deformation at LHeT and after quenching, dislocation rearrangement is considered to have caused this stage. Probably, these treatments had forced the dislocations to move to unstable sites, and during the warmup these dislocations gradually rearranged themselves to the stable sites. This recovery would be too small to be observed in resistivity measurements after quenching.

4.1.2 22 K stage

This stage is characterized by the following facts:

a) the pinning is small and transient,

b) it is not observed in the annealed and slightly deformed specimens after HTB,

c) there is no corresponding recovery in the resistivity measurements.

The cases a) and c) suggest that H atoms arriving at dislocations are small in number and relatively unstable. However, the case b) is difficult to be explained at present.

4.1.3 36 to 50 K stage

This stage has the corresponding stage in electrical resistivity, and was observed after all the treatments in all the specimens with oxide coating except after HTB in the rolled specimens. The reason for being not observed in the rolled ones after HTB is considered to be simply because the concentration of H pinnings on dislocations in this stage is not large enough to suppress the dislocation motion in this case. In the rolled specimens after HTB, the pinning is observed only in the next recovery stage. Also the temperature of this stage tends to increase with increasing deformation. This could be explained in a similar way that the relative concentration of H is smaller in more deformed specimens and pinnings are too weak to hinder the dislocation motion in the early stage of recovery. Furthermore, this stage shifts to higher temperature with increasing amplitude ε_{max} as shown in Fig. 3. This shows again that the pinnings are not strong enough in the early stage of the pinning. From what kind of trapping sites H atoms are detrapped in this stage is not known at present, but H clusters might be a possible one [11].

4.1.4 60 to 80 K stage

According to Hanada, release of H atoms from oxygen (O) and nitrogen (N) occurs at 70 and 90 K, respectively [11]. Therefore, the 60 to 80 K stage observed in the rolled specimens after HTB or quenching is considered to be caused by H released from O and/or N. A large amount of H pinnings may form strong pinners by clustering. This stage was not observed in the annealed or slightly deformed specimen because the dislocation motion has been completely suppressed by the pinning in the 36 to 50 K stage.

4.2 The α-peak in rolled specimens

In a heavily deformed specimens because of the high dislocation density, the relative number of H pinners would be smaller. This would be one reason why we observed the large α-peak in the rolled specimens as shown in Fig. 5. Also, in the rolled specicmens, the effect of HTB is different from that in annealed or slightly deformed speimens, i.e. ΔP at 6 K was very small, ΔP increased only above ≈ 25 K (Fig. 7) and the main recovery of ΔP shifted to higher temperature.

Since dislocation motion that takes place below LHeT is considered to be the kink diffusion along non-screw dislocations, the very small ΔP at 6 K after HTB suggests that the density of non-screw dislocations in the rolled specimen is relatively low. This seems to be not unreasonable since it is well known that screw dislocations are introduced predominantly in b.c.c. metals by low-temperature deformation [15] and RT is considered to be low enough to produce screw dislocations preferentially. Therefore, in the rolled specimens, only the α_1-peak can be observed predominantly which is attributed to the kink diffusion along screw dislocations. Now, as its was shown in Fig. 5 in the previous section, almost no pinning was observed below 80 K after HTB, and the α_1-peaks observed in curves 1′ and 2′ (or 3′) are not affected by the pinning. Therefore, comparing the peak temperatures in two different ε_{max}'s, the activation volume for the peak can be estimated. From the peak shift of less than 1 K and assuming the pre-exponential factor as 10^8 to 10^{10}, the activation volume was estimated to be less than 4 to $6b^3$. This seems acceptable for the kink diffusion [5]. The peak temperature of the α-peak is higher in the rolled specimen than in the slightly deformed specimen. This is probably due to the longer free length of dislocations in the rolled specimens. The α-peak at 37 K observed in the Pd coated specimen is considered to be predominantly α_1 for the same reason as above. Further, the α_2-peak seems to be smaller in the specimen with less H pinners. The reason for this is not known at present.

5. Conclusion

Dislocation pinning below 100 K was studied on high-purity Nb specimens which were given various deformations and contained a few at.ppm of H. The pinning stages were observed after three dislocation breakaway treatments, i.e. high-amplitude vibration, quenching, and deformation at 4.2 K. It was confirmed that these pinnings can be attributed to H, since no pinning was observed in the H free specimen. The modulus recovery below 20 K was attributed to dislocation rearrangement. Other than this, three pinning stages were observed which started at temperatures ≈ 22, 36 to 50, and 60 to 80 K. The ≈ 22 K stage was transient, not present in some specimens and its origin is not clear. The last two stages agree well in temperatures and could be attributed to the same origin as the recovery stages found in electrical resistivity after quenching. The rolled specimen showed characteristic behavior, especially ΔP at 6 K being extremely small after high-amplitude vibrations. This suggests that screw dislocations were predominantly introduced by rolling at RT. The small activation volume found for the α-peak in the rolled specimen would indicate that this α-peak was mostly associated with the kink diffusion along screw dislocations and this is also in agreement with the above suggestion.

Acknowledgement

This work is partly supported by a Grant in Aid for Scientific Research of the Ministry of Education of Japan.

References

[1] R. Hanada, Scripta metall. 7, 681 (1973).
[2] H. Mizubayashi, T. Arai, and S. Okuda, phys. stat. sol. (a) **76**, 165 (1983).
[3] M. Maul, Thesis, Univ. Stuttgart, 1981.
[4] P. Groh and H. Schultz, J. Physique **42**, C5-25 (1981).
[5] N. Kuramochi, H. Mizubayashi, and S. Okuda, phys. stat. sol. (a) **77**, 633 (1983).
[6] J. A. Rodrigues and R. Kirchheim, Scripta metall. **17**, 159 (1983).
[7] G. Funk, M. Maul, and H. Schultz, Proc. VI. Europ. Conf. Internal Friction and Ultrasonic Attenuation in Solids, Lyon 1983, in the press.
[8] S. Okuda and H. Mizubayashi, phys. stat. sol. (a) **16**, 355 (1973).
[9] H. Mizubayashi, S. Okuda, and M. Daikubara, Scripta metall. **13**, 1131 (1979).
[10] G. Pfeiffer and H. Wipf, J. Phys. F **6**, 167 (1976).
[11] R. Hanada, Proc. 2nd Internat. Cong. on Hydrogen in Metals, 1B6, Paris 1977.
[12] R. T. Mattas and H. K. Birnbaum, Acta metall. **23**, 973 (1975).
[13] J. Engelhard, J. Phys. F **9**, 2217 (1979).
[14] A. Gotoh, K. Yamakawa, and F. E. Fujita, Scripta metall. **13**, 349 (1979).
[15] J. W. Christian, Proc. III. Internat. Symp. Reinststoffe in Wissenschaft und Technik, Dresden 1970, Akademie-Verlag, Berlin 1972 (p. 263).

(Received May 15, 1984)

phys. stat. sol. (a) 85, 133 (1984)

Subject classification: 1.4 and 10.2; 20.1; 22.1.2

Zentrum für wissenschaftlichen Gerätebau der Akademie der Wissenschaften der DDR, Berlin[1]) *(a),*
Institut für Halbleiterphysik der Akademie der Wissenschaften der DDR, Frankfurt (Oder)[2]) *(b),*
and Research Institute for Telecommunication, Budapest[3]) *(c)*

Infrared Spectroscopical and TEM Investigations of Oxygen Precipitation in Silicon Crystals with Medium and High Oxygen Concentrations

By

P. GAWORZEWSKI[4]) (a), E. HILD (c), F.-G. KIRSCHT[4]) (b), and L. VECSERNYÉS (c)

Results of IR spectroscopic as well HVTEM investigations are given on the precipitation of oxygen in Si due to heat treatment within the temperature range 600 to 1275 °C. The initial oxygen concentrations of the Si samples are about 9×10^{17} cm^{-3} and about 1.5×10^{18} cm^{-3}, respectively, with different concentrations of carbon within the range 10^{16} cm^{-3} $\leqq [C_s]_0 \leqq 1 \times 10^{17}$ cm^{-3} in each case. It is shown that four different kinds of Si–oxygen precipitates exist at least, in dependence on the temperature of heat treatment. $[O_i]_0$, $[C_s]_0$, and the thermal history of the samples heat treated. It is assumed that there is only heterogeneous nucleation within the whole temperature range. For temperatures $T \geqq 900$ °C the oxygen precipitation occurs by the growth of as-grown microprecipitates. The origin of various IR spectra and corresponding HVTEM microphotographs is interpreted as being caused by differences in the shape and size of the Si–oxygen-precipitates.

Es werden Ergebnisse von infrarotspektroskopischen und transmissionselektronenmikroskopischen Untersuchungen zur Präzipitation von Sauerstoff in Si infolge Wärmebehandlung im Temperaturbereich 600 bis 1275 °C mitgeteilt. Die Ausgangs-Sauerstoffkonzentration der Si-Proben betragen $[O_i]_0 \approx 9 \times 10^{17}$ cm^{-3} bzw. $1,5 \times 10^{18}$ cm^{-3} mit jeweils drei unterschiedlichen Kohlenstoffkonzentrationen im Bereich 10^{16} cm^{-3} $\leqq [C_s]_0 \leqq 1 \times 10^{17}$ cm^{-3}. Es wird gezeigt, daß in Abhängigkeit von der Temperatur der Wärmebehandlung, von $[O_i]_0$, $[C_s]_0$ und der thermischen Vorgeschichte der behandelten Si-Proben mindestens vier verschiedene Arten von Si–Sauerstoff-Präzipitaten nachweisbar sind. Es wird angenommen, daß im gesamten Temperaturbereich nur heterogene Keimbildung vorliegt, wobei bei $T \geqq 900$ °C bevorzugt Wachstum eingewachsener Mikropräzipitate erfolgt. Der Ursprung verschiedenartiger IR-Spektren und damit verknüpfter TEM-Bilder wird im Unterschied von geometrischer Form und Größe der Si–Sauerstoff-Präzipitate gesehen.

1. Introduction

The controlled precipitation of oxygen in Si due to heat treatment (h.t.) is becoming increasingly important for the improvement of quality and yield of microelectronic devices by the possible application of intrinsic gettering [1]. Although the fundamentals of oxygen precipitation are known there remain some questions with respect to nuclea-

---

[1]) Seestraße 82, DDR-1166 Berlin, GDR.

[2]) Walter-Korsing-Str. 2, DDR-1200 Frankfurt (Oder), GDR.

[3]) Gabor Aron utca 65, H-1525 Budapest, Hungary.

[4]) Formerly Institut für Physik der Werkstoffbearbeitung der AdW der DDR, Seestraße 82, DDR-1166 Berlin, GDR.

tion and growth of oxygen precipitates, especially concerning the influence of the as-grown state as well as the influence of the initial concentrations of interstitially solved oxygen, $[O_i]_0$, and substitutionally incorporated carbon, $[C_s]_0$, respectively. There are controversial conceptions concerning the homogeneous [2, 3] or heterogeneous [4 to 7] nature of nucleation. It is usual to identify the precipitates on the basis of their infrared spectra. After low-temperature h.t. (500 °C $\leqq T \leqq$ 750 °C) a broad IR band has been observed with a transmittance minimum near 1100 cm^{-1} [8]. It has been assumed that the appearance of this broad band is connected with the formation of oxygen clusters and SiO_x precipitates ($x < 2$) [8]. After medium-temperature h.t. (850 °C $\leqq T \leqq$ 1050 °C) a double peak band has been observed with minima near 1110 and 1220 cm^{-1} which has been assumed to be connected with cristobalite [8], whereas in [9] the origin of the peak at around 1225 cm^{-1} was interpreted as being plate-like SiO_2 precipitates. A further type of oxygen precipitates has been observed after h.t. at temperatures $T \geqq$ 1050 °C. This type of precipitates has been assumed to be connected with amorphous SiO_2 in [8, 10, 11] and with cristobalite in [12], respectively.

Grown-in precipitates show a weak band at around 1225 cm^{-1}. This type of precipitate has been connected with cristobalite, too [13].

The purpose of the present work is an experimental investigation of the precipitation of oxygen in Si samples involving medium ($[O_i]_0 \approx 9 \times 10^{17}$ cm^{-3}) as well as high ($[O_i]_0 \approx 1.5 \times 10^{18}$ cm^{-3}) oxygen concentrations. We have been interested in the dependence of the precipitate nucleation and growth on the oxygen and carbon content of the Si samples, in the role of the as-grown state, and in the effect of various one-step and combined h.t.'s. For this aim, there have been carried out single and combined h.t.'s at 600, 700, 900, +1000, and 1275 °C. The samples heat-treated have been investigated by means of IR spectroscopy and high voltage electron microscopy.

2. Experimental

The samples were prepared from dislocation-free Czochralski Si crystals with dopant concentrations (boron) of about 10^{15} cm^{-3} and (111) orientation. Sets of Si crystals both with medium and high $[O_i]_0$ and with three different $[C_s]_0$ ranges with $\leqq 1 \times 10^{16}$ cm^{-3}, $\approx 5 \times 10^{16}$ cm^{-3}, and $\geqq 1 \times 10^{17}$ cm^{-3}, respectively, in each set have

Table 1

Specification of the heat treatments

A		B		
pre-h.t.	h.t.	pre-h.t.	h.t. 1	h.t. 2
—	600 °C, 0 to 240 h	—	1275 °C, 1 to 2 h, O_2/HCl	
450 °C, 120 h	600 °C, 0 to 240 h	—	700 °C, 8 to 16 h, N_2	
—	900 °C, 0 to 64 h	—	1000 °C, 2 to 8 h, N_2	
600 °C, 240 h	900 °C, 0 to 64 h	—	700 °C, 8 h	+ 1000 °C 8 h
		1275 °C, 1 h	700 °C, 8 h	
all h.t.'s in N_2		1275 °C, 1 h	1000 °C, 8 h	
ambient		1275 °C, 1 h	700 °C, 8 h	+1000 °C 8 h

been chosen for the experiments. Samples with sizes 8 × 8 × 1.5 mm^3 were prepared from Si slices, polished on both sides, and heat-treated at different temperatures. Two different h.t. complexes marked A and B, respectively, have been used for investigations. The h.t.'s. carried out within these complexes are shown in Table 1. The samples were etched after h.t. in HF solution to remove SiO_2 layers.

IR transmittance spectra of the samples were measured at room temperature by means of a Perkin Elmer 580 infrared spectrophotometer within the wave-number range from 1300 to 450 cm^{-1}. A floating zone Si slice of 1.5 mm thickness was used as reference sample. The concentration of O_i was determined from the magnitude of the 1106 cm^{-1} peak, that of the C_s was obtained from the 606 cm^{-1} peak, using the calibration factors given in [14] and [15], respectively. The initial concentrations $[O_i]_0$ and $[C_s]_0$ of the crystals are presented in Table 2.

Table 2

Initial oxygen and carbon concentrations $[O_i]_0$ and $[C_s]_0$ of the crystals investigated

crystal	$[O_i]_0$ (10^{18} cm^{-3})	$[C_s]_0$ (10^{16} cm^{-3})
a	1.4	2.2
b	1.4	6.0
c	1.5	10.5
y	1.6	10.0
z	1.6	—
d	1.0	—
MR	1.2	—
e	1.0	5.0
f	0.9	8.0

In the case of advanced oxygen precipitation, the interstitial oxygen peak at 1106 cm^{-1} was masked by the band originating from the precipitates. In these cases, the low-wave-number peak at 515 cm^{-1} was used for the determination of $[O_i]$ (see below).

Structural investigations were made by etching with Wright solution followed by HVTEM. For TEM defect recording, bright field images with diffraction vectors of the type ⟨440⟩ near $s = 0$ were taken.

3. Experimental Results

3.1 IR spectra

The effect of h.t. type A on the IR spectra of samples a, c, and d is shown in Fig. 1, 3 to 5. The spectra of the other heat-treated samples are either similar to the presented ones or combinations of them.

In the spectra of the as-grown samples generally four absorption peaks are present at 1225, 1106, 606, and 515 cm^{-1}, Fig. 1, 3 to 7, spectrum 1. The peaks at 1106 and 515 cm^{-1} are due to the vibrational modes of O_i and the band at 606 cm^{-1} to the localized vibrational mode of C_s, respectively. The weak band at 1225 cm^{-1} should be attributed to precipitates formed in the cooling period of crystal growth.

Heat treatments at 600 °C resulted in the formation of precipitates which were characterized by a broad band around 1085 cm^{-1} (type I). The appearance of this band was associated with the decrease of the O_i peak, which was superimposed on that broad band (Fig. 1). Using an interpolated spectrum at around 1106 cm^{-1} as shown by the dotted line in Fig. 1 (5), it was deduced that the intensity ratio of the O_i peaks at

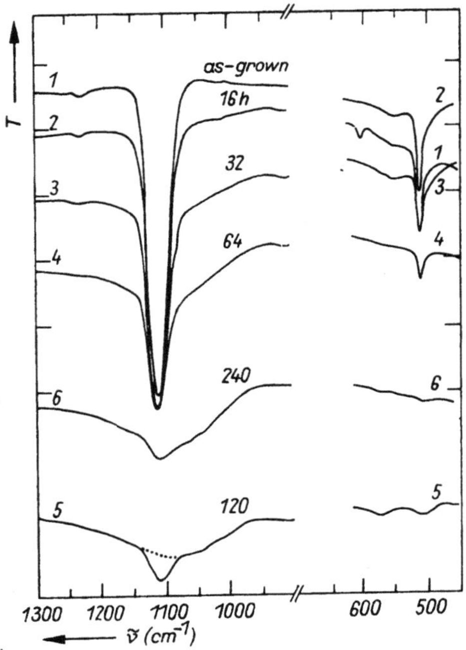

Fig. 1. Transmittance spectra of sample a after h.t. at 600 °C for different durations

1106 and 515 cm^{-1} remained nearly the same for the heat-treated samples as for the as-grown ones. On the basis of that constant intensity ratio between the two O_i peaks, the peak at 515 cm^{-1}, which is not influenced by any band of the Si–oxygen precipitates, can be used for the estimation of $[O_i]$ and consequently for the construction of the band shapes of the precipitates, too.

Fig. 2a shows the reduction of $[O_i]$ during the h.t. at 600 °C. $\Delta[O_i]/\Delta t$ increases with $[O_i]_0$. For samples with $[O_i]_0 \lessapprox 1 \times \times 10^{18}\ cm^{-3}$, there is a strong dependence of $\Delta[O_i]/\Delta t$ on $[O_s]_0$. The reduction of $[O_i]$ is accompanied also by the decrease of $[C_s]$.

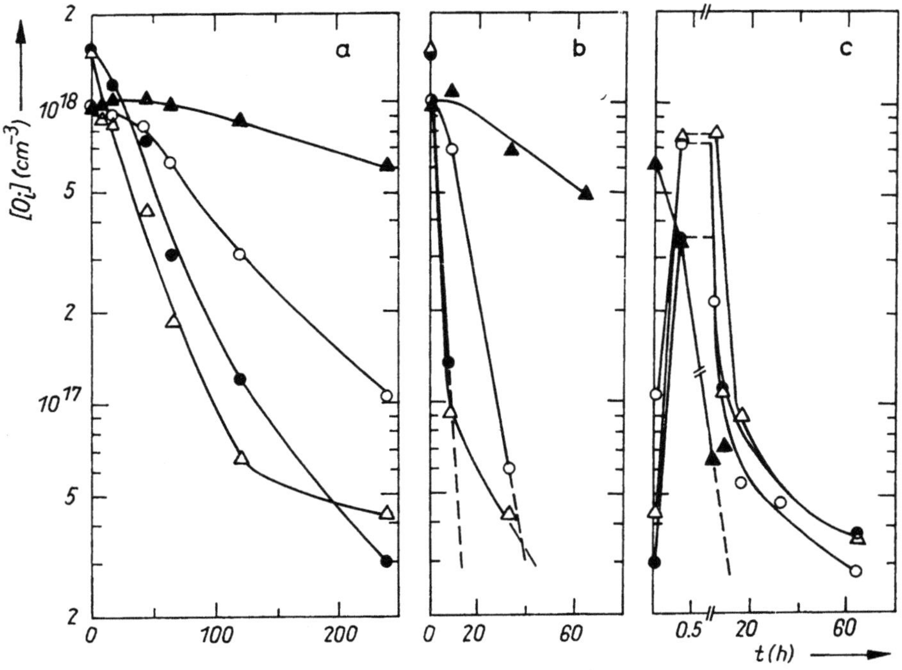

Fig. 2. Oxygen concentration $[O_i]$ in dependence on the duration of heat treatment of different samples (sample a ○, c △, d ▲, e ○) and different heat treatments: a) 600 °C, b) 900 °C, and c) 600 °C for 240 h + 900 °C, respectively

Fig. 3 shows the effect of different h.t.'s of the complex A on the IR spectra of samples a. The h.t. at 600 °C for 240 h resulted in the broad band around 1085 cm^{-1} in the IR spectrum together with the complete disappearance of the C_s band (spectrum 2). If a pre-h.t. at 450 °C for 120 h was applied, the minimum of the broad band shifted to lower wave numbers (spectrum 3). The clusters or precipitates formed at 600 °C were partly dissolved at higher temperatures as shown by an increased O_i peak (spectrum 4). Continuing h.t. at 900 °C, precipitates of new type were formed. Their formation was accompanied by the appearance of a double-peaked band in the spectrum, with minima at about 1225 and 1120 cm^{-1} (type III). Spectrum 5 in Fig. 3 is obviously a mixture of that of the new type precipitates and the broad band of precipitates type I. During a single h.t. at 900 °C, only precipitates characterized by the double-peaked spectrum seemed to be formed (spectrum 6 in Fig. 3).

The results of the h.t. on the carbon-rich sample c are shown in Fig. 4. Spectra 2 and 3 are similar to those in Fig. 3. Spectrum 4 shows that nearly all the precipitates formed at 600 °C for 240 h dissolved due to a subsequent h.t. at 900 °C for 20 min. Prolonged h.t. at 900 °C for 64 h produced the double-peaked band again, together with the reappearance of the C_s band (spectrum 5). Contrary both to the formation of the band type III due to combined h.t. and to the result of the single h.t. at 900 °C in case of samples a and b, this h.t. caused the formation of precipitates characterized by a single band (type II) now and the disappearance of the C_s peak.

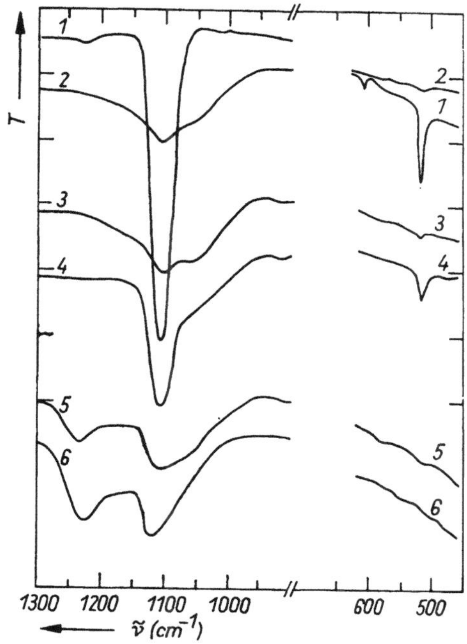

Fig. 3

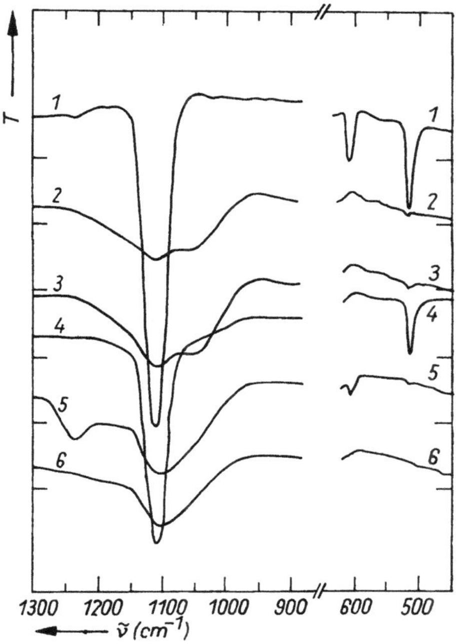

Fig. 4

Fig. 3. Transmittance spectra of sample a after different heat treatments within the heat treatment complex A. Spectrum (1) as-grown; (2) 600 °C, 240 h; (3) 450 °C, 120 h + 600 °C, 240 h; (4) 600 °C, 240 h + 900 °C, 20 min; (5) 600 °C, 240 h + 900 °C, 64 h; (6) 900 °C, 64 h

Fig. 4. Transmittance spectra of sample c after different heat treatments within the heat treatment complex A. Notations as in Fig. 3

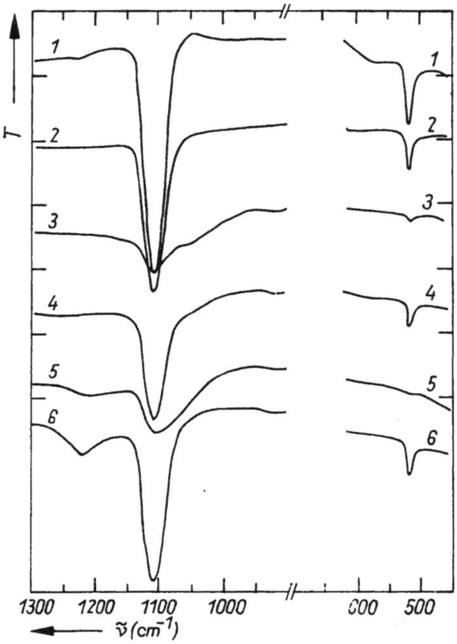

Fig. 5. Transmittance spectra of sample d after different heat treatments within the heat treatment complex A. Notations as in Fig. 3

The influence of h.t. type A on samples with medium $[O_i]_0$ and low $[C_s]_0$ is shown for sample d (Fig. 5). Compared to the $[O_i]_0$-rich samples the decrease of $[O_i]$ due to single h.t.'s both at 600 and 900 °C is very low (spectra 2 to 6; Fig. 2a, 2b). At the same time, the highest rate $\Delta[O_i]/\Delta t$ due to combined h.t. at 600 and 900 °C is observed already after a subsequent h.t. at 900 °C for 16 h. Samples e and f with medium $[O_i]_0$ and medium $[C_s]_0$ show similar IR spectra as the samples of crystal a; spectra of type I appear after low-temperature h.t. and spectra of type III are observed after the medium-temperature one.

Samples of the crystals z, y, and MR were heat-treated according to the h.t. complex B to investigate the influence of pre-h.t. at high temperature (1275 °C) on the formation of oxygen clusters and precipitates at low and medium temperatures.

In the as-grown state sample z did not contain a detectable concentration of C_s as shown in Fig. 6, spectrum 1. After h.t. at 1275 °C for $t \geqq 1$ h, $[C_s] \approx 4 \times 10^{16}$ cm^{-3} was detectable, but $[C_s]$ remained latent for all other h.t. at temperatures $T \leqq 1000$ °C. The h.t. at 1275 °C for 2 h (Fig. 6, spectrum 2) caused a decrease in $[O_i]$ and the appearance of a broad band around 1106 cm^{-1} which was superimposed on the O_i band. Furthermore, an additional band with transmittance minimum at 470 cm^{-1} was observed after that h.t. Investigations of other z samples show that the bands at 1106 and 470 cm^{-1} are connected to each other (type IV precipitate). Heat treatment at 1275 °C resulted in a reduced tendency in the formation of other types of precipitates at subsequent h.t.'s at 700 °C $\leqq T \leqq$ 1000 °C. This effect can be seen comparing spectrum 3 with spectrum 6 (Fig. 6). The concentration of C_s having been released by the pre-h.t. at 1275 °C remained constant due to subsequent h.t.'s. Only combined h.t.'s including a low-temperature pre-h.t. caused a decrease of $[C_s]$ (spectrum 6, 7 in Fig. 6).

Spectra of the $[C_s]_0$-rich sample y are shown in Fig. 7. Heat treatment at 1000 °C produced a spectrum of type II band (spectrum 3 in Fig. 7) similar to the one in the case of sample c. A pre-h.t. at 1275 °C even for 1 h caused such a change of the as-grown state of the crystal y that a subsequent h.t. at 1000 °C produced only a slight decrease of both $[O_i]$ and $[C_s]$. Although $[C_s]_0$ and $[O_i]_0$ were nearly the same for the as-grown state as after the pre-h.t. at 1275 °C for 2 h, $[C_s]$, $[O_i]$, as well as the IR spectra became very different after a subsequent h.t. at 1000 °C (Fig. 7, spectra 3 and 6). On the contrary, the pre-h.t. at 1275 °C did not change essentially the effect of h.t. at lower temperatures resulting in the appearance of the type I band and a decrease of $[C_s]$.

The carbon-poor sample MR showed the double peak spectrum (type III) due to 1000 °C h.t. and a weak tendency for oxygen precipitation at low temperatures was

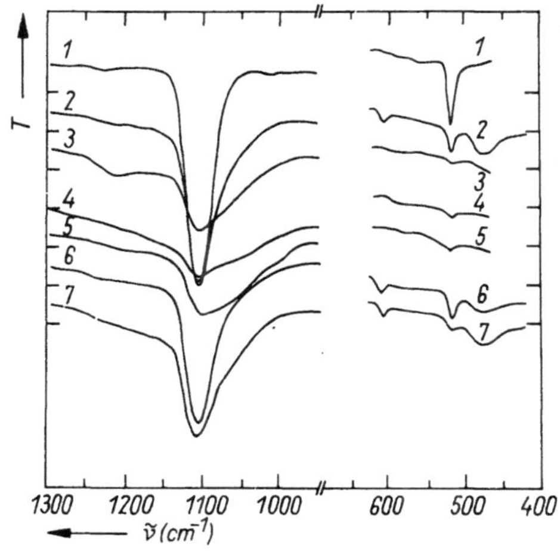

Fig. 6

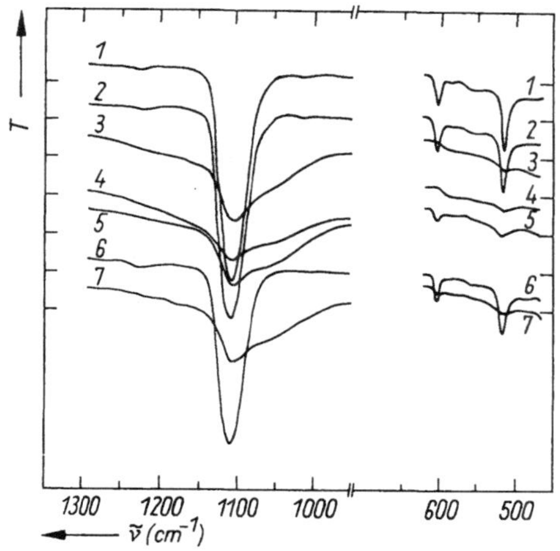

Fig. 7

Fig. 6. Transmittance spectra of sample z after different heat treatments within the heat treatment complex B. Curve (1) as-grown; (2) 1275 °C, 2 h; (3) 1000 °C, 8 h; (4) 700 °C, 16 h; (5) 700 °C, 8 h + 1000 °C, 8 h; (6) 1275 °C, 2 h + 1000 °C, 8 h; (7) 1275 °C, 2 h + 700 °C, 8 h + 1000 °C, 8 h

Fig. 7. Transmittance spectra of sample y after different heat treatments within the heat treatment complex B. Notations as in Fig. 6

observed. The formation of precipitates type III was suppressed by the high-temperature pre-h.t. at 1275 °C. Only after heating the pre-heat-treated samples at low temperature (600 °C) it was possible to produce precipitates type III by the h.t. at 1000 °C, again.

3.2 Structural investigations

Heat treatments at 600 °C for long durations ($t \geqq 100$ h) caused very similar defect formation in all samples with $[O_i]_0 \approx 1.5 \times 10^{18}$ cm^{-3} (Fig. 10a). The average defect density changed from about 5×10^{11} cm^{-3} for samples a and b to about 2×10^{11} cm^{-3} for sample c estimated by structural etching. The average length of the rod-like defects

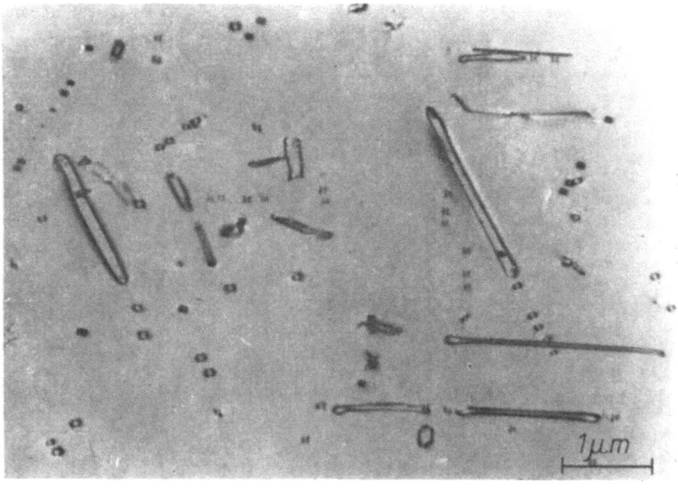

Fig. 8. HVTEM micrograph in sampleb after h.t. 600 °C for 240 h + 900 °C for 64 h

observed amounted to some μm and was slightly higher for sample c than for sample a. After a subsequent h.t. at 900 °C for 20 min it could be shown that rod-like destruction started with an elongation of widened regions observed already after h.t. at 600 °C. As a result, segments of evolving dislocations have been formed with typical dislocation widths of about 50 to 100 nm. A subsequent h.t. at 900 °C for 64 h produced a nearly complete transformation of rod-like defects into perfect dislocation loops and precipitates (Fig. 8). A precipitate density in the order of 10^{11} cm^{-3} and a precipitate size in the range of 50 to 100 nm has been observed. Sample c showed a more homogeneous distribution of precipitates than samples with lower $[C_s]_0$ which contain a large portion of precipitates at traces of former rod-like defects, with distances in the range of 100 to 1000 nm.

Single h.t. at 900 °C showed a similar defect formation only at medium and low $[C_s]_0$. Generally, the precipitates are irregularly shaped (Fig. 10c) and their size is only slightly enlarged compared with precipitates after a preceding h.t. at 600 °C. At the same time, the precipitate density is weakly decreased. A striking difference in precipitation with and without pre-h.t. consists in the stress level around a growing precipitate. Without pre-h.t., a strongly compressed Si lattice is indicated in the surrounding of the precipitates by diffraction contrast. This is in correspondence with the observed early stage of loop punching.

In samples c, the precipitate density is increased due to 900 °C h.t. as compared to samples with lower $[C_s]_0$. The precipitate size is only 10 to 50 nm. Moreover, differently shaped dislocation loops and stacking faults are present with a size of about 1 μm. In comparison with the defect formation due to combined h.t. 600 + 900 °C, the precipitate density is increased (up to 10^{12} cm^{-3}) whereas the precipitate size is reduced.

Samples d showed an advanced stage in defect development after h.t. at 600 °C as compared to samples with higher $[O_i]_0$. After long-time h.t., a strong decomposition of rod-like defects is observed resulting partly in perfect dislocation loops as well as in locally stringed precipitates. These precipitates have typical distances of about 50 nm and sizes of about 10 to 30 nm. Subsequent h.t. at 900 °C for 20 min causes the complete destruction of rod-like defects. Long-time subsequent h.t. at 900 °C results in a defect spectrum which is characterized by a higher portion and a larger variety of dislocation loops. Furthermore, stacking faults with a typical size of about 1 μm are generated (Fig. 9). Single h.t. at 900 °C causes a reduction in precipitate density by about two

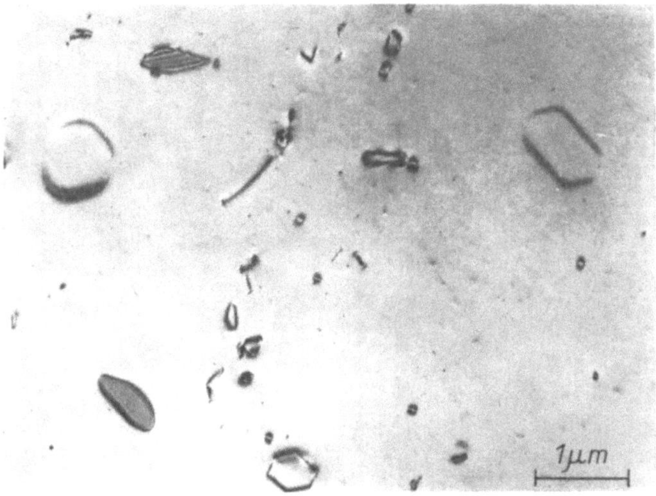

Fig. 9. HVTEM micrograph in sample d after h.t. 600 °C for 240 h + 900 °C for 120 h

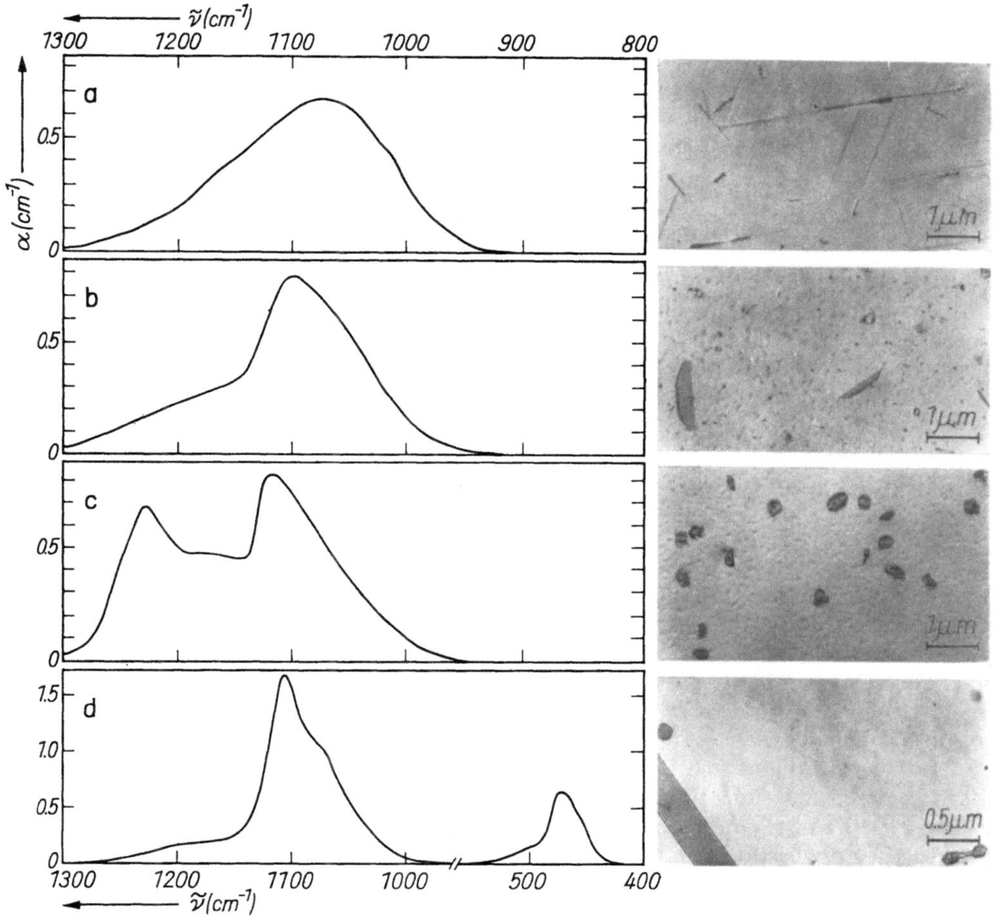

Fig. 10. IR spectra and HVTEM micrographs of the different types of oxygen precipitates observed a) sample a after h.t. at 600 °C for 240 h, b) samplec after h.t. at 900 °C for 64 h, c) sample b after h.t. at 900 °C for 64 h, d) sample z after h.t. at 1275 °C for 2 h

orders of magnitude as compared to h.t. with preceding 600 °C h.t. The edge length of the plate-like precipitates is about 750 nm after h.t. at 900 °C for 64 h.

After h.t. at 1275 °C, in samples z regularly shaped precipitates with a faceted octahedral habitus and a vanishing strain contrast are the dominating defects (Fig. 10d). Long extended stacking faults and isolated dislocations are also present. Precipitates are observed also as lined up configurations along straight dislocation lines. In particular, characteristic double precipitates occur being linked to each other by a "bridge-like" dislocation dipole (Fig. 10d). After subsequent h.t. at low or medium temperatures those defect configurations remain stable, essentially. However, a changed diffraction contrast reveals a higher lattice strain in the surrounding of the precipitates in most cases.

4. Discussion

Comparing the IR spectra of Si–O species present in the samples heat-treated, they seem to be combined from five different types of spectra in dependence on different

$[O_i]_0$–$[C_s]_0$ combinations and temperatures of the h.t.'s:

a) that of the O_i with two peaks at 1106 and 515 cm^{-1}, respectively;
b) spectrum of type I, a broad band around 1085 cm^{-1};
c) spectrum of type II, a broad band near 1100 cm^{-1} but less symmetrical than that of type I;
d) spectrum of type III, a double band with peaks at around 1120 and around 1225 cm^{-1};
e) spectrum of type IV with two bands at 1105 and $\leqq 470$ cm^{-1}, respectively.

It can be seen from Fig. 10 that there is a close connection between these spectra and the different types of defects observed by means of HVTEM.

Spectra of type I are observable after h.t. at temperatures around 600 °C (Fig. 10). The shape of this band and the location of the maximum is only slightly modified by the duration of this h.t. According to [8] the position of the maximum of this broad band depends on the temperature of h.t. within the temperature range 400 to 750 °C. The formation of the broad band is accompanied by the decrease of $[C_s]$. This should mean that C_s is consumed for the formation of oxygen clusters or small precipitates giving rise for the IR spectra measured. For samples with $[O_i]_0 \leqq 1 \times 10^{18}$ cm^{-3} it has been observed that the lower $[C_s]_0$ was, the less the tendency was for both the appearance of the broad band and the decrease of $[O_i]$ due to h.t. at around 600 °C. This indicates the preferentially heterogeneous origin of nucleation of this type of precipitates. Because of the short diffusion length of O_i at these temperatures, those nuclei play the most important role which are available in the highest density. These should be C_s, even in case C_s is less than the detection limit of the IR method. As was found also by other authors [4, 16, 17] a carbon concentration of 10^{16} cm^{-3} is sufficient to initiate heterogeneous nucleation if $[O_i]_0 \geqq 10^{18}$ cm^{-3}.

A further argument for the heterogeneous origin of nucleation at low temperatures is the weak influence of a pre-h.t. at 1275 °C on the shape and intensity of the IR spectrum of type I as well as on the decrease of $[O_i]$ due to a h.t. at around 600 °C. This means that the as-grown state is not of essential importance for both the nucleation and the growth of this type of precipitates.

Carbon as nucleation site for oxygen precipitates at low temperatures is embedded in the growing precipitate and probably, it is part of the stable core of the precipitate. This is indicated by the behaviour of precipitates formed at about 600 °C during subsequent h.t. at higher temperatures (e.g. at 900 °C). Such h.t.'s caused intense changes of the type I band due to dissolution of some of the oxygen clusters and precipitates. On the contrary, carbon which had been consumed during h.t. at 600 °C below the detection limit remained undetectable after a relatively long-time h.t. at 900 °C. Only after very long h.t. at 900 °C, C_s appeared again together with the formation of precipitates type III.

The appearance of the type I band in the IR spectrum is accompanied by the appearance of rod-like defects in the HVTEM patterns, Fig. 10a, as it has been described in the literature different times [18 to 21]. Rod-like defects are dipolar closed loops of an imperfect dislocation with a weak lattice strain [19]. They enclose a certain phase of SiO_x [20], possibly initiated by a bonding rearrangement [21]. Due to subsequent h.t. at higher temperatures (e.g. in the range 900 to 1000 °C) the SiO_x cluster decomposes into individual Si–O–Si groups with the oxygen atoms still being located near their original places, but observable as O_i after short-time h.t. at 900 °C. The second stage of decomposition could involve a redistribution of oxygen and Si self-interstitials, Si_i, by diffusion with a strong component along the dislocation line in ⟨110⟩ direction (pipe diffusion). The flux of Si_i and O_i seems to be antiparallel giving rise to a "demixture" which results in oxygen-rich segments and neighbouring regions with self-interstitial type dislocation loops. With prolonged h.t., single Si–oxygen precipitates are formed (Fig. 8) in the oxygen-rich regions.

The consumption of C_s during 600 °C h.t. does not mean that C_s is necessarily incorporated in rod-like defects. Only a limited portion of the nuclei with an over-critical stress field will initiate the formation of rod-like defects. Therefore, a high density of clusters and microprecipitates containing carbon will also exist and serve as nuclei for further oxygen precipitation outside the decomposed rod-like defects during a subsequent h.t. at 900 °C. The higher the density of carbon-containing clusters and microprecipitates is, the weaker is the influence of pipe diffusion on the decomposition of rod-like defects. Thus, lined-up precipitates are only characteristic of samples a and b (Fig. 8), but not of samples c.

Comparing 900 °C h.t. with and without pre-h.t. at 600 °C in the case of a and b samples, one of the essential differences in defect formation consists in the compressive stresses around the precipitates (compare Fig. 8 with Fig. 10c). The difference in precipitate size and the corresponding difference in misfit strain cannot be responsible for this effect. Probably, in the case of a pure 900 °C h.t. an additional back stress by piled-up self-interstitials in the surrounding of a growing precipitate is the main reason taking into account that the diffusivity of Si_i and O_i are of about the same order of magnitude at this temperature [22]. This situation is strongly changed if sinks for Si_i are present. The elongated perfect dislocation loops in the case of a pre-h.t. at 600 °C act as such sinks thereby extending their widths by climb.

Heat treatments at medium temperatures (850 °C $\leqq T \leqq$ 1100 °C) of oxygen-rich samples with a high content of C_s ($[C_s] \geqq 1 \times 10^{17}$ cm^{-3}) result in the appearance of type II spectra which are connected with small defects of high density (Fig. 10b). This can be seen from the results of the samples c and y.

The decrease of $[C_s]$ due to this h.t. could indicate a preferential role of C_s as nucleus for this type of precipitate. However, a comparison of spectrum 3 and spectrum 6 in Fig. 6 shows that a pre-h.t. at 1275 °C inhibits the formation of type II spectrum due to h.t. at 1000 °C. This pre-h.t. changes the as-grown state of the sample but it does not cause a significant change of $[C_s]$. $[C_s]$ remains nearly constant also during a subsequent h.t. at 1000 °C. This indicates that C_s has only slight importance as direct nucleus for the formation of type II precipitates. C_s plays an indirect role with regard to the formation of microprecipitates with a big variety of sizes at low temperatures. These microprecipitates can grow either during the cooling period of crystal growth or during a low-temperature h.t. High densities of microprecipitates acting as nuclei result in a strong competition concerning the growth of precipitates at higher temperatures. As a result of this process many oxygen precipitates of small size shown in Fig. 10b can grow. The strong decrease of $[C_s]$ could be caused by the change of the substitutional position of carbon atoms due to interaction with each other or with silicon interstitials rather than to being a constituent of the nuclei themselves.

Heat treatments at medium temperatures (850 °C $\leqq T \leqq$ 1100 °C) of samples with high or medium $[O_i]_0$ and $[C_s]_0 < 10^{17}$ cm^{-3} effect the formation of those precipitates which are characterized by a double-peak spectrum (Fig. 10c). The nuclei for the formation of these precipitates are not directly related with C_s. This is concluded from the observations that $[C_s]$ retains nearly its initial value for most samples showing type III spectra, and from the fact that the rate of precipitation depends on h.t. at higher temperatures at one and the same $[C_s]$. Furthermore, it is found that this type of spectrum is present also for carbon-poor samples with $[C_s]_0 \leqq 10^{16}$ cm^{-3}. From the IR measurements it is concluded that high concentrations of C_s even inhibit the formation of these precipitates.

Oxygen clusters and microprecipitates in form of forerunners generated during the cooling period of crystal growth should play the preferential role as nuclei for this type of precipitate. Depending on $[O_i]_0$, $[C_s]_0$, and thermal history of the crystal such nuclei have different thermal stability indicated by the influence of pre-h.t.'s at high temperatures. Partially dissolving the nuclei by those h.t.'s, the formation of new nuclei and the growth of precipitates connected with the type III band is considerably suppressed. But it was observed that the same type of nuclei can also be formed due to h.t. at low temperatures ($T \approx 600$ °C) after a pre-h.t. at $T \geqq 1200$ °C. This is in correspondence with results obtained in [23].

The strong decrease of $[O_i]$ in the case of sample d due to h.t. at 900 °C after h.t. at 600 °C (Fig. 2c) shows the important influence of pre-h.t.'s at low temperatures on carbon-poor Si. Microprecipitates could grow during 600 °C h.t. to a big size on the strength of a small density of competing nucleation sites. They are not destroyed during a subsequent medium-temperature h.t. but grow further giving rise to the strong decrease of $[O_i]$. The resulting advanced state of microprecipitates after 600 °C h.t. can be seen in the HVTEM pattern (Fig. 9).

The nature of precipitates connected with a type III band was discussed in the literature different times [8, 9].

Our measurements do not confirm the assumption that the double band in the IR spectrum is connected with α-cristobalite [8]. Measurements on numerous samples showing the double-peaked spectrum do not reveal the low-wave-number band at about 485 cm^{-1} characteristic of cristobalite at room temperature [24, 25]. From our point of view, the special shape of the precipitate formed at medium temperatures should be responsible for the appearance of the type III spectrum as was pointed out in [9].

The predominant number of these precipitates are of plate-like shape as can be deduced from their contrast behaviour. The nucleation and the growth of plate-like precipitates, and especially of the disc-shaped ones, is favourable with regard to minimize the strain energy between the matrix and the growing precipitate [26]. Plate-like precipitates were observed also by other authors who investigated the oxygen precipitation behaviour at medium temperatures [27, 28, 6]. After [27], these plate-like precipitates with habit plane {100} have thicknesses of some nm. It is known that the position of the strong IR bands of dispersed particles in a matrix strongly depends on their shape because of light scattering phenomena [25]. Therefore, the transmittance minimum can be positioned anywhere between the longitudinal and transverse mode wave number. Hu [9] calculated the absorption spectra of SiO_2 particles of ellipsoidal shape dispersed in a Si matrix. He found that the disc-shaped precipitates with a ⟨100⟩ habit plane produce a strong band at $\tilde{\nu} \approx 1230$ cm^{-1}. This band is shifted up to 1100 cm^{-1} for particles of more and more globular shape. The double-peaked spectrum observed in [9] was interpreted as resulting from the band at 1230 cm^{-1} originating from these SiO_2 platelets, and the 1107 cm^{-1} band due to the retained O_i. However, the disappearance of the O_i band at 515 cm^{-1} showed that O_i exhibits only small concentrations after h.t. at medium temperatures for long durations, which cannot explain the high intensity at 1106 cm^{-1}.

We obtain the same result for an explanation of the 1230 cm^{-1} band considering total reflection from the plate-like particles. Total reflection occurs if $\varepsilon_{pr} - \varepsilon_{Si} \sin^2 \varphi < < 0$, where ε_{pr} and ε_{Si} are the permittivities of the precipitate and of the Si matrix, respectively. φ is the angle between the incident light [111] and the normal to the precipitate plane [100]. At the LO mode wave number ε_{pr} is ≈ 0 and at the TM mode, the imaginary part of ε_{pr} is high. Three-dimensional silicas, crystalline or amorphous, have a range near 1220 cm^{-1} where $\varepsilon \approx 0$ [25, 29]. Using the optical constant of silica glass [29] and angle of incidence $\varphi = 54.7°$ we get a strong transmittance minimum near 1240 cm^{-1} and another one at 1080 cm^{-1}, which is about one order of magnitude smaller. Therefore, Hu's calculation [9] and the total reflection interpretation can only explain the band at about 1230 cm^{-1} as a result of the activation of the LO mode due to the special shape of the precipitates formed at medium temperatures.

The appearance of the band at about 1100 cm^{-1} requires another explanation than originating from the O_i band. We have two possible explanations for the double-peaked type spectrum. The plate-like precipitates are only a few SiO_4 tetrahedra thick as observed in [27, 6, 22] but their lateral dimensions extend to some hundreds nm. A strong band at about 1100 cm^{-1} is characteristic of the three-dimensional silicas [24, 25]. Precipitates of type I, II, and IV also have a single band with a transmittance minimum near 1100 cm^{-1}.

The strong anisotropy between lateral size and thickness of the precipitates can cause intense splitting of the Si–O stretching vibrations just like in the case of layer silicates [25] where the Si–O stretching vibration mode splits into perpendicular and parallel ones.

The perpendicular and parallel vibrations of the layer silicates produce two well-separated absorption bands in the range of the Si–O stretching vibrations. It is known for the layer silicates, too, that the position of the bands can be modified by the shape and size of the particles embedded in a matrix because of scattering phenomena [25].

Another explanation of the double-peaked spectrum is based on the assumption that the two peaks originate from two different kinds of precipitates: The plate-like ones with transmittance minima near 1240 and at 1080 cm^{-1} with very low intensity, and the globular-shaped precipitates with a transmittance minimum at 110 cm^{-1}. Until now, we have not observed a spectrum of heat-treated Si samples with a single peak at 1240 cm^{-1}. Therefore, we assume that both types of precipitates forming the double-peaked spectrum coexist with a tendency of dissolution of globular-shaped precipitates of undercritical size at the expense of the growth of larger ones and of plate-like precipitates. $\alpha_{1100}/\alpha_{1240}$ decrease with the duration of h.t. at temperatures of 900 °C. It has been observed that the time for reaching the equilibrium state concerning oxygen precipitation is very long for medium temperatures for most of the samples (some hundred hours). This observation is in correspondence with investigations in [30]. Equilibrium state for the h.t. durations used is reached only in the case of samples d after combined h.t. 600 + 900 °C but the ratio $\alpha_{1100}/\alpha_{1240}$ is much higher than the one for the other samples heat treated at 600 + 900 °C or at 900 °C. This indicate that plate-like precipitates can coexist with thermally stable globular-shaped precipitates of large size.

Type IV spectra have been observed only for the oxygen-rich sample z having no detectable $[C_s]$ in the as-grown state. Type IV spectra are the only ones showing features of IR spectra of bulk SiO_2 (crystalline or amorphous modifications) [24, 25]. The temperature of formation of oxide precipitates connected with the type IV spectrum is $T \geqq 1100$ °C with a rising tendency for their formation with increasing temperature. This type of high-temperature modification of precipitates was also observed in [11, 23, 31]. The nuclei for these precipitates are not C_s or as-grown microprecipitates connected with C_s because $[C_s]$ is considerably increased by the h.t. at 1275 °C. Furtheron, the HVTEM patterns of samples z heat-treated at 1275 °C doe not show any defects of small sizes except the defects which are connected with type IV oxide precipitates.

The nucleation in samples- z could be initiated by some kind of intrinsic point defect. The observation of double and multiple precipitates (Fig. 10d) suggests that the activation of potential nuclei is easier in the neighbourhood of an already growing precipitate. This should be the consequence of a strong O_i flux to a precipitate site. The formation of an interstitial-type "bridge" dislocation is expected to be caused by the high concentration of Si_i ejected from neighbouring precipitates. At the same time, the "bridges" allow a quick balance of equilibrium conditions with regard to the establishment of stoichiometric composition. From the HVTEM investigations, the portion of precipitated volume was determined. Taking into account the oxygen density in SiO_2 modifications of 4.2×10^{22} cm^{-3} [32] an estimation of the oxygen content involved in the type IV precipitates gives an excellent correspondence with the observed decrease of $[O_i]$ after this h.t. This agrees with the fact that no other kinds of precipitates have been observed.

5. Conclusions

The following conclusions can be deduced from our experimental results in comparison with results in literature:

1. There are acting different types of nuclei of oxygen precipitation in dependence on temperature of h.t. The density of nuclei increases with the decrease of temperature.
2. The defect formation at low temperatures (600 °C $\leqq T \leqq$ 750 °C) is associated with the consumption of C_s as nucleus for the formation of oxygen clusters and small precipitates of SiO_x ($x < 2$) type. These types of clusters and small precipitates are also characterized by their electrical activity [33].
3. There is no evidence for homogeneous nucleation at temperatures $T \geqq$ 600 °C.
4. The higher the temperature of h.t. is within the temperature range 850 °C $\leqq$ $\leqq T \leqq$ 1100 °C, the more important are the type, density, and size distribution of grown-in oxygen clusters and microprecipitates for the further growth of Si–oxygen precipitates. Carbon has no importance as an immediate nucleus at these temperatures. The decrease of $[C_s]$ observed in $[C_s]$-rich samples is connected with the change of position or state of the carbon atoms.
5. Pre-h.t. at temperatures $\geqq$ 1200 °C widely disturbs grown-in oxygen clusters and microprecipitates acting as nuclei for oxygen precipitation. Using low-temperature h.t. ($T \leqq$ 650 °C) the oxygen clusters and precipitates indicated by their IR spectra are nearly the same with and without high-temperature pre-h.t.
6. After a "homogenization" pre-h.t. at high-temperatures nuclei for oxygen precipitation can be formed by many-step h.t. starting at low temperatures.
7. There is a close connection between different types of Si–oxygen precipitates and their corresponding IR spectra and TEM micrographs, respectively. The appearance of a broad band with a transmittance minimum near 1100 cm^{-1} but no other bands at lower wave numbers (type I) is obviously connected with the formation of globular-shaped microprecipitates which are not observable by TEM diffraction contrast because of their small size.

   The appearance of the double-peak spectrum with minima at about 1225 and about 1100 cm^{-1} is connected with the presence of plate-like precipitates.
8. The only type of Si–oxygen precipitates showing properties of bulk silica is the type IV precipitate formed by h.t. at $T \geqq$ 1150 °C and having an octahedral habit.

Acknowledgements

We would like to thank Dr. H. Richter (IHP Frankfurt (Oder)) and Dr. I. Csaba (TKI Budapest) for their interest and encouragement during the course of this work. Useful and stimulating discussions with Dr. D. Wruck (Zentralinstitut für Elektronenphysik der Akademie der Wissenschaften der DDR), Dr. H. Bartsch (Institut für Festkörperphysik und Elektronenmikroskopie der AdW), and Dr. K. Schmalz (IHP

Frankfurt (Oder)) are gratefully acknowledged. The authors are indebted to Mrs. E. Kuhnert and Mrs. R. Kass for their help in sample preparation and optical microscopy. We would like to thank Mr. H. Kessler und Dr. J. Reichel (VEB Spurenmetalle Freiberg) for providing research wafers.

References

[1] T. Y. Tan, E. E. Gardner, and W. K. Tice, Appl. Phys. Letters **30**, 175 (1977).
H. Richter, M. Mai, F.-G. Kirscht, and P. Gaworzewski, Physica (Utrecht) **116B**, 162 (1983).
[2] P. E. Freeland, K. A. Jackson, C. W. Lowe, and J. R. Patel, Appl. Phys. Letters **30**, 31 (1977).
[3] J. Osaka, N. Inoue, and K. Wada, Appl. Phys. Letters **36**, 288 (1980).
[4] S. Kishino, Y. Matsushita, and M. Kanamori, Appl. Phys. Letters **35**, 213 (1979).
[5] A. J. R. de Kock and W. M. van de Wijgert, Appl. Phys. Letters **38**, 888 (1981).
[6] Y. Matsushita, J. Crystal Growth **56**, 516 (1982).
[7] H. F. Schaake, S. C. Baber, and R. F. Pinizzotto, in: Semiconductor Silicon 1981, Ed. H. R. Huff, R. J. Kriegler, and Y. Takeishi, The Electrochem. Soc., Pennington 1981 (p. 273).
[8] K. Tempelhoff, F. Spiegelberg, R. Gleichmann, and D. Wruck, phys. stat. sol. (a) **56**, 213 (1979).
[9] S. M. Hu, J. appl. Phys. **51**, 5945 (1980).
[10] F. Shimura, Y. Ohnishi, and H. Tsuya, Appl. Phys. Letters **38**, 867 (1981).
[11] F. Shimura, J. Crystal Growth **54**, 589 (1981).
[12] K. V. Ravi, J. Electrochem. Soc. **121**, 1090 (1974).
[13] F. Shimura, H. Tsuya, and T. Kawamura, Appl. Phys. Letters **37**, 483 (1980).
[14] K. Graff, E. Grallath, S. Ades, G. Goldbach, and G. Tölg, Solid State Electronics **16**, 887 (1973).
[15] R. C. Newman and R. B. Willis, J. Phys. Chem. Solids **26**, 373 (1965).
[16] A. Murgai, H. C. Gatos, and W. A. Westdorp, J. Electrochem. Soc. **126**, 2240 (1979).
[17] S. Kishino, Y. Matsushita, M. Kanamori, and T. Iizuka, Japan. J. appl. Phys. **21**, 1 (1982).
[18] J. A. Lambert and P. S. Dobson, Phil. Mag. A **34**, 615 (1978).
[19] I. G. Salisbury, J. Microscopy **118**, 75 (1980).
[20] M. Stavola, Ext. Abstracts 163rd Electrochem. Soc. Meeting, San Francisco 1983 (p. 446).
[21] T. Y. Tan, Phil. Mag. A **44**, 101 (1981).
[22] K. Wada, N. Inoue, and J. Osaka, in: Defects in Semiconductors, Ed. S. Mahajan and J. W. Corbett, North-Holland Publ. Co., New York 1983.
[23] F. Shimura and H. Tsuya, J. Electrochem. Soc. **129**, 2089 (1982).
[24] A. G. Vlasov and V. A. Florinskaya, Infrared Spectra of Inorganic Glasses and Crystals, Izd. Khimiya, Leningrad 1972 (in Russian).
[25] V. C. Farmer, The Infrared Spectra of Minerals, Chap. 15, 16, Mineralog. Soc., London 1974.
[26] F. R. N. Nabarro, Proc. Phys. Soc. **52**, 90 (1940).
[27] K. Wada, N. Inoue, and K. Kohra, J. Crystal Growth **49**, 749 (1980).
[28] F. Shimura and H. Tsuya, J. Electrochem. Soc. **129**, 1062 (1982).
[29] V. M. Zolotariev, Optika i Spektroskopiya **29**, 66 (1970).
P. H. Gaskell and D. W. Johnson, J. non-crystall. Solids **20**, 153 (1976).
[30] M. J. Binns, W. P. Brown, J. G. Wildes, R. C. Newman, F. M. Livingston, S. Messoloras, and R. J. Stewart, Appl. Phys. Letters **42**, 525 (1983).
[31] F. A. Ponce, T. Yamashita, and S. Hahn, Ext. Abstracts 163rd Electrochem. Soc. Meeting, San Francisco 1983 (p. 537).
[32] R. W. G. Wyckoff, Crystal Structures, Vol. 1, Chap. IV, Interscience Publishers, New York 1965.
[33] P. Gaworzewski and K. Schmalz, phys. stat. sol. (a) **77**, 571 (1983).

(Received May 17, 1984)

phys. stat. sol. (a) 85, 149 (1984)

Subject classification: 10.1 and 12.1; 21.3

Department of Mining and Metallurgical Engineering, University of Queensland, St. Lucia[1])

Steady-State Creep and Strain Transients for Stress-Dip Tests in Polycrystalline Magnesium at 300 °C

By

D. O. NORTHWOOD[2]) and I. O. SMITH

High purity (99.99%) polycrystalline magnesium is creep tested in air at 300 °C and the stress-dip test used to determine the effective (σ_e) and internal (σ_i) stresses. The stress exponent of steady state creep, and the shape of the primary creep curve, indicate recovery (climb) creep to be rate controlling. There is an instantaneous non-elastic strain for all levels of stress decrement in the stress-dip tests. The absence of incubation periods of zero creep rate in the stress-dip tests, except for one particular value of the stress drop, favour the applicability of glide-recovery rather than straight recovery mechanisms for the high-temperature creep of magnesium. The internal stress varies with the applied stress, σ, increasing with increasing σ. σ_i/σ ratios of about 0.9 agree with previous data for pure metals.

Hochreines (99,99%) ,polykristallines Magnesium wird einem Kriechtest in Luft bei 300 °C unterworfen und der Spannungsabfall-Test benutzt, um die effektiven (σ_e) und die inneren (σ_i) Spannungen zu bestimmen. Der Spannungsexponent des stationären Kriechens und die Form der primären Kriechkurve zeigen, daß das Erholungs(Kletter)-Kriechen geschwindigkeitskontrolliert ist. Es existiert eine instantane nicht-elastische Spannung für alle Spannungsdekremente im Spannungsabfall-Test. Außer für einen bestimmten Wert des Spannungsabfalls spricht das Fehlen von Inkubationsperioden für die Null-Kriechrate im Spannungsabfall-Test eher für die Anwendbarkeit der Gleiterholung als des direkten Erholungsmechanismus für das Hochtemperaturkriechen von Magnesium. Die innere Spannung ändert sich mit der angelegten Spannung σ, wobei sie mit steigendem σ zunimmt. Die σ_i/σ-Verhältnisse von etwa 0,9 stimmen mit früheren Werten für reine Metalle überein.

1. Introduction

Although the deformation mechanisms occurring in f.c.c. metals at elevated temperatures are now understood reasonably well, the creep behaviour of h.c.p. materials has not been fully documented and is the subject of considerable speculation [1]. Generally, however, experiments on h.c.p. metals have shown two trends [1] namely:

(i) The activation energy for creep increases from a value close to self-diffusion at absolute temperatures of ≈ 0.5 to $0.6\ T_m$ to a much higher value at ≈ 0.7 to $0.8\ T_m$ [2 to 5] where T_m is the melting point in K.

(ii) There is a corresponding decrease in the value of the stress exponent, e.g. in Zn from ≈ 8.2 to ≈ 4.7 [2], in Mg from ≈ 5.5 to ≈ 4.0 [3], and in Cd from ≈ 4.3 to ≈ 3.9 [5].

---

[1]) St. Lucia, Queensland 4067, Australia.

[2]) One leave from Department of Engineering Materials, University of Windsor, Windsor, Ontario, Canada N9B 3P4.

Of the h.c.p. metals, magnesium has probably received by far the greatest attention in terms of creep studies. Much of this interest in the creep of Mg and its alloys stems from the extensive use of the alloys for cladding and structural applications in CO_2-cooled nuclear reactors. Crossland and Jones [6] have provided an excellent summary of creep data on magnesium published up to 1972 and this was updated in 1981 by Vagarali and Langdon [1]. Although there has been some general agreement by the different researchers on the general characteristics of creep in magnesium, there are numerous points of disagreement and areas of uncertainty (the latter also being due to a paucity of data in certain areas). For instance the second of the two trends mentioned earlier, i.e. a decrease in the value of the stress exponent with increase in creep temperature, has been found generally not to hold and in fact the reverse trend has been found with the stress exponent increasing from a value of about 5 for temperatures less than 350 °C ($< 0.68\ T_m$) to a value of about 10 at 425 °C ($0.76\ T_m$) [6].

Given these uncertainties in the results and mechanistic interpretations it was decided to study magnesium as part of a wider investigation into the nature of the back stress and rate controlling mechanisms in the creep of pure metals and solid solution alloys. The present paper reports results on the stress exponent of steady-state creep at 300 °C ($0.62\ T_m$) together with data on the effective stresses and internal (back)-stresses as determined from transients during a stress-dip test. Apart from some limited results for Zn–0.4 wt% Ag [7], Zr–1 wt% Nb [8], and α-Zr [9, 10] the present data is, to the authors' knowledge, the only available for internal (back) stresses in h.c.p. metals and alloys obtained during steady-state creep tests.

2. Experimental Details

2.1 Material

The magnesium was of high purity (99.99%) and was received in the form of bar stock. The bar was swaged and drawn to 3.30 mm diameter wire with appropriate intermediate annealing treatments. Creep specimens with a 76 mm gauge length were cut from the wire. These specimens were then annealed for 1 h at 400 °C in a vaccum furnace into which argon was bled. The grain size determined by the linear intercept method was (95 ± 15) μm.

2.2 Creep tests

All creep tests were performed in air at 300 °C with a temperature gradient along the gauge length of less then 1 K. A uniaxial constant stress creep machine with high sensitivity was used. The strain is detected by a pair of LVDT transducers outside the hot zone. Invar rods are used to transmit the strain from the wire grips so that temperature fluctuations have a minimal effect on the strain signal. The temperature in the room was controlled to better than ± 0.5 K and the resulting stability of the strain signal was excellent. Steady-state creep rates down to $10^{-8}\ s^{-1}$ could be measured with confidence and reproducibility. The experimental set-up described in more detail in [11].

The specimens were crept at stresses ranging from 2.5 to 15 MPa until steady-state creep occurred. Stress-dip tests of up to about 50% of the applied stress were performed. Great care was exercised in both unloading and reloading to reduce to a minimum the mechanical vibrations in the system. The strain–time plots presented in this paper were produced by first digitizing the information from the chart recorder and then plotting the data on a flat bed plotter. Care was taken to reproduce the curve as accurately as possible during digitizing to ensure that no smoothing was applied.

3. Experimental Results

3.1 Shape of creep curves

In agreement with Vagarali and Langdon [1] the creep curves were all of the normal type showing an instantaneous strain, extensive primary creep with gradually decreasing strain rate, and then a well-defined steady state stage.

3.2 Stress dependence of steady-state creep

The results are plotted logarithmically in Fig. 1 in the form of the steady-state creep rate, $\dot{\varepsilon}_s$, versus the applied stress, σ. The slope of this line, representing the stress exponent n in the creep equation $\dot{\varepsilon} = A\sigma^n$ was found to be 4.49 by linear regression analysis (correlation coefficient 0.9884). Also plotted on Fig. 1 for comparison purposes are the data of Vagarali and Langdon [1] for Mg with a grain size $d = 80\ \mu m$, where the stress exponent was found to be ≈ 5.2. The creep rates in the present set of experiments are almost two orders of magnitude faster than in the tests of Vagarali and Langdon [1]. Some of this difference is due to the difference in purity of the magnesium used in the two sets of experiments; 99.99% pure in present work compared to 99.94% pure material tested by Vagarali and Langdon.

3.3 Instantaneous strain upon a stress drop

The instantaneous strain upon a stress drop was measured for a wide range of initial stresses and stress drops ranging up to about 50% of the initial stress. The value of this instantaneous strain was compared to the value estimated from the elastic modulus of magnesium at 300 °C, 36.53 GPa. E_{300} for Mg was calculated using the relationships μ (shear modulus) $= (1.92 \times 10^4 - 8.6\ T)$ MPa [12] and $E = 2\mu(1 + \nu)$ where ν = Poissons ratio = 0.28 [13]. Fig. 2 summarises all the results as a plot of the ratio (calculated elastic strain to total instantaneous strain) against the stress drop expressed as a percentage of the initial stress. Even at relatively small stress drops the instantaneous strain was always greater than that expected from purely elastic behaviour. The ratio of the calculated elastic strain to total instantaneous strain reached a limiting value of about 0.4 for stress drops greater than about 30% of the initial stress.

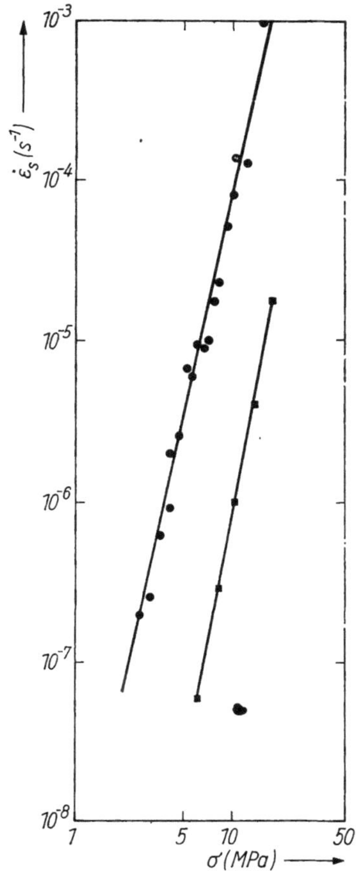

Fig. 1. Dependence of the steady-state creep rate, $\dot{\varepsilon}_s$, on the applied stress, σ, for polycrystalline Mg crept at 300 °C in air. ● present work, ■ after [1]

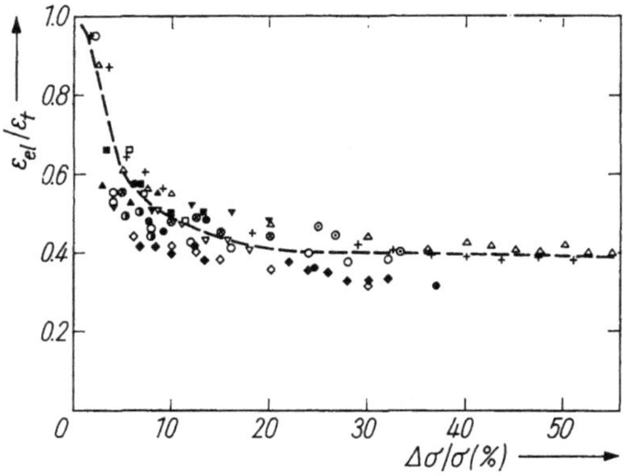

Fig. 2. Ratio of elastic strain, ε_{el}, to total instantaneous strain, ε_t, as a function of stress drop, $\Delta\sigma$, expressed as a percentage on the initial stress. $\sigma = 12$ (⊙), 10 (×), 9 (▽), 8 (◇), 7.5 (◑), 7 (□), 6.5 (●), 6 (◆), 5.5 (+), 5 (○), 4.5 (△), 4 (⊗), 3.5 (▲), 3 (■), 2.5 MPa (▼)

3.4 *Nature of the strain transients upon a stress drop*

Stress reductions of up to about 50% from all initial stress levels showed no evidence of classic incubation times, i.e. a period of zero creep or delay time before creep begins again in a forward manner [14]. Fig. 3a and b show typical results for initial stress levels of 8 and 9 MPa, respectively. The initial strain rate immediately after unloading was either positive for small stress reductions or negative for large stress reductions. Only at one particular value of stress reduction was the creep rate immediately after unloading found to be zero. For lower stress drops, e.g. Fig. 3a for $\Delta\sigma = 0.8$ MPa, the creep transient curves often showed a two- or three-stage behaviour with the creep rate in Stage I being less than that in Stage II, which was also less than that in Stage III. Such behaviour has previously been reported for the h.c.p. Zr–1 wt% alloy [8] and f.c.c. materials Cu [15] and Al–Mg [16, 17].

The value of stress drop that produces a zero creep rate immediately after unloading can be used to determine the internal or back-stress, σ_i, using the relationship

$$\sigma = \sigma_e + \sigma_i \,, \tag{1}$$

where σ_e is the effective stress (stress drop required to produce zero creep rate) and σ the applied stress. It is estimated that σ_e could be determined to ± 0.1 MPa. Values for σ_e and σ_i are given in Table 1. Fig. 4 shows the relationship between normalized internal stress, σ_i/μ, and the normalized applied stress, σ/μ, using a value of $\mu =$

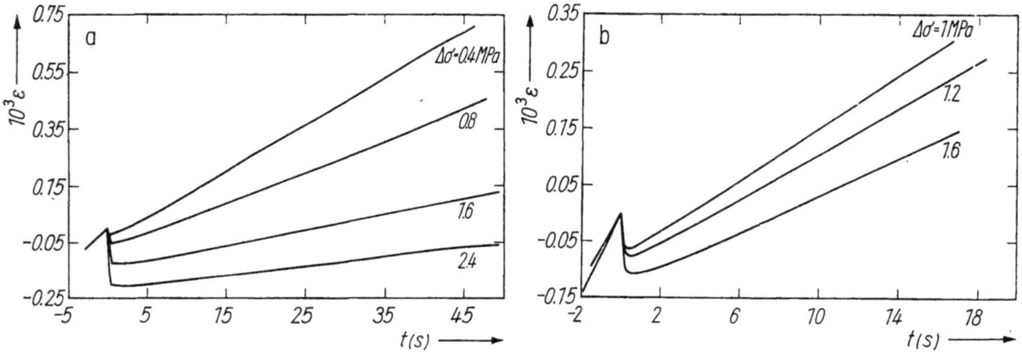

Fig. 3. Typical strain transients in magnesium on unloading after steady-state creep at 300 °C. a) Initial stress 8, b) 9 MPa

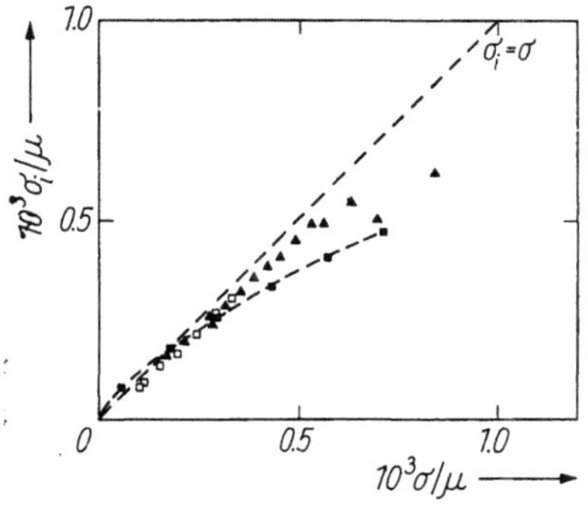

Fig. 4. Relationship between the normalized internal stress, σ_i/μ, and the normalized applied stress, σ/μ. ▲ present result for Mg, ■ calculated using equation (3), □ [71] for Al

Table 1

Experimentally determined values of the effective stress, σ_e, and internal stress, $\sigma_i (= \sigma - \sigma_e)$, for magnesium crept in air at 300 °C

σ (MPa)	σ_e (MPa)	σ_i (MPa)	σ_i/σ
2.5	0.2	2.3	0.92
3	0.2	2.8	0.93
3.5	0.2	3.3	0.94
4	0.3	3.7	0.93
4	0.4	3.6	0.90
4.5	0.4	4.1	0.91
5	0.4	4.6	0.92
5.5	0.4	5.1	0.93
6	0.5	5.5	0.92
6.5	0.6	5.9	0.91
7	0.6	6.4	0.91
7.5	0.5	7.0	0.93
8	1.0	7.0	0.88
9	1.2	7.8	0.87
10	2.8	7.2	0.72
12	3.2	8.8	0.73

= 14.27 GPa at 300 °C [12]. The internal stress, σ_i, can be seen to increase with increasing applied stress, σ. The ratio σ_i/σ decreases with increasing σ from a value of 0.93 ± 0.02 for lower values of σ (< 7.5 MPa) to a value of about 0.72 for higher applied stresses (> 10 MPa).

4. Discussion

The shape of the primary creep curves was, as noted in the results section, always of the normal-type (N-type) with a gradually decreasing strain rate. N-type primary creep curves have also been found by Vagarali and Langdon [1] for magnesium and by Georgy and Kamel for h.c.p. cobalt [18] and are in fact considered to be a characteristic of pure metals in which recovery creep (dislocation climb) is rate controlling [19, 20]. The concept for these early stages of creep is that rapid dislocation multiplication occurs and causes the dislocation density to rise sharply [21]. The creep rate then begins to decay as a direct consequence of strain hardening. As time proceeds, dislocations continue to re-arrange and the density continues to change until steady-state is approached.

The value of the stress exponent (≈ 4.5) for steady-state creep also suggests that dislocation climb is the rate-controlling process. According to the theoretical model for dislocation climb developed by Weertman [22] as subsequently modified for climb involving dipoles [23], the steady-state creep rate, $\dot{\varepsilon}_s$, is given by [24]

$$\dot{\varepsilon}_s = \alpha \frac{D_l}{b^{3.5} M^{0.5}} \frac{\mu \Omega}{kT} \frac{\sigma^{4.5}}{r}, \tag{2}$$

where D_l is the coefficient for lattice self-diffusion, M the number of dislocation sources per unit volume, Ω the atomic volume, and α a constant. The agreement between the theoretically derived value of the stress exponent, 4.5, and our experimentally determined value of 4.49 is excellent. Our value for the stress exponent is also in fairly good agreement with other experimentally determined values for magnesium at temperatures less than about 350 °C where deformation is predominantly by basal slip and stress exponents of about 5 have been found [1, 6]. Limited optical metallography of our creep specimens indicated extensive basal slip with straight and sharply defined slip lines.

As indicated in Fig. 2 there is an instantaneous non-elastic strain in all the stress decrement tests regardless of the level of the stress decrement. These observations are in accord with previous observations for polycrystalline and single crystal Cu [7], Al–Mg alloys [16, 17, 25, 26], and polycrystalline Al [26]. They are however at odds with the observations of Langdon and Yavari [27] for Al–5 wt% Mg at 400 °C where a stress decrement of 25 to 30% was necessary to introduce a non-elastic component to the instantaneous strain, and those of Parker and Wilshire [28] for high-purity Al where the instantaneous contraction was entirely elastic for stress decrements up to about 50%. Although the present results are at odds with those of Langdon and Yavari [27] and Parker and Wilshire [28] they do however indicate that the instantaneous strain becomes more nearly elastic for small stress decrements. The physical explanation for this non-elastic component of the instantaneous strain is still uncertain although the unbowing of dislocation segments, immobilized in a bowed-out configuration, toward a new configuration at reduced stress is a possibility. Also possible is the runback of dislocations piled up against an obstacle under stress [29].

As was noted in the results section there was no evidence of incubation periods of zero creep rate except for one particular stress drop for each value of the initial stress. Rather than periods of zero creep after small stress reductions, we recorded positive strain rates. Data in the literature on a wide variety of metals and alloys are at odds with reports of periods of zero creep [7, 30 to 45], "virtually" zero creep [46, 47], or positive strain rates [8, 10, 16, 17, 48 to 58], after small stress reductions. The only data in the literature for a h.c.p. material, Zn–0.4 wt% Ag, indicated incubation periods of zero creep [7]. Many of these "differences" in behaviour could be due to the relative ability of the creep measurement systems to pick-up and record the changes in strain which can occur in quite short periods of time. Some recent high precision measurements on Al using different recording and measurement systems has shown that there are no incubation periods and a zero creep rate is obtained only for one particular value of stress decrement for each value of initial stress [59].

Our observations for magnesium, showing no evidence of incubation periods of zero creep rate except for one particular stress drop for each value of initial stress, favour the glide-recovery models of high-temperature creep (e.g. Thorpe and Smith [60]) over the recovery models (e.g. Lagneborg [61]). The physical basis of the recovery model is that dislocation glide is possible only when the applied stress is sufficient to break dislocations from their pinning points in the dislocation network. The spacing of these pinning points is controlled by network recovery and this is the rate controlling process. A consequence of this model is the prediction that if a small stress reduction is made, creep will cease for the time that it takes for recovery to increase the network spacing to the extent that the lower applied stress can break the dislocations free, i.e. a delay time (incubation period) will be observed. In the glide-recovery models of high-temperature creep dislocation movement is considered to be thermally activated through an array of short-range obstacles with work-hardening and recovery occuring by dislocations entering cell walls and cell wall coarsening, respectively. In these models the gliding dislocations experience faces from the short-range obstacles, other gliding dislocations, and a longer range stress from the dislocations in the cell wall. The stresses from the last two of these sources constitute the back stress. This type of model predicts that creep will continue, albeit at a reduced rate, after small stress reductions, i.e. no delay time will be observed.

The strain transients, such as that shown for a stress drop of 0.8 MPa for an initial stress of 8 MPa, where there are two or more stages with increasing strain rate indicate that there are changes in the rate controlling mechanisms taking place after certain time intervals. A Stage I–Stage II transition where there is a breakaway in creep rate from Stage I to Stage II has previously been explained by Thorpe and Smith [8] in terms of control of the creep rate switching from one recovery process to another. The switching is due to substuctural changes which occur after a stress decrement. In the model of Thorpe and Smith the primary recovery process is climb and annihilation within the subgrain boundaries, and the secondary recovery process is climb and annihilation of glide dislocations. Note that climb (recovery) remains the rate controlling mechanism but that different "types" of dislocations undergo climb in the two stages.

The internal stress, σ_i, varies with applied stress, σ, increasing with applied stress (Fig. 4 and Table 1). The fact that σ_i increases with increasing applied stress agrees with previous work on Cu [62], Cu–4.04 wt% Co [62], various Al–Mg alloys [16, 49, 55, 63], Al [64], Ni [65], Ni–5 and 66 wt% Co [66], Fe [67, 68], Al–11 wt% Zn [57], and Al–0.4 wt% Li [69]. A fairly constant σ_i with varying σ has however been found for Nimonic 80 A [39], Nimonic 90 [32], a nickel-based superalloy IN 738LC [43], and Al–0.3 wt% Li [70]. The only data for a h.c.p. material, Zn–0.4 wt% Ag, indicates a small increase in the "friction" stress (a component of the internal stress) with increasing applied stress. The general comment that can be made on the data indicating a fairly constant σ_i with varying σ is that the tests were done over a relatively small range of applied stress and given the errors in measuring σ_e by the stress-dip test, any change in σ_i could have been masked by the scatter in values for σ_e.

The internal stress is expected to vary with the applied stress from theoretical considerations. Argon et al. [64] have developed an equation relating σ_i to σ as follows:

$$\frac{\sigma_i}{\mu} = C_i \frac{(1-\nu)}{\pi} (K\theta)^{1/3} \left(\frac{\sigma}{\mu}\right)^{2/3}, \tag{3}$$

where ν is Poisson's ratio, $C_i = \text{const} = 0.317$, $K = \text{const} = 30$ (from the equation relating sub-grain size, d, to stress: $d = K\mu b/\sigma$), and θ is the sub-boundary misorientation.

A plot of σ_i/μ versus σ/μ computed using (3) and data for magnesium at 300 °C ($\mu = 14.27$ GPa [12], $\nu = 0.28$ [13], and $\theta = 1°$) is shown together with the actual experimental results in Fig. 4. The agreement is not particularly good especially at the higher values of σ/μ and would indicate that the stress exponent in the σ/μ term on the right-hand side of the equation is higher than $\frac{2}{3}$ (a log–log plot of σ_i/μ versus σ/μ indicates the stress exponent to be ≈ 0.9). A possible reason for the difference between the experimental values and the calculated values is that the lattice misorientation across a sub-boundary, θ, varies with σ/μ. This is in fact a fairly realistic assumption from a physical standpoint with an increase in misorientation expected for an increase in applied stress. Also included in Fig. 4 are some recent experimental data for polycrystalline Al [71]. Although this data is limited to the lower σ/μ range the data indicates that Al and Mg behave similarly with respect to the dependence of normalized internal stress, σ_i/μ, on the normalized applied stress, σ/μ. This similarity in behaviour is not unexpected since both metals have a close-packed structure (f.c.c. for Al and h.c.p. for Mg) and the stress exponents for steady-state creep are similar, 4.2 for Al [72] and 4.49 in the present work on Mg, indicating recovery (dislocation climb) controlled creep. The ratio σ_i/σ for Mg in the present studies (≈ 0.9) agrees with previous data for pure metals ($\sigma_i/\sigma = 0.9$ to 1.0) and is higher than that found in either Class I alloys ($\sigma_i/\sigma \approx 0.7$) or Class II alloys (0.8 to 0.9) [69].

5. Conclusions

1. The stress exponent of steady-state creep for polycrystalline magnesium at 300 °C, ≈ 4.5, and the shape of the primary creep curve indicate recovery (climb) creep to be rate controlling.

2. An instantaneous non-elastic strain is found in stress-dip tests for all levels of stress decrement.

3. No evidence was found for incubation periods of zero creep rate in the stress-dip tests except for one particular stress drop for each value of the initial applied stress. These observations favour the gliderecovery models of high-temperature creep over the recovery models.

4. The internal stress, σ_i, varies with applied stress, σ, increasing with increasing applied stress. σ_i/σ ratios for Mg (≈ 0.9) agree with previous data for pure metals ($\sigma_i/\sigma = 0.9$ to 1.0).

Acknowledgements

Financial support from the Australian Research Grants Scheme is gratefully acknowledged. The technical assistance of Mr. L. Moerner and Miss K. Daly in conducting the creep experiments is also gratefully acknowledged. One of the authors, D. O. Northwood, would also like to thank the Natural Sciences and Engineering Research Council of Canada for the provision of a Travel Grant and the University of Windsor for granting a sabbatical leave.

References

[1] S. S. VAGARALI and T. G. LANGDON, Acta metall. 29, 1969 (1981).
[2] W. J. McG. TEGART and O. D. SHERBY, Phil. Mag. 3, 1287 (1958).
[3] W. J. McG. TEGART, Acta metall. 9, 614 (1961).
[4] J. E. FLINN and D. E. MUNSON, Phil. Mag. 10, 816 (1964).
[5] J. E. FLINN and S. A. DURAN, Trans. MS AIME 236, 1056 (1966).
[6] I. G. CROSSLAND and R. B. JONES, Metal Sci. J. 6, 162 (1972).
[7] P. W. DAVIES, G. NELMES, K. R. WILLIAMS, and B. WILSHIRE, Metal Sci. J. 7, 87 (1973).
[8] W. R. THORPE and I. O. SMITH, Mater. Sci. Engng. 36, 145 (1978).
[9] M. PAHUTOVA and J. CADEK, Mater. Sci. Engng. 11, 151 (1973).
[10] M. PAHUTOVA, J. CADEK, and P. RYS, Mater. Sci. Engng. 39, 169 (1979).
[11] I. O. SMITH, Australian Atomic Energy Commission Research Project, Progr. Rep. No. 15, January 1971.
[12] L. J. SLUTSKY and C. M. GARLAND, Phys. Rev. 107, 972 (1957).

[13] J. FRIEDEL, in: Dislocations, Pergamon Press, Oxford 1964 (p. 54).
[14] J. B. BILDE-SORENSEN, Phil. Mag. A **38**, 1 (1978).
[15] W. R. THORPE and I. O. SMITH, Phil. Mag. A **40**, 581 (1979).
[16] D. O. NORTHWOOD and I. O. SMITH, Scripta metall. **18**, 521 (1984).
[17] D. O. NORTHWOOD and I. O. SMITH, Mater. Sci. Engng., in the press (1984).
[18] K. H. GEORGY and R. KAMEL, phys. stat. sol. (a) **80**, 95 (1983).
[19] S. KARASHIMA, in: Creep and Fracture of Engineering Materials and Structures, Ed. B. WILSHIRE and D. R. J. OWEN, Pineridge Press, Swansea (U.K.) 1981 (p. 31).
[20] T. G. LANGDON, Proc. VI. Internat. Conf. Strength of Metals and Alloys, Vol. 3, Ed. R. C. GIFKINS, Pergamon Press, 1982 (p. 1105).
[21] C. R. BARRETT, W. D. NIX, and O. D. SHERBY, Trans. ASM **59**, 3 (1966).
[22] J. WEERTMAN, J. appl. Phys. **28**, 362 (1957).
[23] J. WEERTMAN, Trans. ASM **61**, 681 (1968).
[24] J. WEERTMAN, in: Rate Processes in Plastic Deformation of Materials, Ed. J. C. M. LI and A. K. MUKHERJEE, American Society for Metals, Metals Park (Ohio) 1975 (p. 315).
[25] H. OIKAWA and K. SUGAWARA, Scripta metall. **12**, 85 (1978).
[26] L. MOERNER, D. O. NORTHWOOD, and I. O. SMITH, J. Mater. Sci., to be published (1984).
[27] T. G. LANGDON and P. YAVARI, see [19] (p. 71).
[28] J. D. PARKER and B. WILSHIRE, Phil. Mag. A **41**, 665 (1980).
[29] J. C. POIRIER, Acta metall. **25**, 913 (1977).
[30] B. BERGMANN, Scand. J. Metall. **4**, 109 (1975).
[31] P. W. DAVIES and K. R. WILLIAMS, J. Inst. Metals **97**, 337 (1969).
[32] W. J. EVANS and G. F. HARRISON, Scripta metall. **9**, 239 (1975).
[33] W. J. EVANS and B. WILSHIRE, Metal Sci. J. **4**, 89 (1970).
[34] P. W. DAVIES and B. WILSHIRE, Scripta metall. **5**, 475 (1971).
[35] G. NELMES and B. WILSHIRE, Scripta metall. **10**, 697 (1976).
[36] J. D. PARKER and B. WILSHIRE, Metal. Sci. J. **9**, 247 (1975).
[37] J. D. PARKER and B. WILSHIRE, Phil. Mag. **34**, 485 (1976).
[38] J. D. PARKER and B. WILSHIRE, Metal Sci. J. **12**, 453 (1978).
[39] K. R. WILLIAMS and B. WILSHIRE, Metal Sci. J. **7**, 176 (1973).
[40] P. L. THREADGILL and B. WILSHIRE, in: Creep Strength in Steel and High Temperature Alloys, The Metals Society, London 1972 (p. 8).
[41] Y. TAKAHASHI and T. YAMANE, J. Mater. Sci. **16**, 397 (1981).
[42] Y. TAKAHASHI, K. NAKAGAWA, and T. YAMANE, Z. Metallk. **71**, 572 (1980).
[43] P. J. HENDERSON and M. MCLEAN, Acta metall. **31**, 1203 (1983).
[44] C. OYTANA, P. DELOBELLE, and A. MERMET, J. Mater. Sci. **14**, 549 (1979).
[45] M. MCCLEAN, Proc. Roy. Soc. **A373**, 93 (1980).
[46] S. K. MITRA and D. MCCLEAN, Proc. Roy. Soc. **A295**, 288 (1966).
[47] A. ROSEN, J. Inst. Metals **99**, 111 (1971).
[48] C. N. AHLQUIST and W. D. NIX, Scripta metall. **3**, 679 (1969).
[49] C. N. AHLQUIST and W. D. NIX, Acta metall. **19**, 373 (1971).
[50] M. PAHUTOVÁ, K. KUCHAŘOVÁ, and J. ČADEK, phys. stat. sol. (a) **6**, K33 (1971).
[51] J. C. GIBELING and W. D. NIX, Metal Sci. J. **11**, 453 (1977).
[52] J. C. GIBELING and W. D. NIX, see [20] (p. 613).
[53] K. MINO, M. KITAGAWA, A. OHOTOMO, and Y. SAIGA, Trans. Iron Steel Inst. Jpn. **20**, 826 (1980).
[54] W. BEERE and I. G. CROSSLAND, Acta metall. **30**, 1891 (1982).
[55] K. KUCHAROVA, I. SAXL, and J. CADEK, Acta metall. **22**, 465 (1974).
[56] J. C. GIBELING and W. D. NIX, Acta metall. **29**, 1769 (1981).
[57] W. BLUM, J. HAUSSELT, and G. KÖNIG, Acta metall. **24**, 293 (1976).
[58] M. PAHUTOVA, J. CADEK, and P. RYS, Mater. Sci. Engng. **36**, 185 (1978).
[59] L. R. CUTLER, W. R. THORPE, and I. O. SMITH, to be published Scripta metall. (1984).
[60] W. R. THORPE and I. O. SMITH, phys. stat. sol. (a) **52**, 487 (1979).
[61] R. LAGNEBORG, Metal Sci. J. **3**, 161 (1969).
[62] B. WILSHIRE, in: Creep in Structures, Ed. A. R. S. PORTER and D. R. HAYHURST, Springer-Verlag, Berlin 1981 (p. 72).
[63] D. O. NORTHWOOD and I. O. SMITH, phys. stat. sol. (a) **83**, 227 (1984).
[64] A. S. ARGON, F. PRINZ, and W. C. MOFFATT, see [19] (p. 1).

[65] H. Oikawa and S. Karashima, Metallurg. Trans. **5**, 1179 (1974).
[66] V. Sagar Dwivedi, Scripta metall. **13**, 833 (1979).
[67] A. Orlova, M. Pahutova, and J. Cadek, Phil. Mag. **25**, 865 (1972).
[68] M. Pahutova, A. Orlova, K. Kucharova, and J. Cadek, Phil. Mag. **28**, 1099 (1973).
[69] D. O. Northwood and I. O. Smith, Strain–Time Transients for Stress Change Tests in an Al–0.4 wt% Li Solid Solution Alloy, Paper to be presented at Fall TMS-AIME Meeting, Detroit, September 17 to 21, 1984.
[70] L. R. Cutler and I. O. Smith, to be published.
[71] D. O. Northwood and I. O. Smith, Unpublished data, University of Queensland, 1984.
[72] D. O. Northwood, L. Moerner, and I. O. Smith, J. Mat. Sci., to be published (1984).

(Received May 29, 1984)

phys. stat. sol. (a) 85, 159 (1984)

Subject classification: 10.1; 8; 12.1; 21.1

Institute of Chemistry, Academy of Sciences of the USSR, Ural Scientific Centre, Sverdlovsk[1])

Structural Vacancies in Nonstoichiometric Compounds at High Pressure

Thermodynamic Model

By

A. I. GUSEV

A thermodynamic model is proposed which describes the effect of high pressure on the equilibrium structural vacancy concentration of nonstoichiometric compounds possessing wide regions of homogeneity. The implicit pressure and temperature dependence of the equilibrium structural vacancy concentration is found in general form for any nonequivalent crystal sublattice of a compound.

Предложена термодинамическая модель, описывающая влияние высокого давления на равновесную концентрацию структурных вакансий в нестехиометрических соединениях, обладающих широкими областями гомогенности. В общем виде для любой неэквивалентной кристаллической подрешетки соединения найдена неявная зависимость равновесной концентрации структурных вакансий от давления и температуры.

1. Introduction

High-melting NaCl-structure compounds such as transition metal carbides, nitrides, and oxides posses wide regions of homogeneity. Within these regions they may have structural vacancies in either the nonmetallic or metallic sublattice, or in both at a time. For some of the compounds concerned (for example, such as vanadium and zirconium carbides and vanadium monoxide), no stoichiometric compositions with all the sublattices being free from structural vacancies have been synthesized so far. A possible way to produce practically defect-free compounds (or compounds with crystal sublattice defect relations incapable of realization using conventional synthesis methods) is to apply high pressures in combination with high temperatures. (A defect-free compound is one containing no structural vacancies; the defect concentration, or imperfection, is the structural vacancy content of the crystal).

Since temperature and pressure involved in the formula for the free energy of formation of a compound are opposite in sign, an increase in pressure, just as a decrease in temperature, should in the general case lead to a lower equilibrium structural vacancy concentration for the conditions given. However, the synthesis of practically defect-free (devoid of structural vacancies) compounds in the region of temperatures close to absolute zero appears to be unfeasible due to primarily kinetic reasons. To synthesize a compound with a crystal lattice containing no structural vacancies it is obviously an acceptable procedure to use simultaneously high temperatures assuring sufficiently high rates of diffusion processes in the solid state and high pressures reducing the structural vacancy concentration. It should be noted that the theoretical

---

[1]) Pervomaiskaya 91, 620219 Sverdlovsk, USSR.

equilibrium structural vacancy concentration depends on temperature and pressure in an exponential manner, i.e. may be as small as desired but never goes to zero. In this context it is necessary to consider the possibility of synthesizing under certain conditions not absolutely defect-free compounds but compounds possessing a structural vacancy concentration as small as desired, i.e. practically defect-free ones.

Depending on the relation of sublattice defects in the basic material and on which relation between defect concentrations one wishes to obtain, the synthesis may be performed with or without the addition of some components of a given basic material. If the synthesis involving the combined use of pressure and temperature is effected without adding components the formation of a defect-free (with respect to one or several sublattices) compound will occur due to the migration of a certain number of atoms of one component or another from the surface of the crystal and due to the collapse of vacancies in the corresponding sublattices. In this case conditions can also be established under which not all of the vacancies vanish completely. It is thus possible to synthesize a compound with crystal sublattice vacancy contents which are not realized under normal conditions. If relevant components are added to a defective compound subject to compression, the formation of a defect-free compound occurs due to the migration of the atoms of the components added inside the crystal and due to these atoms filling the vacancies in the corresponding sublattices. Like in the previous case, it is possible here to create conditions under which not all the vacancies will be filled.

A number of experimental and theoretical papers are available which deal with the effect of high pressure on materials possessing structural vacancies (e.g., on compounds such as TiO, VO, and NbO which contain structural vacancies in the nonmetallic and metallic sublattices simultaneously). Thus, it has been demonstrated experimentally [1, 2] that when titanium monoxide is compressed under a pressure of 60 kbar and more and at temperatures above 1500 K the structural vacancy concentration can be decreased substantially. An investigation of the effect of high pressure on the crystal structure of extreme compositions in the region of homogeneity of TiO_x at $T = 300$ K has revealed a very insignificant lattice parameter variation [3]. A statistical thermodynamical analysis of the equilibrium vacancy concentration as a function of pressure P and temperature T, performed in [4], has shown that, subject to the conditions specified in [1], no defect-free TiO should arise. However, that analysis was carried out involving a number of arbitrary assumptions. The Grüneisen constant γ was taken to be negative, although the value of γ is positive for all solid substances. The values of the formation energy of a vacancy in the titanium and oxygen sublattices were taken to be equal and composition-independent [4], although the behaviour of the Ti- and O-sublattice defects in the region of homogeneity of TiO_x indicates that these energies differ in magnitude and vary with composition. A calculation of the formation energy of a vacancy in the titanium and oxygen sublattices of titanium monoxide has confirmed their dependence on composition TiO_x and their substantial difference in magnitude [5].

The present paper proposes a thermodynamic model to describe nonstoichiometric compounds containing structural vacancies. This model permits the calculation of the pressure and temperature required to synthesize, under thermodynamic-equilibrium conditions, similar compounds with any predetermined crystal sublattice defect relation, including practically defect-free compounds.

2. Thermodynamic Model

Examine a defective compound possessing nonequivalent crystal sublattices, in each of which the relative number of structural vacancies (defect concentration) per "molecule" is equal to $n_i = N_{\square i}/N_i$, and the relative number of occupied sites is $m_i = M_i/N_i$, where $N_{\square i} + M_i = N_i$ is the total number of i-sublattice sites; hence $n_i + m_i = 1$. Let the Gibbs free energy of this defective compound be equal to $\Delta G(n)$. Then the variation of the Gibbs free energy in the case of passing from a defective compound to a compound with a given crystal sublattice defect relation ("crystal sublattice defect relation" is a particular combination of the relative numbers of structural vacancies present in each crystal sublattices, i.e. a particular set of values $n_1, n_2 \dots n_i$, characterizing the compound under investigation) may be represented as

$$\Delta G(n) = \Delta G + \sum \int \Delta H_i \, dn_i - T\Delta S_c + \sum \iint \Delta V_i \, dn_i \, dP + \Delta F \,, \qquad (1)$$

where ΔG is the Gibbs free energy of a crystal with a given sublattice defect relation, ΔH_i the formation energy of vacancies in the i-sublattice, ΔS_c the change in the configurational entropy, ΔV_i the change in volume during the formation of structural vacancies in the i-sublattice, ΔF the change in the vibrational free energy of the crystal (ΔH_i and ΔV_i are continuous functions of n_i).

If some components are added to a defective compound subject to compression, the formation of a compound with a lower structural vacancy content or of a totally defect-free compound occurs due to the added-component atoms migrating inside the crystal and filling the vacancies in the corresponding sublattices. The change in volume here will be

$$\sum \int \Delta V_i \, dn_i = V + V_{add} - V_f \,, \qquad (2)$$

where V is the volume of the defective crystal, V_{add} the volume of the component added, V_f the volume of the crystal produced as a result of P–T treatment. The volume will vary not only due to a change in defect concentration but also due to the compression of the material under pressure, as well as due to thermal expansion. Let v_i be the composition-independent volume of the i-component of the compound, $v_{\square i}$ the volume of an i-sublattice vacancy, b_i a proportionality coefficient indicating which portion of the structural vacancies of the i-sublattice should be filled as a result of the P–T treatment, $x = V_P/V_0$ is the compression (V_0 is the volume of the material when $P = 0$, V_P is the volume of the compressed material), α the thermal expansion coefficient.

Calculation in terms of one "molecule" in this case yields

$$V = \sum \left[v_i(1 - n_i) + \int v_{\square i} \, dn_i\right] , \qquad V_{add} = \sum b_i v_i n_i \,,$$
$$V_f = x(1 + 3\alpha T) \sum \left[v_i - v_i n_i(1 - b_i) + (1 - b_i) \int v_{\square i} \, dn_i\right] . \qquad (3)$$

For solid substances the reduced pressure P/K (with K being the bulk modulus) dependence of compression x is described, according to [6], by the expression

$$\frac{P}{K} = \frac{3}{A - 1}\left[\left(\frac{1}{x}\right)^{(A+3)/3} - \left(\frac{1}{x}\right)^{4/3}\right], \qquad (4)$$

where $A = (3K' - 7)$ ($K' = \partial K/\partial P$ when $P = 0$). According to [6], for various solid materials $(\partial K/\partial P)_{P=0}$ varies over a quite narrow interval and the mean value of this quantity amounts to 4.8, the respective value for A being 7.4. With these values of the parameters (4) gives a good fit to the compression of various solid materials (except for solid inert gases) with $P/K \leqq 2$. For $P/K \leqq 0.25$ (4) may be shown to transform to a linear function of compression versus reduced pressure

$$x = 1 - 0.8\frac{P}{K}. \qquad (5)$$

In view of (5) the total change in the volume of the crystal as calculated in terms of one ,,molecule'' of the compound will be

$$\Sigma \int \Delta V_i \, dn_i = \Sigma \, [v_i(1 - n_i + n_i b_i) + \int v_{\Box i} \, dn_i] - \\ - \left(1 - 0.8 \frac{P}{K}\right) (1 + 3\alpha T) \, \Sigma \, [v_i - v_i n_i (1 - b_i) + \\ + (1 - b_i) \int v_{\Box i} \, dn_i] \, . \tag{6}$$

If the compression is effected without adding any component the number of crystalline atoms remains unchanged, the volume of the crystal upon P–T treatment is

$$V_f = \left(1 - 0.8 \frac{P}{K}\right)(1 + 3\alpha T) \, \Sigma \, [v_i(1 - n_i) + (1 - b_i) \int v_{\Box i} \, dn_i] \, , \tag{7}$$

and the change in the volume of the crystal as a result of compression is

$$\Sigma \int \Delta V_i \, dn_i = \Sigma \, [v_i(1 - n_i) + \int v_{\Box i} \, dn_i] - \\ - \left(1 - 0.8 \frac{P}{K}\right) (1 + 3\alpha T) \, \Sigma \, [v_i(1 - n_i) + \\ + (1 - b_i) \int v_{\Box i} \, dn_i] \, . \tag{8}$$

Expression (6) differs from (8) in that it contains an additional term, which is equal to

$$\left[1 - \left(1 - 0.8 \frac{P}{K}\right) (1 + 3\alpha T)\right] \Sigma \, b_i v_i n_i \, . \tag{9}$$

The change in vibrational free energy may be written in the form

$$\Delta F = \Delta E - T \Delta S \, . \tag{10}$$

To describe the thermal vibrations one may adopt the Einstein model and represent the crystal as a set of harmonic linear oscillators with the mean energy $\overline{E}$,

$$\overline{E} = \frac{kT^2}{Z} \left(\frac{\partial Z}{\partial T}\right)_V = \frac{h\nu}{2} \, \frac{\exp\left(\frac{h\nu}{kT}\right) + 1}{\exp\left(\frac{h\nu}{kT}\right) - 1} \, , \tag{11}$$

where Z is the statistical sum of states, k the Boltzmann constant, h the Planck constant. In this case the change in energy ΔE in a three-dimensional crystal at the transition from a defective state to a defect-free one will be

$$\Delta E = \frac{3}{2} \, \Sigma \, m_i h \nu_n \, \frac{\exp\left(\frac{h\nu_n}{kT}\right) + 1}{\exp\left(\frac{h\nu_n}{kT}\right) - 1} - \frac{3}{2} \, \Sigma \, m_i^0 h \nu \, \frac{\exp\left(\frac{h\nu}{kT}\right) + 1}{\exp\left(\frac{h\nu}{kT}\right) - 1} \, , \tag{12}$$

where m_i and m_i^0 stand for the number of atoms in a defective crystal and in a crystal resulting from P–T treatment, respectively, ν_n and ν are the atomic vibration frequencies for a defective crystal and for a crystal resulting from P–T treatment.

The total change in the vibrational non-configuration entropy ΔS may be found, as has been shown in [7], in the form

$$\Delta S = \sum \int \Delta S_i \, dn_i \,, \tag{13}$$

where ΔS_i is the change in the non-configuration entropy during the formation of i-sublattice vacancies.

The change in configuration entropy is determined by the equation

$$\Delta S_c = k \sum \ln \frac{(m_i + n_i)!}{m_i! n_i!} \tag{14}$$

to be subsequently expanded according to the Stirling approximation.

When an equilibrium concentration of structural vacancies in the i-sublattice for a given pressure and for a given temperature is reached, the Gibbs free energy $\Delta G(n)$ will be minimum and, accordingly, $\partial \Delta G(n)/\partial n_i = 0$. Proceeding from this situation it is possible to determine the temperature and pressure dependence of the equilibrium structural vacancy concentration.

Substitution of (6), (10), (12) to (14) or (8), (10), (12) to (14) into (1) and differentiation of the latter with respect to n_i yields

$$\frac{\partial \Delta G(n)}{\partial n_i} = \Delta H_i - T \frac{\partial \Delta S_c}{\partial n_i} + \frac{\partial \sum \iint \Delta V_i \, dn_i \, dP}{\partial n_i} + \frac{\partial \Delta E}{\partial n_i} - T \Delta S_i \,, \tag{15}$$

where

$$\frac{\partial \Delta S_c}{\partial n_i} = -k \ln \frac{n_i}{m_i}, \tag{16}$$

$$\frac{\partial \sum \iint \Delta V_i \, dn_i \, dP}{\partial n_i} = \int v_{\Box i} \left[1 - \left(1 - 0.8 \frac{P}{K}\right)(1 + 3\alpha T)(1 - b_i)\right] dP - \\ - \int v_i (1 - b_i) \left[1 - \left(1 - 0.8 \frac{P}{K}\right)(1 + 3\alpha T)\right] dP \,, \tag{17}$$

$$\frac{\partial \Delta E}{\partial n_i} = \frac{3}{2} m_i h \frac{\partial \nu_n}{\partial n_i} \left\{ \frac{\exp\left(\frac{h\nu_n}{kT}\right) + 1}{\exp\left(\frac{h\nu_n}{kT}\right) - 1} - \frac{2h\nu_n}{kT} \frac{\exp\left(\frac{h\nu_n}{kT}\right)}{\left[\exp\left(\frac{h\nu_n}{kT}\right) - 1\right]^2} \right\}. \tag{18}$$

Allowance for $h\nu_n = k\theta$ and

$$h \frac{\partial \nu_n}{\partial n_i} = k \frac{\partial \theta}{\partial n_i} = k \frac{\partial \theta}{\theta} \frac{V}{\partial V} \frac{\theta}{V} \frac{\partial V}{\partial n_i} = k \frac{\theta}{V} \frac{\partial \ln \theta}{\partial \ln V} \frac{\partial V}{\partial n_i} = -k\theta\gamma \frac{1}{V} \frac{\partial V}{\partial n_i}, \tag{19}$$

where θ is the characteristic Einstein temperature, gives

$$\frac{\partial \Delta E}{\partial n_i} = -\frac{3}{2} m_i \gamma k \theta \, \frac{\frac{\partial v_{\Box i}}{\partial n_i} n_i + v_{\Box i}}{\sum (v_i m_i + v_{\Box i} n_i)} \left\{ \frac{\exp\left(\frac{\theta}{T}\right) + 1}{\exp\left(\frac{\theta}{T}\right) - 1} - \frac{2\theta}{T} \frac{\exp\left(\frac{\theta}{T}\right)}{\left[\exp\left(\frac{\theta}{T}\right) - 1\right]^2} \right\}. \tag{20}$$

Substituting (16), (17), (20) into (15) and solving it for $\ln \frac{n_i}{m_i}$ we find

$$\ln \frac{n_i}{m_i} = -\frac{\Delta H_i}{kT} + \frac{\Delta S_i}{k} - \frac{1}{kT} \int v_{\square i} \left[1 - \left(1 - 0.8\frac{P}{K}\right)(1 + 3\alpha T)(1 - b_i)\right] \mathrm{d}P + \\ + \frac{1}{kT} \int v_i(1 - b_i) \left[1 - \left(1 - 0.8\frac{P}{K}\right)(1 + 3\alpha T)\right] \mathrm{d}P + \\ + \frac{3}{2} m_i \gamma \frac{\theta}{T} \frac{\frac{\partial v_{\square i}}{\partial n_i} n_i + v_{\square i}}{\sum (v_i m_i + v_{\square i} n_i)} \left\{ \frac{\exp\left(\frac{\theta}{T}\right) + 1}{\exp\left(\frac{\theta}{T}\right) - 1} - \frac{2\theta}{T} \frac{\exp\left(\frac{\theta}{T}\right)}{\left[\exp\left(\frac{\theta}{T}\right) - 1\right]^2} \right\} \tag{21}$$

for the case when some components are added to a defective compound subject to P–T treatment. If the latter is performed without adding any components the last but one term in (21) should be written as

$$\frac{1}{kT} \int v_i \left[1 - \left(1 - 0.8\frac{P}{K}\right)(1 + 3\alpha T)\right] \mathrm{d}P\,. \tag{22}$$

Other parameters being known, the derived equation (21) admits a self-consistent solution for n_i.

3. Numerical Calculations

As an example, we will calculate the thermodynamic-equilibrium conditions of synthesis for zirconium and vanadium carbides, which are close to stoichiometry. Referred to one "molecule" and suitable for the 1500 to 3000 K temperature range, the structural vacancy formation energy in the carbon sublattice $\Delta H = (0.524 - 0.520n)$ eV and the dependence of the vibrational entropy on the defect concentration of the carbon lattice $\Delta S = (0.88 - 31.84n) \times 10^{-5}$ eVK^{-1}, which are required for this calculation, were determined earlier for ZrC_{1-n} [7]. According to the most trustworthy phase diagram for the Zr–C system [8], the zirconium carbide composition which is closest to stoichiometry is $ZrC_{0.98}$. For a zirconium carbide close to stoichiometry the characteristic Einstein temperature θ is equal to 750 K [9]. The value of the Grüneisen constant γ for most of the carbides ranges between 1 and 2 [10], in the present calculation the mean γ value of 1.5 has been adopted. To estimate the volume of a carbon vacancy in zirconium carbide the following assumptions have been made. The volume of one "molecule" of stoichiometric ZrC with NaCl structure equals, on the one hand, one fourth of the volume of the unit cell and, on the other hand, the sum of the volumes of a zirconium atom, v_{Zr}, and of a carbon atom, v_C. The volume of an atom is proportional to R^3, where R is the atomic radius. Hence, if a is the period of a unit cell of a stoichiometric compound and $R_{Zr}/R_C = q$, then

$$v_{Zr} = \frac{a^3 q^3}{4(1 + q^3)}, \qquad v_C = \frac{a^3}{4(1 + q^3)}. \tag{23}$$

For some composition $ZrC_{1-\Delta n}$

$$V = v_{Zr} + (1 - \Delta n)\, v_C + \Delta n v_{\square C} = v_{Zr} + v_C + \Delta n(v_{\square C} - v_C) = \frac{a_n^3}{4}. \tag{24}$$

Since $v_{Zr} + v_C = a^3/4$ then $v_{\square C} - v_C = (a_n^3 - a^3)/4$; in the limit $\Delta n \to 0$, $a_n \to a$, therefore $v_{\square C} - v_C = \partial(a_n^3)/4\, \partial n$ and

$$v_{\square C} = \frac{1}{4}\left[\frac{\partial(a_n^3)}{\partial n} + \frac{a^3}{1 + q^3}\right]. \tag{25}$$

Using the data [11] on the variation of the lattice period of zirconium carbide in the region of homogeneity we have arrived at the result

$$v^{\mathrm{ZrC}}_{\square\mathrm{C}} = (2.494 - 2.684n) \times 10^{-30}\ \mathrm{m}^3 .$$

For vanadium carbide the upper boundary of the region of homogeneity is $VC_{0.87}$ [11]. The temperature dependence of the Gibbs free energy $\Delta G_i = \Delta H_i - T\Delta S_i$ for the formation of structural vacancies in the carbon sublattice of vanadium carbide for a temperature interval from 300 to 1300 K, as calculated for one "molecule", has the form [12]

$$\Delta G^{\mathrm{VC}}_{\square\mathrm{C}} = 0.415 + 1.574 \times 10^{-3}T + 7.553 \times 10^{-8}T^2 - \\ - 0.17 \lg T - 0.458 \times 10^{-3}T \lg T + 4.525T^{-1}\ \mathrm{eV} .$$

For the vanadium carbide $VC_{0.87}$ the characteristic Einstein temperature θ is equal to 1050 K [9]. The volume of a carbon vacancy in vanadium carbide has been calculated with the use of structure data [13, 14], just as is the case with zirconium carbide, and amounts to

$$v^{\mathrm{VC}}_{\square\mathrm{C}} = (3.586 - 19.45n - 0.56n^2 + 6.95n^3) \times 10^{-30}m^3 .$$

The values of the bulk modulus K for the carbides $VC_{0.87}$ and $ZrC_{0.98}$ are equal to 3.98×10^{14} and 2.65×10^{14} Pa, respectively [14, 15], and the respective values of the mean linear thermal expansion coefficient for these carbides are 9×10^{-6} and $7 \times 10^{-6}\ \mathrm{K}^{-1}$ [15].

For the structural vacancies to be filled the synthesis of defect-free zirconium and vanadium carbides should obviously be performed involving a supply of the carbon deficit in these carbides.

Assuming reasonable values of the quantity n and temperature T, calculations based on (21) were carried out to determine the values of the pressure required to synthesize the compositions ZrC_{1-n} and VC_{1-n} when using $ZrC_{0.98}$ and $VC_{0.87}$ as initial carbides (see Table 1).

Table 1

Pressure P required to synthesize ZrC_{1-n} and VC_{1-n} with a given n at a given temperature T under thermodynamic-equilibrium conditions

n	zirconium carbide ZrC_{1-n}				vanadium carbide VC_{1-n}			
	filling factor b	pressure P (10^{11} Pa) T (K)			filling factor b	pressure P (10^{11} Pa) T (K)		
		1500	2000	2500		800	1000	1300
5×10^{-2}					0.615	30	70	120
1×10^{-2}	0.500	130	280	410	0.923	80	110	160
1×10^{-3}	0.950	310	520	730	0.992	140	190	260
1×10^{-4}	0.995	500	770	1050	1.0	210	270	370
1×10^{-5}	1.0	690	1030	1370	1.0	280	360	480
1×10^{-6}	1.0	880	1280	1680	1.0	350	450	600

As the calculation results show, with increasing pressure the structural vacancy concentration decreases quickly, asymptotically approaching zero. The initial vacancy concentrations are equilibrium concentration for $P = 0$ and for the temperatures chosen, i.e. the upper boundaries of the region of homogeneity coincide practically with the corresponding phase boundaries on the phase diagrams of the Zr–C and V–C-systems [8, 13]. For example, calculation according to (21) at $P = 0$ and $T = 1300$ K for the vanadium carbide yields $n = 0.135$, i.e. in accordance with the calculation, vanadium carbide $VC_{0.865}$ is in thermodynamic equilibrium under these conditions, which is practically in agreement with the phase diagram.

It follows from (21) that the pressure is determined chiefly by the relation between ΔH_i and ΔS_i, i.e. by the quantity ΔG_i since $\partial \Delta E/\partial n_i$ is small in absolute magnitude compared to ΔG_i. Taking account of the fact that the free energies of formation of structural vacancies in transition metal carbides and nitrides are close in magnitude [7, 12, 16], it may be expected that approximately equal conditions will be required for the synthesis of these compounds in a practically defect-free state. Indeed, comparison of the results obtained shows that the synthesis of zirconium and vanadium carbides possessing equal crystal lattice imperfections requires approximately equal conditions: for ZrC_{1-n} with $n = 10^{-3}$ the equilibrium temperature and the equilibrium pressure are equal to 1500 K and 3.07×10^{13} Pa, respectively, whereas for VC_{1-n} with $n = 10^{-3}$ a pressure of 2.58×10^{13} Pa is required at $T = 1300$ K.

Thus, simultaneous application of high pressures and high temperatures permits synthesis of compounds with crystal sublattice defect relations that are not realized under normal conditions. The temperature and pressure dependence, found in general form (taking into account the volume variation of a substance due to a change in defect content, due to compression under pressure and due to thermal expansion), of the equilibrium vacancy concentration of any nonequivalent sublattice of a compound, makes it possible to calculate the conditions of P–T treatment for any compound containing structural vacancies.

References

[1] A. Taylor and N. J. Doyle, Colloq. Internat. Centre Nation Rech. Sci. **188**, 261 (1970).
[2] M. D. Banus and T. B. Reed, in: The Chemistry of Extended Defects in Non-Metallic Solids, North-Holland Publ. Co., Amsterdam 1970 (p. 488).
[3] Y. Syono, T. Goto, J. Nakai, Y. Nakagawa, and H. Iwasaki, J. Phys. Soc. Japan **37**, 442 (1974).
[4] H. Iwasaki, Japan. J. appl. Phys. **10**, 1149 (1971).
[5] A. I. Gusev, Zh. fiz. Khim. **54**, 773 (1980).
[6] F. F. Voronov, Dokl. Akad. Nauk SSSR **226**, 1052 (1976).
[7] A. I. Gusev, Teplofiz. vysok. Temp. **17**, 1232 (1979).
[8] R. V. Sara, J. Amer. Ceram. Soc. **48**, 243 (1965).
[9] A. S. Bolgar, A. G. Turchanin, and V. V. Fesenko, Termodinamicheskie svoistva karbidov, Izd. Naukova Dumka, Kiev 1973.
[10] C. Jun and P. Shaffer, J. less-common Metals **23**, 367 (1971).
[11] H. Bittner and H. Goretzky, Monatsh. Chem. **93**, 1000 (1962).
[12] A. I. Gusev, Zh. fiz. Khim. **57**, 1382 (1983).
[13] E. K. Storms and R. J. McNeal, J. phys. Chem. **66**, 1402 (1962).
[14] G. V. Samsonov, G. Sh. Upadhaya, and V. S. Neshpor, Fizicheskoe materialovedenie karbidov, Izd. Naukova Dumka, Kiev 1974.
[15] A. I. Gusev, N. A. Ivanov, G. P. Shveikin, and P. V. Geld, Izv. Akad. Nauk SSSR, Ser. neorg. Mater. **12**, 1391 (1976).
[16] A. I. Gusev, Zh. neorg. Khim. **25**, 629 (1980).

(Received April 6, 1984)

phys. stat. sol. (a) 85, 167 (1984)

Subject classification: 11 and 20.1; 10.2; 22.5

General Physics Institute, Academy of Sciences of the USSR, Moscow[1]) *(a) and Moscow State University (b)*

Effect of Ionizing Radiation on Optical Properties of Thallium Halides in the 0.36 to 15 μm Range

By

V. G. Artyushenko (a), V. V. Voitsekhovskii (a), L. S. Kornienko (b), I. S. Lisitskii (a), and A. O. Rybaltovskii (b)

Absorption coefficients measured at 5 to 6 and 10.6 μm by CO- and CO_2-laser calorimetry of TlCl, KRS-6 (TlCl–TlBr), TlBr, and KRS-5 (TlBr–TlI) crystals are found to increase upon irradiation with ^{60}Co γ-rays. It is found that the radiation resistance of the crystals substantially increases with the halogen number. The dose dependence of γ-induced absorption spectra from 0.36 to 15 μm is measured for TlCl. The results of the measurements are explained by the formation of Tl aggregate centres.

Методом лазерной калориметрии измерен рост коэффициентов поглощения на длинах волн CO- и CO_2-лазеров в кристаллах TlCl, KPC-6(TlCl–TlBr), TlBr и KPC-5 (TlBr–TlI) с дозой γ-облучения ^{60}Co. Установлено, что радиационная стойкость кристаллов существенно возрастает с увеличением порядкового номера галогена. В TlCl измерена дозная зависимость спектров наведенного поглощения в диапазоне 0,36 до 15 μm. Полученные результаты объясняются образованием при облучении агрегатных центров Tl.

1. Introduction

Together with the evolution of powerful lasers and fibre optics for the middle IR range, materials were investigated having low losses in this spectral region. For a number of applications these materials should be resistant to ionizing irradiation. Recently the dependence of absorption for NaCl and KCl only at 10.6 μm upon the irradiation dose with ^{60}Co γ-rays [1, 2] and with high-energy electrons [3] was measured. It was found by the authors of these papers that the absorption at 10.6 μm increased upon irradiation, and that was associated with the occurrence of F aggregate centres. Among the thallium halides the TlCl resistance to X-ray irradiation in the visible region was investigated [4]. In those investigations the radiation colour centres (RCC) were not observed in pure TlCl upon X-ray irradiation at 300 and 70 K.

In our paper we present new results obtained for thallium halide crystals by way of measurements of the induced absorption in the range 0.36 to 15 μm, particularly at CO and CO_2 laser wavelengths, depending on the dose of γ-irradiation.

2. The Experimental Procedure

TlCl, KRS-6 (TlCl–TlBr), TlBr, and KRS-5 (TlBr–TlI) crystals for this study were grown by the Stockbarger method. The samples with dimensions $10 \times 10 \times 50$ mm^3 and $2.5 \times 10 \times 20$ mm^3 were cut and polished. In the samples of $10 \times 10 \times 50$ mm^3 the volume absorption coefficients were measured at CO (5 to 6 μm) and CO_2 (10.6 μm) lasers wavelengths by the laser calorimetry method [5]. The transmission spectra were measured by a Fourier spectrometer Bruker IFS-115s in the range 2.5 to 15 μm. The

[1]) ul. Vavilova 38, 117942 Moscow, USSR.

induced absorption (IA) spectra were measured at 0.2 to 2.5 μm with the spectrometer SF-8 in the samples with thickness 2.5 mm. The irradiation was made by the ^{60}Co source with γ-rays of energy 1.2 MeV and the dose rate 450 Gy/s.

3. Results and Discussion

The irradiation of 10^2 to 10^5 Gy at room temperature changed the absorption coefficients measured at the laser wavelengths for no more than a factor 2 in all the samples of thallium halides. The changes could be associated with RCC occuring after γ-irradiation for which uncontrolled impurities in the crystal matrix [6, 7] were responsible. It was known that the initial absorption in the investigated samples (1 to 3) × $\times\ 10^{-4}\ cm^{-1}$ was associated mainly with the absorbing impurities but not with the fundamental optical losses, which could be as low as 10^{-8} to $10^{-6}\ cm^{-1}$ at 5 to 11 μm [8].

The irradiation at 300 K by large doses 10^5 to 10^7 Gy significantly influenced the measured absorption coefficients (Fig. 1). The crystals (two samples of each thallium halide) were irradiated and measured five times one after another. Note that one cycle of irradiation and the following measurement lasted for nearly a week. Fig. 1 shows that the absorption at 5 to 6 and 10.6 μm increases upon irradiation and the radiation resistance significantly increases in the series of crystals TlCl, KRS-6, TlBr, KRS-5.

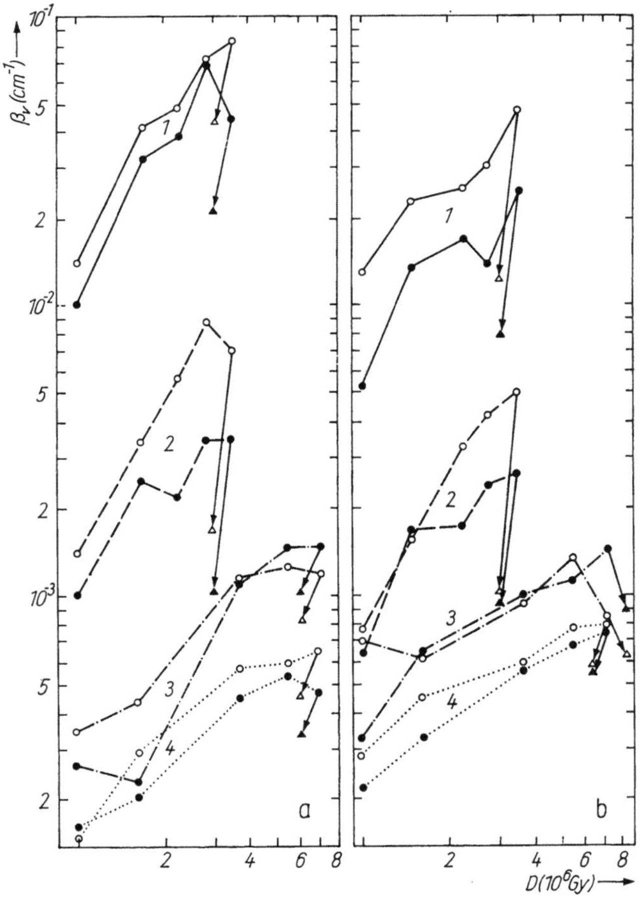

Fig. 1. a) 5 to 6 μm and b) 10.6 μm volume absorption coefficients for No. 1 (○) and No. 2 (●) samples of TlCl (1) KRS-6 (2), TlBr (3), KRS-5 (4) crystals as a function of the irradiation dose. The β_v reduction 10 weeks after the irradiation is shown by the arrows. $\Delta\beta_v/\beta_v \approx 20\%$

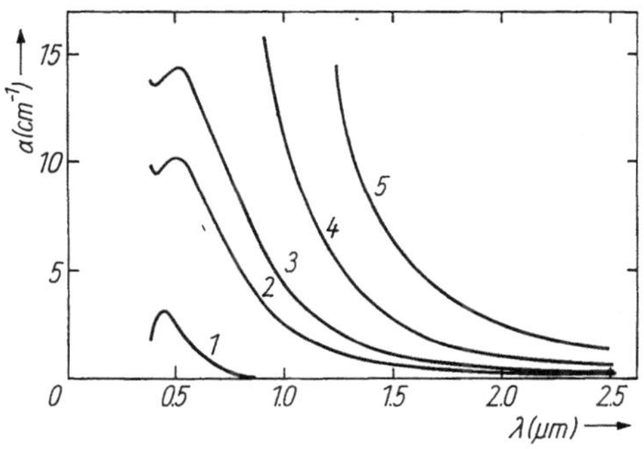

Fig. 2

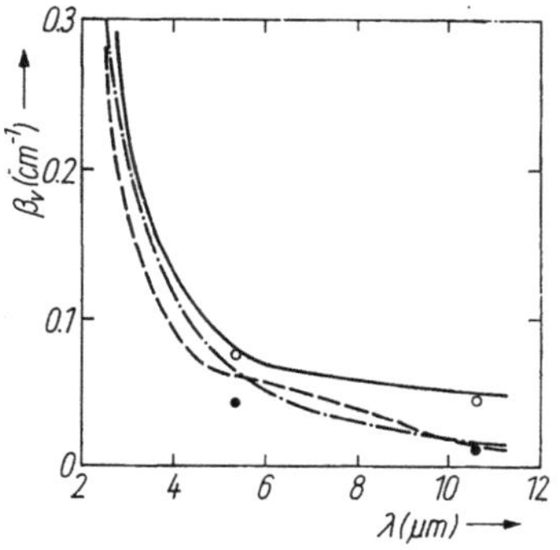

Fig. 3

Fig. 2. Induced absorption spectra in γ-irradiated TlCl crystals at 300 K: (1) 10^5, (2) 4×10^5, (3) 10^6, (4) 2.2×10^6, and (5) 4×10^6 Gy

Fig. 3. IA spectra of two TlCl crystal samples irradiated by 4×10^6 Gy (——— and – – –) λ^2 curve (—·—) and IA of these samples (○, ●) obtained by laser calorimetry

In the TlCl crystal the absorption increases by more than two orders upon irradiation of 3×10^6 Gy and in the KRS-5 crystal only by 2 to 3 times upon irradiation of 7×10^6 Gy. The variation of the absorption coefficient values in different samples of the same crystal is due to the variation of uncontrolled impurities and structure defects on which the RCC occurs first of all. As the RCC are formed mostly in TlCl crystals, we have investigated their optical characteristics in detail.

The γ-irradiation of TlCl crystals at room temperature with the dose $\geq 10^5$ Gy caused grey colour in the visible range. The colour density increased with the dose, and radiation decoration of structure defects, such as microcracks and damaged surface layer, was observed. Sample annealing at 500 to 550 K preceding the irradiation resulted in a more uniform RCC distribution. This could be associated with the thermal relief of inner strains and with the cure of a number of structure defects.

Fig. 2 shows the IA spectra for TlCl crystals irradiated with various doses at room temperature. IA increases with the irradiation dose in the range 0.36 to 2.5 μm and its spectrum consists of a band the maximum of which shifts from 0.44 μm at a dose of 10^5 Gy to 0.52 μm at 10^6 Gy, and of a short-wave tail, which can be related to the band with the maximum in the range < 0.36 μm as well as to the electron absorption edge shifted upon the γ-irradiation [9]. The spectrum analysis shows that the IA increase with dose is determined mainly by the increase of the short-wave tail. In order to clear up the nature of the tail it is necessary to measure the transmission spectrum at wavelengths shorter than 0.36 μm, that is in the range where the fundamental absorption is high. The intensity increase of the band 0.44 to 0.52 μm slows down with dose. Note, that mainly its long-wave tail grows. The IA spectrum in the middle IR range (2.5 to 12 μm) is measured for two samples of TlCl crystals only for the dose 4×10^6 Gy (Fig. 3). It is clear from the figure that the absorption coefficients at 5 to 6 and 10.6 μm measured by the laser calorimetry method coincide well with the results obtained by the Fourier spectrometer. However, it is necessary to note, first, that for $\lambda = 2.5$ μm IA measured by the transmission spectra of 2.5 mm thick samples exceeds IA measured for 50 mm long samples by a few times. Second, the IA spectrum extrapolation from the range 0.36 to 2.5 μm into the range 2.5 to 12 μm provides higher absorption in the middle IR range. The main reason for this is, in our mind, that the surface absorption for thin samples has become more important.

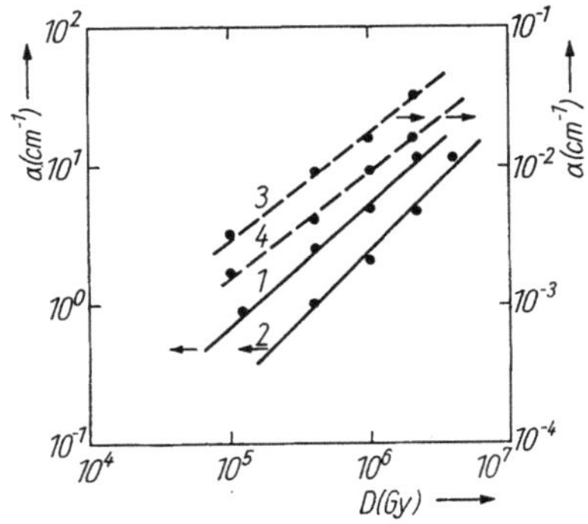

Fig. 4. IA value as a function of the irradiation dose for TlCl samples in the points λ = (1) 1.0, (2) 1.3, (3) 5 to 6, and (4) 10.6 μm

The investigation of IA increase upon irradiation dose in definite points of the spectrum, i.e. λ = 1.0, 1.3, 5 to 6, and 10.6 μm, shows that for these wavelengths the increase has the form $\sim D^n$, where $n = 0.8$ to 1.0 for these irradiation doses (Fig. 4). Thus, we can suppose that IA in the middle IR range is mostly associated with the same RCC as those determining IA in the near IR range. This supposition is confirmed by the fact that IA thermobleaching in TlCl samples in the near and middle IR ranges occurs in the same temperature interval. It should be pointed out that the effective bleaching temperature in the range 0.36 to 2.5 μm depends on the irradiation dose and varies from 500 K at 7×10^5 Gy to 560 K at 2×10^6 Gy.

At room temperature the RCC reformation rate — that is aggregation, desintegration, etc. — is much slower than at the temperature of the effective bleaching. But IA changes significantly in the irradiated samples during a few months of investigations. Fig. 1 shows the reduction of IA coefficients in all the samples of thallium halide crystals ten weeks after the maximum irradiation. But the character of the IA spectrum in the middle IR range in TlCl and KRS-6 does not change. As the sample transparency increases also in the visible range in KRS-6 samples irradiated with the dose 4×10^6 Gy and kept at room temperature during a month, block boundaries and dislocations decorated by dark microinclusions have been definitely observed and after about three months the samples become practically transparent. During that period TlCl samples also become practically transparent. For a comparison we should note that the induced absorption in thallium halide crystals decreases upon irradiation and in alkali-halide crystals at 10.6 μm it increases [1]. According to our data, the volume absorption coefficients at 10.6 μm in KCl and NaCl increase during ≈ 1.5 months (at dose 10^6 Gy), and after that decrease during a few months to their initial values. If the supposition that the absorption peak at 10.6 μm is associated with the aggregate F-centres in these crystals [1, 2] is correct, then the IA reduction with time can be associated with the dissociation of these centres.

TlCl crystal irradiation with the dose 10^6 Gy at 80 K does not lead to their colouring in the visible range, but after heating up to room temperature, they possess the characteristic colour with practically the same IA spectrum as in the case of irradiation at 300 K but with a dose which is a few orders smaller. This fact shows that in TlCl crystals at 80 K γ-irradiation transforms the initial RCC reformed by heating up to 300 K into colour centres characteristic of irradiation at room temperature.

The measurement results of the optical losses induced by γ-irradiation in thallium halide crystals can be clearly explained by the occurrence of colloid centres. There are some data on the investigation of colloid centres in alkali halide crystals cited in [10, 11] and in AgCl crystals cited in [12], concerning the absorption band shape and the location of these centres. By analogy with [10 to 12] the peak in the TlCl IA spectrum at 0.46 μm and its shift up to 0.52 μm upon the higher irradiation doses can be explained by the occurrence of Tl colloid particles, tending to growth with increasing dose. The increase the annealing temperature of IA with the dose can be also explained by the colloid growth. Besides, in the samples with high IA a Tyndall scattering cone is observed when a He–Ne laser beam (0.63 μm) is transmitted, owing to the presence of scattering colloid particles [13]. The decoration of microcracks and spoilt surface layers observed in the samples occurs, as shown in [10], by way of colloid formation mostly on structure defects. At low temperatures ($\approx$ 80 K) the formation of colloid particles is practically impossible and occurs only when the irradiated samples are heated up to 300 K. Colloid formation associated with polyvalent thallium ions [14] has, in our mind, the following radiolysis mechanism:

$$3\,\mathrm{TlCl} \xrightarrow{\gamma} \mathrm{Tl_2Cl_3} + \mathrm{Tl} \xrightarrow{\gamma} \mathrm{TlCl_3} + 2\,\mathrm{Tl}\,. \quad (1)$$

Another mechanism can possibly take place together with the first one,

$$2\,\mathrm{TlCl} \xrightarrow{\gamma} 2\,\mathrm{Tl} + \mathrm{Cl_2}\,. \quad (2)$$

There are some reasons to suppose that the second mechanism can be realized only in the spoilt surface layer, as we have observed no EPR V_k-centre spectra from which the halogen molecule is usually formed [15] in TlCl samples irradiated and investigated at 80 K.

We consider that if resistivity of thallium halides to radiation is associated with such secondary RCC as colloid particles, then the increase of the resistivity is due to the corresponding decrease of the primary RCC accumulation rate. The nature of primary RCC in thallium halides is not yet clear.

Acknowledgement

The authors want to thank E. M. Dianov and E. E. Kudryavtseva for helpful discussions.

References

[1] H. G. Lipson, P. Ligor, and J. J. Martin, phys. stat. sol. (a) **37**, 547 (1976).
[2] A. V. Gektin, T. A. Charkina, and N. V. Shiran, Kvantovaya Elektronika **8**, 2237 (1981).
[3] T. J. Magee, N. Johnson, and J. Peng, phys. stat. sol. (a) **30**, 81 (1975).
[4] R. W. Christy and J. D. Dimock, Phys. Rev. **141**, 806 (1966).
[5] V. G. Artyushenko, E. M. Dianov, and E. N. Nikitin, Kvantovaya Elektronika **5**, 1065 (1978).
[6] P. I. Berezhinskĭĭ, L. M. Belokrinitskaya, I. F. Pryadko, and I. N. Khalimonova, Fiz. tverd. Tela **23**, 3677 (1981).
[7] P. D. Alekseev and G. I. Baranov, Fiz. tverd. Tela **22**, 1213 (1980).
[8] J. A. Harrington, Opt. Spectra **2**, 39 (1981).
[9] F. Seitz, Rev. mod. Phys. **23**, 328 (1951).
[10] V. I. Gotlib, Ya. Zh. Kristapsen, K. K. Shvarts, and Yu. A. Ekmanis, Radiat. Fiz. (Riga) **7**, 143 (1973).
[11] Yu. A. Ekmanis and I. S. Radchenko, phys. stat. sol. (b) **115**, 389 (1983).
[12] W. Kaiser, Z. Phys. **132**, 497 (1952).
[13] A. A. Blistanov, V. V. Geraskin, N. S. Kozlova, and S. A. Zubareva, Fiz. tverd. Tela **18**, 2279 (1976).
[14] V. N. Fadeev and P. I. Fedorov, Zh. neorg. Khim. **9**, 2028 (1964).
[15] T. N. Nurahmetov and A. A. Elango, Trudy Inst. Fiz. Akad. Nauk Eston SSR **49**, 1 (1979).

(Received July 13, 1984)

phys. stat. sol. (a) 85, 173 (1984)

Subject classification: 18.2.1; 22.8.2

Czechoslovak Technical University, Deptartment of Physics K 2020, Prague (a)[1])
and Institute of Physics, Czechoslovak Academy of Sciences, Prague (b)[2])

Yttrium Iron Garnet Films Substituted by Gd and Mn

By

J. ČERMÁK (a), R. NOVÁK (a), P. NOVÁK (b), and M. NEVŘIVA (b)

Mn substituted YIG and (Y, Gd)IG systems are studied in order to investigate mechanisms of the magnetic anisotropy of Mn-containing garnet layers. The phase diagrams for YIG system substituted by Mn ions are presented for given Y concentration. Results obtained for the growth parameters, Gd and Mn segregation, and temperature dependence of magnetic anisotropy are compared with appropriate models and discussed.

Для исследования влияния марганца на анизотропию были подготовлены тонкие пленки $Y_3Fe_5O_{12}$ содержающие марганец и гадолиний. Фазообразование из раствора-расплава $PbO-B_2O_3$ определено для постоянной концентрации иттрия. Получены результаты для сегрегации гадолиния и марганца. Температурная зависимость магнитной анизотропии сопоставлена с теорией и дискутирована.

1. Introduction

In YIG films containing octahedrally coordinated Mn^{3+} ions a special type of non-cubic anisotropy appears which is connected with a strong Jahn-Teller effect. In a simplified approach this effect leads to a tetragonal deformation of each Mn^{3+} ion octahedral surrounding [1, 2]. As there are three possibilities of tetragonal axis directions, three states of each Mn^{3+} ion must be considered. In an ideal YIG structure the three states are equivalent but this is not the case in epitaxial films where the lattice constant misfit produces additional deformation. The states corresponding to the elongated or contracted octahedron due to the deformation exhibit a lower respectively higher energy compared with the states of the non-deformed system. Consequently Mn^{3+} local Jahn-Teller distortions will be ordered due to the misfit strain. With low Mn^{3+} concentration and at absolute zero temperature the ordering is complete, at higher temperatures the ordering is only partial. Inhomogeneities of strain caused by lattice defects (vacancies, impurities including also Mn^{3+} themselves at higher concentrations, etc.) also lower the ordering.

A summary of the theoretical prediction of non-cubic anisotropy is given for individual crystallographic directions of Mn-containing films with respect to the misfit orientation in Table 1. These values indicate the fact the Mn-substituted garnet films could provide a suitable material for submicron bubble memories. In addition Mn substitution produces only little damping and therefore a high bubble mobility can be expected, especially for (110) films with orthorhombic anisotropy.

As follows from Table 1 the proper value of the lattice constant misfit is necessary to obtain the required magnetic anisotropy. Usually a positive misfit, i.e. the film is under pressure due to its larger lattice constant than that of the substrate, is obtained by Gd substitution into dodecahedral sites where only a low damping is introduced by Gd ions.

---

[1]) Suchbátarova 4, 16607 Prague 6, Czechoslovakia.

[2]) Na Slovance 2, 18040 Prague 8, Czechoslovakia.

Table 1
Theoretically predicted values of anisotropy for $Y_{3-x}Gd_xFe_{5-y}Mn_yO_{12}$ garnet films

plane	(001)		(111)		(110)	
sign μ	pos.	neg.	pos.	neg.	pos.	neg.
$T = 0$ K						
K_u/y	−1.7	3.2	−1.3	1.8	−3.1	2.6
K_i/y	—	—	—	—	3.1	−0.5
$T = 300$ K						
$K_u/y\ \lvert\mu\rvert$	−47	52	−23	24	−36	20
$K_i/y\ \lvert\mu\rvert$	—	—	—	—	14	−20

Values of K_u and K_i are in 10^6 J/m³ units; $\mu = 1 - a_f/a_s$, where a_f and a_s are the lattice constants of film and substrate, respectively. For K_u and K_i calculation the data of Bates and Wardlaw [3] were used. Room temperature values of anisotropy were evaluated using the results of Breed et al. [4].

2. Experimental

Epitaxial Mn^{3+} substituted (Y, Gd)IG films were grown by a conventional LPE isothermal method from 16 different $PbO-B_2O_3$ based melts in order to find a region of primary garnet crystallization and to obtain low defect films with given composition of Gd and Mn. During the film growth the substrates were rotated at a constant rate of 100 to 150 rpm. The growth temperatures ranged from 880 to 950 °C at supercooling of 1.2 to 57 K. Corresponding growth rates fell within the limits 0.03 to 2.20 μm/min. GGG substrates of 22 mm in diameter were cut in (111), (110), and (100) planes.

The film compositions were analysed by a Jeol JXA-50A electron microprobe at 15 kV using YIG, GGG, $MnFe_2O_4$, and PbSe single crystals as standards. The relative accuracy of calculated concentrations was better than ±6% except for Pb where the accuracy was ±15%.

Lattice constant misfits were measured by double-crystal X-ray diffraction and the misfit homogeneity was checked by double-crystal X-ray topography [5].

The temperature dependence of the magnetic anisotropy was studied by means of the ferromagnetic resonance method at 25.4 GHz within the range 4.2 to 300 K. The anisotropy constant, K_u, was calculated using the known temperature dependence of the magnetization of pure YIG [6].

3. Results and Discussion

A survey of the melt compositions and corresponding liquidus temperatures is given in Table 2. The R-coefficients used for the melt characterization are defined by the following atomic ratios:

$$\begin{aligned} R_1 &= Fe_2O_3/(Y_2O_3 + Gd_2O_3)\,, \\ R_2 &= 2\ Fe_2O_3/MnCO_3\,, \\ R_3 &= PbO/2\ B_2O_3\,, \\ R_4 &= \text{garnet (atomic)/melt (atomic)}\,, \\ R_5 &= (2\ Fe_2O_3 + MnCO_3)/2\ (Y_2O_3 + Gd_2O_3)\,. \end{aligned} \tag{1}$$

The following definitions of the R-coefficients, similar to those in [7, 8], are very advantageous for the description of the fields of primary crystallization of the individual

Table 2

Survey of selected melt compositions, the corresponding liquidus temperatures, T_L, and designations of primary crystallizing phases (g — garnet, o — orthoferrite, and h — hexaferrite)

melt	R_1	R_2	R_3	R_4	R_5	T_L (°C)	phase
49	20.3	20.3	7.43	0.211	42.7	933.5	g
53	20.3	9.07	7.43	0.220	45.1	950.0	g
54	20.3	3.90	7.43	0.241	51.1	925.0	h
56	18.9	3.90	7.52	0.226	47.4	>1100	h
57	17.6	3.90	7.59	0.214	44.3	923.0	o + g
58	16.5	3.90	7.66	0.202	41.5	920.0	o
59	18.7	4.41	7.61	0.218	45.8	>1100	h
60	18.7	4.41	7.64	0.202	45.8	?	o
61	20.0	9.23	7.64	0.218	44.3	>1100	h
64	24.6	20.4	7.43	0.181	25.8	894.0	g
65	17.5	20.4	7.43	0.183	18.3	952.0	g

phases:

$$\begin{aligned} R_Y &= 2\,Y_2O_3/(2\,B_2O_3 + PbO)\,,\\ R_{Gd} &= 2\,Gd_2O_3/(2\,B_2O_3 + PbO)\,,\\ R_{YG} &= R_Y + R_{Gd}\,,\\ R_{Fe} &= 2\,Fe_2O_3/(2\,B_2O_3 + PbO)\,,\\ R_{Mn} &= MnCO_3/(2\,B_2O_3 + PbO)\,. \end{aligned} \qquad (2)$$

The phase relations for the investigated garnet system are illustrated in Fig. 1 and the corresponding melt compositions using the R-coefficients according to formulae (2) are summarized in Table 3.

The R_Y and R_{YG} coefficients were nearly constant for all melts and equal to that published in [4]. The Mn concentration in the melt was gradually changed and in this manner the required Mn concentrations in the film ($y = 0.15$, 0.32, and 0.50) were obtained.

The characteristic Gd content (for zero growth rate) in the films grown from the melts No. 64 and 65 was 0.50 and 0.76, respectively, and satisfied the requirement of a positive misfit between the lattice constants of the grown film and the substrate.

The relative distribution coefficients of Gd and Mn were calculated according to the

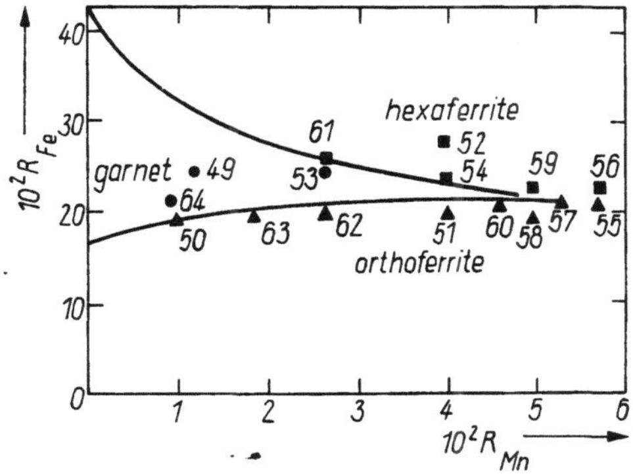

Fig. 1. R_{Fe}–R_{Mn} diagram of the garnet system $(Y, Gd)_3(Fe, Mn)_5O_{12}$

Table 3

Survey of selected melt compositions expressed by the R_Y, R_{Gd}, R_{YG}, R_{Fe}, and R_{Mn} coefficients (× 100)

melt	R_Y	R_{Gd}	R_{YG}	R_{Fe}	R_{Mn}
49	1.10	0	1.20	24.3	1.20
53	1.20	0	1.20	24.3	2.68
54	1.20	0	1.20	24.3	6.23
56	1.18	0	1.18	22.3	5.72
57	1.17	0	1.17	20.7	5.30
58	1.16	0	1.16	19.2	4.93
59	1.17	0	1.17	21.8	4.95
60	1.06	0	1.06	19.8	4.49
61	1.20	0	1.20	24.0	2.60
64	0.615	0.208	0.823	20.3	0.993
65	0.951	0.208	1.16	20.3	0.993

following relations [9]:

$$K\ (\mathrm{Gd}) = \frac{x_0}{3 - x_0} \frac{[\mathrm{Y}]}{[\mathrm{Gd}]} \tag{3}$$

and

$$K\ (\mathrm{Mn}) = \frac{y_0}{3 - y_0} \frac{[\mathrm{Fe}]}{[\mathrm{Mn}]}, \tag{4}$$

where x_0 and y_0 are the numbers of atoms per formula unit incorporated into the film at zero growth rate and [Gd], [Y], [Fe], and [Mn] are the cation fractions of the corresponding elements in the melt.

The solubility of the investigated garnet systems in the melt was interpreted according to van Erk's model [10]. The garnet concentration, L, was evaluated using the definition

$$L = \frac{[\mathrm{Y}]^3\,[\mathrm{Fe}]^5}{\left(1 - \frac{x}{3}\right)^3 \left(1 - \frac{y}{3}\right)^3}. \tag{5}$$

The standard formation enthalpy ΔH was calculated in compliance with van Erk's theory [10] from

$$\frac{\Delta H}{RT_{\mathrm{L}}} - \frac{\Delta S}{R} = \ln L\,, \tag{6}$$

Table 4

Distribution coefficients and concentration of garnet in the melt, L, expressed using equation (5)

melt	T_L	$\ln L$	K (Gd)	K (Mn)
49	933.5	−19.30	—	0.747
53	950.0	−19.18	—	0.757
57	923.0	−19.77	—	0.633
64	894.0	−21.54	0.334	0.689
65	952.0	−19.86	0.876	1.056

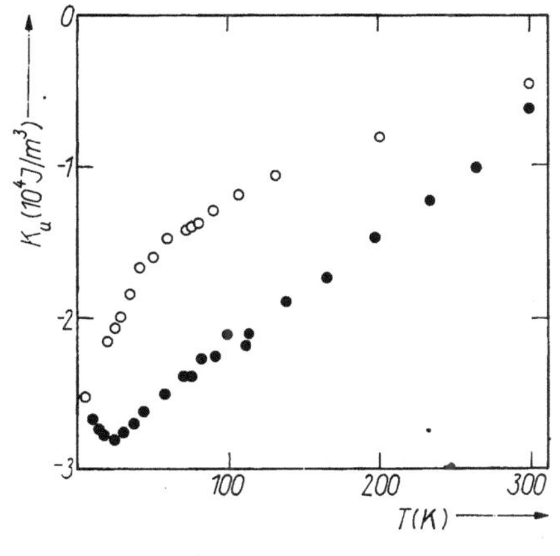

Fig. 2. Temperature dependence of K_u in Gd and Mn substituted YIG films; y = ○ 0.15, ● 0.32

ΔS being the standard formation entropy and R the universal gas constant. The results are summarized in Table 4. The value of the formation enthalpy for Gd and Mn-substituted YIG and $PbO–B_2O_3$ based flux is —392 kJ/mol. This is in good agreement with the values published for pure YIG by van Erk [10] (—335 to —350 kJ/mol).

Experimental dependences of magnetic anisotropy on temperature for two different Mn concentrations are presented in Fig. 2. Comparison of these results with the theory [2] is rather complicated because no information about strain variations due to defects in the films was available. However, qualitative agreement between theory and experiment was obtained. Moreover, the films with a predicted perpendicular anisotropy exhibit a stripe domain structure having a period of 1.2 μm.

Acknowledgements

The authors are indebted to Dr. S. Krupička for fruitful discussions, to Dr. J. Šimšová for performing the electron probe microanalysis, and to Dr. J. Kub for the lattice constant misfit measurement.

References

[1] P. Novák, Czech. J. Phys. **B21**, 1197 (1971).
[2] P. Novák, M. Maryško, S. Krupička, J. Kub, J. Šimšová, J. Čermák, and M. Nevřiva, phys. stat. sol. (a) **80**, K213 (1982).
[3] C. A. Bates and R. S. Wardlaw, J. Phys. C **13**, 3609 (1980).
[4] D. J. Breed, J. M. Robertson, H. A. Algra, B. A. H. van Bakel, W. de Geus, and J. P. H. Heynen, Appl. Phys. **24**, 163 (1981).
[5] O. Nittono and S. Shimizu, J. Crystal Growth **45**, 476 (1978).
[6] J. Čermák, J. Kub, M. Maryško, M. Nevřiva, P. Novák, L. Půst, J. Šimšová, and W. Schüppel, Internat. Symp. Jahn-Teller Effect, Tue-P6, Liblice (1983).
[7] J. M. Robertson, W. Tolksdorf, and H. D. Jonker, J. Crystal Growth **27**, 241 (1974).
[8] M. Nevřiva, R. Novák, and J. Čermák, J. Crystal Growth **66**, 221 (1984).
[9] C.-P. Klages and W. Tolksdorf, J. Crystal Growth, to be published.
[10] W. van Erk, J. Crystal Growth **46**, 539 (1979).

(Received June 18, 1984)

phys. stat. sol. (a) 85, 179 (1984)

Subject classification: 18.2.1; 19; 22.8.2

Institute of Physics, Czechoslovak Academy of Sciences, Prague[1])

Method for Measurement of Magnetostriction Constants in (001) Thin Films Using FMR

By

L. PŮST and Z. FRAIT

FMR formulas for (001) cubic films with an applied uniaxial or isotropic in-plane stress σ are derived for various film orientations toward the external magnetic field H. The resonance field shift δH caused by σ is explicitly evaluated including quadratic terms ($\sim \sigma^2$) and the magnetocrystalline anisotropy effect. Measurements of δH performed on LPE garnet thin films (composition $Y_3Fe_{4.87}Mn_{0.13}O_{12}$) are well described by the theory.

Es werden FMR-Formeln für eine kubische Schicht unter dem Einfluß einer in der Schichtebene wirkenden einachsigen oder isotropen Spannung σ für verschiedene Orientierungen der Schicht im äußeren Magnetfeld abgeleitet. Die durch σ verursachte Verschiebung δH des Resonanzfeldes wird berechnet, wobei quadratische Glieder ($\sim \sigma^2$) und der Einfluß der magnetokristallinen Anisotropie einbezogen werden. An dünnen LPE-Granatschichten (Zusammensetzung $Y_3Fe_{4,87}Mn_{0,13}O_{12}$) durchgeführte Messungen von δH werden durch die Theorie gut beschrieben.

1. Introduction

Ferromagnetic resonance (FMR) measurements in thin films with applied external stress enable the determination of the magnetostriction constants (λ_{100} and/or λ_{111}); usually a simple experimental configuration with external magnetic field H and applied stress σ parallel to principal crystallographic directions is used and the dependence of the resonance field on σ is considered in a simple linear approximation [1 to 5], while the possible deviations from the linear dependence between resonance field shift and applied stress is not considered for stress values reached in experiments. Recently a more detailed theoretical analysis of FMR in (111) thin films with applied stress was published [6]. The aim of this paper is to evaluate resonance conditions and expressions for the resonance field shift due to an applied stress up to the quadratic approximation for various directions of applied stress and static magnetic field, for the case of thin cubic epitaxial (001) films, and to compare them with experimental data.

Resonance conditions are derived from the general expression

$$\left(\frac{\omega}{\gamma}\right)^2 = \frac{1}{(M_s \sin \vartheta)^2} \left(F_{\varphi\varphi} F_{\vartheta\vartheta} - F_{\varphi\vartheta}^2\right) , \qquad (1)$$

where ω is the angular frequency of microwave radiation, γ the spectroscopic splitting ratio, φ and ϑ are spherical coordinates of the saturation magnetization vector M_s; $F_{\varphi\varphi}$, $F_{\vartheta\vartheta}$, and $F_{\varphi\vartheta}$ are the second partial derivatives of the free energy of the system evaluated at the equilibrium direction of magnetization (i.e. for $F_\varphi = F_\vartheta = 0$).

A dimensionless formulation has a great advantage in simplifying the problem [6]. Using $\Omega = \omega/(\gamma\mu_0 M_s)$ and $\Phi = F/(\mu_0 M_s^2)$ we obtain the dimensionless resonance

[1]) Na Slovance 2, 18040 Prague 8, Czechoslovakia

condition

$$\Omega^2 = \frac{1}{(\sin\vartheta)^2}\,(\Phi_{\varphi\varphi}\Phi_{\vartheta\vartheta} - \Phi^2_{\varphi\vartheta})\,. \tag{2}$$

Here the free energy Φ includes the Zeeman, uniaxial anisotropy, magnetostatic, magnetocrystalline anisotropy, and stress terms; they will be evaluated in the following sections.

2. Resonance Conditions

2.1 Parallel configuration

2.1.1 General in-plane direction of H_e and σ

We shall investigate the case of an external magnetic field H lying in the film plane xy (see Fig. 1) and making an angle $\bar{\varphi}$ with the x-axis. An uniaxial stress σ is oriented in the film plane with angle φ_s toward the x-axis.

The Zeeman term has the form

$$\Phi^z = -\eta \sin\vartheta \cos(\varphi - \bar{\varphi})\,, \tag{3}$$

where $\eta = H/M_s$ is the dimensionless magnetic field. The uniaxial anisotropy pointing normally to the film plane and the magnetostatic part of the free energy can be written as

$$\Phi^{um} = \frac{m_e}{2}\cos^2\vartheta\,, \tag{4}$$

where $m_e = 1 - 2K_u/(\mu_0 M_s^2)$ is the dimensionless effective magnetization, K_u the constant of uniaxial anisotropy ($K_u > 0$ and $K_u < 0$ corresponds to the case of the easy axis and easy plane, respectively). The magnetocrystalline anisotropy energy is

$$\Phi^a = (b_1 + b_2\cos^2\vartheta)\sin^4\vartheta\sin^2\varphi\cos^2\varphi + b_1\cos^2\vartheta\sin^2\vartheta\,, \tag{5}$$

where $b_i = K_i/(\mu_0 M_s^2)$ are the dimensionless magnetocrystalline anisotropy constants, K_i the anisotropy constants.

The magnetoelastic energy will be expressed in terms of the magnetostriction constants λ_{100} and λ_{111} as

$$\Phi^s = -3S_0(\alpha_x^2\gamma_y^2 + \alpha_y^2\gamma_y^2 + \alpha_z^2\gamma_z^2) - 6S_1(\alpha_x\alpha_y\gamma_x\gamma_y + \alpha_y\alpha_z\gamma_y\gamma_z + \alpha_z\alpha_x\gamma_z\gamma_x)\,, \tag{6}$$

where

$$S_0 = \frac{\sigma\lambda_{100}}{2\mu_0 M_s^2}\,,\qquad S_1 = \frac{\sigma\lambda_{111}}{2\mu_0 M_s^2}\,, \tag{7}$$

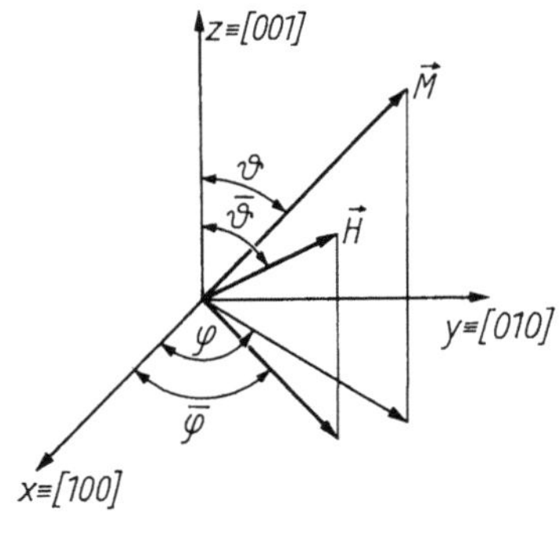

Fig. 1. The coordinate system

and α_i and γ_i are the direction cosines of M_s and the tensile stress σ, respectively. Formula (6) is strictly valid only in the zero limit of λ_{100} and λ_{111}. The total dimensionless energy of the system Φ is the sum of (3), (4), (5), and (6).

We suppose that the external magnetic field is high enough to maintain the single-domain state in the film with the equilibrium direction of $\overline{M}_s$ in the film plane. Because of the symmetry in the film plane $\vartheta = \pi/2$, but $\delta_\varphi = \varphi - \overline{\varphi}$ is a small, generally non-zero value. Then $\sin \delta_\varphi \doteq \delta_\varphi$, $\cos \delta_\varphi \doteq 1 - \delta_\varphi^2/2$ and all terms containing third powers of S_i and/or b_i can be neglected,

$$S_i^3 \approx 0, \qquad S_i^2 b_j \approx 0, \qquad S_i b_j^2 \approx 0, \qquad b_j^3 \approx 0 . \tag{8}$$

The first partial derivatives of Φ can be written as

$$\Phi = 0, \qquad \Phi_\varphi = z_0 + \delta_\varphi(\eta + z_1) + \dots , \tag{9a}$$

where

$$z_0 = \frac{b_1}{2} \sin 4\overline{\varphi} - 3S_1 \cos 2\overline{\varphi} \sin 2\varphi_s + 3S_0 \sin 2\overline{\varphi} \cos 2\varphi_s , \tag{9b}$$

$$z_1 = 2b_1 \cos 4\overline{\varphi} + 6S_1 \sin 2\overline{\varphi} \sin 2\varphi_s + 6S_0 \cos 2\overline{\varphi} \cos 2\varphi_s . \tag{9c}$$

Respecting the quadratic approximation in S_i and b_i (8) the condition $\Phi = 0$ yields

$$\delta_\varphi = -\frac{z_0}{\eta}\left(1 - \frac{z_1}{\eta}\right). \tag{10}$$

Using the above-mentioned substitutions and approximations we obtain

$$\Phi_{\varphi\vartheta} = 0 , \tag{11a}$$

$$\Phi_{\varphi\varphi} = \eta + 2b_1 \cos 4\overline{\varphi} + 6(S_1 \sin 2\overline{\varphi} \sin 2\varphi_s + S_0 \cos 2\overline{\varphi} \cos 2\varphi_s) + \frac{(B_1 + 6Q)(31B_1 + 42Q)}{8\eta} , \tag{11b}$$

$$\Phi_{\vartheta\vartheta} = \eta + m_e + \frac{3}{2} b_1 + \frac{b_2}{4} + \cos 4\overline{\varphi}\left(\frac{b_1}{2} - \frac{b_2}{4}\right) + 3S_0 + 3(S_1 \sin 2\overline{\varphi} \sin 2\varphi_s + S_0 \cos 2\overline{\varphi} \cos 2\varphi_s) + \frac{1}{2\eta} B_1\left(\frac{7}{4} B_1 - B_2\right) + Q(15B_1 - 6B_2) + 27Q^2 , \tag{11c}$$

where

$$B_i = b_i \sin 4\overline{\varphi} , \tag{12a}$$

$$Q = S_0 \sin 2\overline{\varphi} \cos 2\varphi_s - S_1 \cos 2\overline{\varphi} \sin 2\varphi_s . \tag{12b}$$

When evaluating magnetostriction constants only the resonance field shift (in the dimensionless form expressed as $\delta\eta = \eta - \eta_0$, where η_0 is dimensionless resonance field without σ) caused by an applied stress σ is measured, this shift being small compared to the resonance field. In the course of measurements the uniaxial stress σ is applied in the film plane in one direction determined by the sample shape, this direction being usually parallel to the main crystallographic directions ⟨100⟩ or ⟨110⟩. In the following sections we shall express analytically the resonance field shift for these special directions of σ as the first two terms in the power series of σ (in the dimensionless expression σ is included in S_0 and S_1, (7)). The linear term (in many cases the dominant one) proportional to σ includes also the influence of magnetocrystalline anisotropy via b_1 and b_2.

2.1.2 Uniaxial stress σ parallel to [100]

σ parallel to [100] corresponds to $\varphi_s = 0$. As $\sin 2\varphi_s = 0$ the resonance field in this case is a function of S_0 (corresponding to λ_{100}) only. The resonance field η will be expressed as a sum of the resonance field without stress η_0 and the stress-dependent term $\delta\eta = \eta - \eta_0$. Using substitutions

$$\omega' = \sqrt{\left(\Omega^2 + \frac{m_e^2}{4}\right)}, \qquad \alpha = \frac{m_e}{2\omega'} \tag{13}$$

and approximation (8), the shift of the resonance field can be written as

$$\delta\eta = 3S_0[-2 + (3 + \alpha)\sin^2\bar{\varphi} + Z_b] + 9S_0^2 \sin^2\bar{\varphi} \times \\ \times \left[\frac{1 - \alpha^2}{2\omega'} - \cos^2\bar{\varphi}\left(\frac{14\alpha}{\eta_0} + \frac{21 - \alpha^2}{2\omega'}\right)\right], \tag{14a}$$

$$Z_b = \sin^2 2\bar{\varphi} \sin^2\bar{\varphi}\left[b_1\left(\frac{38\alpha}{\eta_0} + \frac{51 - 3\alpha^2}{2\omega'}\right) - b_2\left(\frac{7 + \alpha^2}{4\omega'}\right)\right] - \\ - \sin^2 2\bar{\varphi}\left[b_1\left(\frac{12}{\omega'} + \frac{19\alpha}{\eta_0}\right) - b_2\frac{1}{\omega'}\right]. \tag{14b}$$

In the case of external magnetic field parallel to the applied stress (parallel to [100]) the field shift can be expressed simply as

$$\delta\eta = -6S_0, \tag{15}$$

but at other magnetic field directions (especially along the [010] direction) the influence of magnetocrystalline anisotropy and of the quadratic term must be estimated. In many materials

$$b_1, b_2 \ll \omega', \tag{16a}$$

so that the influence of magnetocrystalline anisotropy on $\delta\eta$ determined by Z_b can be neglected. Condition (16a) can be then expressed as

$$\frac{K_1}{M_s}, \quad \frac{K_2}{M_s} \ll \frac{\omega}{\gamma}. \tag{16b}$$

The deviations from the linear dependence between $\delta\eta$ and the applied stress characterized mostly by the quadratic terms complicate the data evaluation. The quadratic term proportional to S_0^2 is negligible in comparison with the linear term proportional to S_0 when

$$S_0 \ll \omega \tag{17a}$$

corresponding to

$$\frac{\sigma_{max}\lambda_{100}}{2M} \ll \frac{\omega}{\gamma}. \tag{17b}$$

(16) and (17) are valid for small values of the applied stress σ_{max}, high microwave frequency ω, and/or high magnetization value M_s of the film.

Neglecting the magnetocrystalline and quadratic terms the resonance field shift due to the applied uniaxial stress σ parallel to the crystallographic direction [100] can be written as

$$\mu_0\,\delta H = [-6 + (9 + 3\alpha)\sin^2\bar{\varphi}]\frac{\sigma\lambda_{100}}{2M_s}, \tag{18}$$

where $\bar{\varphi}$ is the angle between H and σ. This simplified expression was also used in Section 3 for experimental data evaluation.

2.1.3 Uniaxial stress σ parallel to [110]

Using (2), (11) to (13) the resonance field shift can be expressed as

$$\delta\eta = 3S_0 \frac{\alpha - 1}{2} \left[1 - \frac{1+\alpha}{4\omega'} \sin^2 2\bar{\varphi}(6b_1 + b_2)\right] + 3S_1 \sin 2\bar{\varphi} \times$$
$$\times \left\{-\frac{3+\alpha}{2} + b_1 \left[\frac{19}{\eta_0} + \frac{12}{\omega'} - \left(\frac{19}{\eta_0} + \frac{51 - 3\alpha^2}{4\omega'}\right) \sin^2 2\bar{\varphi}\right] + \right.$$
$$\left. + \frac{b_2}{8\omega'} [(7 + \alpha^2) \sin^2 2\bar{\varphi} - 8]\right\} + 9S_0^2 \frac{1 - \alpha^2}{8\omega'} -$$
$$- 9S_0S_1 \sin 2\bar{\varphi} \frac{1 - \alpha^2}{4\omega'} + 9S_1^2 \frac{1 - \alpha^2}{8\omega'} - \cos^2 2\bar{\varphi} \left(\frac{7\alpha}{2\eta_0} + \frac{21 - \alpha^2}{8\omega'}\right). \tag{19}$$

The nonlinear terms are negligible when besides (17) also an analogous condition for λ_{111} is fulfilled,

$$S_1 \ll \omega, \qquad \frac{\sigma_{max}\lambda_{111}}{2M_s} \ll \frac{\omega}{\gamma}. \tag{20}$$

If the influence of magnetocrystalline anisotropy can be neglected according to (16) the resonance field shift due to the applied stress σ parallel to [110] is

$$\mu_0\, \delta H = -\frac{3\sigma}{4M_s} [(1 - \alpha)\, \lambda_{100} + (3 + \alpha)\, (1 - 2 \sin^2 \bar{\varphi}')\, \lambda_{111}], \tag{21}$$

where $\bar{\varphi}'$ is the angle between σ and $H(\bar{\varphi}' = 0$ for $H \parallel [110])$.

2.1.4 Isotropic stress

An isotropic tensile stress σ applied in the film plane (corresponding to two uniaxial stresses perpendicular to each other) is also used for magnetostriction constant measurement [4, 5]. The solution is similar, only instead of (9) we obtain

$$z_0 = \frac{b_1}{2} \sin 4\bar{\varphi}, \qquad z_1 = 2b_1 \cos 4\bar{\varphi}, \tag{22}$$

and instead of (11),

$$\Phi_{\varphi\varphi} = \eta + 2b_1 \cos 4\bar{\varphi} + \frac{31B_1^2}{8\eta}, \tag{23a}$$

$$\Phi_{\vartheta\vartheta} = \eta + m_e + \frac{3b_1}{2} + \frac{b_2}{4} + \frac{\cos 4\bar{\varphi}(2b_1 - b_2)}{4} + 6S_0 + \frac{(7B_1^2 - 4B_1B_2)}{8\eta_0}. \tag{23b}$$

The resonance field shift can be expressed as

$$\delta\eta = 3S_0(\alpha - 1) \left[1 - \frac{1+\alpha}{4\omega'} (6b_1 + b_2) \sin^2 2\bar{\varphi}\right] + 9S_0^2 \frac{1 - \alpha^2}{2\omega'}. \tag{24}$$

Neglecting magnetocrystalline anisotropy and quadratic terms we get

$$\mu_0\, \delta H = -3(1 - \alpha) \frac{\sigma\lambda_{100}}{2M_s}. \tag{25}$$

2.2 *Perpendicular configuration*

The resonance condition for FMR with external magnetic field H perpendicular to the thin film can be deduced in a similar way as in Section 2.1. Uniaxial tensile stress σ (with angle φ_s between σ and [100] axis) yields the resonance condition

$$\Omega^2 = \eta_0^2 - 6\eta_0 S_0 + 9(S_0^2 - S_1^2) \sin^2 2\bar{\varphi}_s \,, \tag{26a}$$

where

$$\eta_0 = \eta - m_e + 2b_1 \,. \tag{26b}$$

The resonance field shift due to the applied stress is

$$\delta\eta = 3S_0 + \frac{9}{2\Omega} (S_0^2 \cos^2 2\varphi_s + S_1^2 \sin^2 2\varphi_s) \,. \tag{27}$$

For isotropic stress in the film plane we obtain

$$\Omega = \eta - m_e + 2b_1 - 6S_0 \,, \tag{28}$$

$$\delta\eta = 6S_0 \,. \tag{29}$$

Expressions (26), (28), and (29) are exact, while (27) is evaluated up to a quadratic approximation in S_i and b_i (8).

3. Experimental Details

FMR measurements were performed at 35.76 GHz (Q-band) on a LPE garnet thin film (composition $Y_3Fe_{4.87}Mn_{0.13}O_{12}$, film thickness 2.1 μm) grown on a (001) GGG substrate. The saturation magnetization value was estimated according to the composition as $\mu_0 M_s = 0.177$ T using (5) in [7]. The following values of material parameters were obtained: spectroscopic splitting factor $g = 2.012$, effective magnetization

$$\mu_0 M_{\text{eff}} = m_e \mu_0 M_s = \mu_0 M_s - \frac{2K_u}{M_s} = 0.205 \text{ T} \,, \tag{30}$$

and effective anisotropy field

$$\mu_0 H_a = \frac{2K_1}{M_s} = -3.8 \text{ mT} \tag{31}$$

corresponding to the first magnetocrystalline constant $K_1 = -540$ J/m³. The second magnetocrystalline constant K_2 was negligible. The corresponding dimensionless quantities are $\Omega = 7.178$, $m_e = 1.157$, $\omega' = 7.201$, $\alpha = m_e/(2\omega') = 0.0803$, and $b_1 = -0.021$.

The resonance measurements were performed using a microwave spectrometer (built by one of us, Z.F.) described in [8].

Samples in the form of rectangular plates (12 × 2 × 0.41 mm³) with the longer dimension parallel to the [100] crystallographic direction were fixed at the end of the waveguide section in a special sample holder and the reflected microwave signal was detected and analysed. The sample holder for FMR measurements with H parallel to the sample surface is illustrated in Fig. 2. The end of the circular waveguide (WG) is covered by thin (0.01 mm) Al foil (AL) with an aperture of 1.8 mm in diameter, by means of which the central part of measured thin film (TF) is exposed to microwave radiation. The sample (SP) supported by edges of the waveguide (distance $L = 7.5$ mm) is strained by a special fine-thread screw. From the angle of its revolution the changes of deformation V of the sample can be determined with an accuracy of 1 μm.

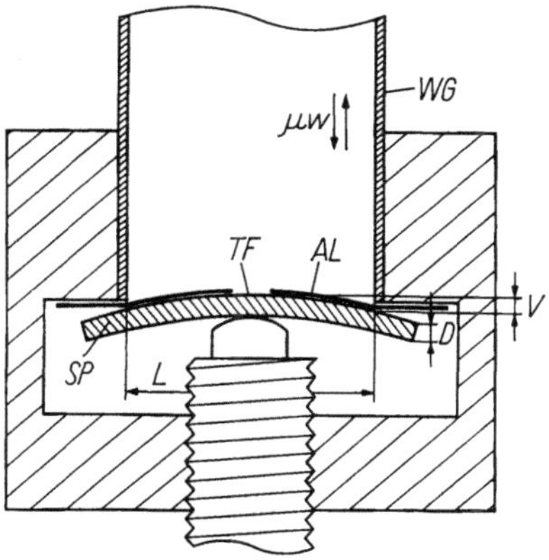

Fig. 2. Sample holder for FMR measurements with H parallel to the thin film. Description of the used symbols is in the text

Another sample holder was constructed for measurements with magnetic field H perpendicular to the sample surface. In this case the sample was again pressed to the aluminia thin foil diaphragm with circular aperture of 1.8 mm in diameter which covered a rectangular window (8.5 mm long) milled in the narrow side wall of a rectangular Q-band shorted waveguide section. By turning this holder by 90° it is also possible to measure FMR in parallel configuration (H perpendicular to the uniaxial stress σ only).

The tensile stress value σ in the central part of the film (see Fig. 2) was for both holders calculated from the formula

$$\sigma = Y \frac{6VD}{L^2}, \tag{32}$$

where $D = 0.41$ mm is the thickness of the sample and Y Young's modulus ($Y = 208$ GPa) [3].

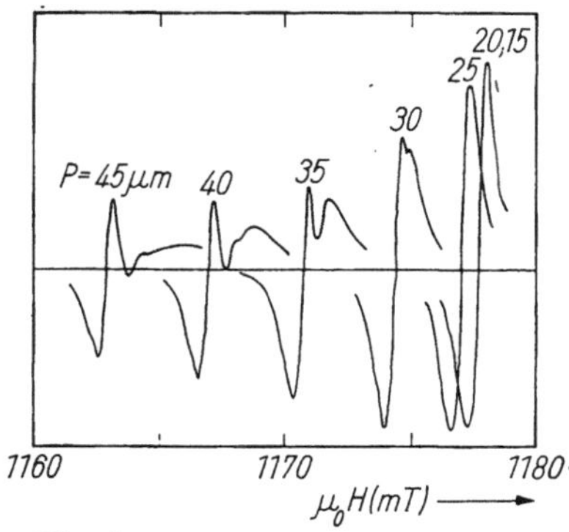

Fig. 3

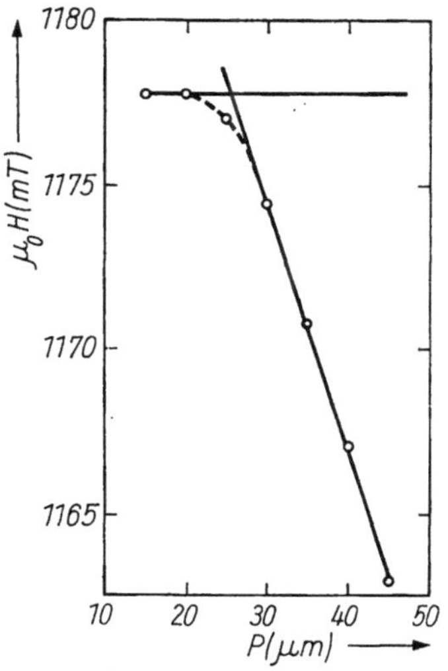

Fig. 4

Fig. 3. FMR signals measured at various positions P of the straining screw. Microwave frequency 35.76 GHz, H parallel to σ

Fig. 4. The dependence of FMR fields on position P of the straining screw (data from Fig. 3)

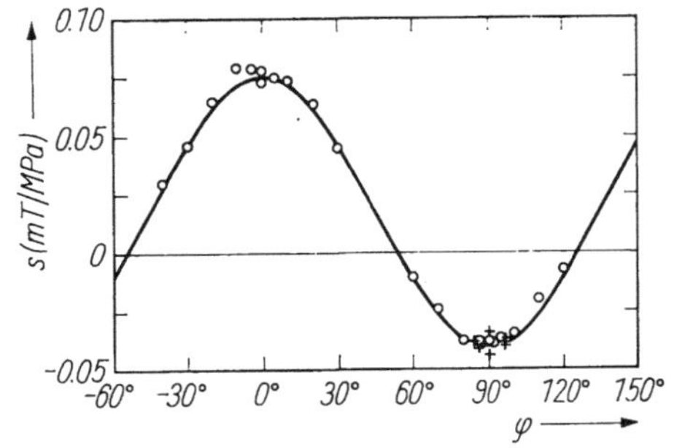

Fig. 5. Sensitivity $s = \Delta(\mu_0\, \delta H)/\Delta\sigma$ of the resonance field shift on the stress σ applied in the film plane parallel to the [100] crystallographic direction. External in-plane magnetic field H makes an angle φ with σ (○, + are data obtained using the "parallel" and "perpendicular" sample holder, respectively)

We have used deformation values up to $V_{max} \approx 30$ µm, which corresponds approximately to $\sigma_{max} \approx 200$ MPa. For $\lambda_{100} \approx 4 \times 10^{-6}$ and $\sigma = \sigma_{max}$ the quantity S_0 reaches its top value of 0.0035, which yields $S_0/\omega \approx 0.0005$. According (17) we can neglect terms quadratic in $\delta\eta$ and can consider the resonance field shift δH as a linear function of applied stress σ. As $b_1/\omega' \approx 0.003$ the influence of magnetocrystalline anisotropy can be also neglected and the simplified expression (18) can be used for data evaluation.

In Fig. 3 an example of FMR derivative absorption signals measured at various positions P of the straining screw is plotted. By extrapolation of the data obtained at higher deformations, the zero value of V corresponds to $P = 26$ µm. The direct determination of screw position corresponding to $V = 0$ is difficult because of the nonlinearity at very small values of V (see Fig. 4). For each such set of signals we determined the slope $\Delta(\mu_0\, \delta H)/\Delta V$ and by using (32) we calculated the sensitivity $s = \Delta(\mu_0\, \delta H)/\Delta\sigma$ of the resonance field shift to the stress σ.

Using the sample holder illustrated in Fig. 2 we measured s at various directions of external magnetic field H in the film plane, determined by the angle φ between H and [100]. The result is plotted in Fig. 5 along with the theoretical curve calculated according to (18) for $\lambda_{100} = 3.6 \times 10^{-6}$. The values of s obtained with the second ("perpendicular") sample holder (for H parallel to the film also plotted in Fig. 5) agree well with the same λ_{100}.

4. Discussion

The resonance field shift δH (the quantity which can be directly measured in experiment) caused by an applied stress is approximately a linear function of $(\sigma\lambda_i/M_s)$, $i = 100$ or 111, but deviations of various origin from the simple linear function can occur. In this analysis we supposed that the magnetostriction effect can be described by (6), and we determined the nonlinearity directly from the resonance condition (1), this nonlinearity being represented mainly by quadratic terms proportional to σ^2. For uniaxial σ parallel to H lying parallel to $\langle 100\rangle$ in the film plane and isotropic σ with H perpendicular to the film only simple linear expressions (15) and (29) for δH follow from (1). The coefficients at $(\sigma\lambda_i/M_s)$ are independent of M_s, ω, K_1, and K_2 for these two simple situations, as for the case of in-plane uniaxial stress with H perpendicular to the film (27); however, for other directions of H and σ the coefficients at $(\sigma\lambda_i/M_s)$ depend on the above-mentioned parameters and also the nonlinear terms must be generally considered. As follows from (17) and (20) the influence of quadratic terms ($\sim \sigma^2$) is generally negligible for high microwave frequency ω, high magnetization M_s, and small values of stress σ. If the quadratic terms cannot be neglected, it is necessary to fit the data using a computer and the more complicated nonlinear expressions (14), (19), (23), and/or (26).

The influence of magnetocrystalline anisotropy on δH was calculated and is included in the appropriate expressions in this paper. According to (16) this influence is negligible for high values of M_s and ω.

If the uniaxial stress σ (the direction of which is determined by the longer dimension of the rectangular sample) is applied along the $\langle 100 \rangle$ direction, the resonance field shift δH can be described by only one magnetostriction constant λ_{100} (as follows from (14) and (26)) at $\varphi_s = 0$. This situation was used also in the measurements presented in this paper. Similarly in the case of isotropic stress σ only λ_{100} need to be considered as follows from (23) and (28). In the case of other in-plane directions of the uniaxial stress σ (including the $\langle 110 \rangle$ direction) both λ_{100} and λ_{111} constants must be taken into account and both constants can also be evaluated.

We measured the resonance field shift δH for H perpendicular to the thin film (27) and for various directions of H parallel to the film ((14) and (18)). The angular dependence of δH for parallel configuration is well described by (18) as is illustrated in Fig. 5. The value of $\lambda_{100} = 3.6 \times 10^{-6}$ describes well not only the in-plane values of δH, but also the measurement in the case of H perpendicular to the film.

The magnetostriction constants of $Y\,Mn_xFe_{5-x}O_{12}$ garnet were measured by Dionne [9, 10] and Breed et al. [7]. By interpolating their data to $x = 0.13$ we obtain $\lambda_{100} = 6.6 \times 10^{-6}$ [9, 10] and 5.6×10^{-6} [7]. The difference between these data and the value of $\lambda_{100} = 3.6 \times 10^{-6}$ which fits our FMR measurements is larger than the inaccuracy of our λ_{100} value estimated (from inaccuracy and scattering of the FMR data) to about $\pm 10\%$. A similar discrepancy was obtained by Hoekstra et al. [3] using the sample bending method; they found a large (20 to 50%) difference between the strain calculated according to (32) and the real strain present in the film, which was found lower than the calculated value. The discrepancy may be caused by an inacuracy in determination of the length L (see Fig. 2 and/or (32)) of the bent part of the sample. In our calculations we suppose L as the distance between the opposite edges of the sample holder (see Fig. 2), but as the bending is only very small the real length L may be longer due to a slight unevenness of surface at the place where the samples are supported. The error caused by the value of L can be relatively large as the strain in the film is proportional to $1/L^2$.

5. Conclusions

The resonance field shift δH caused by an applied stress can be generally considered as linear function of $(\sigma\lambda_i/M_s)$ at high microwave frequency, high magnetic field, and small values of stress. At two special directions of H and σ the field shift δH is exactly a simple linear function of σ (uniaxial $\sigma \parallel H \parallel \langle 100 \rangle$ and isotropic σ with H perpendicular to the film). Using uniaxial σ parallel to $\langle 110 \rangle$ both constants λ_{100} and λ_{111} can be evaluated when δH is measured for various (at least two) directions of H.

Acknowledgements

The authors wish to thank Dr. M. Maryško for fruitful discussions and Mr. Suk for preparation of the samples.

References

[1] A. B. Smith, Rev. sci. Instrum. **39**, 378 (1968).
[2] G. F. Dionne, J. appl. Phys. **41**, 2264 (1970).
[3] B. Hoekstra, F. van Dovern, and J. M. Robertson, Appl. Phys. **12**, 261 (1977).
[4] G. P. Vella-Coleiro, Rev. sci. Instrum. **50**, 1130 (1979).
[5] X. Wang, C. S. Kraft, and M. H. Kryder, IEEE Trans. Magnetics **18**, 1295 (1982).

[6] L. Půst, phys. stat. sol. (a) 78, 741 (1983).
[7] D. J. Breed, A. B. Voermans, P. Q. J. Nederpel, and B. A. H. van Bakel, J. appl. Phys. **54**, 1519 (1983).
[8] L. Půst and Z. Frait, phys. stat. sol. (b) **122**, 535 (1984).
[9] G. F. Dionne, IEEE Trans. Magnetics 7, 715 (1971).
[10] G. F. Dionne and J. F. Goodenough, Mater. Res. Bull. 7, 749 (1972).

(Received July 10, 1984)

phys. stat. sol. (a) 85, 189 (1984)

Subject classification: 18.2.1; 22.8.2

Institute of Physics, Warsaw Technical University[1])

On the Inhibition of Dynamic Conversion in Bubble Garnet Domains by a Thin Midplane Layer[2])

By

J. J. Żebrowski and A. Sukiennicki

The effect of an additional, thin layer placed at the midplane of a bubble garnet film on the velocity and structure of a domain wall is studied by numerical means. The force-free and the pinning boundary conditions are assumed at the midplane layer and their effect on the dynamics is discussed. It is found that such a layer may be used for the inhibition of dynamic conversion of bubble domain wall structure. An indirect comparison with conventional techniques of conversion suppression is given.

L'effet sur la dynamique des parois des domaines d'une couche ultramince placée symétriquement entre les deux surfaces d'une couche du grenat à bulle est étudié par des méthodes numériques. Deux types des conditions aux limites à la couche ultramince sont considerées: les conditions neutres et les conditions de pinning. Il est conclu que la couche placée comme ça entre les deux surfaces extérieurs de la couche du grenat peut être utilisée à l'inhibition de la conversion dynamique de la structure des parois des domaines à bulle. Une comparaison indirect avec les techniques conventionels d'inhibition de la conversion est donnée.

1. Introduction

One of the most fundamental problems of bubble device technology is the inhibition and general control of the dynamic conversion of the state of the bubble wall. Usually ion implantation or additional LPE garnet layers are used for this purpose. Hagedorn [1] suggested that dynamic conversion may be inhibited by a thin, nonmagnetic layer placed at the midplane of the garnet film.

The purpose of this paper is to study by theoretical means the effect of an additional, thin layer, placed at the midplane of the film, may have on the velocity and structure of a bubble garnet domain wall.

2. Theory

The equations of motion of the twisted, flat domain wall, as derived by Slonczewski [2], were solved numerically for the case of a garnet film with a midplane layer using the DuFort-Frankel explicit scheme [3, 4]. The basic idea expressed in [1] was to introduce a thin, midplane layer as a barrier hindering the propagation of Bloch loops. In our calculation a thin midplane layer is used to disable the interaction between the structure of the wall in the upper half of the film and that in the lower film half. The existence of the midplane layer was simulated by treating the lower and upper halves of the film separately. The two halves were distinguished by the sign of the surface stray field given by the logarithmic formula described elsewhere [1].

---

[1]) ul. Koszykowa 75, 00-662 Warszawa, Poland.

[2]) Financial support provided by the Institute of Physics, Polish Academy of Sciences, Warsaw.

Since the two halves of the wall are separated by the midplane layer their position in space — after a certain time of motion — may become different. Any difference in the position of the upper and lower halves of the wall should give rise to an additional magnetic stray field. In this calculation no attempt was made to take into account the effect of such a field on the motion of the wall. It was assumed that in a real experimental situation the true position of the wall will correspond to the average of the calculated positions of the upper and lower halves of the domain wall.

In our experience [5, 6] a markedly different behaviour of domain walls may be obtained if different boundary conditions are used. Below, two cases are considered: a) at the midplane layer the force-free boundary conditions are used [3, 4] and b) the layer is the source of a pinning boundary condition leaving the wall free to move but constraining the azimuthal angle of the magnetization at the midplane layer to the value the angle had at the beginning of the motion. Note that case b) is different from that considered in [5, 6] as there the pinning acted at the outer surface of the film whereas here it is present in the midplane.

The material parameters used in the following calculations were the same as in [4] and [5]. They were: exchange constant $A = 0.81 \times 10^{-7}$ erg/cm, magnetization $4\pi M_s = 140$ G, damping coefficient 0.156, gyromagnetic ratio $\gamma = 1.73 \times 10^7$ 1/Oe s, film thickness $h = 1.4$ μm.

3. Results

3.1 *Force-free boundary conditions*

The calculated dependence of the velocity of the wall on the drive field is shown in Fig. 1 by a solid line. The dashed line in Fig. 1 depicts the velocity of the wall calculated for the same material parameters but for a film without the midplane layer [3] referred to below as simple film. For drives less then 10 Oe the velocity rises linearly with the drive. A minimum of the velocity versus drive curve was obtained at 12.5 Oe. For drives larger than 15 Oe the velocity of the wall drops sharply again and

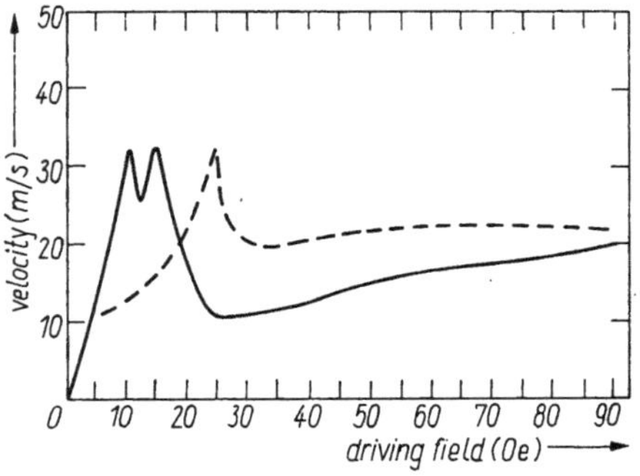

Fig. 1

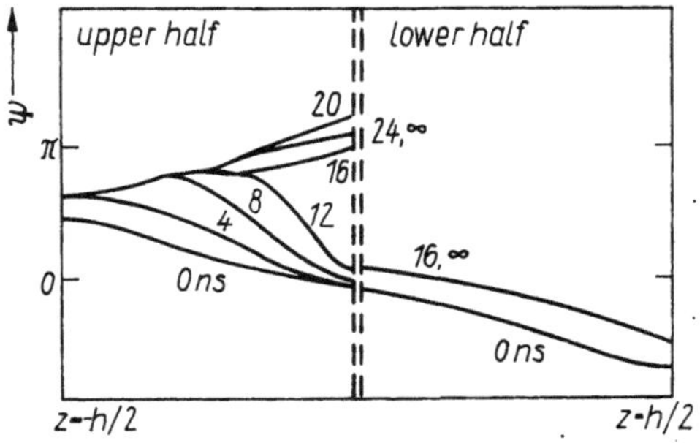

Fig. 2

Fig. 1. Force-free boundary conditions. Solid curve: velocity of the wall versus drive; dashed line: same dependence for a homogeneous film

Fig. 2. Force-free boundary conditions: evolution of the wall structure with time for the 10 Oe drive. Dashed lines indicate position of midplane layer

slowly rises for drives above 25 Oe. Here, the velocity at high drives is lower than obtained for the simple film. On the other hand, the linear mobility region is now twice larger.

The shape of the velocity versus drive curve is closely related to the occurrence of a dynamic wall structure. Thus, for drives less than 10 Oe, the wall structure in the lower half of the wall is Walker-like, i.e. the magnetization rotates by only a rather small angle in a correlated way (see CDOM in [7] and also discussion in [3] and [4]). In the upper half of the wall such a behaviour was observed only for the 2 Oe drive. For drives 10 Oe and less, a horizontal Bloch line was generated close to the upper film surface, moved towards the midplane layer and punched-through there (Fig. 2). From then on the wall structure in the upper half of the film and in the lower one are basically the same except for the direction of the twist due to the local value of the sign of the surface stray field. Consequently, once the horizontal Bloch line punches-through in the upper half, both halves of the wall assume a Walker-like structure and move with the same velocity. A Walker-like structure in the domain wall instead of a propagating horizontal Bloch line or of dynamic Bloch line stacking means that dynamic conversion may not occur. Note that a twisted domain wall cut in two by the midplane layer moves with a larger velocity (Fig. 1) than such a wall in a simple film for the same (low) drive.

For drive field values between 10 and 15 Oe, the structure in the lower half of the wall was also Walker-like as for low drives but, in the upper half, after a Bloch line had punched-through at the midplane layer, another was generated and moved in the opposite direction. The continued generation and punch-through of Bloch lines is the cause for the minimum in the velocity curve in Fig. 1.

For drives larger than 15 Oe, the structure called dynamic Bloch line stacking in the case of a simple film [4]. This is the structure which causes the sharp drop and the following slow rise of the velocity with increasing drive (solid curve in Fig. 1). Due to the fact that the intervening layer is placed at the midplane of the film, the type of structure obtained for large drives in the present calculation basically does not differ from the one obtained in [4]. However, the fact that the structure in the two halves of the wall evolves independently gives rise to a different dependence of the velocity on the drive from that obtained in [4].

3.2 Pinning boundary conditions

It has been shown elsewhere [5, 6] for the case of a simple film that the use of a pinning boundary condition at one of the surfaces of the film leads to a high drive stationary motion, i.e. to motion of the wall without change of structure regardless of the value of the drive. In the present calculation, if pinning was assumed at the midplane layer, stationary motion was also obtained. Somewhat surprisingly, in the present case, the changes of wall structure ceased much earlier for a given drive than in the case of a simple film.

For drives less than 30 Oe, stationary motion was obtained either immediately at the beginning of the motion (the lowest drives) or after a single horizontal Bloch line residing in the upper half of the wall had been generated and had stacked against the pinning surface. Since at the same time no significant structure was created in the lower half of the wall, the velocities of the two parts of the wall are much more different than in the case of the force-free boundary conditions. This may be seen in Fig. 3, where the velocity of the upper and lower halves is plotted as a function of the drive (solid curves, the dashed line depicts the results of [5]). Note that for both halves the velocity is strongly dependent on the drive. As stated in the theoretical section above, the stray field due to the difference in position of the upper and lower parts of

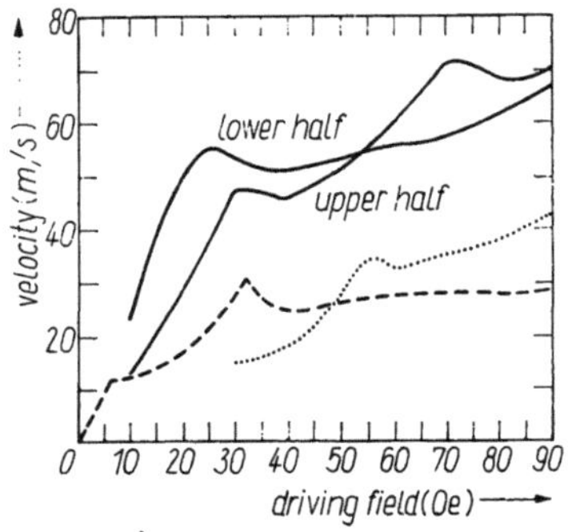

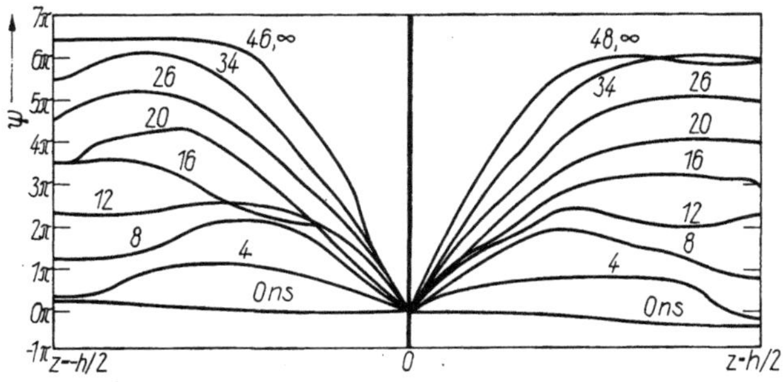

Fig. 3 Fig. 4

Fig. 3. Pinning boundary conditions. Solid curves: velocity of the two halves of the wall versus drive after stationary motion has set in (these curves do not cross or touch each other); dotted curve: velocity during the dynamic Bloch line stacking phase of motion; dashed line: velocity of the wall in a homogeneous film with pinning assumed at upper film surface

Fig. 4. Pinning boundary conditions: evolution of the wall structure with time for the 70 Oe drive. Thick solid line indicates position of the midplane layer

the wall has been neglected in the calculations. It is to be expected that this field will change the value of the drive acting on a given part of the wall and, consequently, in a real film the separation of the two parts of the wall would be negligible.

For drives larger than 30 Oe stationary motion was preceded by a brief period of dynamic Bloch line stacking. An example of the evolution with the time of the structure of the wall for the 70 Oe drive may be seen in Fig. 4. From 46 ns (upper half) or from 48 ns (lower half of the wall) the structure remained unchanged (stationary motion). During dynamic Bloch line stacking the velocity of the wall (dotted curve in Fig. 3) was low, less than for low drives prior to stationary motion and less than during stationary motion at low drives. Note also that in the case treated here the velocity of the wall during dynamic Bloch line stacking depends on the value of the drive. Such a dependence for the case of a simple film (dashed line in Fig. 3) was weak [4 to 6].

4. Conclusions

In the case of force-free boundary conditions, a midplane layer inhibits dynamic conversion of a bubble wall but only if the drive does not exceed a certain value. For larger drives dynamic conversion will occur through dynamic Bloch line stacking. On the other hand, the presence of the midplane layer increases the range of drive fields for which the Walker-like structure of the wall was obtained, by about two times.

Our results for the case of a pinning midplane layer seem more encouraging. For low drives stationary motion is obtained with a very simple wall structure. Thus, dynamic conversion is inhibited. For high drives, dynamic Bloch line stacking occurs but only for a limited time. The final structure of the wall is more complicated than for low drives and so a certain degree of dynamic conversion of the state of the wall is possible. But such conversion is limited by the early onset of stationary motion. The calculated velocity of the wall in the case of a pinning midplane layer may be twice higher than for a simple film with pinning at the outer surface.

It is expected that a real film grown with a midplane layer may not exhibit a pinning effect with the infinite breaking energy assumed in the calculations. On the other hand, the early onset of stationary motion indicates that with pinning at the midplane of the film less energy is needed to arrest changes of the structure of the wall. Thus, such techniques as ion implantation with subsequent annealing or the growth of an LPE layer with in-plane anisotropy, when applied to the formation of the midplane layer, may be more effective in the inhibition of dynamic conversion than when used at the outer surface of the film.

References

[1] F. B. Hagedorn, J. appl. Phys. **45**, 3129 (1974).
[2] J. C. Slonczewski, J. appl. Phys. **44**, 1759 (1973).
[3] B. E. MacNeal and F. B. Humphrey, IEEE Trans. Magnetics **15**, 1272 (1979).
[4] J. J. Żebrowski and A. Sukiennicki, J. appl. Phys. **52**, 4176 (1981).
[5] J. J. Żebrowski, IEEE Trans. Magnetics **18**, 1346 (1982).
[6] J. J. Żebrowski, J. appl. Phys., to be published.
[7] B. Kos and A. Sukiennicki, J. appl. Phys. **50**, 7865 (1979).

(Received April 27, 1984)

phys. stat. sol. (a) 85, 195 (1984)

Subject classification: 17; 11; 21

Faculty of Science, Department of Physics, Alexandria University, Egypt (a) and Sektion Physik der Wilhelm-Pieck-Universität Rostock[1]) *(b)*

Energy–Depth Relation of Electrons in Bulk Targets by Monte-Carlo Calculations

By

M. GABER (a) and H.-J. FITTING (b)

Monte-Carlo calculations are used to calculate the energy of penetrating electrons as a function of the depth in thick targets of Ti, Fe, Cu, As, In, and Au. It is shown that the mean energy ratio $\bar{E}(z)/E_0$ decays exponentially with depth z and depends on the backscattering coefficient η_B of the bulk material and the maximum range $R(E_0)$ of the primary electrons with initial energy E_0. Thereby a normalized plot $\bar{E}/E_0$ as a function of the reduced depth z/R becomes possible.

Mittels Monte-Carlo Rechnungen wird die Energie eindringender Elektronen als Funktion der Tiefe in dicken Targets Ti, Fe, Cu, As, In und Au bestimmt. Es kann gezeigt werden, daß die mittlere Energierate $\bar{E}(z)/E_0$ mit der Targettiefe z exponentiell abfällt und vom Rückstreukoeffizienten η_B des massiven Materials und der maximalen Eindringtiefe $R(E_0)$ der Primärelektronen mit der Anfangsenergie E_0 abhängt. Dadurch gelingt eine normalisierte Darstellung $\bar{E}/E_0$ als Funktion der reduzierten Targettiefe z/R.

1. Introduction

The interaction of keV electrons with solid targets is of considerable interest in many fields of electron probe microanalysis, X-ray emission spectroscopy, Auger electron spectroscopy, electron beam induced currents, and cathode luminescence as well as in electron lithography. E.g. in electron probe microanalysis some corrections must be done to convert the measured X-ray intensity into concentration values of sample elements [1]. The calculation of these correction factors requires an exact knowledge of the depth distribution of X-ray generation in thick targets. Information about the energy dissipation of electrons with target thickness z can be used to calculate the spatial distribution function of X-ray production.

The well-known Bethe expression scarcely can be applied directly to calculate the variation of the electron mean energy with the target thickness z, since the Bethe law is related to the real path length of the straggling electron.

An according expression based on measurements was already given by Whiddington [2] with the variation of the most probable energy E_m of transmitted electrons in dependence on their initial energy E_0 and the penetration depth z,

$$E_0^2 - E_m^2 = Bz\,. \tag{1}$$

However, the mean energy $\bar{E}$ of scattered electrons in dependence on the target depth cannot be expressed by a formula like (1) as has been shown in [3]. For that reason all forward- and backscattered electrons within the target must be taken into account in order to obtain the intensity of energy transfer and excitation in bulk materials.

---

[1]) Universitätsplatz 3, DDR-2500 Rostock, GDR.

The present paper describes Monte-Carlo results for the mean electron energy $\bar{E}(z)$ and its variation with the depth z taking into account the multiple scattering through the depth z.

2. Some Computation Details

The Monte-Carlo calculations are based on Moliere's multiple scattering theory [4]. Furthermore the Bethe energy loss equation is used to calculate the energy of electrons at each scattering point. More details about the present Monte-Carlo model are given in [5, 6]. Some results concerning the backscattering of electrons and the depth ionization distribution for one- and two-element targets were already published in [6, 7]. For the computation the samples had thicknesses greater than the Bethe range, that means, all incident electrons will be either absorbed in the target or backscattered through the surface. The number of histories which is used for each target is different, but lies between 1500 and 3000 electrons.

The mean energy data and their variation with depth were determined by the Monte-Carlo program for Ti, Fe, Cu, As, In, and Au targets with primary electron energies in the range between $E_0 = 7$ and 30 keV.

3. Energy–Depth Relation

The ratio of electron penetrating through the depth z in solid targets has been measured extensively by the thin film method and could be expressed by an empirical expression for the transmission coefficient [8],

$$\eta_T(z, E_0) = \exp\left[-4.605\left(\frac{z}{R(E_0, \varrho)}\right)^{p(\eta_B)}\right], \tag{2}$$

where the depth dependence is normalized by the reduced penetration depth z/R. R is the maximum range R reached by 1% of the impinging primary electrons: $\eta_T(z = R) = 0.01$;

$$R = 90\varrho^{-0.8}E_0^{1.3} \quad \text{for} \quad E_0 < 10\ \text{keV}\,, \tag{3a}$$

$$R = 45\varrho^{-0.9}E_0^{1.7} \quad \text{for} \quad E_0 > 10\ \text{keV} \tag{3b}$$

with R in nm, the target material density ϱ in g/cm³, and the initial energy E_0 of primary electrons in keV. The transmission parameter p is correlated to the normal incident electron backscattering fraction η_B [8],

$$p(\eta_B = 0.1) \approx 2.4\,, \qquad p(\eta_B = 0.2) \approx 2\,, \qquad p(\eta_B = 0.3) \approx 1.7\,,$$

$$p(\eta_B = 0.4) \approx 1.5\,, \qquad p(\eta_B = 0.5) \approx 1.3\,.$$

According energy depth relations computed in the present work indicate a general behaviour of an exponential slope of the mean energy $\bar{E}$ of penetrating electrons with depth z,

$$\bar{E}(z) = CE_0 \exp\left(-k(E_0)\, z\right). \tag{4}$$

Fig. 1 shows typical results for the target element Fe and initial energies of primary electrons $E_0 = 10$, 15, 20, 25, and 30 keV. We can state that the points lie close to a series of straight lines for not too large z. Obviously up to depths $z < 0.6R$ when the mean energy $\bar{E}$ has been decreased to about 50% of the initial one the $\bar{E}$–z slope can be approximated by a linear decrease,

$$\bar{E}(z) = CE_0(1 - q(E_0)\, z)\,. \tag{5}$$

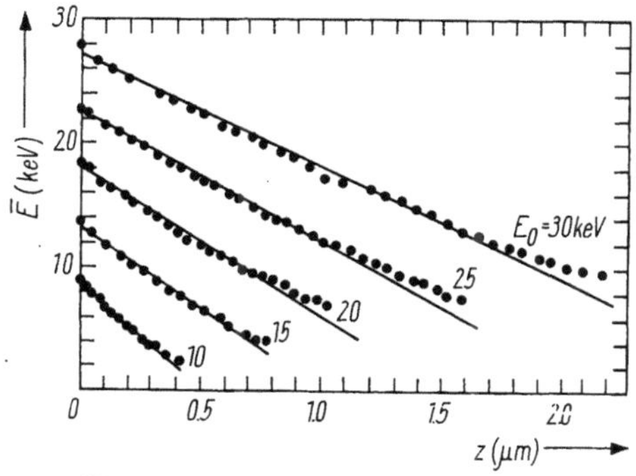

Fig. 1

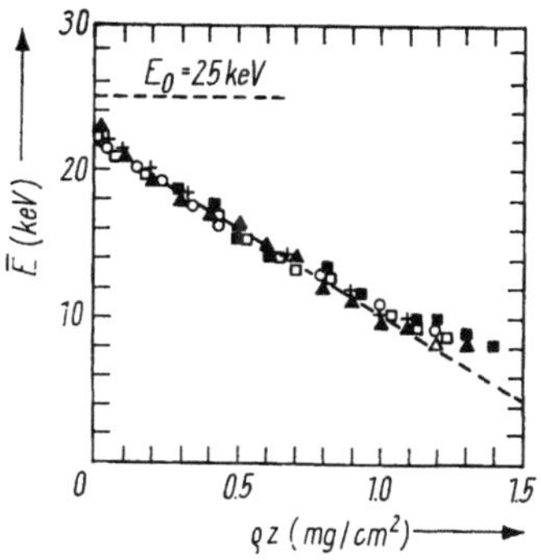

Fig. 2

Fig. 1. Mean energy $\bar{E}$ of penetrating electrons as a function of the depth z in Fe targets (E_0 initial energy of primary electrons)

Fig. 2. Mean energy $\bar{E}$ of penetrating electrons versus mass thickness (ϱz) of several target elements: + Ti, ▲ Cu, □ As, ○ In, ■ Au

In order to describe the material and energy dependence of the mean energy ratio at the surface,

$$C = \frac{\bar{E}(z = 0)}{E_0}, \tag{6}$$

and the approximation parameter $k(E_0, \varrho)$ and $q(E_0, \varrho)$ we have tried a plot over the commonly used mass thickness (ϱz), Fig. 2. It is shown that for a large number of elements and one initial energy such attempt becomes successful. A full normalization including only the maximum range R has been carried out in Fig. 3 where all material and energy parameters are involved by $R(E_0, \varrho)$ according to (3a) and (3b). In that way the presentation $\bar{E}/E_0$ in dependence on the reduced depth z/R (Fig. 3) can by applied to different target materials and primary electron energies. The analytical approxima-

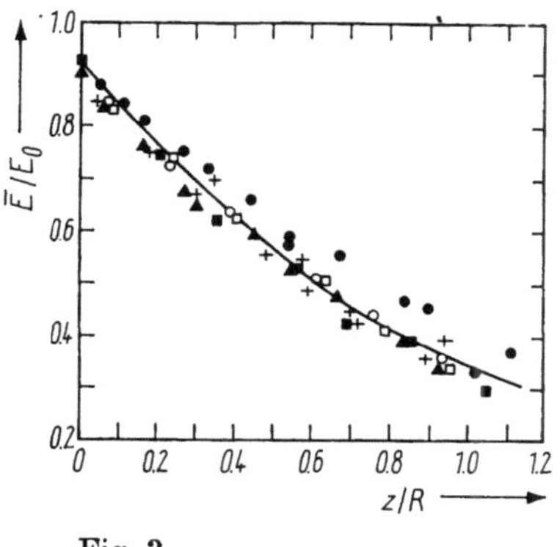

Fig. 3

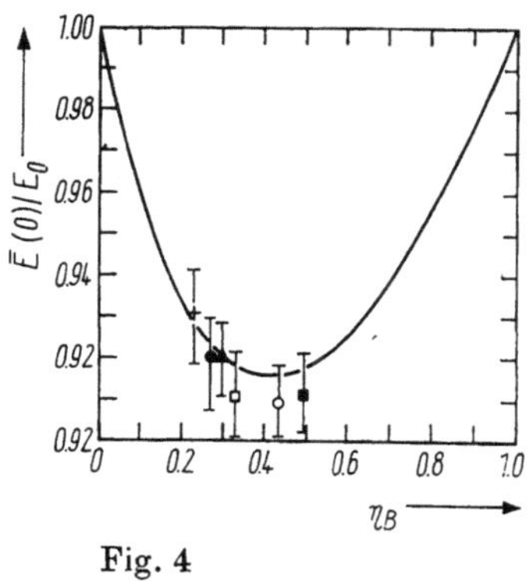

Fig. 4

Fig. 3. Normalized plot of the mean energy fraction $\bar{E}/E_0$ as a function of the reduced depth z/R in serveral target elements: + Ti, ● Fe, ▲ Cu, □ As, ○ In, ■ Au, and for different electron initial energies $E_0 = 7$ to 30 keV

Fig. 4. Mean energy at the surface $C = \bar{E}(z = 0)/E_0$ in dependence on the electron backscattering fraction η_B according to (11) (full line) and computed values: + Ti, ● Fe, ▲ Cu, □ As, ○ In, ■ Au

tion can by expressed over the whole range $z \leqq R$ by an exponential slope,

$$\frac{\bar{E}}{E_0} = C \exp\left(-\frac{z}{R}\right), \tag{7}$$

i.e. the mean energy at the maximum range $z = R$ amounts to 37% of that value at the surface $z = 0$. Taking into account that only 1% of the impinging electrons passes through this depth $z = R$ this large amount of the remaining energy seems somewhat surprising but is in good agreement with measured values [3, 8]. Comparison with (4) yields $k(E_0) \cdot R(E_0) \approx 1$.

The linear approximation (5) can be described by

$$\frac{\bar{E}}{E_0} = C\left(1 - q\frac{z}{R}\right) \quad \text{for} \quad z < 0.6R \tag{8}$$

with the slope parameter $q = q(E_0) \cdot R(E_0) \approx 0.25$.

On the other hand, the mean energy at the surface $C = \bar{E}(z = 0)/E_0$ is not equal to E_0 because of the contribution of backscattered electrons having a mean energy less than E_0. The present work has shown that for a certain target material the variation of C with initial energy E_0 is very small, while the variation of C with atomic number is larger. An expression to calculate C can be deduced from the mean energy $\bar{E}_B(z = 0)$ of backscattered electrons at the surface of a bulk target. Experimental measurements by Fitting [3] have shown that this energy $\bar{E}_B$ is given by

$$\bar{E}_B \approx E_0(0.5 + 0.5\eta_B) \tag{9}$$

with η_B as the backscattering coefficient including both the energy and material dependence of the target. Then the mean energy at the surface can be written, under consideration of (9),

$$\bar{E}(z = 0) = \frac{N_0 E_0 + N_B \bar{E}_B}{N_0 + N_B} = \frac{E_0 + \eta_B \bar{E}_B}{1 + \eta_B}, \tag{10}$$

where N_0 is the number of electrons incident on the target, N_B is the number of backscattered electrons, and $\eta_B = N_B/N_0$ the backscattering fraction. Combining (10) with (9) one obtains

$$C = \frac{\bar{E}(z = 0)}{E_0} = \frac{1}{1 + \eta_B} + \frac{\eta_B}{2}. \tag{11}$$

This connection between the mean energy at the surface (factor C in (4) to (8)) and the backscattering fraction is demonstrated in Fig. 4. The spread of the computed values is caused by the statistical error of the Monte-Carlo calculations. For (7) C can be chosen according to (11), but for the linear approximation (8) a somewhat lower value for C (some per cent) has to be applied in order to obtain the best fit, e.g. see Fig. 1, 2, and 3. Both descriptions for the mean energy– depth relation can help in analytical or numerical calculation of the spatial electron beam excitation in solid targets.

References

[1] R. H. PACKWOOD and J. D. BROWN, X-Ray Spectrometrie **10**, 138 (1981).
[2] R. WHIDDINGTON, Proc. Roy. Soc. **A88**, 554 (1914).
[3] H.-J. FITTING, J. Phys. D 8, 1480 (1975).
[4] G. MOLIERE, Z. Naturf. **3a**, 78 (1948).
[5] M. GABER, Dissertation A, TU Dresden 1980.
[6] M. GABER and P. JUGELT, Isotopenpraxis **16**, 185 (1980).
[7] M. GABER and P. JUGELT, Isotopenpraxis **17**, 336 (1981).
[8] H.-J. FITTING, phys. stat. sol. (a) **26**, 525 (1974).

(Received May 7, 1984)

phys. stat. sol. (a) 85, 199 (1984)

Subject classification: 20.1; 22.1.2

Department of Physics, National University of Singapore[1])

Temperature Dependence of Interband Optical Absorption of Silicon at 1152, 1064, 750, and 694 nm

By

E. H. SIN, C. K. ONG, and H. S. TAN

The optical absorption coefficients of Si are calculated for 1152, 1064, 750, and 694 nm at elevated temperatures. The optical absorption processes employed in the calculation model are the indirect transitions of the valence band electrons to the set of lowest conduction band minima and involving the absorption or emission of acoustical and optical phonons. The calculated results are found to be in excellent agreement with the available experimental data up to a temperature of about 1000 K. Comparison is also made with the results calculated with the expressions employed earlier by Jellison and Modine. The present calculation model and the calculated results for Si should prove to be valuable references to many laser annealing heat flow calculations and solar cell analyses.

Die optischen Absorptionskoeffizienten von Si bei 1152, 1064, 750 und 694 nm werden für ausgewählte Temperaturen berechnet. Die in dem Rechnungsmodell einbezogenen optischen Absorptionsprozesse sind die indirekten Übergänge der Valenzbandelektronen zu dem Satz der tiefsten Leitungsbandminima und schließen die Absorption und Emission von akustischen und optischen Phononen ein. Es wird gefunden, daß die berechneten Ergebnisse in ausgezeichneter Übereinstimmung mit verfügbaren experimentellen Werten sind bis zu Temperaturen von etwa 1000 K. Es wird auch ein Vergleich mit Ergebnissen durchgeführt, die mit den früheren von Jellison und Modine benutzten Ausdrücken berechnet werden. Das neue Rechnungsmodell und die berechneten Ergebnisse für Si könnten sich als nützliche Angaben für viele Laserausheilwärmefluß-Rechnungen und Solarzellenanalysen erweisen.

1. Introduction

Pulsed and continuous-wave laser annealing of silicon has been studied extensively in recent years. In particular, the most active theoretical and experimental studies are those on pulsed laser annealing processes. For nanosecond pulsed lasers, the thermal evolution of the annealing processes has been found to be described very well by the solution of the following one-dimensional heat-flow equation when the possibility of phase changes is allowed [1]:

$$\varrho c \frac{\mathrm{d}T}{\mathrm{d}t} = \alpha(T)\, I(x, t) + \frac{\partial}{\partial x}\left(k(T) \frac{\mathrm{d}T}{\mathrm{d}x}\right), \tag{1}$$

where c is the heat capacity and ϱ the mass density of Si; $T = T(x, t)$ is the temperature, $\alpha(T)$ the temperature-dependent optical absorption coefficient, and $k(T)$ the thermal conductivity. The surface of Si is chosen to be at $x = 0$. $I(x, t)$ is the absorbed radiation power density at depth x and is given by

$$I(x, t) = (1 - R)\, I(0, t) \exp(-\alpha x)\,, \tag{2}$$

where R is the reflectivity at the surface.

[1]) Kent Ridge, Singapore 0511, Singapore.

To solve (1), a numerical scheme generally has to be employed. Owing to a lack of information of temperature-dependent optical and thermal properties [2], most of the earlier computations were performed with these quantities treated as constants [2, 3], while in the other computations the temperature-dependent expressions for optical absorption were often inaccurate [4]. Therefore, the maximum melt depths and crystal regrowth rates obtained were questionable. Furthermore, without an accurate knowledge of the temperature dependence of the optical properties of solid phase Si, these earlier computations would also produce inaccurate estimates of the threshold energies and the instants at which the surface of the material would start to melt. On the other hand, the optical absorption coefficient and reflectivity are not easy to determine experimentally at high temperature due to the growth of an oxide layer [5]. A theoretical approach that allows the estimation of these parameters at high temperature should therefore be helpful.

In this paper, we have calculated the temperature dependence of the absorption coefficient of Si with a simple theoretical expression containing a parameter to be determined from the experimental data measured at a particular temperature, usually room temperature. The expression was taken to be valid up to the melting point of Si. Calculations were performed for various wavelengths of radiations that were of interest to those working on laser annealing of Si.

2. Calculation Model

The two major mechanisms for optical absorption are the fundamental absorption and the free-carrier absorption. The magnitude of the free-carrier absorption coefficient of silicon can be estimated with the classical equations [6]. It was found that for pure Si the free-carrier absorption is less than a few per cents of the fundamental absorption even at temperature close to the melting point of Si. Therefore, the contribution by the free carriers to optical absorption of pure Si may be neglected.

In the fundamental absorption processes of Si, incident photons of energy close to the band-gap energy of Si would excite electrons from the valence band to the conduction band. Since Si is an indirect-gap semiconductor, the absorption of a photon is usually accompanied by the emission or absorption of a phonon.

As suggested by Macfarlane et al. [7], there are four types of phonons involved in the indirect transitions and the coupling constants for the longitudinal acoustic (LA), transverse acoustic (TA), transverse optical (TO), and longitudinal optical (LO) phonons involved in the transitions have the following ratios:

$$C_{\mathrm{LA}} \approx 5.5 C_{\mathrm{TA}} \approx 4 C_{\mathrm{TD}} \approx 4 C_{\mathrm{LO}} \,. \tag{3}$$

Hence, for Si the absorption coefficient due to the various indirect transitions is given by

$$\alpha(h\nu, T) = A \sum_{j=1}^{4} P_j \left[\frac{(h\nu - E_{\mathrm{g}}(T) + E_{\mathrm{p}j})^2}{\exp(E_{\mathrm{p}j}/kT) - 1} + \frac{(h\nu - E_{\mathrm{g}}(T) - E_{\mathrm{p}j})^2}{1 - \exp(-E_{\mathrm{p}j}/kT)}\right], \tag{4}$$

where the indices 1, 2, 3, 4 are associated with the LA, TA, TO, and LO phonons, respectively; $E_{\mathrm{p}j}$ is the energy of type j phonons with $\boldsymbol{k}$-vector in the $\langle 100 \rangle$ directions, T the temperature of the sample, $E_{\mathrm{g}}(T)$ the energy gap between the minimum of the conduction band and the maximum of the valence band. The proportionality constants P_j calculated from the coupling constant relations (3) are $P_1 = 1$; $P_2 = 0.033$, $P_3 = 0.063$, and $P_4 = 0.063$. The energy of the phonons, $E_{\mathrm{p}j} = k\Theta_j$, is given by Macfarlane et al. [7] as $\Theta_1 = 212$, $\Theta_2 = 670$, $\Theta_3 = 1050$, and $\Theta_4 = 1420$ K. The value of A is often chosen such that (4) will give the experimental value of absorption coefficient at room temperature for a given incident photon frequency. A is then assumed constant for the absorption coefficient calculation at other temperatures.

In (4) the argument within each set of parentheses have to be positive. We used the empirical formula given by Thurmond [8] for the temperature dependence of the energy gap, which is

$$E_g(T) = E_g(0) - \frac{BT^2}{C + T}, \tag{5}$$

where $B = 4.73 \times 10^{-4}$ eV K^{-1}, $C = 635$ K, $E_g(0)$ the energy gap at 0 K is taken as 1.155 eV as suggested by Jellison and Lowndes [9].

In (4) we have considered only the indirect transitions between the top of the valence band and the set of first-lowest conduction band minima. Hence, it was not suitable for calculating the absorption coefficient for wavelengths smaller than 650 nm where the absorption contribution from the indirect transitions to the set of second-lowest conduction band minima must be included. We also neglect the absorption term due to indirect transitions to exciton states, because its contribution which is proportional to $(h\nu - E_g(T) + E_{pj})^{1/2}$ becomes insignificant as $E_g(T)$ becomes small compared with $h\nu$ at elevated temperature. In principle, due to spin–orbit coupling the top of the valence band should be considered as consisting of an upper $P_{2/3}$ quadruplet and a lower $P_{1/2}$ doublet with a separation of 0.044 eV [10]. However, the splitting which is equivalent to a temperature of 500 K can also be ignored at higher lattice temperatures. Therefore, the splitting of the valence band has been neglected in (4).

3. Results

3.1 $\lambda = 1152$ and 1064 nm

The near-infrared line ($\lambda = 1152$ nm) of a He–Ne laser has recently been employed in several time-resolved transition studies of pulsed laser annealing of Si. An accurate knowledge of the temperature-dependent α at this wavelength is essential to reliable analyses of the data obtained. From (4) A was found to be 6200 if the absorption coefficient calculated at 300 K was to be the experimental value of 0.54 cm^{-1} [7, 11].

The calculated absorption coefficients with (4) from 300 to 1600 K together with Jellison and Lowndes' experimental data [9] are shown in Fig. 1. (Jellison and Lown-

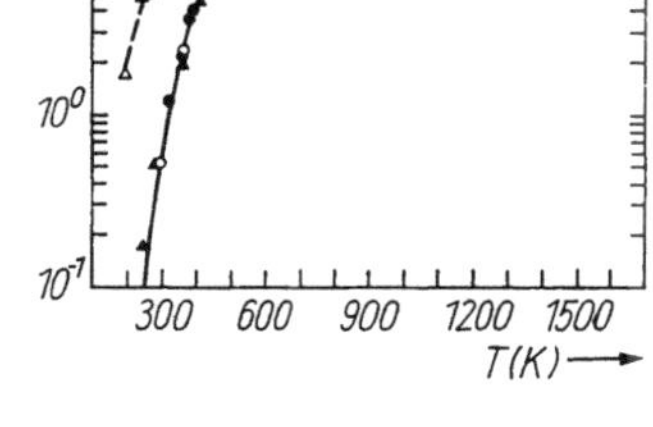

Fig. 1. Temperature dependence of optical absorption coefficient of Si at $\lambda = 1152$ (solid curve) and 1064 nm (broken curve) with $A = 6200$ and 5403, respectively. The experimental data from Jellison and Lowndes (●), Svantesson and Nilsson (○), Weakliem and Redfield (▲), and Macfarlane et al. (△) are also shown

des have calibrated their data to give $\alpha = 0.54\ \text{cm}^{-1}$ at 300 K.) There is a good agreement between the calculated values and the experimental data up to the temperature of 1100 K. Above this temperature, our results are lower than the experimental data. This is mainly due to the fact that (4) does not include the contributions to α from the other indirect and direct transitions that become significant as the band gap becomes sufficiently narrow at high temperatures. The assumption of a rigid band structure by having A constant should also fail at high temperature.

Another near-infrared line which is of interest to many workers in laser annealing studies of Si is the $\lambda = 1064$ nm emission of the Nd:YAG or Nd:glass pulsed lasers. Although the Si absorption coefficient at this wavelength is very small at room temperature, it increases very rapidly with temperature. Therefore, the $\lambda = 1064$ nm radiation from these pulsed lasers is capable of producing a thin molten layer on Si.

In requiring α to have the measured value [13] of $9.5\ \text{cm}^{-1}$ at $T = 294$ K and $\lambda = 1064$ nm, the value A of (4) was found to be 5403. α calculated from (4) for the temperature range from 200 to 1600 K is also shown in Fig. 1. It is in good agreement with the experimental data reported by various groups [7, 13]. No experimental data above the temperature of 500 K is available for comparison. However, it is believed that the calculated results should be good up to a temperature of about 1000 K. Above this temperature, the calculated α would be smaller than the actual values of α for the same reasons associated with the case with $\lambda = 1152$ nm. Nonetheless, the calculated and actual values of α should have the same order of magnitude.

The relative importance of the four types of phonons involved in the interband transitions were also investigated. An equation similar to (4) but involving only the two acoustic phonons was employed in the calculation of α. Different values of A were required to fit the room temperature experimental data of α. However, the values of α up to 1600 K calculated with the equation that involves only the two acoustic phonons show no significant deviation from those values calculated with (4). This is expected as the two optical phonons are associated with much higher temperatures while their coupling constants in the absorption processes are relatively small. Further calculations also indicated that both the acoustic phonons are equally important in explaining the temperature-dependent behaviour of α observed experimentally.

Similar temperature-dependent absorption calculations have also been performed by Jellison and Lowndes [9] for λ of 1152 and 1064 nm. Their calculations employed the two empirical absorption expressions of Macfarlane et al. [7] that are associated with the TA and LA phonons, respectively. Their empirical absorption expressions can be recognized as to consist of the absorption contributions from indirect interband transitions and indirect transitions to exciton states. The calculated results were very much the same as ours, thus confirming the neglect of the exciton contribution to the absorption in our calculations is valid. On the other hand, our two-phonon and four-phonon calculations also confirmed that the neglect of the two-high-temperature optical phonons in their calculations is also valid.

3.2 $\lambda = 750$ nm

Recently, the usefulness of the Q-switched alexandrite laser has been demonstrated for the doping, gettering, and removal of ion-implantation damages of Si wafers [14]. For industrial application the high repetition rate of the laser is particularly attractive for the rapid processing of Si wafers. The fundamental wavelength of the alexandrite laser is 750 nm. The optical absorption coefficient at 298 K and 750 nm was measured to be $1.45 \times 10^3\ \text{cm}^{-1}$ by Weakliem and Redfield [11]. By substituting these values into (4) the constant A was found to be 4019. The absorption coefficient was then calculated for temperatures ranging from 300 to 1600 K and the results are shown in Fig. 2.

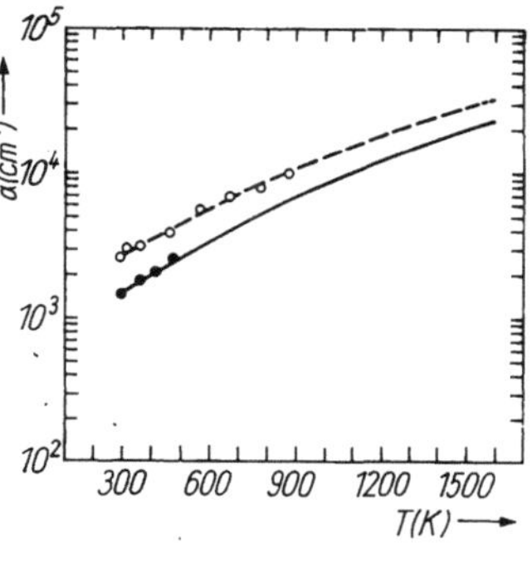

Fig. 2. Temperature dependence of optical absorption coefficients of Si at $\lambda = 750$ (solid curve) and 694 nm (broken curve) with $A = 4019$ and 4493, respectively. The experimental data from Jellison and Modine (○) and Weakliem and Redfield (●) are also shown

The agreement with the limited amount of experimental data [11] is again good. On the other hand, the $\alpha(T)$, values calculated at this wavelength with Jellison and Lowndes' expressions [9] are about 50% greater than the values calculated with (4) over the whole temperature range.

3.3 $\lambda = 694$ nm

By and large, the most often used pulsed laser in the laser annealing studies of Si is the ruby laser with the emission radiation of 694 nm. Therefore, it is especially important to provide an accurate knowledge of $\alpha(T)$ at this wavelength for various theoretical analyses. At 297 K and 694 nm, α was measured to be 2.6×10^3 cm^{-1} by Jellison and Modine [12]. Again, by substituting these values into (4), A was found to be 4493. $\alpha(T)$ calculated at elevated temperatures is also shown in Fig. 2. The agreement with the experimental data [12] is also found good. The $\alpha(T)$ values calculated with the expressions employed by Jellison and Lowndes [9] are also found to be about 30% greater than the experimental data and the results calculated with (4).

4. Conclusion

The present study has demonstrated that (4) is very useful in calculating the temperature-dependent optical absorption $\alpha(T)$ of Si for radiation of wavelengths up to 690 nm. The $\alpha(T)$ values calculated are in excellent agreement with the experimental data up to a temperature of about 1000 K. Also, it was found that the $\alpha(T)$ calculated by considering those indirect transitions involving only the two acoustic phonons shows insignificant deviations from the $\alpha(T)$ calculated with (4). At a wavelength significantly shorter than 690 nm or at temperature higher than 1000 K, the calculated $\alpha(T)$ values are found to be smaller than the actual values. It is obvious that the absorption processes involving the optical phonons did not help in explaining the larger $\alpha(T)$ measured for temperatures above 1100 K. It is likely that the discrepancy is due to the fact that (4) does not include other absorption contributions such as the indirect interband transitions from the valence band to the set of second-lowest conduction band minima which are at the Brillouin zone edge on the $\langle 111 \rangle$ axes. This second set of indirect transitions is expected to contribute to the optical absorption as the band gap becomes sufficiently narrow at high temperatures. For radiation of $\lambda < 360$ nm direct optical transitions should also be included in the calculations of $\alpha(T)$.

References

[1] P. BAERI and S. U. CAMPISANO, in: Laser Annealing of Semiconductors, Ed. J. M. POATE and J. W. MAYER, Academic Press, 1982 (p. 75).
[2] P. BAERI, S. U. CAMPISANO, G. FOTI, and E. RIMINI, J. appl. Phys. **50**, 788 (1979).
[3] R. F. WOOD and G. E. GILES, Phys. Rev. B **23**, 23 (1981).
[4] D. J. GODFREY, A. C. HILL, and C. HILL, J. Electrochem. Soc. **128**, 1798 (1981).
[5] G. E. JELLISON, JR. and F. A. MODINE, J. appl. Phys. **53**, 3745 (1982).
[6] R. J. ELLIOT and A. F. GIBSON, An Introduction to Solid State Physics and Its Application, MacMillan Co., New York 1974 (p. 266).
[7] G. G. MACFARLANE, T. D. MCLEAN, J. E. QUARRINGTON, and V. ROBERT, Phys. Rev. **111**, 1245 (1958).
[8] C. D. THURMOND, J. Electrochem. Soc. **122**, 1133 (1975).
[9] G. E. JELLISON, JR. and D. H. LOWNDES, Appl. Phys. Letters **41**, 594 (1982).
[10] S. ZWERDLING, K. J. BUTTON, B. LAX, and L. M. ROTH, Phys. Rev. Letters **4**, 173 (1960).
[11] H. A. WEAKLIEM and D. REDFIELD, J. appl. Phys. **50**, 1491 (1979).
[12] G. E. JELLISON, JR. and F. A. MODINE, Appl. Phys. Letters **41**, 180 (1982).
[13] K. G. SVANTESSON and N. G. NILSSON, J. Phys. C **12**, 3837 (1979).
[14] J. M. THOMAS and T. K. MCNAB, Solid State Technol. **25**, 7, 7A (1982).

(Received June 1, 1984)

phys. stat. sol. (a) 85, 205 (1984)
Subject classification: 11; 13; 22.2.4

Sektion Physik der Humboldt-Universität zu Berlin[1]) *(a)*
and Institut für Elektronenmikroskopie und Festkörperphysik
der Akademie der Wissenschaften der DDR, Halle (Saale)[2]) *(b)*

Kilovolt Electron Energy Loss Distribution in GaAsP

By
G. OELGART (a) and U. WERNER (b)

Using the electron beam induced barrier current and Monte Carlo simulation the depth-dose function as well as the distribution function for the generation source for electron–hole pairs perpendicular to the injection direction are determined. The lateral dose distributions show measured and calculated a part I formed by the focused electron beam penetrating into the solid target and a part II due to the progressive spreading of the electron beam during the interaction with the target. Spatial dose distributions are computed for the typical case of scanning electron microscopy and for the electron microprobe analysis.

Die Tiefendosiskurve und die Verteilung der Elektron–Loch-Paargeneration senkrecht zur Einschußrichtung werden unter Ausnutzung des elektronenstrahlinduzierten Stromes sowie mit Monte-Carlo-Rechnungen ermittelt. Die experimentell und rechnerisch erhaltenen Lateralverteilungen bestehen aus einem Teil I, der durch das Eindringen des fokussierten Elektronenstrahles in den Festkörper hervorgerufen wird und einem Teil II, der sich durch die Aufspreizung des Elektronenstrahles infolge der Wechselwirkung mit dem Festkörper ausbildet. Für den typischen Fall der Rasterelektronenmikroskopie und der Elektronenstrahlmikroanalyse werden die räumlichen Dosisverteilungen angegeben.

1. Introduction

A determination of the generation volume associated with electron penetration into solid materials is required especially with regard to the increasing use of quantitative scanning electron microscopy and electron microprobe analysis. A lot of measurements have been carried out in order to determine the depth–dose distribution of the generation volume of different materials (e.g. [1 to 5]). Little attention, however, has been paid to the aspects of the lateral spread of electrons during the electron scattering in solids. The aim of the present paper is to describe this lateral dose distribution and the factors which influence it by way of example of the distribution in GaAsP in the energy range of 10 to 45 keV. This enables the extension of the assertions concerning the depth–dose distributions of GaAsP given in [5]. Furthermore, the measurements are to be supported by using Monte Carlo simulations of the electron–target interaction. Based on the simple diffusion model of Archard [6] and the results of the Monte Carlo calculation a first approximation of an empirical description of the spatial dose distributions is formulated for GaAsP.

2. The Monte Carlo Simulations

Similar to the model applied by Curgenven and Duncumb [7] the present model of the Monte Carlo simulation [8] describes the angular spreading of the incident electron beam by using a simple analytical scattering formula. But different to [7], the screened

[1]) Invalidenstr. 42, DDR-1040 Berlin, GDR.
[2]) Weinberg, DDR-4020 Halle (Saale), GDR.

one is modified instead of the simple Rutherford scattering cross-section. The probability of an electron dp to be scattered through angle ϑ into a solid angle element $d\Omega$ is evident from the screened cross-section,

$$\frac{dp}{d\Omega} = \frac{\alpha(1 + \alpha)}{\pi(1 - \cos\vartheta + 2\alpha)^2}. \tag{1}$$

For single scattering the screening parameter α can be written as

$$\alpha = \alpha_0 = 0.00434 \frac{Z^{2/3}}{E}, \tag{2}$$

where Z stands for the atomic number and E for the electron energy (E in keV). In generalizing the screening parameter by setting

$$\alpha = \alpha_0 \nu^{1.30} \tag{3}$$

relation (1) presents the angular distribution resulting not only from single scattering but from multiple scattering, too. The parameter ν is the mean number of single scattering events which an electron suffers along its path. In this way ν single scatterings by real atoms can be reduced to one single scattering by a fictive atom, which is characterized by α.

The Bethe range covered by the electrons is usually subdivided into 60 segments of equal length and, moreover, the first segment is subdivided into subsegments approximately equal to the mean free path between two single scattering acts. The angular deflection of an electron per segment or subsegment is determined statistically via the probability of a scattering through an angle ϑ given by (1) to (3). Using a simple power function which approximates Bethe's energy loss law,

$$-\frac{dE}{ds} = 24.6 \frac{\varrho}{A} \frac{Z^{4/5}}{E^{2/3}}, \tag{4}$$

the retardation of an electron along a segment is calculated. The symbol s indicates the travelled path, ϱ the density and A the relative atomic mass (E in keV, s in μm, and ϱ in g/cm³). The resulting simple character allows the model to run quickly on a 16 bit microcomputer. Comparison of results of the presented Monte Carlo model with experimental results of Neubert and Rogaschewski [9] shows the efficiency of the model (see [8]).

3. The Depth–Dose Distribution

By measurements on angle-lapped specimens, as shown in Fig. 1a, in [5] an universal depth distribution for the range of primary energy 10 keV $\leq E_{pr} \leq$ 45 keV was found.

Fig. 2 compares the gained universal depth–dose distribution (curve) and results of the applied Monte Carlo simulation (points) for 20 keV. Energies different from

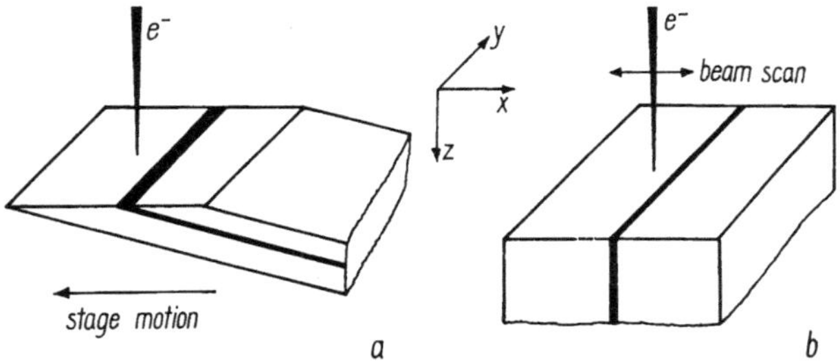

Fig. 1. Geometry of specimens

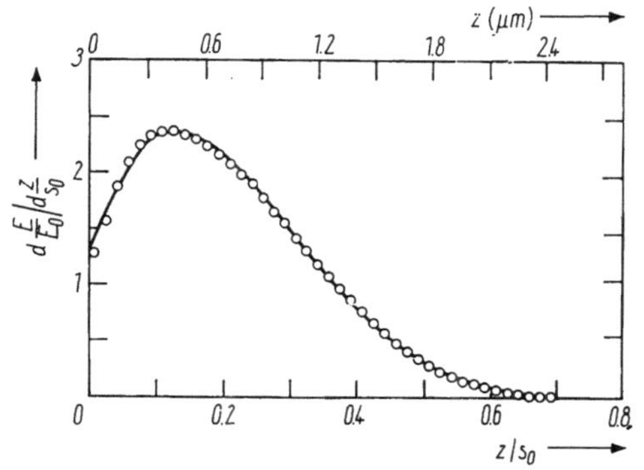

Fig. 2. Comparison in $GaAs_{0.35}P_{0.65}$ between the empirical description [5] of the measured depth-dose distribution (curve) and the simulated one (points) resulting from 10^5 histories of electrons ($E_{pr} = 20$ keV). (At the abscissa read E_{pr} instead of E_0)

20 keV cause slight deviations of the calculated distribution from the measured one. These are due to a slightly different energy dependence of the simulated scattering process; the depth of the maximum calculated exhibits an energy dependence of $E_{pr}^{1.71}$ instead of $E_{pr}^{1.65}$ as it was experimentally found in [5]. Monte Carlo calculations performed for 20 keV and the atomic number range of $24 \leqq Z \leqq 32$ shows that in this range the results vary only slightly. Accordingly, in this range relations found in [5] are almost valid, provided that the distance of the maximum under the surface a is multiplied by ϱ_{GaAsP}/ϱ.

4. Experiments

For determination of the generation volume the electron beam induced barrier current (BC) was measured. The measuring arrangement utilized in detecting the short-circuit barrier current (BC) was described in [5]. In contrast to [5] the specimens used here were not angle-lapped ones (Fig. 1a) but perpendicular to the p–n junction cleaved structures (Fig. 1b). The measurements were carried out by scanning of the electron beam in x-direction. The effect of scanning along x is then to alter the distance between the point of impact of the electron beam and the space charge region, where the electron–hole pairs generate the BC. The lateral distribution of the generation source can be obtained in this manner if the space charge region and the diffusion length of minority carriers are small compared with the lateral extension of generation volume.

A computer controls the step-by-step scanning of the structure investigated by the electron beam [10].

Freshly cleaved surfaces were used for the measurements in order to prevent (i) a strong recombination of excess carriers at the scanned surface especially for low primary energies and (ii) the bending of the space charge region at the surface. (This effect was experimentally obtained on aged surfaces for low primary energies.) The specimens used in the experiments here were prepared from different epitaxial discs $GaAs_{0.35}P_{0.65}$ ($\varrho = 4.4$ g/cm³) which have an effective atomic number $Z_{eff} = \sum_i p_i Z_i = 28.7$ (p_i in wt%). The donor (Te) concentrations are in the region 8×10^{17} cm$^{-3} \leqq N_{Te} \leqq 3 \times 10^{18}$ cm^{-3}. The p–n junction was prepared by Zn diffusion. The penetration depth of the junction was about 15 μm.

5. Measurements

Fig. 3a shows a record of the position dependence of the BC ($E_{pr} = 40$ keV, current absorbed by the specimen $I_{abs} = 2 \times 10^{-11}$ A). The needle-like appearance in the immediate vicinity of the p–n junction is characteristic of measurements at high primary energies on material with short diffusion lengths. The needle-like shape of the BC signal surrounding the p–n junction must be understood as a hint for the complicated form of the lateral distribution function of electron–hole pairs. The position dependence of the BC was tested for various specimen-absorbed currents (in the range 5×10^{-12} A $\leqq I_{abs} \leqq 5 \times 10^{-9}$ A). For currents $I_{abs} \leqq 10^{-10}$ A the position dependence of the BC was nearly independent of the value of I_{abs} at the primary energies used here. The change of the position dependence of the BC with increasing $I_{abs} > > 10^{-10}$ A can be caused by (i) increasing diameter of the electron probe, (ii) Coulomb interaction of generated carriers, and (iii) increasing diffusion length of minority carriers attended with the increasing injection level in the solid. The experiments discussed here were carried out with specimen-absorbed currents $I_{abs} < 10^{-10}$ A in order to prevent the former described effects.

By scanning of a surface perpendicular to the p–n junction with electron beam (see Fig. 1b) the diffusion lengths of minority carriers were obtained from the position dependence of the BC ($E_{pr} = 10$ keV). For the determination normalized BC-values less than some percent were used.

Fig. 4a shows the measurements of the total barrier current as a function of the position for the primary energies in the range 10 keV $\leqq E_{pr} \leqq$ 45 keV measured on the specimen D1 ($N_D = 3 \times 10^{18}$ cm^{-3}; $L = 0.08$ μm; cf. [5]). In contradiction to the experiments in [5] the BC shown in Fig. 4a is proportional to the energy transfer in discs (thickness 0.2 μm) perpendicular to the surface. The maximum signal is generated if the point of impact of the electron beam coincide with the space charge region. As found for the depth distribution function the maximum signal increases with decreasing energy except for the measurement for 10 keV. For $E_{pr} = 10$ keV the signal decreases due to the recombination of excess carriers at the surface. The distance of the depth distribution maximum to the surface is only 0.13 μm and consequently comparable with the diffusion length. Neglecting the surface recombination the

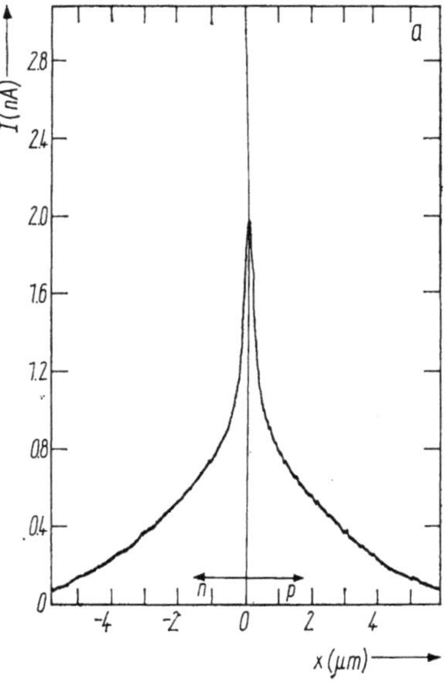

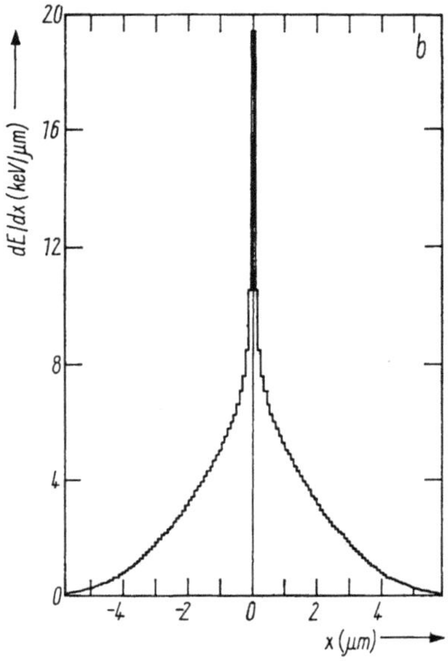

Fig. 3. Lateral dose distribution ($E_{pr} = 40$ keV) in $GaAs_{0.35}P_{0.65}$. a) measured on specimen D1, $L_e = L_h = 0.08$ μm, $I_{abs} = 2 \times 10^{-11}$ A, b) calculated by way of 3×10^4 electron trajectories for an infinitely small beam diameter

Monte Carlo calculation provides the maximum signal increase with decreasing energy in the whole energy region investigated here.

Fig. 4b shows the measurements on specimen D2 ($N_D = 8 \times 10^{17}$ cm^{-3}; $L_h = 0.35$ μm) for the same experimental conditions. In contradiction to Fig. 4a the signal is highest for $E_{pr} = 45$ keV. The signal decreases with decreasing energy due to the increasing recombination of excess carriers at the surface. The distance of the depth distribution maximum to the surface is 0.4 μm for 20 keV and comparable with the diffusion length $L_h = 0.35$ μm in specimen D2. The poorer local resolution in this experiment is clearly visible for small values of x especially for high primary energies. Experiments, as shown in Fig. 4a and Fig. 4b, were carried out on a set of specimens with different values of diffusion lengths in order to test the influence of the minority carrier diffusion on the position dependence of $I(x)$ for different primary energies of electrons. The experiments discussed in the following sections were carried out on specimens D1 ($L_h = 0.08$ μm) prepared from the epitaxial disc used in [5]. Except for the immediate vicinity of the point of impact the diffusion of minority carriers has no remarkable effect on the experimentally obtained position dependence of BC for D1. The maximum total signal for specimen D2 is ten times higher than the maximum BC for specimen D1. Taking into account the creation energy of one electron–hole pair we conclude that the value of maximum total BC obtained on specimen D1 is generated by only 1% of the electron–hole pairs (excited by the electron beam $E_{pr} = 45$ keV).

In Fig. 5 the normalized position dependence of the BC is shown in the vicinity of the p–n junction measured on specimen D1. The outstanding feature of the energy transfer measured in layers perpendicular to the surface of the specimen is the part nearly independent of the primary energy near the impact point of electron beam. This part generates the needle-like appearance of the function $I(x)$ for high primary energies (see also Fig. 3a). Whereas the former discussed experiments (Fig. 4a, b) were carried out on an electron beam microanalyser (SEMQ, working distance 0.6 cm) the results in Fig. 5a were obtained by measurements with a scanning electron microscope (JSM U3, working distance 1.4 cm).[3] The comparison of the experimentally obtained data shows the influence of the electron-optical design on the energy inde-

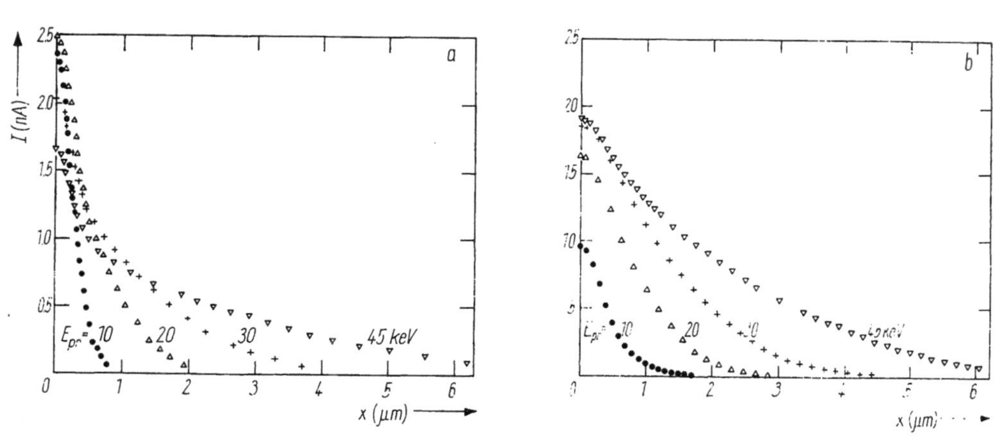

Fig. 4. Total BC in GaAsP vs. position in the energy range $10 \leqq E_{pr} \leqq 45$ keV ($I_{abs} = 2 \times 10^{-11}$ A) a) measured on specimen D1; $L_h = 0.08$ μm, b) measured on specimen D2; $L_h = 0.35$ μm

[3]) These experiments were carried out by Dr. H. Menniger, Zentralinstitut for Elektronenphysik der AdW, Berlin, GDR.

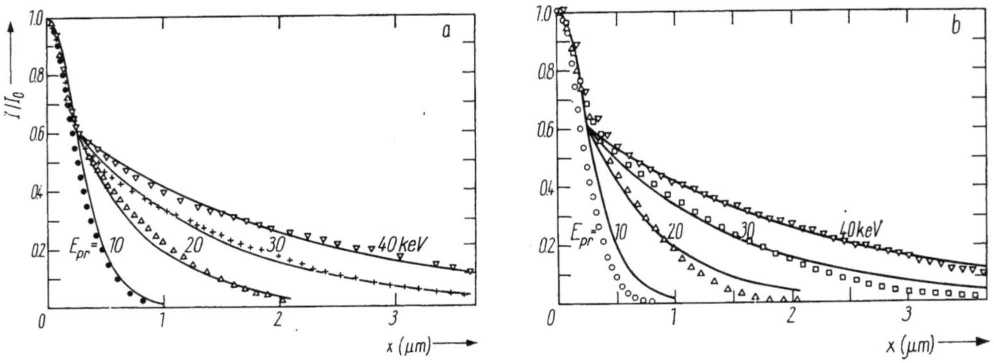

Fig. 5. Position dependence of the normalized BC in GaAsP; comparison of the empirical relation (7, 8) (solid lines) and a) experimental points measured on specimen D1 ($L = 0.08$ μm, $I_{abs} = 2 \times 10^{-11}$ A), b) Monte-Carlo simulation of 5×10^4 trajectories of electrons incident within a finely focused beam (points)

pendent part of the function $I(x)$. Contrary to this fact $I(x)$ is nearly independent of the electron-optical design for greater distances between the point of impact and the space charge region.

The experimentally obtained position dependence of the BC can be described by an energy independent Gaussian distribution in the neighbourhood of the impact point of the electron beam. For greater distances the position dependence of the BC can be fitted by an exponential function with an exponent dependent on the energy,

$$I(x) = I_0 \begin{cases} \exp\left[-\left(\frac{x}{0.38}\right)^2\right] & \text{for} \quad 0 \leqq |x| \leqq 0.27 \\ 0.6 \exp\left[-\left(\frac{|x| - 0.27}{h}\right)\right] & \text{for} \quad 0.27 < |x| \end{cases} \tag{5}$$

with

$$h = 4.64 \times 10^{-3} E_{pr}^{1.65} \qquad (x, h \text{ in } \mu\text{m}; E_{pr} \text{ in keV}) . \tag{6}$$

Fig. 5a shows the curves calculated from (5) and (6) in good agreement with the experimentally obtained results in the whole energy region.

6. Comparison with Monte-Carlo Calculations

The measured and the calculated lateral dose distributions (Fig. 3) show that the distributions can be divided into two parts: Part I is formed by a needle-like peak which directly results from the focused electron beam penetrating into the solid target. On the other hand, the progressive spreading of the electron beam due to the interaction with the target produces part II.

The needle-like peak (part I) in Fig. 3b calculated for a finely focused electron beam crucially complicates the experimental determination of the lateral dose distribution. With respect to the width of this peak the lateral resolution of the measurement is not sufficient. For this reason, part I of the measured lateral dose distribution is influenced not only by the scattering and energy loss of the incoming electrons but decisively by the diffusion of the electron–hole pairs, the width of the space charge region, and probably also by the distribution of the electrons within the electron probe. These influences affect the measurement of the lateral dose distribution in such a way that part I in Fig. 3a shows a lower and broader maximum.

The last-mentioned influences, which affect the shape of the measured part I, are independent of the electron–target interaction. That is why part I in Fig. 5a is not influenced by the incident energy. Contrary to this, part II caused by the spreading of the incident beam increases proportionally to the electron range. This behaviour of the BC is illustrated in Fig. 5b by the simulation of the lateral spread of an incident finely focused electron beam (cf. [11]): The generated BC is computed on the assumption that the rate of pair creation is proportional to the rate of energy dissipation. The surface recombination is not considered. However, the simulation takes into account the diffusion of the minority carriers and the effect of the space charge region by the convolution of the lateral dose distribution. If the calculation only considers the diffusion length $L_e = L_h = 0.08$ μm and the actual width of the space charge region $W = 0.03$ μm the agreement is worse than shown in Fig. 5b. This seems to be due to the fact that scarcely calculable factors influence the position dependence as e.g. an ill-defined focused electron probe or errors resulting from the Monte Carlo calculation of small-angle scattering. Such factors which are hard to estimate are roughly taken into consideration by increasing W for the Monte Carlo simulation of Fig. 5b up to 0.2 μm.

An exponential slope of part II has been found experimentally. As the lateral spread of electrons cannot be larger than their depth penetration, this slope can only be valid for a BC not too low. We suppose the measured slope of lower BC values to be influenced by deviations of the electron probe from an accurately focused electron beam. Accordingly, in Fig. 5b the Monte Carlo-simulated BC with its stronger decrease displays a more realistic behaviour regarding an accurately focused beam.

The realistic description of the Monte Carlo model, verified by Fig. 2 and 5b, also guarantees a realistic result by computing the spatial dose distribution. Fig. 6 shows two extreme cases of the spatial dose distribution projected on the plane of the paper: (i) typical of scanning electron microscopy (Fig. 6a), and (ii) typical of electron microprobe analysis (Fig. 6b). The relative doses q marking the equidose lines in Fig. 6 differ by a factor of about 1/3. For representing the distribution a normalization has been chosen so that for an extremely small probe the equidose line characterizing the region of incidence is set unity. In Fig. 6a this small region near the impact point exhibits a relatively strong absorbed dose directly caused by the incident finely focused electron beam. The further two equidose lines ($q = 0.3, 0.1$) indicate in which way the electron beam penetrates into the target and is spread out. Starting from the

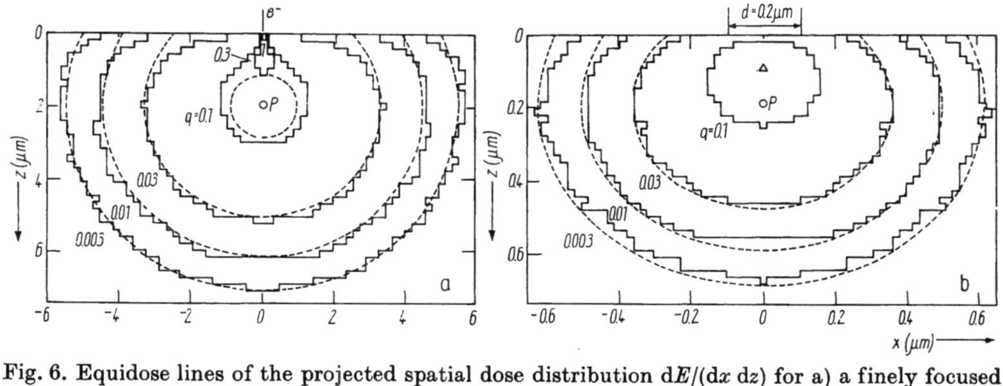

Fig. 6. Equidose lines of the projected spatial dose distribution $dE/(dx\,dz)$ for a) a finely focused beam of 40 keV and b) a beam of 10 keV 0.2 μm in diameter.
The histograms present the result of the Monte Carlo calculation of 5×10^4 histories of electrons and the dashed curves follows from the analytical description (7)

third line ($q = 0.03$) the lines have an almost circular shape with the point P as the centre. After Archard's model [6] this point can be regarded as an isotropic point source situated in the diffusion depth. Fig. 6b reveals that due to the relatively large beam diameter the region directly caused by the incident beam shows a weaker absorbed dose. Furthermore, the maximum of the projected spatial dose distribution (designated by Δ) is shifted from the surface in the direction of the point source. Obviously, this maximum is due to the higher absorbed dose resulting from electrons which wander in this depth mainly parallel to the surface. Apart from some details the lines indicating doses more distant from the region of incidence resemble the corresponding lines of Fig. 6a. Considering these facts and the fact that the shape of lines in Fig. 6a is hardly energy-dependent we can conclude that the spatial distribution displays the same behaviour as the lateral one: Part of the distribution is strongly sensitive whereas the other is hardly sensitive with reference to the beam diameter; they are part I and II, respectively.

7. Analytical Description

As pointed out above, the spatial dose distribution can be roughly explained by Archard's simple diffusion model: The energy absorbed while the electrons penetrate down to the point source forms the equidose lines with $q = 1, 0.3, 0.1$ (Fig. 6a). On the other hand, the electrons travelling away from the point source as in a diffusion cause the lines with $q = 0.03, 0.01, 0.003$ (Fig. 6a). This part resulting from the diffusion process shall be described analytically on the basis of the Monte Carlo results. At first Fitting [4] has shown that the depth–dose distribution can be approximately derived from a spherically symmetric volume distribution given by a three-dimensional Gaussian function. Different from [4] the spherical symmetry is abandoned and the centre of the Gaussian function is placed at Archad's point source P instead of at the depth of the maximum of the depth–dose. Resulting from these assumptions the projection of the part II of the spatial dose distribution on the plane of the paper can be analytically described by

$$\frac{\mathrm{d}E}{\mathrm{d}x\,\mathrm{d}z} = \frac{0.83E_{\mathrm{pr}}}{2\pi\sigma_x\sigma_z} \exp\left[-\frac{x^2}{2\sigma_x^2} - \frac{(z - z_\mathrm{D})^2}{2\sigma_z^2}\right] \quad \text{for} \quad z > 0 \tag{7}$$

with

$$\sigma_x^2 = \sigma_\mathrm{s}^2 + \sigma_\mathrm{c}^2 + \sigma_\mathrm{b}^2 \quad \text{and} \quad \sigma_z^2 = \sigma_\mathrm{s}^2 .$$

Here σ_x and σ_z signify the standard deviations of the distribution in x- and z-direction, respectively. The quantity σ_s marks the standard deviation of the spherically symmetric approach. Taking into account that there is no real point source due to the spreading to the incident beam the correction σ_c is introduced. Provided the distribution of primary electrons in the focused spot also follows a Gaussian function the standard deviation of this distribution is considered to be linked with the beam diameter by $\sigma_\mathrm{b} = 0.60d$ [12]. The diffusion depth z_D, determining the position of the centre of the Gaussian function (7), has been computed with $Z_\mathrm{eff} = 28.7$ after relation (21) given in [13]. For the employed GaAsP the parameters used in (7) can be estimated by

$$\sigma_\mathrm{s} = 0.17s_0\,, \qquad \sigma_\mathrm{c} = 0.070s_0\,, \quad \text{and} \quad z_\mathrm{D} = 0.17s_0\,. \tag{8}$$

Considering GaAsP as a mixed-element target the Bethe range results from (4) in

$$s_0 = \frac{E_{\mathrm{pr}}^{5/3}}{41\varrho \sum\limits_i (p_i Z_i^{4/5}/A_i)}\,, \tag{9}$$

where p_i denotes the weight percentage of the i-th element (s_0 in μm, E_{pr} in keV, and ϱ in g/cm³). Because of the normal incidence $\sigma_y = \sigma_x$ is valid. Thus, the actual spatial distribution may be easily obtained by multiplying the projected distribution (7) by the one-dimensional Gaussian function of y. Fig. 6 demonstrates that the analytical description (7) reflects the influence of the beam diameter as well as of the primary energy (normalization factor in Fig. 6 $1.1E_{pr}/\sigma_z^2$).

We suppose that part I, too, may be approximated by a Gaussian function, similarly to the approximation used by Donolato [14]. For GaAsP, however, the analytically described part II undoubtedly represents the main part of the dose distribution; moreover, this part yields the extension of the generation volume. Only near the surface this description differs from the Monte Carlo result (Fig. 6) owing to the electron backscattering. In order to get the atomic number dependence of the Gaussian approximation of part II Archad's model may be used as a key.

References

[1] A. F. Makhof, Fiz. tverd. Tela **2**, 2161 (1960).
[2] T. E. Everhart and P. H. Hoff, J. appl. Phys. **42**, 5837 (1971).
[3] T. Matsukawa, R. Shimizu, K. Harada, and T. Kato, J. appl. Phys. **45**, 733 (1974).
[4] H.-J. Fitting, H. Glaefeke, and W. Wild, phys. stat. sol. (a) **43**, 185 (1977).
[5] G. Oelgart and H. Scholz, phys. stat. sol. (a) **75**, 547 (1983).
[6] G. D. Archard, J. appl. Phys. **32**, 1505 (1961).
[7] L. Curgenven and P. Duncumb, Tube. Investments Res. Rep. No. 303 (1971).
[8] U. Werner and J. Heydenreich, Ultramicroscopy, to be published.
[9] G. Neubert and S. Rogaschewski, phys. stat. sol. (a) **59**, 35 (1980).
[10] G. Oelgart, H. Scholz, and R. Stegmann, Exper. Tech. Phys. **28**, 485 (1980).
[11] B. Akamatsu, J. Henoc, and P. Henoc, J. appl. Phys. **52**, 7245 (1981).
[12] O. C. Wells, Scanning Electron Microscopy, McGraw-Hill Publ. Co., New York 1974 (p. 72).
[13] U. Werner, H. Bethge, and J. Heydenreich, Ultramicroscopy 8, 417 (1982).
[14] C. Donolato, phys. stat. sol. (a) **65**, 649 (1981).

(Received May 29, 1984)

phys. stat. sol. (a) 85, 215 (1984)

Subject classification: 13.4 and 14.3; 22.2.1

Department of Electrical and Electronic Engineering, University of Nottingham[1])

Electrical Measurements on Silver-Diffused GaAs

By

G. A. ADEGBOYEGA[2])

Silver-doped samples of p-type GaAs are obtained from diffusions at 1000 °C. The doping levels are found from preliminary experiments with radioactive silver. *C–U*, Hall effect, and conductivity measurements are carried out on samples at 300 K. A reasonably good correspondence is found between silver and hole concentrations. Analysis of the resulting data shows that silver is most probably a double acceptor in GaAs.

Silberdotierte Proben von p-leitendem GaAs werden durch Diffusion bei 1000 °C erhalten. Die Dotierungsniveaus werden aus vorhergehenden Experimenten mit radioaktivem Silber bestimmt. *C–U*-, Halleffekt- und Leitfähigkeitsmessungen werden bei 300 K an den Proben durchgeführt. Eine vernünftige Übereinstimmung zwischen der Silber- und Löcherkonzentration wird gefunden. Die Analyse der Werte zeigt, daß Silber höchstwahrscheinlich ein Doppelakzeptor in GaAs ist.

1. Introduction

Most previous reports [1 to 4] on the diffusion of silver in GaAs were aimed at determining the atomic diffusion profiles and very little or no data were presented on the electrical effects of silver in GaAs. Recently, Tuck and Adegboyega [5] showed that silver acts as an acceptor in GaAs. In this paper, we present a more detailed report and discussion of the electrical measurements carried out on silver-diffused GaAs.

2. Experimental Details

The sample preparation and diffusion details have been described elsewhere [5], but may be briefly outlined as follows. A slice of n-type ($n \approx 2 \times 10^{16}$ cm^{-3}) GaAs was cleaned and put into a silica ampoule together with a small amount of radioactive silver and arsenic (when needed). The loaded ampoule was evacuated to below 10^{-4} N/m^2 and sealed. It was then put into a diffusion furnace preset to 1000 °C for specific periods. At the end of each period, the sample was removed and the diffusion profile was plotted by using sectioning and counting techniques.

The above diffusions were then duplicated using non-radioactive silver dopant. The non-radioactive diffused specimen were each cleaved into two parts and used for electrical measurements. One part was used for the *C–U* measurements and the other for the Hall effect and conductivity measurements. Capacitance–voltage (*C–U*) measurements were made on diffused samples by means of a mercury probe [6]. This was combined with serial sectioning in order to produce a carrier concentration profile. Typical profiles are shown in Fig. 1 together with corresponding atomic distributions. *C–U* measurements were also carried out on the as-grown n-type starting material and on other n-type (e.g. 2×10^{18} cm^{-3}) commercially available GaAs for comparison purposes.

---

[1]) Nottingham NG7, 2RD, U. K.

[2]) Present address: Department of Electronic and Electrical Engineering, University of Ife, Ile-Ife, Nigeria.

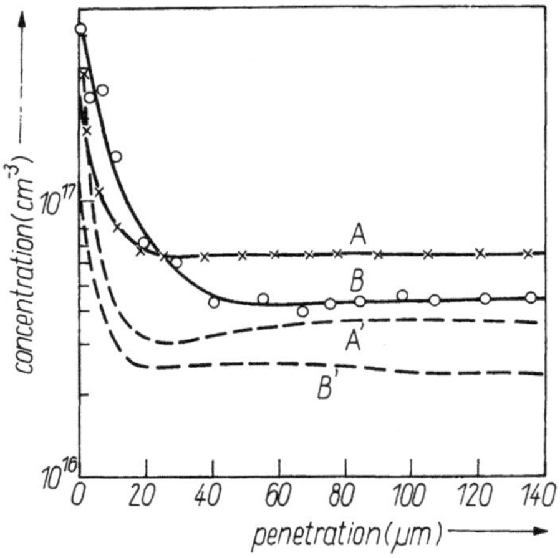

Fig. 1. Carrier concentration profiles for GaAs diffused at 1000 °C for 1 h. Curve A without, curve B with arsenic overpressure. A′ and B′ give corresponding atomic distribution

One of the advantages of the shape of the diffusion curves of Fig. 1 is that the profiles are essentially flat for most of their penetration depth. It was therefore possible to producc homogeneously doped, diffused specimens of known doping concentration by simply etching off about 50 μm from specimen surfaces after diffusion. Standard conductivity and Hall effect measurements were then performed on the resulting specimens. Hall effect measurements were also performed on the n-type starting material as well as on other commercially available n-type GaAs.

3. Results and Discussion

C–U and Hall effect measurements showed samples to be electrically p-type after diffusion and it was established that the p-type conductivity was not due to thermal conversion. The results of Hall effect and conductivity measurements are shown in Table 1. For comparison purposes, the tracer atomic concentration and carrier concentrations obtained from *C–U* and Hall effect measurements are shown in Table 2. This table shows a reasonably good correspondence between silver and hole concentrations. A close look at the two tables shows that measurements on as-grown n-type materials gave results which were close to nominal values. However, there is a discrepancy between the carrier concentration values resulting from Hall effect and *C–U* measurements on diffused p-type specimens.

Table 1
Hall-effect measurement

specimen	N (cm^{-3})	ϱ (Ω cm)	mobility ($cm^2\ V^{-1}\ s^{-1}$)
n-type wafer used for diffusion ($n \approx 2 \times 10^{16}\ cm^{-3}$)	1.78×10^{16}	0.13	2667
n-type as -grown ($n \approx 2 \times 10^{18}$)	1.92×10^{18}	0.0015	2167
p-type Ag-diffused	2.28×10^{16}	0.66	415
p-type Ag diffused + As	1.37×10^{16}	0.88	559

Table 2

Comparison of radiotracer, C–U and Hall measurements

specimen	radiotracer atom (cm^{-3})	C–U concentration (cm^{-3})	Hall-effect concentration (cm^{-3})
n-type as-grown ($n \approx 2 \times 10^{16}$ cm^{-3})	—	2.29×10^{16}	1.78×10^{16}
n-type as-grown ($n \approx 2 \times 10^{18}$ cm^{-3})	—	2.03×10^{18}	1.92×10^{18}
p-type Ag-diffused	3.6×10^{16}	6.2×10^{16}	2.28×10^{16}
p-type Ag-diffused + As	2.6×10^{16}	4.5×10^{16}	1.37×10^{16}

Theoretically, if a concentration of N_a acceptors is diffused into a substrate material with an initial concentration of donors N_d, then the net concentration of acceptors is $N_a > N_d$ should be $p = N_a - N_d$. Based on this simple expression, if silver atoms are assumed to be single acceptors and if all acceptor atoms are completely ionized, then the Hall-effect measurements should give net acceptor concentration values of 1.82×10^{16} cm^{-3} and 0.82×10^{16} cm^{-3} for the silver-diffused and silver + arsenic-diffused specimens, respectively. These values fall short of the experimental figures of 2.28×10^{16} cm^{-3} and 1.37×10^{16} cm^{-3}, respectively. The corresponding C–U acceptor concentration values should give 1.31×10^{16} cm^{-3} and 0.31×10^{16} cm^{-3}, respectively. These values fall very short of the experimental values of 6.20×10^{16} and 4.50×10^{16} cm^{-3}, respectively.

Using the same simple expression and still assuming complete ionization, if Ag atoms act as double acceptors, then the expected Hall results should give 5.42×10^{16} cm^{-3} and 3.42×10^{16} cm^{-3} for the Ag-diffused and Ag + As-diffused specimens, respectively. The corresponding C–U results should give 4.91×10^{16} cm^{-3} and 2.91×10^{16} cm^{-3}. Under the assumption of Ag atoms being double acceptors, the Hall effect results fell short of theoretical values, while C–U results are a little bit above the expected value. A comparison of the above theoretical and experimental values suggests that Ag is most probably a double acceptor in GaAs.

It is interesting to note that Tuck and Adegboyega [5] have proposed a probable diffusion mechanism for Ag in GaAs in which most of the silver exists substitutionally on Ga sites while a small proportion occurs interstitially. Silver atoms on Ga sites would act as double acceptors, while the interstitial silver would act as a single donor, silver being a monovalent atom. The observed acceptor concentrations in this work seem to agree with the proposed mechanism. Possible reasons for the discrepancy between Hall-effect and C–U values are given below.

Shishiyanu and Boltaks [7] have reported a deep acceptor level of 0.11 eV above the top of the valence band due to silver in GaAs. The low acceptor concentration values obtained in the Hall-effect measurements may therefore be due to incomplete ionization of silver atoms if the energy levels due to silver are fairly deep. Whether this statement is correct or not can only be proved by an experimental determination of the energy levels of Ag in GaAs, for instance by measuring the Hall effect in a broader temperature range. In the case of the C–U measurements, in the surface layer as a result of the band bending down, the acceptor levels should be completely ionized. This might be the reason for the high acceptor concentration values obtained for the C–U results.

4. Conclusions

C–U- and Hall-effect measurements have shown silver to be an acceptor in GaAs. While Hall-effect results gave low acceptor concentration levels which were attributed to incomplete ionization due to deep acceptor levels of Ag in GaAs, C–U results gave high acceptor concentration levels which have been connected to the band bending and hence complete ionization at the surface layer of the GaAs. The results of these measurements in conjunction with the simple theory of charge compensation have led to the conclusion that silver is probably a double acceptor in GaAs.

References

[1] V. Rybka, M. Yoseli, and M. Aoki, J. Phys. Soc. Japan **17**, 1812 (1962).
[2] B. I. Boltaks and F. S. Shishiyanu, Soviet Phys. — Solid State **5**, 1680 (1964).
[3] T. D. Dzhafarov and G. S. Kulikov, Soviet Phys. — Solid State **12**, 1237 (1970).
[4] B. I. Boltaks, T. D. Dzhafarov, G. S. Kulikov, and V. M. Mikhelashvili, Soviet Phys. — Solid State **13**, 2631 (1971).
[5] B. Tuck and G. A. Adegboyega, J. Phys. D **13**, 433 (1980).
[6] P. R. Jay and B. Tuck, phys. stat. sol. (a) **46**, 199 (1978).
[7] F. S. Shishiyanu and B. I. Boltaks, Soviet. Phys. — Solid State **8**, 1053 (1966).

(Received March 20, 1984)

phys. stat. sol. (a) 85, 219 (1984)

Subject classification: 13.4 and 14.3.3; 22.4.3

Institute of Radio Engineering and Electronics, Czechoslovakian Academy of Sciences, Prague[1]) *(a)*
and Department of Mathematics and Physics, Charles University, Prague[2]) *(b)*

Analysis of DLTS Curves of Aggregated Deep Level Impurities

By

K. Žďánský (a) and N. T. Thuc Hien[3]) (b)

Rate equations of the occupancy of aggregated impurities including two interaction parameters are solved. The DLTS curves, in particular their widths, heights, and shifts for various interaction strengths are calculated. Further, the filling factor as a function of the filling pulse duration is calculated for various interactions. Finally, the DLTS on n-CdTe Schottky junctions are measured; the dominant peak is interpreted as a deep level aggregate and its parameters are determined.

На основе дифференцияльного уравнения с двумя параметрами взаимодействия была описана временная зависимость захвата электронов на примесных скоплениях. Крывые релаксационной спектроскопии глубоких уровней (DLTS), именно их ширины, высоты и смещения вычисленны для разных сил взаимогействия. Фактор заполнения в зависимости от продолжительности заполнающего импульса вычислен для разных взаимодействий. На образцах CdTe n-типа проводимости с переходом Шотки были измерены крывые DLTS; доминирующий пик интерпретируется как скопление глубоких уровней. Параметры наиденого примесного скопления былы определены.

1. Introduction

Deep level transient spectroscopy (DLTS) is a powerful tool for investigating deep level impurities in semiconductors [1] provided that the impurities are of a small concentration and that they are randomly distributed in the lattice; hence the capacitance transient of a junction is exponential. A detailed analysis of the DLTS curves corresponding to non-exponential transients due to high concentration of impurities in a p–n junction has been published in our recent paper [2]. To describe the capture and emission processes on aggregates of deep level impurities a potential barrier model was introduced some time ago [3] and it was used recently for the interpretation of the DLTS spectra [4]. An exact solution of the rate equation describing the occupancy of aggregated impurities including the thermal emission only has been derived in our recent paper [2]. It is the aim of this paper to bring a detailed analysis of the rate equations for the filling and emptying periods in DLTS and to apply it to the 0.75 eV level in CdTe.

2. Theory

The rate equation for the occupation number n_T of aggregated deep acceptors in an n-type semiconductor reads as follows [4]:

$$\frac{dn_T}{dt} = \frac{N_T - n_T}{\tau_1} e^{-q\varphi_1/kT} - \frac{n_T}{\tau_2} e^{q\varphi_2/kT} , \tag{1}$$

[1]) Lumumbova 1, 18251 Praha 8, Czechoslovakia.
[2]) Ke Karlovu 3, 12000 Praha 2, Czechoslovakia.
[3]) On leave from the Department of Physics, University of Hanoi, Vietnam.

where N_T is the concentration of deep acceptors, τ_1^{-1}, τ_2^{-1} are the capture and the emission rates of an isolated deep acceptor, φ_1 is the total Coulomb potential in the place of a deep acceptor due to the other charged (filled with electrons) acceptors of the aggregate, and $q\varphi_2$ is the rise of the deep acceptor energy due to the Coulomb field; q is the absolute value of the electron charge, k the Boltzmann constant, and T the absolute temperature.

The values of φ_1 and φ_2 depend on the occupation number n_T. In the case of a line aggregate the dependence is linear; let us introduce the following relations:

$$q\varphi_1 = E_1 \frac{n_T}{N_T}, \tag{2a}$$

$$q\varphi_2 = E_2 \frac{n_T}{N_T}, \tag{2b}$$

where the first and the second interaction energies E_1 and E_2 are the maximum values of $q\varphi_1$ and $q\varphi_2$.

After the filling period the capture rate τ_1^{-1} approaches zero (in the depletion region of a junction) and (1), using (2b), can be rewritten for the occupation factor

$$n(t) = \frac{n_T}{N_T} \tag{3}$$

as follows:

$$\frac{dn(t)}{dt} = -\frac{n(t)}{\tau_2} e^{n(t)E_2/kT}. \tag{4}$$

The solution of (4) has been shown in our recent paper [2] to have the following form:

$$n(t) = a\, e^{-g(b_2 a)}\, e^{g[b_2 n(t)]}\, e^{-t/\tau_2}, \tag{5}$$

where a is the filling factor,

$$a = \frac{n_0}{N_T} \tag{6}$$

with n_0 being the initial value of n_T. Further, the factor

$$b_2 = \frac{E_2}{kT} \tag{7}$$

is introduced and the function $g(z)$ is defined by the infinite series

$$g(z) = z - \frac{z^2}{2 \cdot 2!} + \frac{z^3}{3 \cdot 3!} - \dots \tag{8}$$

During the filling period both terms on the right-hand side of (1) have to be considered in general. While for isolated centres the capture process is always many orders of magnitude faster than the emission process, this rule may be violated for an aggregate as pointed out in [4]. However, in favourable cases, when the effect of φ_1 and φ_2 does not overcome it, the rule may be still maintained as it is indeed the case for the 0.75 level in CdTe discussed below. Thus in the following we assume that the second term on the right-hand side of (1) can be neglected during the filling period. Equation

(1), using (2a) can be rewritten for the filling factor a (6) as follows:

$$\frac{da}{dt_p} = \frac{1 - a}{\tau_1} e^{-aE_1/kT} . \tag{9}$$

where t_p is the duration of the filling pulse. The solution of (9) can be calculated in the same way as the one of (4) and with the initial condition ($a = 0$ if $t_p = 0$) it reads as follows:

$$a = 1 - e^{-g(b_1)} e^{g[b_1(1-a)]} e^{-t_p/\tau_1'} . \tag{10}$$

where

$$b_1 = \frac{E_1}{kT}, \tag{11}$$

and

$$\tau_1' = \tau_1 e^{b_1} \tag{12}$$

have been introduced.

3. DLTS Curve

Equation (5) represents a non-exponential transient for the occupation factor $n(t)$. Assuming the differential type of detection [1] the DLTS curve is a signal proportional to the difference of $n(t)$ at two subsequent times t_1 and t_2 displayed versus the temperature T. The temperature dependence enters into the DLTS curve through the temperature-dependent values of τ_2 and b_2 in (5). However, in the vicinity of the DLTS peak the temperature dependence of b_2 is less pronounced than that of τ_2, thus the DLTS curve can be analyzed with a constant b_2 in the first approximation. When doing so it is advantageous to define a new DLTS-like curve

$$y(x) = n(t_1) - n(t_2) \tag{13}$$

as a function of

$$x = \ln\left(\frac{\tau_2}{\tau_x}\right), \tag{14}$$

where τ_x^{-1} is the rate window [1],

$$\tau_x = \frac{(t_1 - t_2)}{\ln (t_1/t_2)} . \tag{15}$$

Let us point out that the function $y(x)$ depends on the ratio t_1/t_2 (or equivalently on t_1/τ_x), i.e. different (t_1, t_2) pairs constrained with a constant t_1/t_2 give the same function $y(x)$. In the limit of $b_2 = 0$ the transient $n(t)$ in (5) is exponential and the maximum of $y(x)$ occurs at $x = x_m = 0$. For non-zero b_2 the maximum of $y(x)$ occurs at a non-zero positive x_m, the height of the peak is decreased and the linewidth is broadened.

Let us show how the activation energy E_T of an isolated deep acceptor can be determined. The emission rate τ_2^{-1} is given by the following relation [1]:

$$\tau_2^{-1} = \lambda T^2 e^{-E_T/kT} . \tag{16}$$

where λ is a parameter depending on the electron capture cross-section. Inserting (16) into (14) and substituting x_m for x and the corresponding temperature T_m for T we get

$$\ln T_m^2 \tau_x = \frac{E_T}{kT_m} - \ln \lambda - x_m . \tag{17}$$

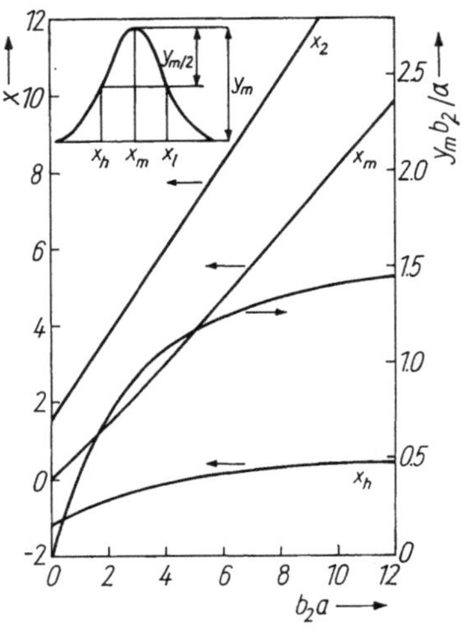

Fig. 1. Plots of the parameters x_{m}, x_{l}, and x_{h} vs. b_2a and the plot of $y_{\mathrm{m}}b_2/a$ vs. b_2a. In the insert the meaning of the parameters is shown

The energy E_T can be determined from (17) by using the Arrhenius plot of $T_2^{\mathrm{m}}\tau_x$ verus $(kT_m)^{-1}$ provided that x_m is constant. Indeed it is constant if the ratio t_1/t_2 and the filling factor a are constant. To ensure a constant filling factor a either a long filling pulse duration can be used (which makes $a \approx 1$) or, if the capture cross-section is not large enough, the filling pulse duration can be varied at each (t_1, t_2) pair to get a constant height of the DLTS peak which ensures a constant value of a. It should be stressed that with this procedure the value of E_T is determined in the first approximation; the approximation does not take into account changing of b_2 with T as pointed out above.

Now let us show how the parameter b_2 (or the second interaction energy E_2) can be determined. As the linewidth of the DLTS curve depends on b_2 we may suggest to use it for determination of b_2. However, the linewidth can be influenced by other effects like the dependence of the parameter λ in (16) on the electric field [5] or the positional variation of E_T in alloy crystals [6]. The positional variation of E_T can appear also in strained crystals. Thus it is advisable to use also some other relations for the determination of b_2. The most useful seems to be the interrelation of the DLTS peak height and its shift when the filling is varied. While in the final step the dependence of b_2 on T has to be taken into account, it is advantageous in the first step to determine b_2 approximately by using the $y(x)$ curve defined in (13). For that purpose we have to evaluate the shift of the peak x_{m}, the height of the peak y_{m} and the full linewidth $|x_l - x_h|$ which is given by two values of x at one half of the height y_{m}. For numerical evaluation it is useful to express the relation between the ratio $y(x)/a$ and the product b_2a from (5) and (13). By using this relation the curves of x_{m}, x_{l}, x_{h}, and $y_m b_2/a$ versus b_2a for $t_1/\tau_x = 0.4$ have been evaluated and plotted in Fig. 1.

The graphs in Fig. 1 can be used for determination of b_2 in the following way. First the value of b_2 is estimated from the linewidth $|x_l - x_h|$ which can be calculated from the experimental DLTS curve when the value of E_T has been determined. Both the low-temperature $|x_l - x_m|$ and the high-temperature $|x_{\mathrm{h}} - x_{\mathrm{m}}|$ half-linewidths can be compared with the experiment and their interrelation checked. As is obvious from the graphs the high-temperature half-linewidth broadens much faster with b_2 than the low-temperature one. In the next step the curve of the fractional height $y_m[a]/y_{\mathrm{m}}[1]$ verus the shift $(x_{\mathrm{m}}[a] - x_{\mathrm{m}}[1])$ can be constructed from the graphs in Fig. 1 for the estimated value of b_2 and compared with the curves evaluated from the experiment. The best fit can be found by changing the value of b_2 in a small range. Final corrections of both values E_T and E_2, including the temperature dependence of b_2 according to (7), have to be done by finding a new Arrhenius plot and a new curve of fractional height versus the shift of the DLTS line in the temperature scale.

4. Filling Process

If the values of E_T and E_2 are known the filling factors can be evaluated from the measured DLTS heights while the pulse duration t_p is varied. In the first approximation the graph of $y_m b_2/a$ plotted in Fig. 1 can be used for the evaluation of a from y_m. Once the dependence of a on t_p is determined, the parameters E_1 and τ_1 can be found by using (10), (11), and (12). For isolated impurities ($b_1 = 0$) it is convenient to plot $-\ln(1 - a)$ versus t_p which gives a straight line with the slope equal to τ_1^{-1}. The plot is less convenient for aggregates when the value of b_1 increases as can be seen in Fig. 2. With increasing value of b_1 the curves are more and more non-linear and take logarithmic-like shapes. Thus we have plotted the dependence of the filling factor a on the value of the parameter $\xi = \ln(t_p/\tau_1)$ as shown in Fig. 3. As is obvious, the curves are nearly linear in a range of three orders of magnitude or greater if $b_1 \geqq 10$. The evaluation of these curves by using (10) is less accurate when increasing the value of b_1 (and decreasing the value of a) because the result is given by a difference of large numbers which nearly cancel each other. For $b_1 = 20$ the accuracy of twelve decimal figures during the calculation is insufficient for getting any meaningful result. Thus we have found an approximate formula which can be applied with a satisfactory accuracy when $b_1 \gg 1$. The formula is derived in the Appendix and it reads as follows:

$$\xi = \ln \frac{t_p}{\tau_1'} = -b_1 - \ln b_1 + \ln \left(\frac{e^{b_1 a}}{1 - a} - 1 \right). \tag{18}$$

By using (18) it can also be shown that the dependence of a on ξ is nearly linear; if $a \gg 1/b_1$ the one in (18) can be neglected and as far as $b_1 a$ dominates over $\ln(1 - a)$ the dependence (18) is nearly linear. The curves of a versus ξ shown in Fig. 3 can be used to determine the values of b_1 and τ_1' from capture rate measurements. The curves shift towards lower values of ξ and get smaller slopes with increasing value of b_1. The shift ξ_0 corresponding to $a = 1 - e^{-1}$ is approximately given by the following power relation:

$$\xi_0 = 0.69 b^{0.88} , \tag{19}$$

and the slope at the same value of a is approximately given by the following linear relation:

$$\frac{d\xi_0}{da} = b_1 + e . \tag{20}$$

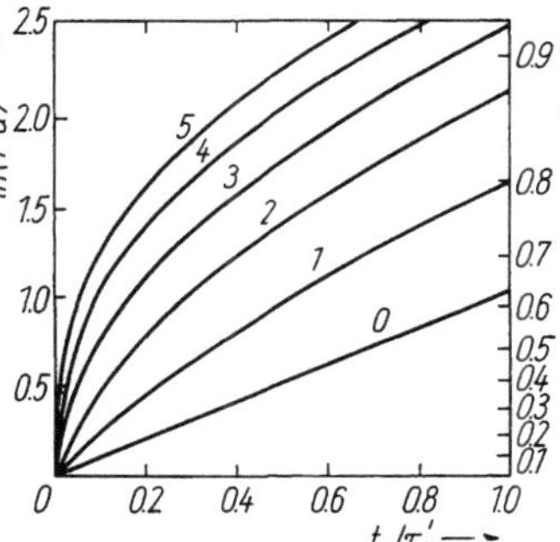

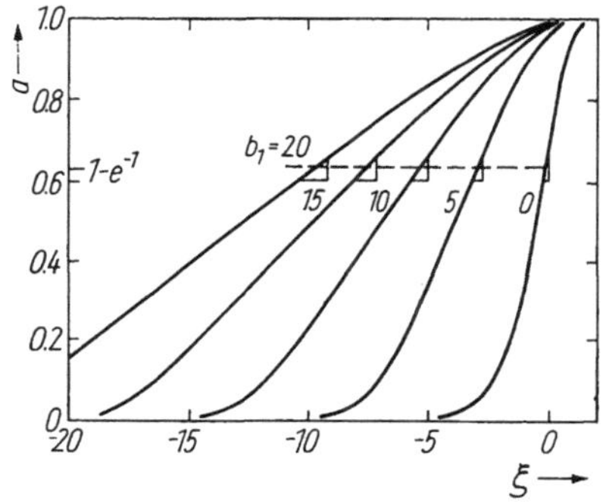

Fig. 2

Fig. 3

Fig. 2. Plots of $-\ln(1 - a)$ vs. the filling pulse duration t_p for various values of the interaction parameter b_1

Fig. 3. Plots of the filling factor a vs. $\xi = \ln(t_p/\tau_1')$ for various values of the interaction parameter b_1. The shift ξ_0 corresponding to $a = 1 - e^{-1}$ is indicated by the dashed line and the slopes $\Delta\xi_0/\Delta a$ are marked by thin lines

Equations (19) and (20) can be used for an estimate of the values of b_1 and τ_1' from experiment and further by using (11) and (12) the first interaction energy E_1 and the capture rate τ_1^{-1} can be estimated.

Final corrections of both parameters E_1 and τ_1^{-1} have to include a change of the DLTS peak position T_m with the filling pulse duration t_p.

5. Experimental Results and Discussion

All DLTS measurements were done on the deep level spectrometer made in the Hungarian Academy of Sciences. In the spectrometer the lock-in type of detection is used with the ratio of the cut-off time to the pulse period held constant and equal to 0.05. As the calculation presented above has been done for the differential type of detection we have checked how far it can be used for the lock-in detection. We have found that the DLTS lines calculated for the lock-in detection (with the time ratio 0.05) are nearly coincident with the DLTS lines calculated above (for $t_1/\tau_x = 0.4$) in the upper halves of the lines both for an exponential or for non-exponential transients.

The samples were prepared by the following procedure. On a plate cut from a bulk of n-type CdTe grown from the melt a Schottky junction was prepared by the chemical deposition of gold after polishing and etching in aqueous NaCN solution. The ohmic contact on the opposite side of the plate was formed by soldering indium. The plate was cleaved in small pieces and for the DLTS measurements the pieces were clipped under metal strips welded on TO-18 headers. The typical break-down voltage of the prepared diodes is greater than 5 V.

All samples show several DLTS peaks when measured in the reduced bias regime. The dominant peak appears at about 300 K with the pulse period 0.2 s. The position of the peak and its linewidth depend on the filling pulse duration thus manifesting a non-exponential transient. As the high-temperature part of the peak is partly overlapped with another peak only the low-temperature half-linewidth could be measured accurately.

The activation energy was determined from the Arrhenius plot of $T_m^2\tau_x$ versus $1/kT_m$ while the repetition frequency was varied from 500 Hz to 1 Hz.

In the first approximation it was equal to 0.7 eV. To determine the second interaction energy, the plot of the fractional DLTS height versus its shift was made as shown in Fig. 4. The upper part of the plot was fitted simultaneously with the Arrhe-

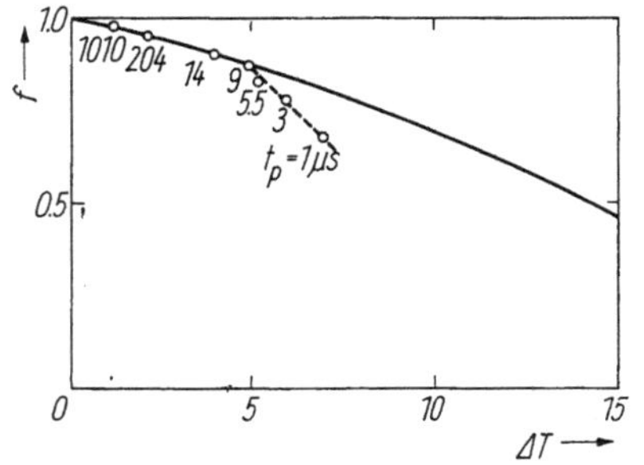

Fig. 4

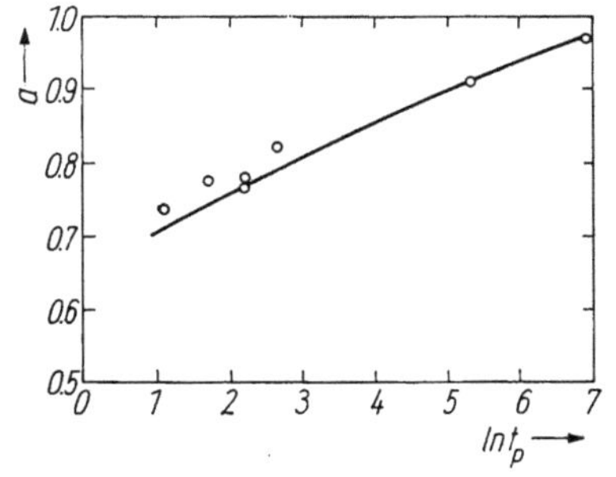

Fig. 5

Fig. 4. Plot of the fractional DLTS height $f = y_m[a]/y_m[1]$ vs. its shift $\Delta T = T_m[a] - T_m[1]$. The experimental values are indicated by circles labelled with the used filling pulse duration t_p in µs. The solid line shows the calculated values

Fig. 5. Plot of the filling factor a vs. $\ln t_p$. The experimental values are indicated by circles and the solid line shows the calculated values

nius plot while the temperature dependence of the parameter b_2 (7) was taken into account; in this way the activation energy $E_T = 0.75$ eV, the second interaction energy $E_2 = 0.08$ eV and $\ln \lambda = 16.6$ were determined. Knowing the value of λ the electron capture cross-section σ can be calculated assuming a non-degenerate level of the impurity [7]. Taking the electron effective mass of CdTe equal to 0.16 in units of the free electron mass [8], we get the value of $\sigma = 3 \times 10^{-14}$ cm^2. In the lower part of the graph in Fig. 4 the experimental DLTS heights deviate from the calculated ones towards smaller values. This effect is caused by a distortion of the short-time filling pulse during transmission from the generator to the sample; in this way the effective amplitude of the pulse is diminished and the measured DLTS height is lowered.

The filling factor a was calculated from the measured DLTS heights while the filling pulse duration t_p was varied in the range from 3 ms to 9 μs. For shorter values of t_p (up to 1 μs) the filling factor a was calculated from the shift of the DLTS line. The plot of a versus $\ln t_p$ is shown in Fig. 5. By fitting this plot with the calculated curve the first interaction energy $E_1 = 0.407$ eV and the electron capture rate $\tau_1^{-1} = 10^7$ s^{-1} were determined. From the electron capture rate the electron capture cross-section σ can be calculated if the electron concentration is known [9]. The electron concentration determined by C–V measurements is equal to 6×10^{14} cm^{-3}. With this value the cross-section is equal to $\sigma = 1.5 \times 10^{-15}$ cm^2. This value is one order of magnitude smaller than the value determined from the emission process. The reason of this difference is probably due to the effect of the electric field; the latter value of σ is measured on the non-biased junction while the former one is measured on a junction with 5 V bias voltage.

The measured low-temperature DLTS half-linewidth at full filling ($a = 1$) is equal to 36.2 K while the calculated value is 25.8 K. An explanation of this discrepancy in terms of electric field broadening and/or the strain broadening as pointed out above may be suggested.

For a simplified estimate the first interaction energy E_1 can be taken as the energy of a free electron in the Coulomb potential when all but one of the impurities on the line are charged. If we assume a straight line the potential in its middle can be calculated as follows:

$$U = \frac{2q}{\varepsilon d}\left(1 + \frac{1}{2} + \frac{1}{3} + \dots + \frac{1}{N}\right), \tag{21}$$

where $2N + 1$ is the number of impurities on the line, ε the dielectric constant, and d the distance between the neighbours. The sum within the brackets in (21) diverges when N goes to infinity, thus the finite length of the line has to be taken into account. At large values of N the sum approaches the value of $\ln (2N)$. If we assume that the line length is of the order of 1 mm and we use $\varepsilon = 9.6$ for CdTe we get the distance between neighbours on the straight line equal to 9 nm. For shorter lines, like a dislocation loop, the value of d will be smaller, thus the estimated value should be understood as the upper limit of d. To estimate the lower limit we may consider a circle of the length of about 10 nm. Then a similar calculation as above gives 1.4 nm as the lower limit of d. The value of E_2 depends on the chemical nature of impurities which is not known but the fact that E_2 is much smaller than E_1 is reasonable.

Acknowledgements

One of us (N.T.T.H.) would like to thank Dr. I. Toušková for supplying CdTe crystals and for helping with preparation of the samples. We are also grateful to Dr. R. Kužel for his support of our investigation of CdTe and for helpful and stimulating discussions with Dr. Z. Šroubek.

Appendix

The differential equation (9) in the text above can be solved by separating the variables and integrating as follows:

$$-\mathrm{Ei}(-b_1) + \mathrm{Ei}[-b_1(1-a)] = \frac{t_p}{\tau_1'}, \tag{A1}$$

where the exponential integral is

$$\mathrm{Ei}(x) = \int\limits_{-\infty}^{x} \frac{e^z}{z} \, dz \qquad \text{for} \quad x < 0 . \tag{A2}$$

The exponential integral can be written [10] as

$$\mathrm{Ei}(-x) = e^{-x} \left[-\frac{1}{x} + \int\limits_0^\infty \frac{e^{-t} \, dt}{(x+t)^2} \right] \quad \text{for} \quad x > 0 . \tag{A3}$$

The following inequality holds for the second term on the right-hand side of (A3):

$$\left| \int\limits_0^\infty \frac{e^{-t} \, dt}{(x+t)^2} \right| < \left| \frac{1}{x^2} \int\limits_0^\infty e^{-t} \, dt \right| = \left| -\frac{1}{x^2} \right| . \tag{A4}$$

Thus (A3) can be rewritten for $x \gg 1$ as follows:

$$\mathrm{Ei}(-x) \approx \frac{-e^{-x}}{x} . \tag{A5}$$

Using (A5) in (A1) we can write for $b_1 \gg 1$ and for a not too close to one,

$$\frac{e^{-b_1}}{b_1} \left(\frac{e^{b_1 a}}{1-a} - 1 \right) = \frac{t_p}{\tau_1'} . \tag{A6}$$

By taking the logarithm of (A6) (18) in the text is obtained.

References

[1] D. V. Lang, J. appl. Phys. **45**, 3023 (1974).
[2] K. Žďánský and N. T. Thuc Hien, phys. stat. sol. (a) **81**, 353 (1984).
[3] S. R. Morrison, Phys. Rev. **104**, 619 (1956).
[4] G. Ferenczi and L. Dózsa, Crystal Res. Technol. **16**, 203 (1981).
[5] A. Mircea, D. Pons, and S. Makram-Ebeid, Lecture Notes in Physics **122**, 69 (1980).
[6] P. Omling, L. Samuelson, and H. G. Grimmeiss, J. appl. Phys. **54**, 5117 (1983).
[7] A. Mintonneau, Philips Res. Rep. **31**, 244 (1976).
[8] B. R. Nag, Electron Transport in Compound Semiconductors, Springer Series in Solid State Science, Vol. 11, Springer-Verlag, Berlin 1980 (p. 30).
[9] C. H. Henry and D. V. Lang, Phys. Rev. **15**, 989 (1977).
[10] I. M. Ryshik and I. S. Gradstein, Tables of Series, Products and Integrals, Deutscher Verlag der Wissenschaften, Berlin 1957 (p. 291).

(Received April 26, 1984)

phys. stat. sol. (a) 85, 227 (1984)

Subject classification: 10.2 and 13.4; 14.3; 22.1.1

School of Engineering, Okayama University[1])

Origin of a Shallow Acceptor in Quenched Germanium

By

Y. KAMIURA and F. HASHIMOTO

The origin of a quenched-in shallow acceptor $(SA)_1$ at $E_V + 9$ meV is studied by investigating the effect of copper and iron dopings via their diffusion on the density of this acceptor. Hall-effect measurements are done in the range 5 to 150 K to determine the energy and density of the SA_1 level. A defect pair model consisting of mobile and immobile constituents is proposed for the SA_1 acceptor. The iron interstitial as the mobile constituent suggested recently by Broeckx et al. is not favored by the present doping experiments. Neither is a vacancy plausible in view of the authors recent electron-irradiation experiments. It is likely due to another unknown interstitial impurity. The immobile constituent, which is left behind when the SA_1 acceptor disappeared at 320 °C, has an acceptor level near $E_V + 40$ meV. It is very probably attributable to substitutional copper as suggested by Broeckx et al., but the other possibility that it is due to any other impurity or defect giving rise to a similar acceptor level is still remaining.

Der Herkunft des durch Abschrecken erzeugten flachen Akzeptors (SA_1) bei $E_V + 9$ meV wird nachgegangen, indem der Einfluß von Kupfer- und Eisendotierungen, eingebracht durch Diffusion, auf die Dichte des Akzeptors untersucht wird. Um Dichte und Lage des SA_1-Zustandes zu bestimmen, wird der Halleffekt zwischen 5 und 150 K gemessen. Es wird ein Defektpaar-Modell vorgeschlagen, das aus beweglichen und unbeweglichen Partnern besteht. Als bewegliche Spezies wurde kürzlich interstitielles Eisen von Broeckx et al. vermutet, was aber durch die Dotierungsexperimente nicht gestützt wird. Neuere Experimente der Autoren mit Elektronenbestrahlung geben auch keinen Hinweis für die Leerstelle als bewegliche Spezies. Vermutlich handelt es sich um ein unbekanntes interstitielles Fremdatom. Der unbewegliche Partner, der nach Verschwinden des SA_1-Akzeptors bei 320 °C zurückbleibt, hat ein Akzeptorniveau bei $E_V + 40$ meV. Er ist sehr wahrscheinlich mit substitutionell gelöstem Kupfer identisch, wie von Broeckx et al. vermutet, aber die andere Möglichkeit, daß er von einem anderen Fremdatom oder Defekt mit ähnlichem Akzeptorniveau herrührt, bleibt offen.

1. Introduction

It has been known that quenching of germanium crystals from high temperatures close to the melting point produces several unknown shallow acceptors [1 to 3]. Although their properties such as ionization energies and annealing behaviour have been studied by means of low-temperature Hall measurements [1] and far-infrared photoconductivity measurements [2, 3], their nature is not understood yet. Recently we found [4] that one of these acceptors whose level is located at $E_V + 9$ meV is closely related to copper, which accidentally diffuses into the bulk of samples during the heat treatment. Our finding indicated that annealing around 300 °C caused this acceptor (SA_1) to disappear leaving at equal additional density of substitutional copper, strongly suggesting that the SA_1 acceptor is a defect complex containing at least one copper atom. The subsequent investigation of Broeckx et al. [5] revealed that the copper or nickel diffusion prior to the quenching greatly affected the ionization energy and thermal stability of the SA_1 acceptor. From these results, they drew

[1]) 700 Okayama, Japan.

the conclusion that various kinds of SA_1 acceptors were produced by the heat treatments, and were attributable to impurity complexes involving fast diffusing transition metals which enter the samples during the heat treatments. In addition they proposed an impurity pair model consisting of substitutional copper (Cu^s) and interstitial iron (Fe^i) for the normal SA_1 acceptor produced by the quenching without intentional impurity doping, because iron has been known as a principal contaminant in quenching experiments of silicon [6, 7]. According to their model, the annealing near 300 °C makes the SA_1 acceptor dissociate, so that the Fe^i atom disappears to precipitate, leaving the Cu^s atom behind. However, they presented no positive evidence for the correlation of SA_1 with iron. Their identification was based only upon the speculation that iron may contaminate germanium as well as silicon during the heat treatment. Perhaps the most direct verification of the Cu^s–Fe^i pair model can be given by the iron and copper doping experiments where one lets those impurities diffuse into samples during the heating prior to the quenching and compares the concentration of SA_1 acceptors with that for the normal quenching without the intentional doping. The photoconductivity measurement is not suitable to do this kind of experiment, because from this technique it is difficult to obtain any information about the concentrations of centers. We have therefore performed low-temperature Hall measurements to evaluate the concentrations of SA_1 acceptors.

In this paper we report the effect of impurity doping on the SA_1 concentration. All experimental efforts were concentrated on clarifying whether or not the impurity doping produces the SA_1 acceptors in larger concentrations than for the normal quenching. A discussion is made on the origin of the SA_1 acceptor.

2. Experimental Procedure

Two undoped germanium crystals with net acceptor densities of 2×10^{12} (P6) and 5×10^{12} (P9) cm^{-3} were used as the starting material. They were cut into rod samples with a dimension of $1 \times 2 \times 15$ mm^3.

Heating of samples without the intentional doping and subsequent quenching was performed in the same manner as described in our previous reports [8, 9]. In short, samples were chemically etched, rinsed, heated in a nitrogen atmosphere at 900 °C for 3 h, and then quenched into ethylene glycol at room temperature. This procedure will be referred to hereafter as the normal quenching.

Impurity diffusion experiments were done in two ways.

(i) Evaporation diffusion: Impurity metals were evaporated onto samples in a vacuum. Samples were then heated at 900 °C for 3 h and quenched in the same way as the normal quenching.

(ii) Gas phase diffusion: Samples were mounted together with small blocks of impurity metals on a quartz boat in a furnace. They were separated from each other, but were so closely located that an ambient nitrogen gas stream could carry enough vapor of the impurities to the sample surface. They were heated at 900 °C for 3 h, and were quenched in the same way as the normal quenching.

The gettering action of liquid gallium [9] was used to reduce the density of copper already diffused-in. Etched samples were coated with a thin layer of gallium. They were pre-heated at 700 °C for 1 h and then heated at 900 °C for 3 h, and finally quenched. The pre-heating at 700 °C was an important procedure to accomplish effective gettering of copper [9].

Annealing was done by immersing samples in molten salt (KNO_3:$NaNO_3$ = 1:1) kept at 320 °C [4].

Samples subjected to the above heat treatments were prepared for Hall measurements. Gold wires were spot-welded to etched samples as electrical contacts. This kind of contacts played an essential role in evaluating accurate changes in level densities resulting from the annealing [4], because the contacts were not melted at the annealing temperature. Hall measurements were carried out in a field of 3.7*kG* in the region 5 to 150 K. Hole densities, p, were calculated using Morin's data [10] on the ratio of Hall mobility and conductivity mobility as a function of temperature so that constant hole densities in the exhaustion region were obtained.

The energies and densities of levels were determined by fitting the experimental data to the theoretical curve calculated using the multilevel statistics. The analytical procedure is as follows. We consider p-type germanium containing m kinds of acceptors and a small amount of compensating donors. The hole density, p, is expressed by the equations

$$p + N_D = \sum_{i=1}^{m} \frac{N_i}{1 + \beta_i \exp\left[(E_i - E_F)/kT\right]} \tag{1}$$

and

$$p = N_v \exp\left(-\frac{E_F}{kT}\right), \tag{2}$$

where E_i, N_i, and β_i are the ionization energy, the density, and the degeneracy factor of the i-th acceptor level, respectively, $N_v = 1.087 \times 10^{15} T^{3/2}$, N_D the density of compensating donors, E_F the Fermi level, k the Boltzmann constant, and T temperature. The value of β_i was assumed to be four for all the acceptor levels considered here. An equation of $(i + 1)$-th order in p was derived from (1) and (2), and was numerically solved at a given temperature and for given values of fitting parameters (E_i, N_i, N_D). Similar calculations were done for all the temperatures where the experimental data had been taken, and were compared with the data. This procedure had been repeated after making small changes in the values of the fitting parameters until the sum of squares of differences between experimental and calculated values of p reached a minimum.

3. Results

Before presenting the results obtained in the present experiments, it is useful to point out the following properties of the SA_1 acceptor:

(i) It has an acceptor level at $E_v + 9$ meV [1, 2, 4].

(ii) It is completely annihilated by the annealing at 320 °C for 10^3 min, being replaced by a substitutional copper atom [4].

These properties will be used to identify the shallow acceptors produced in the present experiments as the SA_1 acceptor.

Fig. 1 shows the result of iron doping together with the data of the normal quenching for comparison. In both cases, three kinds of acceptor levels are involved in the curves: SA_1, another shallow acceptor (SA_2) at $E_v + 13$ meV, and the lowest level of Cu^s. The two levels, SA_1 and SA_2, are so closely located in the band gap that their level parameters cannot be determined accurately by the fitting. We therefore made the fitting first for the data after the annealing at 320 °C, where the SA_1 acceptors disappear while the SA_2 acceptors are stable [1, 2, 4], in order to determine the level parameters of SA_2 and Cu^s. We then fitted the as-quenched data to determine the energy and density of SA_1, and the density of Cu^s. The detailed procedure is described in [4]. The level parameters determined in this way are given in Table 1. The annealing at 320 °C completely annihilated SA_1 acceptors, and replaced them by the same density of Cu^s.

Table 1

Energies E and densities N of levels determined by the fitting of the curves in Fig. 1 to Fig. 3. The energy of the Cu^s level was fixed at 41 meV. The densities of compensating donors were always 10^{12} to 10^{13} cm^{-3}

figure	curve	SA_1		SA_2		Cu^s
		E (meV)	N (10^{14} cm^{-3})	E (meV)	N (10^{14} cm^{-3})	N (10^{14} cm^{-3})
1	a	8	14	12	2.8	21
	b	9	0.85	12	4.6	5.1
	c	—	0	12	4.6	5.9
2	a	8	15	12	4.8	18
	b	—	0	12	4.8	33
3	a	8	15	14	0.27	14
	b	—	0	14	0.27	31

It is confirmed that the SA_1 acceptor in the iron-diffused sample is the same center as the normal SA_1 acceptor because of their complete agreement in level energy and annealing behavior. However, the density of the SA_1 acceptor was obviously reduced by the iron doping. This is quite contradictory to the Cu^s–Fe^i model.

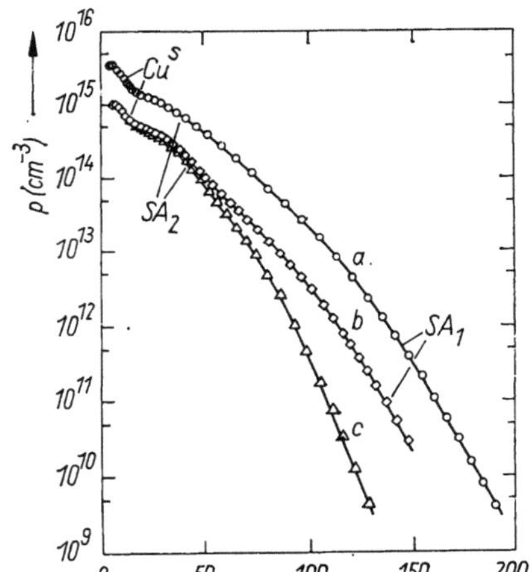

Fig. 1

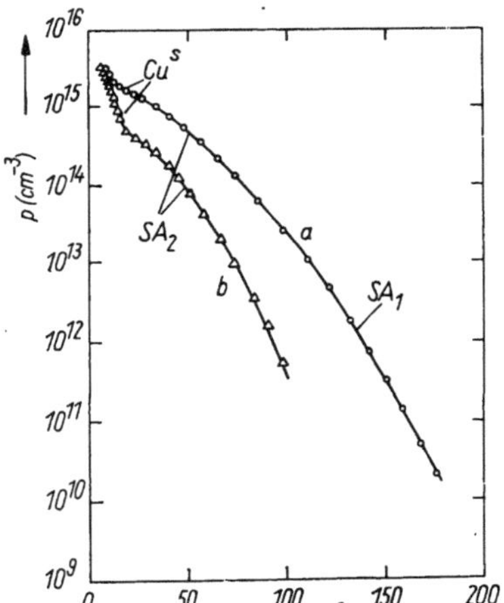

Fig. 2

Fig. 1. Comparison of the result of iron doping with that of the normal quenching. The hole density p is plotted against the reciprocal temperature T. (a) Normally quenched (P6-39-1); (b) iron-doped by evaporation diffusion (P6-82B8); (c) subsequently annealed at 320 °C for 1046 min (P6-82B8). Solid curves represent the results of the theoretical calculation considering the presence of three acceptor levels (SA_1, SA_2, and Cu^s). The level parameters determined by the fitting are listed in Table 1

Fig. 2. Effect of the double doping with copper and iron by the evaporation diffusion (P6-82B20). (a) Copper-diffused and then iron-diffused; (b) subsequently annealed at 320 °C for 1000 min. The level parameters are listed in Table 1

Next we carried out copper doping experiments. The results were the same as already reported [1], that is, the SA_1 acceptors are completely absent and only substitutional copper was dissolved to its solubility limit ($\approx 4 \times 10^{16}$ cm^{-3}). This is also a contradictory result to the model and to our observation about the annealing behavior of the SA_1 acceptor. We will discuss it in the next section.

The third kind of experiments are concerned with the double doping with iron and copper. A sample was first doped with copper, confirming that no SA_1 acceptors were introduced. The sample was then doped with iron. The result is shown as curve a in Fig. 2, where shallow acceptors like SA_1 appear after the iron doping, and simultaneously the copper density has been reduced from 4×10^{16} to 1.8×10^{15} cm^{-3}. The subsequent annealing at 320 °C for 1000 min completely annihilated those acceptors, and yielded the same density of Cu^s, indicating that they are certainly the normal SA_1 acceptors. We do not interpret this result as an evidence supporting the Cu^s–Fe^i model, because the density of the appearing SA_1 acceptors is the same as after the normal quenching. Further experiments on the double doping supported our argument. When the iron doping was done first and then the copper doping followed it, no SA_1 acceptors were produced at all in contrast to the first double doping in the reverse order of doping steps. Moreover, neither did SA_1 acceptors appear when one face of a sample was plated with iron and the opposite face with copper and then the sample was heated and quenched. In the latter two cases, only copper was introduced in densities of (1 to 3) $\times 10^{16}$ cm^{-3} which are almost at its solubility limit. We interpret the appearance of SA_1 in the first (copper and then iron) double doping (Fig. 2) in terms of the gettering action of iron on copper [9]. In fact, after the iron doping, the copper density was diminished to the same level as after the normal quenching. If our interpretation is correct, other gettering agents are expected to have the same effect. We chose liquid gallium as a gettering agent to test our assumption, because gallium was reported to extract copper effectively from the interior of germanium [9], and moreover gallium is available in high purity and does not diffuse into germanium. A copper-doped sample, which contained only Cu^s atoms with a density of 4×10^{16} cm^{-3}, was etched and coated with gallium. It was then pre-heated at 700 °C, heated at 900 °C, and quenched. The result was what had been expected as shown in Fig. 3, where it is verified that the gettering of copper by gallium also gave rise to the appearance of SA_1 acceptors, reducing the copper density to 1.4×10^{15} cm^{-3}. The subsequent annealing annihilated the SA_1 acceptors, which were replaced by the same density of Cu^s atoms. It is therefore established that the result in Fig. 2 does not mean any role of in-diffused iron in the formation of SA_1, but can be explained merely by the gettering action of iron on copper at the sample surface. The reason why the appearance

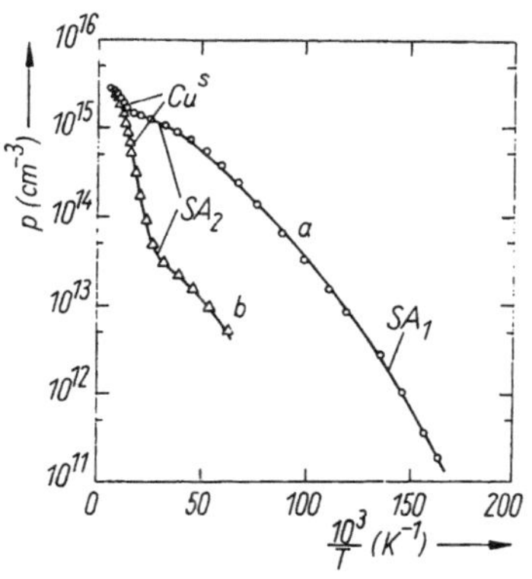

Fig. 3. Effect of gallium coating on the gettering of copper and the appearance of SA_1 acceptors (P9-82B6). (a) Copper-diffused by evaporation diffusion and then copper-gettered by gallium coating; (b) subsequently annealed at 320 °C for 1000 min. The level parameters are listed in Table 1

of SA_1 is caused by the reduction of the copper density will be discussed in the next section.

We finally performed further experiments on the iron doping and the iron–copper double doping by the gas phase diffusion. Iron and copper were supplied to the sample surface only via the gas phase, so that any gettering actions at the metal–germanium interface were avoided. Our iron-doping experiment (P9-83B36) was not so successful. We could not carry out measurements below 20 K because of bad electrical contacts of the sample. However, the result clearly indicated that the total density of shallow acceptors, probably SA_1 and SA_2, was only about 1×10^{14} cm^{-3}, which was by an order of magnitude smaller than that for the normal quenching (curve a of Fig. 1). The result of the double-doping experiment indicated that no SA_1 acceptors were introduced, while only substitutional copper was detected with a density of $1.3 \times \times 10^{16}$ cm^{-3}. These results show quite the same tendency as those of the iron and iron–copper doping experiments by the evaporation diffusion.

4. Discussion

Broeckx et al. [5] interpreted our annealing data on the SA_1 acceptor [4] as the dissociation of a pair consisting of a Cu^s atom and an unknown interstitial impurity, which is mobile at the annealing temperature. After the dissociation the interstitial precipitates leave a Cu^s atom behind. Furthermore, they proposed an iron interstitial for the mobile constituent in view of its role in the contamination of heat-treated silicon. However, our results of iron-doping experiments provided no positive evidence for the possibility of iron as one of the constituents of the SA_1 acceptor. Although we have no evidence for the actual diffusion of iron into the interior of samples during our diffusion experiments, we argue that much more iron might have diffused in than in the case of the normal quenching if iron contamination also occurred for the heat treatment of germanium as speculated by Broeckx et al. [5]. We therefore conclude that the SA_1 acceptor involves no iron atom.

An alternative assignment of the mobile constituent to a vacancy may be possible, because it is known to be a principal thermal defect at high temperatures [11, 12] and to be mobile below room temperature [13]. However, our recent results of electron irradiation experiments on copper-diffused germanium showed that no SA_1 acceptors were produced by the irradiation [14]. We therefore assume that the mobile constituent of the SA_1 acceptor is an unknown impurity other than iron.

Our copper-doping experiments indicated that the SA_1 acceptors were completely absent after the copper diffusion. A straightforward interpretation may be that no copper atoms are incorporated in the SA_1 acceptor. However, our previous annealing experiments [4] clearly indicated that the annealing transformed one SA_1 acceptor into one Cu^s atom. Moreover, our present experiments on the iron doping (curve b in Fig. 1) showed that the gettering action of iron resulted in a reduction in the SA_1 density as well as in the Cu^s density in comparison with those for the normal quenching (curve a in Fig. 1). These results suggest that the SA_1 acceptor contains at least one copper atom.

One interpretation to explain these conflicting results is to assume that the SA_1 acceptor involves one or more copper atoms, and that the copper diffusion leads to the reaction of SA_1 acceptors with copper atoms to make larger complexes which are electrically neutral. According to this picture, the copper-related SA_1 acceptor, SA_1 (Cu), discovered by Broeckx et al. [5] may be assigned to a relatively small and electrically active complex consisting of the normal SA_1 acceptor and a copper atom. The copper density ($\approx 10^{16}$ cm^{-3}) after the copper diffusion might be high enough to transform most of SA_1 acceptors into large neutral complexes and only a small part of them into SA_1 (Cu) acceptors. This could explain why our Hall measurements did not detect the SA_1 (Cu) acceptors and the photoelectric spectroscopy study of Broeckx et al. [5] did because of its high sensitivity [15]. In this picture, if the copper density is reduced to the low level as that for the normal quenching, larger complexes may dissociate into SA_1 acceptors according to the mass action principle. This is consistent with our present results showing the appearance of SA_1 acceptors in the samples where copper was diffused first and then was gettered by iron or gallium (curves a in Fig. 2 and Fig. 3).

An alternative interpretation to remove the contradiction between the data is to attribute the $E_v + 40$ meV level appearing as the SA_1 acceptor disappears not to substitutional copper but to any other impurities or defects giving rise to a similar level. One of the most probable candidates may be gold in view of its level position, diffusion coefficient, and solubility [16]. However, the lowest level of gold has donor character. It can easily be shown that any donor level at $E_v + 40$ meV appearing after the annealing is inconsistent with our data. If the $E_v + 40$ meV levels had donor character, they would compensate the levels of the SA_2 acceptors at low temperatures after the annealing. In this case, if the SA_1 density is larger than the SA_2 density before the annealing, the SA_2 levels would completely be absent in p versus $10^3/T$ curves after the annealing. However, such a situation has never occurred in our experiments (Fig. 2 and Fig. 3 of the present paper, and Fig. 1 of [4]). In addition, our results of the fitting indicated that the total density of compensating donors was always less than 10^{13} cm^{-3} (Table 1). We therefore conclude that the $E_v + 40$ meV level which the SA_1 acceptor leaves behind during the annealing has acceptor character.

5. Conclusions

The SA_1 acceptor is considered to be a defect pair consisting of mobile and immobile defects and/or impurities. The iron interstitial which had been proposed as the mobile constituent by Boreckx et al. was not favored by the present doping experiments. Neither is a vacancy plausible in view of our recent electron irradiation experiments. It is likely to be due to an unknown interstitial impurity other than iron. The immobile constituent, which is left behind when the SA_1 acceptor disappears at 320 °C, has an acceptor level near $E_v + 40$ meV. It is very probably attributable to substitutional copper, but the other possibility that it is due to any other impurity or defect giving rise to a similar acceptor level is still remaining. This equation should be solved by further studies by means of DLTS or infrared absorption spectroscopy.

Acknowledgements

We would like to thank Messrs. S. Yoneyama and T. Isoi for their help in our experiments.

References

[1] Y. Kamiura, F. Hashimoto, T. Takada, Y. Sakaji, and T. Hattori, Inst. Phys. Conf. Ser. No. 59, 187 (1981).
[2] Y. Kamiura, J. Broeckx, P. Clauws, and J. Vennik, Solid State Commun. **38**, 883 (1981).
[3] J. Broeckx, Y. Kamiura, P. Clauws, and J. Vennik, Solid State Commun. **40**, 149 (1981).
[4] Y. Kamiura, F. Hashimoto, H. Sugiyama, and S. Yoneyama, Japan. J. appl. Phys. **21**, L357 (1982).
[5] J. Broeckx, Y. Kamiura, P. Clauws, and J. Vennik, Solid State Commun. **43**, 499 (1982).
[6] Y. H. Lee, R. L. Kleinhenz, and J. W. Corbett, Appl. Phys. Letters **31**, 142 (1977).

[7] E. WEBER and H. G. RIOTTE, Appl. Phys. Letters **33**, 433 (1978).
[8] F. HASHIMOTO and Y. KAMIURA, Japan. J. appl. Phys. **13**, 762 (1974).
[9] Y. KAMIURA, F. HASHIMOTO, S. YONEYAMA, and T. NOBUSADA, phys. stat. sol. (a) **82**, 135 (1984).
[10] F. J. MORIN, Phys. Rev. **93**, 62 (1954).
[11] A. SEEGER and K. P. CHIK, phys. stat. sol. **29**, 455 (1968).
[12] M. HIRATA and M. KIRITANI, Physica **116B**, 623 (1983); Proc. 12th Internat. Conf. on Defects in Semiconductors, Ed. C. A. J. AMMERLAAN, North-Holland, Amsterdam 1983 (p. 623).
[13] R. E. WHAN, Phys. Rev. **140**, A690 (1965).
[14] Y. KAMIURA, F. HASHIMOTO, S. YONEYAMA, and N. FUKUOKA, Japan. J. appl. Phys. **23**, L126 (1984).
[15] SH. M. KOGAN and T. M. LIFSHITS, phys. stat. sol. (a) **39**, 11 (1977).
[16] A. G. MILNES, Deep Impurities in Semiconductors, John Wiley & Sons, New York 1973 (pp. 36 to 38).

(Received June 29, 1984)

phys. stat. sol. (a) 85, 235 (1984)

Subject classification: 14.3.1 and 20.3; 13.4; 22.4.2

The Electrical Engineering Laboratories, The University of Manchester[1]) *(a) and Department of Physics, The University of Hull*[2]) *(b)*

Photoluminescence and Electrical Properties of Vapour Phase Epitaxial ZnSe Grown on GaAs

By

P. LILLEY (a), M. R. CZERNIAK[3]) (a), and J. E. NICHOLLS (b)

Vapour phase epitaxial ZnSe is grown on GaAs substrates under various conditions and the layers characterized by low temperature photoluminescence and electrical measurements. Growth conditions are found to have a strong influence on the resistivity of the layers (0.1 to 10^9 Ωcm) and this variation is considered in terms of autodoping from the substrate by gallium ions. Photoluminescence of the near band gap region shows excitonic structure and shallow donor–acceptor pair lines which are discussed in terms of the electrical properties of the layers.

Epitaktisches ZnSe aus der Dampfphase wird auf GaAs-Substraten unter verschiedenen Bedingungen hergestellt und die Schichten mittels Tieftemperaturphotolumineszenz und elektrischen Messungen charakterisiert. Es wird gefunden, daß die Wachstumsbedingungen einen starken Einfluß auf den Widerstand der Schichten haben (0,1 bis 10^9 Ωcm) und diese Änderung wird mit Selbstdotierung aus dem Substrat durch Galliumionen diskutiert. Die Photolumineszenz im Bereich um die Bandlücke zeigt die Exzitonenstruktur und flache Donator-Akzeptor Paar-Linien die mit den elektrischen Eigenschaften der Schichten diskutiert werden.

1. Introduction

Thin films of epitaxial ZnSe have potential applications for blue light emitting diodes and have been prepared by vapour phase deposition, liquid phase epitaxy, molecular beam epitaxy, and more recently by organo-metallic CVD. However, in order to achieve efficient diodes in the blue region a deeper understanding of the opto-electronic properties is necessary in order to achieve both low-resistivity semiconducting layers, n- and p-type, and strong near band edge emission compared with the usually dominant deep centre emission.

Vapour phase epitaxy (VPE) is one method by which ZnSe can be deposited onto substrates of GaAs or Ge, with mass transport and deposition taking place in a continuously refreshed flow of H_2 wherein impurities condense in different parts of the growth apparatus or are removed from the quartz reactor in the exhaust-gases [1]. Low-temperature photoluminescence is a convenient technique for assessing the overall impurities of the material that in turn are likely to control the opto-electronic properties. In this paper we report such characterization of the layers and also electrical measurements.

2. Experimental

A schematic diagram of the growth system used in the present work is shown in Fig. 1 where vaporisation of a solid or powder ZnSe source at 950 °C in H_2 flow gives sufficient mass transport to yield a growth of ≈ 1 μm/h at 550 °C. Deposition occurs in the

1) Manchester M13 9PL, Great Britain.

2) Hull HU6 7RX, Great Britain.

3) Currently at Philips Research Laboratories, Redhill, Surrey, Great Britain.

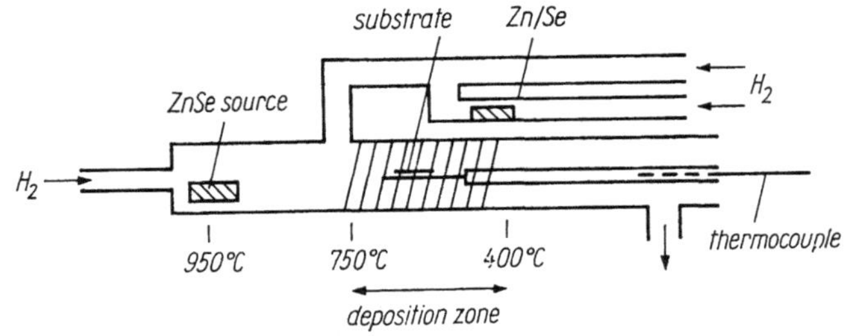

Fig. 1. The quartz growth system

quartz reactor tube within the temperature range 750 to 500 °C with a growth rate that increases exponentially with increasing temperature. A low substrate temperature is preferred to minimise possible contamination from the substrate but a compromise with growth rate is necessary, 550 °C providing 0.8 μm/h. Two types of ZnSe were used as source material in the reactor: ZnSe powder (Merck, supplied by BDH, Ltd) with impurities quoted as Cu (0.5 ppm) and Fe (1.00 ppm) and (as shown by mass spectrometry) to have Cd of a few ppm, and polycrystalline ZnSe of the type used for laser windows with known impurities of Cu (0.4 ppm) and Fe (1.5 ppm). Metallic zinc for providing the zinc overpressure (supplied by Johnson-Matthey, Ltd.) had both Cu and Fe in the region of 1 ppm.

In the present work the standard conditions for growth on GaAs substrates in the reactor were as shown in Table 1. Photoluminescence spectra were recorded at 2 K,

Table 1

"Standard" conditions used for growth of ZnSe epilayers on GaAs subtrates. These conditions resulted in layers 1 to 2 μm thick

source temperature	970 °C
substrate temperature	550 °C
transport flow (H_2)	250 cm^3/min
secondary flow (H_2)	1300 cm^3/min
Zn source temperature	530 °C
Zn transport flow (H_2)	250 cm^3/min
substrate material	GaAs (100)
substrate etchant	1 H_2SO_4:5 H_2O_2:1 H_2O
growth time	90 min

with the sample immersed in overpumped liquid helium, by exciting the layers with 40 mW argon-ion laser radiation at 351/364 nm. The spectra were subsequently corrected for instrumental response and plotted as photoluminescence intensity per unit wavelength interval (Fig. 2a to c). Some differences are apparent in the edge region between the high- and low-resolution spectra caused by excitation of different parts of the layer. Semi-insulating Cr-doped GaAs, with a resistivity greater than 10^6 Ωcm, was used as substrates for the layers used for electrical measurements. Contacts of In were applied to these layers by heating in Ar at 250 °C for 2 min and values of carrier concentrations and mobilities were assessed from van der Pauw and Hall measurements. Table 2 lists some of the electrical properties at room temperature of the ZnSe layers grown under different conditions and shows that the resistivity varies over many orders of magnitude whereas the mobilities are fairly constant.

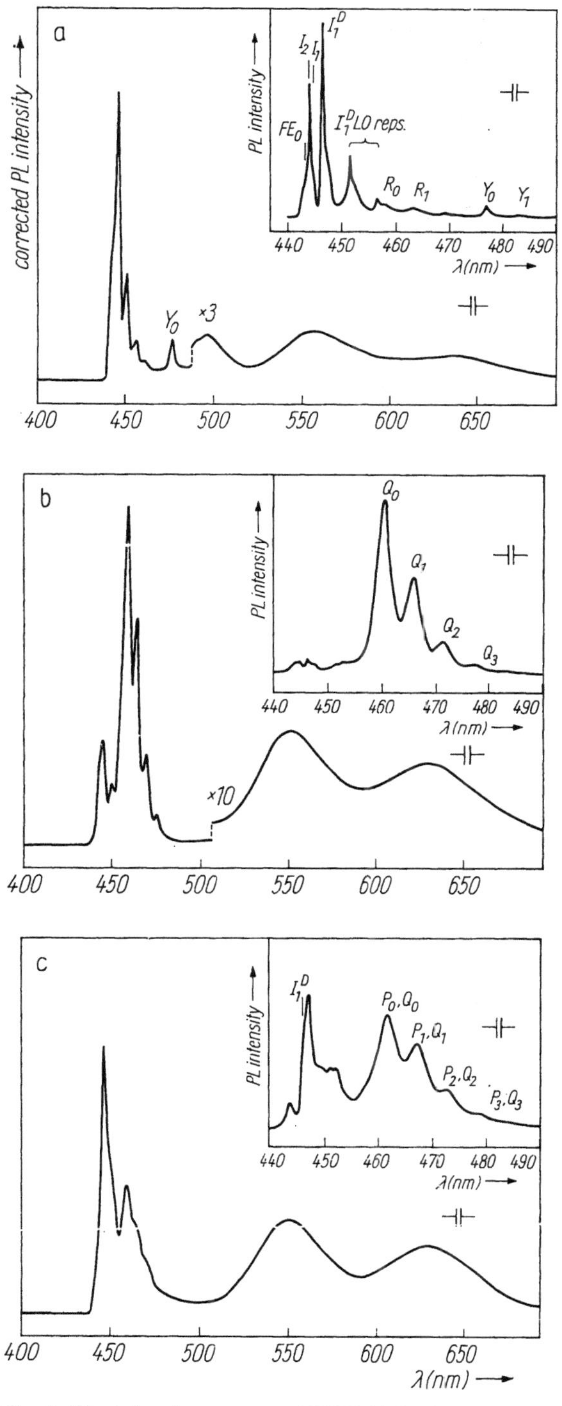

Fig. 2. Photoluminescence of VPE ZnSe layers that have been grown a) under the "standard" conditions of Table 1, b) in a flux of excess zinc, and c) in a reduced zinc flux, recorded with the sample at 2 K and with 40 mW, 351/364 nm laser excitation

Table 2

Electrical properties measured at room temperature of ZnSe epilayers grown under different conditions

	ϱ (Ωcm)	n (cm^{-3})	μ (cm^2/Vs)	substrate
standard conditions	0.1–10	10^{15}–10^{18}	140–160	GaAs
excess Zn overpressure	1–10^3	10^{14}–10^{16}	150–170	GaAs
reduced Zn overpressure or Se overpressure	10^5–10^9	—	100–150	GaAs
buffer layers	10^3–10^4	10^{12}–10^{14}	100–200	GaAs
standard conditions	$> 10^8$	—	—	quartz, Si, Ge

3. Photoluminescence of the Epilayers

Epilayers grown under the standard conditions of Table 1 showed room temperature photoluminescence that consisted of a band at 458 nm, 5 to 10 nm wide, and strong emission in the red region of the spectrum. The intensity ratio of the peak of the red emission to the blue emission was typically 5 or more. Fig. 2a shows the photoluminescence from one of these layers at 2 K. This spectrum is typical of many layers grown under the standard conditions (Table 1) and in the near band edge (NBE) region is dominated by the deep I_1^D exciton emission and phonon replicas which are associated with a bound exciton (BE) transition at a Cu acceptor [2, 3]. In Fig. 2a the I_1^D-line is shifted by about 2.5 meV to shorter wavelengths compared to the same line in bulk crystalline material [2]: in broad agreement with the shifts observed by other workers in ZnSe epilayers [4, 5]. Based on the position with respect to this I_1^D BE transition, the strong exciton line observed at higher energies is assigned to the I_2D^0X BE transition with adjacent weaker FE and I_1A^0X BE transitions. Strong I_2 transitions have also been observed in OMCVD epilayers [4, 5]. The exciton linewidths ($\approx$ 3 meV) are also broader than observed in crystalline ZnSe and this has been attributed to strain in the epilayers [4] and in our samples prevents the I_2 BE lines being resolved. The energy difference between the I_2D^0X and the I_1^D-lines is slightly less than in bulk material [6]; this, and the exciton linewidth, prevents identification of the donor species. However, in view of the later discussion, it is likely that the excitons are trapped at Ga donors. Weak R-series DAP emission and Y-series emission is also usually observed in our epilayers. The R emission has been attributed to DAP recombination between lithium interstitial and lithium substitutional ions [7] while the impurity causing the Y-emission is unknown but appears to be a persistent impurity in our layers and also in OMCVD ZnSe [4] where it is believed to result from free exciton recombination in the vicinity of an impurity or defect, possibly a transition metal ion, and where it was found to correlate with the first phonon replica of the FE transition. However this phonon replica of the FE transition is not usually observed in our epilayers.

At longer wavelenths three broad band emissions, which are characteristic of our epilayers, are observed around 495, 550, and 635 nm although relative intensities vary from sample to sample. The position of the 635 nm band also varies from sample to sample and is probably the result of two overlapping bands that are commonly observed in bulk ZnSe, the self-activated luminescence, donor–V_{Zn} associate [8], at 620 nm and the Cu-R luminescence, donor–Cu_{Zn} associate [9, 10] at 640 nm. The origins of the 495 and 550 nm bands are unknown but both are frequently observed in both OMCVD and VPE material. The band at 550 nm has been attributed by Dean et al.

[4] to the Cu-G emission in their OMCVD material and such an interpretation is attractive in the present work as the existence of both the I_1^D-lines and the Cu-R distant DAP bands show copper contamination of the layers. However, in bulk ZnSe the Cu-G band has been identified nearer 530 nm [10, 11] and an alternative interpretation may be more appropriate. The band may be the same as that observed by Bouley et al. [12] and attributed by them to an alkaline impurity complex. The appearance of the 495 nm band is noted in spectra published by Bouley et al. [12], Blanconnier et al. [13], and Dean et al. [4] and appears to be a characteristic band in epitaxial material. In our own samples this band is found to be stronger under cathodoexcitation rather than photoexcitation and is also stronger in thinner epilayers which suggests that the band may originate from the metalurgical interface region. However it is interesting to note that the band is also observed in our epilayers grown on Si and Ge as well as on GaAs.

3.1 Effects of ZnSe source material

The sharp exciton lines and the weaker shallow DAP and deep emission bands suggest that the epilayers grown under the standard conditions of Table 1 are relatively free of high concentrations of impurities. However in order to investigate the source of impurities, changes in the photoluminescence of the layers were observed when different ZnSe sources were used. The photoluminescence of Fig. 2a was observed when the epilayers were grown from a polycrystalline, sintered source of ZnSe, of the type used for laser windows, and after the source had been used for several runs. In contrast, when the sintered polycrystalline source material was first used, some differences were observed from Fig. 2a although the general features were the same. The photoluminescence showed a stronger I_2 BE emission compared to the I_1^D-line, an additional P-DAP series, attributed [7] to (Al_{Zn}–Na_{Zn}) pairs, that were stronger than the accompanying R-series and the absence of the Y-series line and 495 nm band. When the polycrystalline source was replaced with Merck ZnSe powder the photoluminescence spectrum was the same as Fig. 2a in the near band edge region but no emission was observed from the Y-series and the deep emission bands were much weaker.

The photoluminescence from the polycrystalline source material was also recorded both before and after use in the growth system. Before use it showed well resolved emissions from all of the emission lines and broad bands that were observed in the expitaxial material, except for the absence of the P-DAP series and an additional strong Q-DAP series, attributed to (Al_{Zn}–Li_{Zn}) pairs [7]. After just one growth cycle the polycrystalline source material became scintered and only the I_1^D-series, the Y-series, and weak I_1 and I_2 exciton lines were observed, which probably indictes considerable leaching of impurities from the surface during growth of the epilayers, although, after further use, the emission from the source did not change again.

3.2 Layers grown with different zinc overpressures

Other epilayers were grown under conditions that we describe as non-standard. Of the different parameters listed in Table 1 we found that changing the zinc flux produced the greatest effects on the photoluminescence at 2 K. The zinc flux was varied by moving the zinc source to different positions in the growth system where it was at a different temperature. Epilayers grown with increased zinc flux showed the photoluminescence of Fig. 2b; the photoluminescence in the NBE region is dominated by the Q-series DAP bands with much weaker, broadened exciton emission. Comparison with Fig. 2a shows similar 550 and 635 nm bands but an absence of the 495 nm band. Layers grown without the zinc source, so that the zinc pressure throughout the system was due entirely to the ZnSe source material, gave the photoluminescence spectrum of Fig. 2c. Similar broad bands at 550 and 630 nm are observed but the NBE emission is dominated by overlapping bands, that probably result from both the P and Q series, and broadened bands in the exciton region. The I_1^D-line is apparent on the side of a broad component while one of the transitions around 450 nm may result from the S band reported by Blaconnier et al. [13] which was attributed to DAP recombination.

Replacement of the zinc elemental source by one of selenium enabled layers to be prepared in a Se-rich atmosphere, with the selenium held at 150 °C. The principal differences between these layers and the standard ones was the greatly increased intensity of all three deep emission bands, and the loss of any resolvable structure in the NBE region.

4. Discussion

4.1 Photoluminescence

In preparing ZnSe epilayers for use as blue LEDs the most important parameter is the room temperature ratio of the blue emission observed in the region of 460 nm to the broad band emission. The blue emission at room temperature has been accounted for by a free exciton–electron interaction model [14, 15] with the recombination involving 2 LO phonons. The ratio of the peak of the broad band emission to blue emission is typically 5 or more in our VPE samples grown at 550 °C which compares unfavourably with a ratio of around 0.002 to 0.01 that can now be obtained with OMCVD [4]. However recent work by us on a modified VPE apparatus that enables growth at the much lower temperature of 300 °C, similar to that used in OMCVD systems, produces comparable low ratios [16]. For the blue emission to dominate, pure epilayers are required and while the exciton structure observed from our layers grown under standard conditions shows apparently that the layers are relatively free of impurities, the shallow DAP and deep emissions show that some contamination of the layers is occuring at this growth temperature, either from the sources (ZnSe or metallic Zn), the quartz, or the substrate.

The dominant exciton line that is observed in all our layers is the I_1^D, known to result from copper contamination [2, 3]. Although the same line was also observed in the ZnSe source materials, suggesting that the source contributes to the copper contamination, there is little reduction in the I_1^D-series after repeated use of the polycrystalline source material in the growth system and it is likely that the quartz provides a further source of this contamination. Other persistent impurities in the epilayers are Li and Na which result in the shallow DAP bands and it appears that these elements are quickly leached from the ZnSe polycrystalline source material as after use the source shows a marked reduction in these lines and the first runs with new source material show enhanced exciton emission from the epilayers. However subsequent runs show no change in the photoluminescence of the source material but the continued presence of these bands in the epilayers shows that the quartz is likely to be a further source of contamination of these elements. Epilayers grown with excess zinc overpressure are dominated by the Q-series, associated with lithium acceptors [7]. Lithium is a contaminate of zinc metal and the higher temperature of the zinc source probably results in the appearance of the band. At lower zinc pressures and with selenium overpressures both the P-series, associated with Na acceptors [7], and the Q-series are observed. The reason why the P-series is favoured at lower zinc pressures is not clear but could indicate out diffusion rates of the different metal ions from the silica, or incorporation rates of the ions into the layers, to be dependent of zinc pressure.

4.2 Electrical properties

In Table 2 it is seen that the resistivity of the ZnSe layers grown under different conditions varies over many orders of magnitude, but that the mobility of the layers are all within the range 100 to 200 $m^2 s^{-1} V^{-1}$. This is the same range as found by Leigh et al. [17] in ZnSe epilayers and attributed to ionized impurity scattering. All of our layers were n-type with activation energies in the range 10 to 20 meV whereas the calculated hydrogenic level is 29 meV; similar low energies have been reported by Stutius [18] for layers grown on GaAs by OMCVD who suggested impurity conduction to be the dominant transport process at low temperatures in heavily doped material and hopping conductivity in lightly doped layers.

Epilayers grown under the standard conditions on GaAs substrates from either the elemental sources, used polycrystalline source material or Merck powder have resistivities $\leqq 10\ \Omega$cm although when the polycrystalline source material is first used the resistivity is $\approx 10^3\ \Omega$cm. This probably results from more impurities being leached from the source when first used, as evidenced by the significant reduction in impurity related photoluminescence of the source material after being used for the first time. Growth under non-standard conditions with excess or reduced zinc overpressure or with selenium overpressure increases the resistivity by three or four orders of magnitude. However, when the epilayers are grown on substrates of quartz, Si, or Ge with the standard conditions of Table 1, high-resistivity material was produced with resistivities $> 10^8\ \Omega$cm.

The dramatic differences in the resistivity of the layers grown under different conditions on GaAs, and particularly the high resistivity of the layers grown on Ge, suggest that autodoping by the substrate is a major source of contamination of the layers and is the major factor controlling the n-type resistivity at the temperature of 550 °C at which our layers were grown. Contamination from the substrate has also been observed in homoepitaxy layers grown by CVD at 900 °C [3]. These changes in resistivity are almost certainly the result of Ga diffusion from the substrate entering the layers as shallow donors and recently Dean et al. [4] have shown the presence of Ga as a major contaminant on OMCVD layers, even at the much lower growth temperature 350 to 400 °C.

In many of our layers grown on GaAs there is a general correlation between the resistivity of the layers and the presence or absence of exciton emission and shallow DA pair bands. Layers with a resistivity $\lesssim 1000\ \Omega$cm show photoluminescence like that in Fig. 2a which is from a layer of resistivity 0.1 Ωcm. These layers show well resolved exciton structure, dominated by the I_2- and I_1^D-lines, with weaker shallow DA pair emission although in some of these layers the shallow DA pair emission is too weak to be observed. Since the resistivity of our layers seems to be principally determined by autodoping from the substrate, the I_2 BE emission that is observed in all of our lower resistivity layers is expected to result from excitons trapped at gallium donors. However we find little correlation between the intensity of the I_2 BE emission and the resistivity. Similar results were also observed by Stutius [18] who found only small differences in intensity of the I_2 BE emission in undoped, lightly doped, and degenerately doped material and this probably results from exciton broadening at donor concentrations above which the exciton wavefunction overlaps neighbouring donors. Consequently the intensity of the I_2 BE emission cannot be used as a guide to the resistivity of the layer. In layers with resistivities greater than about 1000 Ωcm there is either an absence of exciton structure or it is very weak. Instead the near band gap emission is dominated by the shallow DA pair bands. This can be seen in Fig. 2b and c which are recorded from layers with resistivity of 10^3 and $10^7\ \Omega$cm, respectively.

In these layers compensation of Ga donors by shallow acceptors is probably responsible both for the increased resistivity and the shallow DA pair recombination.

In the present work autodoping from the substrate was confirmed by growing layers at 650 °C directly onto GaAs and onto GaAs on which a thin buffer layer had been first deposited at 500 °C. In the former case the resistivity of the layer was $< 0.2\ \Omega$cm, in the latter 600 Ωcm. In both cases the layers showed similar I_2 BE emissions.

5. Conclusions and Summary

VPE ZnSe has been grown on GaAs substrates under various conditions with the growth system described above. It was found that the growth conditions listed in Table 1 produced an acceptable growth rate of 1 μm/h and optimum near band edge emission with respect to the broad band emission beyond 500 nm although the ratio of these emissions at room temperature was found to be poor compared to the best OMCVD material now available [4, 5]. This results from the greater contamination at the higher growth temperatures used in VPE. The low-temperature photoluminescence indicates that impurity doping originates both from the ZnSe source material and the quartz reactor system itself. Large variations in resistivity and carrier concentrations are found for samples grown under different conditions. It is concluded that autodoping by Ga from the substrate is the dominant mechanism controlling the electrical properties of the layers at the growth temperatures used in the present work even though autodoping does not appear to strongly influence the photoluminescence of the layers.

Acknowledgements

We are pleased to express our gratitude to Dr. K. Lewis of R.S.R.E. Malvern for supplying the ZnSe polycrystalline material, and one of us (M.R.C) to the S.E.R.C. for a research studentship.

References

[1] P. LILLEY, P. M. R. KAY, and C. N. W. LITTING, J. Mater. Sci. **10**, 1317 (1975).
[2] R. N. BHARGAVA, J. Crystal Growth **59**, 15 (1982).
[3] C. WERKHOVEN, B. J. FITZPATRICK, S. P. HERKO, R. N. BHARGAVA, and P. J. DEAN, Appl. Phys. Letters **38**, 540 (1981).
[4] P. J. DEAN, A. D. PITT, M. S. SKOLICK, P. J. WRIGHT, and B. COCKAYNE, J. Crystal Growth **59**, 301 (1982).
[5] W. J. STUTIUS, J. Crystal Growth **59**, 1 (1982).
[6] P. J. DEAN, D. C. HERBERT, C. J. WERKHOVEN, B. J. FITZPATRICK, and R. N. BHARGAVA, Phys. Rev. B **23**, 4888 (1981).
[7] R. N. BHARGAVA, R. J. SEYMOUR, B. J. FITZPATRICK, and S. P. HERKO, Phys. Rev. B **20**, 2407 (1979).
[8] J. E. NICHOLLS, J. J. DAVIES, B. C. CAVENETT, J. R. JAMES, and D. J. DUNSTAN, J. Phys. C **12**, 361 (1979).
[9] F. J. BRYANT and P. S. MANNING, J. Phys. Chem. Solids **35**, 97 (1974).
[10] G. JONES and J. WOODS, J. Lum. **9**, 389 (1974).
[11] J. L. PATEL, J. J. DAVIES, and J. E. NICHOLLS, J. Phys. C **14**, 5545 (1981).
[12] J. C. BOULEY, P. BLANCONNIER, A. HERMAN, P. GED, P. HENOC, and J. P. NOBLANC, J. appl. Phys. **46**, 3549 (1975).
[13] P. BLANCONNIER, J. F. HOGREL, A. M. JEAN-LOUIS, and B. SERMAGE, J. appl. Phys. **52**, 6895 (1981).
[14] K. ERA and D. W. LANGER, J. Lum. **1/2**, 514 (1970).
[15] X. W. FAN and J. WOODS, J. Phys. C **14**, 1863 (1981).
[16] P. LILLEY, Crystal Growth, to be published.
[17] W. B. LEIGH, P. BESOMI, and B. W. WESSELS, J. appl. Phys. **53**, 532 (1982).
[18] W. STUTIUS, J. appl. Phys. **53**, 284 (1982).

(Received April 3, 1984)

phys. stat. sol. (a) 85, 243 (1984)

Subject classification: 10.2 and 13.4; 14.3; 22.4.3; 22.4.4

Electronic Materials Division, National Institute for Materials Research, CSIR, Pretoria[1])

Piezoelectrically Induced Charge Distributions around Dislocations in CdTe and HgCdTe

By

H. Booyens and J. H. Basson

The piezoelectrically-induced equilibrium polarization charge distributions associated with dislocations in CdTe are determined. It is shown that there exists a minimum energy configuration of these dislocation with an associated unique line polarization charge density. The results of these calculations are extended to a consideration of the implications for photosensitive HgCdTe.

Die mit Versetzungen verknüpfte piezoelektrisch induzierte Gleichgewichtspolarisationsladungsverteilung in CdTe wird bestimmt. Es wird gezeigt, daß eine Konfiguration dieser Versetzungen mit minimaler Energie existiert mit einer damit verbundenen bestimmten Polarisationslinienladungsdichte. Die Ergebnisse dieser Berechnungen werden für die Diskussion von photoempfindlichem HgCdTe erweitert.

1. Introduction

CdTe has been receiving increasing attention due to its application as a substrate material for HgCdTe infrared devices [1]. Additionally the material has found application in the field of nuclear detection [2] and solar cell technology [3]. One of the limiting factors in the use of CdTe as a substrate material lies in the high defect density typically encountered in this material [4, 5]. Furthermore, defects such as dislocations have been shown to have deleterious effects on the optical and electrical properties of many wide bandgap semiconductors [6]. Many of these problems are extended to HgCdTe epitaxial layers grown on semi-insulating CdTe substrates which are used for infrared detector applications. In the light of the above it seems appropriate that the properties of dislocations in these materials should be studied in greater depth.

CdTe and HgCdTe are group II–VI compounds having the zincblende structure. As such, these materials are noncentrosymmetric and exhibit the piezoelectric effect according to which an applied stress will result in an electric displacement field. The nature of the field is determined by the orientation of the crystal and the detailed stress field applied.

In this paper the case of dislocations in high resistivity CdTe is considered and the piezoelectrically induced volume charge distributions and associated line charges are calculated. These results are discussed and extended to a consideration of photoconductive HgCdTe.

2. The Piezoelectric Effect in CdTe

The piezoelectric effect in an elastic crystalline material may be expressed in the following form:

$$\tau_{ij} = C_{ijkl}\sigma_{kl} - e_{lij}E_l \,, \tag{1}$$

[1]) POB 395, Pretoria.

$$D_j = e_{jkl}\sigma_{kl} + \varepsilon_{kl}E_l \,, \tag{2}$$

where D_i and E_i are components of the electric displacement and electric field, respectively, while τ_{ij} and σ_{ij} are elements of the stress and strain fields, respectively. C_{ijkl} are elements of the elastic constant tensor, e_{lij} of the elastoelectric constant tensor, and ε_{ij} those of the dielectric constant tensor.

The solution of equations (1) and (2) in the presence of a dislocation is based on the method used by Saada [7] and Faivre and Saada [8]. As in the work of Booyens and Vermaak [9] the problem is formulated in the following coordinate system:

$$\bar{x}_1 = \frac{1}{\sqrt{2}}[110]\,, \qquad \bar{x}_2 = [001], \qquad \bar{x}_3 = \frac{1}{\sqrt{2}}[1\bar{1}0]\,,$$

where $\bar{x}_3$ is in the direction of the dislocation line. In this coordinate system the piezoelectric constant tensor assumes the following form in reduced notation:

$$[e_{ijk}] = e\begin{pmatrix} 0 & 0 & 0 & 0 & 0 & 1 \\ 1/2 & 0 & -1/2 & 0 & 0 & 0 \\ 0 & 0 & 0 & -1 & 0 & 0 \end{pmatrix}. \tag{3}$$

Allowing for the anisotropic nature of the elastic properties of CdTe, we transform the elastic constant tensor to the above coordinate system to obtain,

$$C_{ijkl} = \begin{bmatrix} 1/2(C_{11}+C_{12}+2C_{44}) & C_{12} & 1/2(C_{11}+C_{12}-2C_{44}) & 0 & 0 & 0 \\ C_{12} & C_{11} & C_{12} & 0 & 0 & 0 \\ 1/2(C_{11}+C_{12}-2C_{44}) & C_{12} & 1/2(C_{11}+C_{12}+2C_{44}) & 0 & 0 & 0 \\ 0 & 0 & 0 & C_{14} & 0 & 0 \\ 0 & 0 & 0 & 0 & 1/2)C_{11}-C_{12}) & 0 \\ 0 & 0 & 0 & 0 & 0 & C_{44} \end{bmatrix} \tag{4}$$

in reduced notation. Furthermore, it is assumed that the structure is isotropic, as regards the dielectric effect [10], so that

$$\varepsilon_{ij} = \varepsilon\delta_{ij} \qquad (i, j = 1, 2, 3)\,. \tag{5}$$

The relevant constants for CdTe are given in Table 1.

Table 1
Physical constants of CdTe

	value	units
lattice constant [10]	0.6481	nm
elastic constants [12]		
C_{11}	5.351×10^{11}	dyn cm^{-2}
C_{12}	3.681×10^{11}	dyn cm^{-2}
C_{44}	1.994×10^{11}	dyn cm^{-2}
elastoelectric constant [11]		
e_{14}	9.114×10^{3}	e.s.u. cm^{-2}
dielectric constant [10]		
ε	10.3	

In the absence of external forces on the crystal, one has that

$$\frac{\partial \tau_{ij}}{\partial x_j} = 0 \tag{6}$$

while for the case of a crystal with no free charge

$$\frac{\partial D_j}{\partial x_j} = 0 \,. \tag{7}$$

Application of these two conditions to (1) and (2) results in

$$C_{ijkl} \frac{\partial^2 U_k}{\partial x_j \, \partial x_l} + e_{lij} \frac{\partial^2 \Phi}{\partial x_j \, \partial x_l} = 0 \qquad (i = 1, 2) \tag{8}$$

and

$$e_{jkl} \frac{\partial^2 U_k}{\partial x_j \, \partial x_l} - \varepsilon \frac{\partial^2 \Phi}{\partial x_l{}^2} = 0 \,, \tag{9}$$

where U_k and Φ are the elastic displacement and electric potential, respectively. Furthermore, the cylindrical symmetry associated with the dislocation reduces the problem to two dimensions.

According to Saada [7], one may now assume a solution of the form

$$U_k = A_k \ln (x_1 + s x_2) \qquad (k = 1, 2, 3) \,, \tag{10}$$

where $\Phi = U_3$.

Equation (10), combined with (8) and (9), results in the following set of simultaneous equations:

$$\begin{aligned}
&\left(\tfrac{1}{2} (C_{11} + C_{12} + 2C_{44}) + s^2 C_{44}\right) A_1 && + (C_{12} + C_{44}) \, s A_2 + e s A_3 = 0 \,, \\
&(C_{12} + C_{44}) \, s A_1 \quad + (C_{11} s^2 + C_{44}) \, A_2 + \tfrac{1}{2} e A_3 && = 0 \,, \\
&e s A_1 \quad + \tfrac{1}{2} e A_2 \quad - \varepsilon (1 + s^2) \, A_3 && = 0 \,.
\end{aligned} \tag{11}$$

The condition for non-triviality of the above equations results in a sixth-order polynomial equation in s with three sets of mutually conjugate roots

$$s_n = P_n + i q_n \,. \tag{12}$$

From equation (11) the following ratios are determined:

$$\frac{A_{2n}}{A_{1n}} = \omega_n + i v_n \,, \tag{13}$$

$$\frac{A_{3n}}{A_{1n}} = f_n + i d_n \,, \tag{14}$$

where $A_{1n} = \alpha_n + i \varepsilon_n$ is determined from the boundary conditions

$$\begin{aligned}
&\int_c \frac{\partial U_k}{\partial x_i} \, \mathrm{d}x_i = b_k \,, \\
&\textstyle\iint_s \tau_{ij} n_i \, \mathrm{d}s = 0 \,, \\
&\int_c \frac{\partial \Phi}{\partial x_i} \, \mathrm{d}x_i = 0 \,, \\
&\textstyle\iint_s D_i n_i \, \mathrm{d}s = 4 \pi \lambda L \,.
\end{aligned} \tag{15}$$

c is any closed curve around the dislocation and s any closed surface. λ is the line charge density and L the dislocation length.

The equations for α_n, ε_n as a function of the Burgers vector $\bar{b} = b_1\bar{x}_1 + b_2\bar{x}_2$ and the line charge λ are given as follows:

$$\begin{pmatrix} \varepsilon_1 \\ \varepsilon_2 \\ \varepsilon_3 \\ \alpha_1 \\ \alpha_2 \\ \alpha_3 \end{pmatrix} = \begin{pmatrix} 7.65 \times 10^{-7} & 0 & -6.43 \times 10^{-9} \\ -7.96 \times 10^{-2} & 0.1 & 3.22 \times 10^{-9} \\ -7.96 \times 10^{-2} & -0.1 & 3.22 \times 10^{-9} \\ 0 & 5.38 \times 10^{-6} & 0 \\ 7.5 \times 10^{-2} & 1.78 \times 10^{-2} & 2.09 \times 10^{-9} \\ -7.5 \times 10^{-2} & 1.78 \times 10^{-2} & -2.09 \times 10^{-9} \end{pmatrix} \begin{pmatrix} b_1 \\ b_2 \\ \lambda \end{pmatrix}. \tag{16}$$

The line charge density is determined using the condition that the radial electric field is zero under minimum energy conditions. Under these conditions, no charge is attracted or repelled by the dislocation line. This condition is met when

$$\sum_{n=1}^{3} (\alpha_n f_n - \varepsilon_n d_n) = 0 \tag{17}$$

corresponding to a critical line charge density of

$$\lambda_c = (9.49 \times 10^2)\, b_1 \tag{18}$$

in electrostatic units.

The corresponding charge density may be determined from $U_3 = \Phi$ using Poisson's equation

$$\varrho = -\frac{1}{4\pi} \nabla^2 U_3 \,. \tag{19}$$

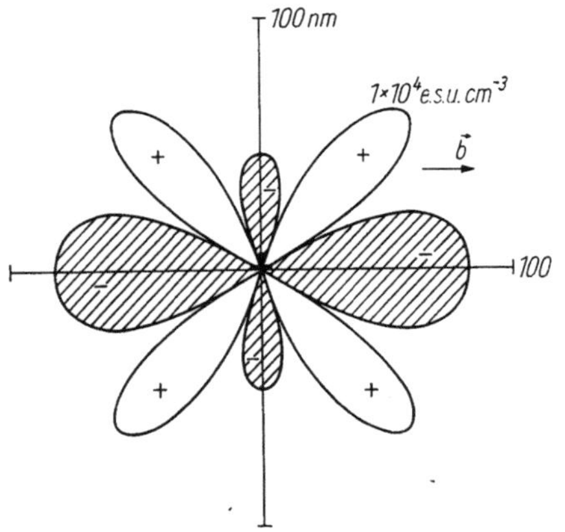

Fig. 1

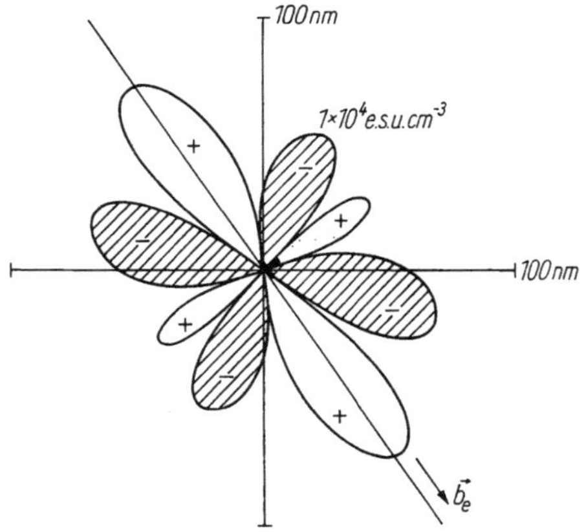

Fig. 2

Fig. 1. Contours of constant polarization charge density around a pure edge dislocation in semi-insulating CdTe. The signs in the lobes denote the sign of the charge and the line charge is chosen equal to the critical value

Fig. 2. Contours of constant polarization charge density around a 60° dislocation in semi-insulating CdTe. The signs in the lobes denote the sign of the charge and the line charge is chosen equal to the critical value

As is the case in most crystals with the zincblende structure, dislocations in CdTe glide on {111} planes and have Burgers vectors of the ⟨110⟩ type [12]. Two types of dislocations are considered here: the pure edge dislocation with line in the $[1\bar{1}0]$ direction and $\bar{b} = (a/2)\,[110]$ and the 60° dislocation with line in the $[1\bar{1}0]$ direction and $\bar{b} = (a/2)\,[1\bar{0}1]$.

In Fig. 1 a contour plot of the polarization charge density distribution around an edge dislocation is shown for the specific case of $\lambda = \lambda_c$ which represents the most stable situation. In this case $b_1 = a/\sqrt{2}$ and $b_2 = 0$ with $a = 0.6481$ nm the lattice parameter of CdTe.

A similar plot is shown in Fig. 2 for a 60° dislocation where, in this case, $b_1 = a/2\sqrt{2}$ and $b_2 = -a/2$.

3. Discussion and Conclusion

The above results strictly apply only in the absence of free carriers which could screen the piezoelectrically induced potential of the dislocations [8]. Consequently, the results would apply only to material of very low doping or which is compensated.

Considering the fact that this material may have very high dislocation densities [13, 14], the effect on charge carriers by piezoelectrically induced fields is expected to be significant. Majority charge carriers are expected to screen the piezoelectrically induced effects at distances greater than $5r_0$ as assessed for Te by Faivre and Saada [8], where

$$r_0 = \left(\frac{\varepsilon k T\, 4\pi}{n e^2}\right)^{1/2} \tag{20}$$

in electrostatic units.

In the case of CdTe material with a doping concentration of 10^{14} cm^{-3} this screening length would be of the order of 10 μm. For heavily dislocated material with dislocation densities $> 10^6$ cm^2 this screening length would be of the same order or larger than the average dislocation separation. Consequently the major portion of the crystal is expected to be under the influence of the piezoelectrically induced fields, which could adversely affect minority carriers and could reduce carrier mobility due to scattering. Similar considerations would apply to the depletion regions of photovoltaic devices, where the free carrier concentration does not screen the dislocation potential.

Although the lack of necessary information as regards the piezoelectric constants for HgCdTe precludes a detailed calculation for this material, the above consideration may nevertheless be extended on the assumption that the qualitative results remain the same. Photoconductive grade $Hg_{1-x}Cd_xTe$ ($x = 0.2$) has a typical carrier concentration of the order of 10^{14} cm^{-3} at 77 K. Consequently, the results obtained above for CdTe will also qualitatively apply to HgCdTe at 77 K. HgCdTe epitaxial layers are typically grown on CdTe substrates with a misfit of 2.5×10^{-3} and consequently misfit dislocations with an average separation of approximately 0.04 μm [13] may be expected to be introduced. The above results indicate that this misfit dislocation array would lead to the introduction of a piezoelectric charge distribution in the epitaxial layer. This is expected to adversely affect the mobility of charge carriers in the epilayer and to have deleterious effects on photo-generated carriers. The same basic results also apply to bulk HgCdTe where typical dislocation densities of the order of 10^6 cm^{-2} are encountered [14].

In conclusion, the piezoelectrically induced charge effects due to dislocations in CdTe have been calculated. The results have been extended to a discussion of the probable effects of dislocations on charge carrier properties in both CdTe and HgCdTe.

References

[1] M. Chu and C. C. Wang, J. appl. Phys. **51**, 2255 (1980).
[2] R. Triboulet, Y. Marfaing, A. Cornet, and P. Siffert, J. appl. Phys. **45**, 2759 (1974).
[3] R. Triboulet and Y. Marfaing, J. Electrochem. Soc. **120**, 1260 (1973).
[4] M. Klimkiewicz, J. Auleytner, and T. Warminski, Crystal Res. Technol. **16**, 175 (1981).
[5] I. V. Kurilo and V. I. Kuchma, Inorg. Mater. **18**, 479 (1982).
[6] H. Kressel, Semiconductors and Semimetals, Vol. 16, Academic Press, New York 1981.
[7] G. Saada, phys. stat. sol. (b) **44**, 717 (1971).
[8] G. Faivre and G. Saada, phys. stat. sol. (b) **52**, 127 (1972).
[9] H. Booyens and J. S. Vermaak, J. appl. Phys. **50**, 4302 (1979).
[10] Landolt-Börnstein, New Series, Vol. 17b, Physics of II–VI and I–VII Compounds, Semimagnetic Semiconductors, Springer-Verlag, Berlin 1982.
[11] D. Berlincourt, H. Jaffe, and L. R. Shiozawa, Phys. Rev. **129**, 1009 (1963).
[12] G. Lu and D. J. H. Cockayne, Physica (Utrecht) **116B**, 646 (1983).
[13] T. W. James and R. E. Stoller, Appl. Phys. Letters **44**, 56 (1984).
[14] S. Cole, A. F. W. Willoughby, and M. Brown, J. Crystal Growth **59**, 370 (1982).

(Received June 19, 1984)

phys. stat. sol. (a) 85, 249 (1984)

Subject classification: 16 and 20.3; 22.4.4

Department of Materials Science,
School of Engineering, Tôhoku University, Sendai[1])

Effect of the Stoichiometry Control on the Photoelectrical Properties of ZnS_xSe_{1-x}

By

K. MOCHIZUKI and K. SUZUKI

The effect of the stoichiometry deviation y of $Zn_{1+y}(S_{0.25}Se_{0.75})$ single crystals on the photoelectrical properties is examined. The stoichiometry is controlled through the vapour growth process by regulating the sum of chalcogen partial pressure $P_{VI}(= P_{S_2} + P_{Se_2})$, and by keeping the partial pressure ratio $P_r(= P_{S_2}/P_{Se_2})$ constant in each vapour growth experiment. With increasing P_{VI}, the emission intensity of the broad band observed at 485.5 nm (named A-peak) increases, but the intensities of the band edge emissions, I_2, and free excitonic peak, E_x, decrease. The dark electrical conductivity at room temperature of the grown crystals increases from 10^{-14} to 10^{-12} $(\Omega cm)^{-1}$ with increasing P_{VI}. From these results, it is found that the deviation y from stoichiometry is controlled effectively and the results can be explained by the increase of native defects such as zinc vacancy V_{Zn} with increasing P_{VI}.

Es wird der Einfluß einer Stöchiometrieabweichung y von $Zn_{1+y}(S_{0.25}Se_{0.75})$-Einkristallen auf die photoelektrischen Eigenschaften untersucht. Die Stöchiometrie wird über den Wachstumsprozeß aus der Dampfphase gesteuert, indem die Summe der Chalkogenpartialdrücke $P_{VI}(= P_{S_2} + P_{Se_2})$ reguliert und das Verhältnis der Partialdrücke $P_r(= P_{S_2}/P_{Se_2})$ bei jeder Dampfphasenzüchtung konstant gehalten wird. Mit wachsendem P_{VI} nimmt die Emissionsintensität der breiten Bande, die bei 485,5 nm (A-Maximum) beobachtet wird, zu, jedoch nehmen die Intensitäten der Bandkantenemissionen I_2, und das Maximum der freien Exzitonen ab. Die elektrische Dunkelleitfähigkeit der gewachsenen Kristalle bei Zimmertemperatur nimmt von 10^{-14} bis 10^{-12} $(\Omega cm)^{-1}$ mit wachsendem P_{VI} zu. Aus diesen Ergebnissen wird geschlossen, daß die Stöchiometrieabweichung y effektiv gesteuert wird, und die Ergebnisse lassen sich durch das Anwachsen von Eigendefekten, z.B. Zinklücken V_{Zn}, mit steigendem P_{VI} erklären.

1. Introduction

It has been known that the deviation from stoichiometry of II–VI binary compounds influences seriously the photoelectrical properties. However, for the case of II–VI ternary system, efforts for controlling the stoichiometry have not been made except our case. In 1978, one of the authors proposed a principle to control the stoichiometry of $Cd_{1+y}(S_xSe_{1-x})$ without changing the composition x [1].

Now, attention is paid to ZnS, ZnSe, and ZnS_xSe_{1-x} for making blue emitting diodes [2 to 6]. However, to control the stoichiometry in ZnS_xSe_{1-x} still remains an important and unresolved problem.

In this study, the principle proposed was applied to the ternary $Zn_{1+y}(S_{0.25}Se_{0.75})$ and the effects of the deviation from stoichiometry on the photoelectrical properties were examined.

The stoichiometry was mainly controlled during the vapour growth process by regulating the sum of chalcogen partial pressure P_{VI} $(= P_{S_2} + P_{Se_2})$, keeping the partial pressure ratio $P_r(= P_{S_2}/P_{Se_2})$ constant in each growth experiment.

[1]) Sendai, Miyagi 980, Japan.

Since the partial pressure control of sulfur and selenium was achieved by introducing small amounts of them into a closed growth tube, efforts were made to minimize the vapour pressure over the source specimen as far as possible.

The vapour transport rate, photoluminescence, reflection spectra, and dark electrical conductivities of the grown crystals were measured as a function of controlled P_{VI}. From these results, the effects of the deviation from stoichiometry on the photoelectrical properties were discussed.

2. Experimental Procedure

The proposed principle to control the stoichiometry of $Zn_{1+y}(S_{0.25}Se_{0.75})$ grown crystals is based on the following equation introduced by Handelman and Kaiser [7]:

$$\frac{x_{ZnS}}{x_{ZnSe}} = \frac{K_{ZnSe}}{K_{ZnS}} \left(\frac{P_{S_2}}{P_{Se_2}}\right)^{1/2}. \tag{1}$$

Here, P_{S_2} and P_{Se_2} are the partical pressure of sulfur and selenium, x_A and K_B denote the mole fraction and dissociation constant of A- and B-components of the mixed crystal, respectively. As the dissociation constant for a fixed growth temperature can be estimated from a recent report [8], the stoichiometry with a fixed composition x must be controlled by regulating P_{VI}, holding the P_r constant in each growth run. The amounts of sulfur and selenium necessary for controlling the partial pressure were calculated by applying the ideal gas law.

The schematic construction of the apparatus for the vapour growth is shown in Fig. 1. The source chamber (s) contains $ZnS_{0.25}Se_{0.75}$ polycrystal of about 0.2 g. The source crystals were prepared from ZnS and ZnSe synthesized crystals which were vapour transported under the partial pressure of Zn corresponding minimum total pressure [9]. This vapour transport keeps the sublimation pressure over the source specimen as small as possible. The ZnS and ZnSe were synthesized by the chemical reaction of pure Zn (Cominco 6 N), Se (Mitsubishi $>$ 5 N), and S (Wako 6 N).

The source and growth temperatures were kept at 1061 and 1051 °C, respectively. After the crystal growth, the quartz ampoules were water-quenched and the transport rate was measured.

The samples with size of about 1 × 2 × 4 mm³ were cut from the grown single crystals along the cleavage planes for the photoluminescence measurements. The photoluminescence spectra were measured at 4.2 K using a Xe lamp as the exciting source. A W lamp was used as a exciting source for reflection spectra. The detection system consists of grating monochromator with a reciprocal dispersion of 1.5 nm/mm and HTVR-292 photomultiplier. Electrical measurements were performed by using In

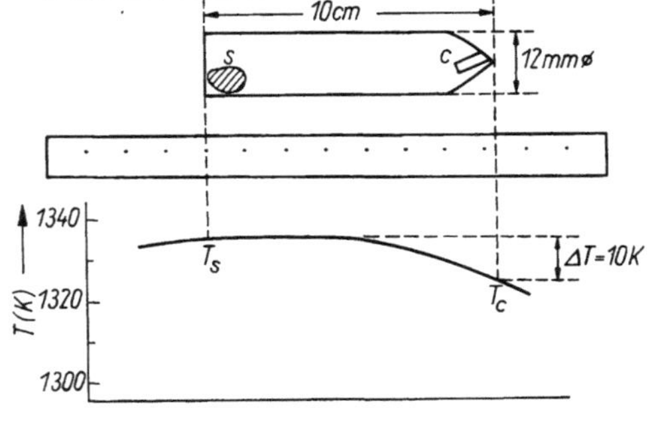

Fig. 1. Schematic construction of the apparatus used for the single crystal growth of $Zn_{1+y}(S_{0.25}Se_{0.75})$ and temperature distribution. s source chamber which contains polycrystalline $ZnS_{0.25}Se_{0.75}$; c growth chamber with conic shape

electrodes and the contacts were confirmed to be ohmic. The photoconductivity was measured by irradiating the samples with light near 365 nm from a high-pressure mercury lamp.

3. Results

Under the temperature difference $\Delta T = 10$ K between source and growth chamber, two or three single crystals with well-developed faces grew together in the growth chamber. The composition of the inner parts of the grown crystals was analyzed by an electron probe microanalyser and the deviation from the target composition $x = 0.25$ was ± 0.01 at%.

Fig. 2 shows the transport rate as a function of P_{VI} under P_r constant ($= 0.076$) in each growth run. As seen in the figure, the transport rate is proportional to the $-\frac{3}{2}$-th power of P_{VI}. The solid line is the calculated transport rate limited by the diffusion process in the vapour phase and this agrees well to the experimental results.

Fig. 3a shows the typical photoluminescence spectra at 4.2 K of the samples grown under P_{min} and $P_{VI} = 1.33 \times 10^3$ Pa. P_{min} means the minimum sum of the chalcogen partial pressure ($P_{VI} = 120$ Pa) and the vapour growth without introducing sulfur and selenium into the growth tube. The features of the photoluminescence are as follows:

1. The photoluminescence consists of four parts: a sharp peak at band edge, two small peaks next of the sharp peak, and two broad peaks at 485.5 nm (named A-peak) and 610 nm (named B-peak).
2. A sharp peak at the band edge is very intense and other peaks are weak.
3. The broad peak at 485.5 nm becomes observable for the crystal grown under $P_{VI} = 1.33 \times 10^4$ Pa.

Fig. 3b shows the detailed photoluminescence and reflection spectra at band edge. As clear in the figure, a sharp peak consists of two peaks, I_2 and E_x.

Fig. 4 shows the typical photoluminescence and reflection spectra with increasing P_{VI}.

The features of these spectra are as follows:

1. The peaks A and B are observed for all the crystals grown under controlled P_{VI}.
2. A broad peak at about 430 nm appears instead of two peaks observed for the crystals grown under P_{min} condition and $P_{VI} = 1.33 \times 10^4$ Pa.
3. The reflection spectra become unclear with increasing P_{VI}.

Fig. 5 shows the emission intensity of A- and I_2-peak. The emission intensity of I_2-peak decreases with increasing P_{VI} whereas that of A-peak increases with increasing P_{VI}.

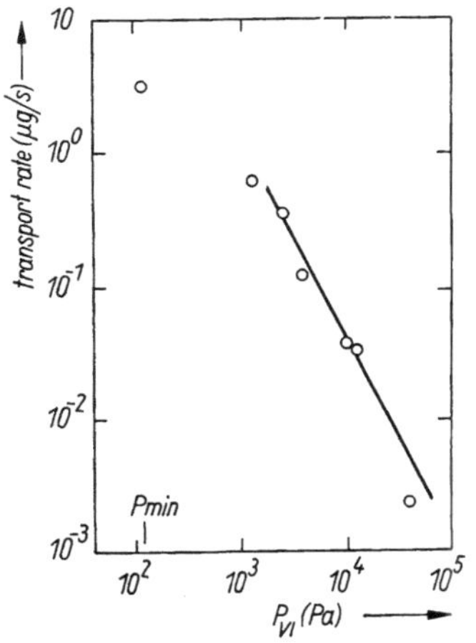

Fig. 2. Vapour transport rate in $ZnS_{0.25}Se_{0.75}$ as a function of P_{VI} at 1329 K, $\Delta T = 10$ K. Solid line shows the calculated transport rate determined by the diffusion process in the vapour phase

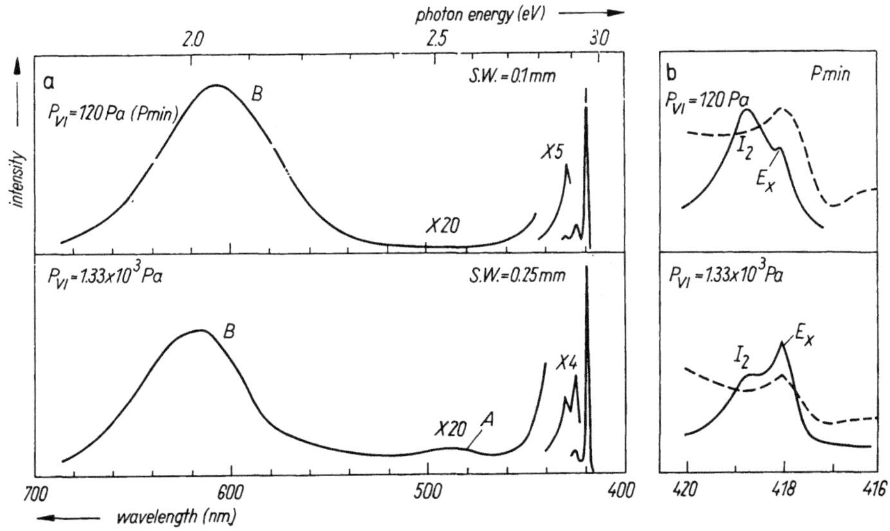

Fig. 3. a) Photoluminescence spectra at 4.2 K of $ZnS_{0.25}Se_{0.75}$ single crystals grown under P_{min} = = 120 and P_{VI} = 1.33×10^3 Pa; b) detailed photoluminescence and reflection spectra at band edge. Symbol S.W. means slit width

Fig. 6 shows the dark and photoconductivity at room temperature as a function of P_{VI}. The dark conductivity increases with increasing P_{VI} by two orders of magnitude, but the photoconductivity shows the opposite dependence.

Fig. 7 and 8 show the photoluminescence at several temperatures of A- and B-peaks and the temperature dependence of the emission intensity of these peaks. The

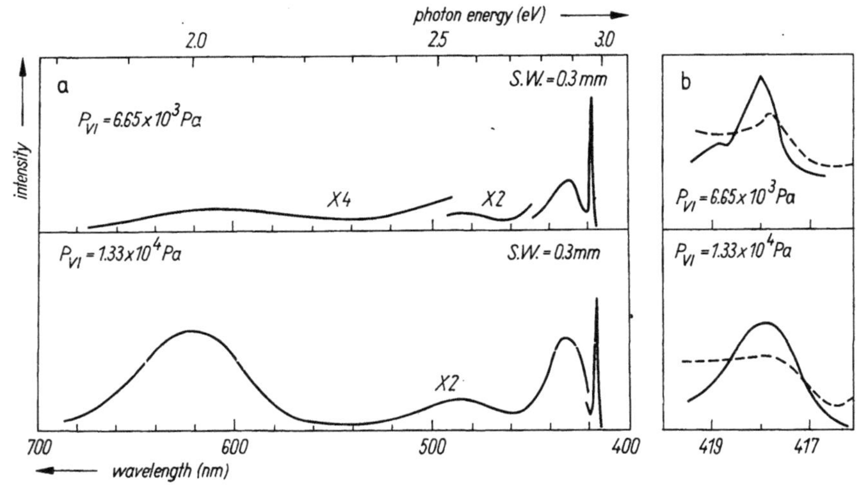

Fig. 4. a) Typical photoluminescence at 4.2 K of $ZnS_{0.25}Se_{0.75}$ single crystals grown under different P_{VI}; b) detailed photoluminescence and reflection spectra at band edge. Symbol S.W. means slit width

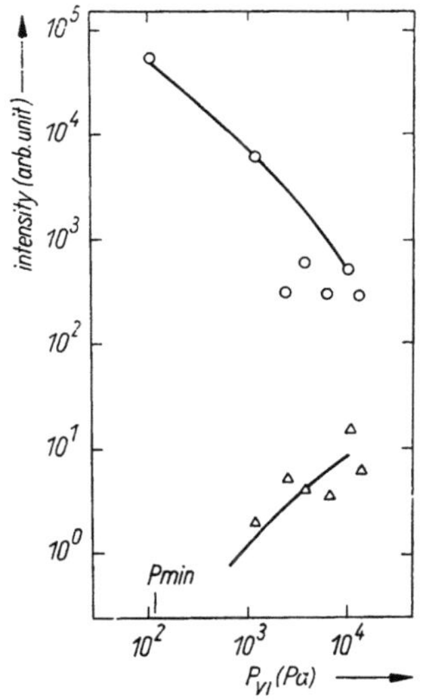

Fig. 5

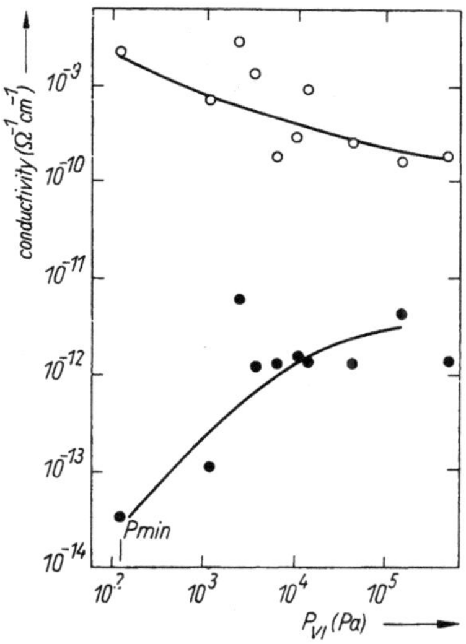

Fig. 6

Fig. 5. Photoluminescence intensity of A- (△) and I_2-peaks (○) as a function of P_{VI}

Fig. 6. Dark (●) and photoelectrical (○) conductivity at room temperature as a function of P_{VI}. Photoconductivity was measured by irradiating the sample with 365 nm line of a Hg lamp

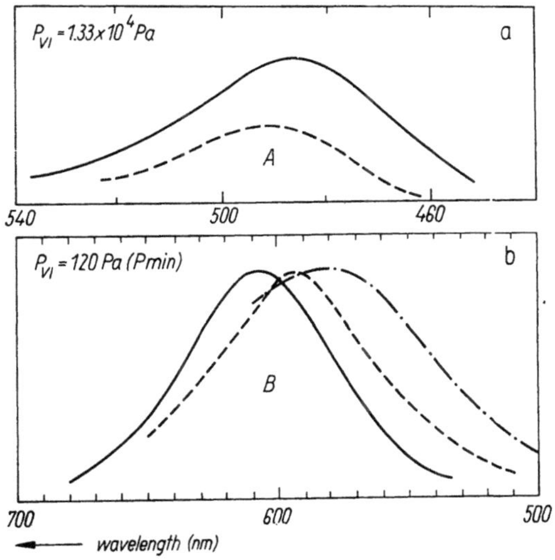

Fig. 7

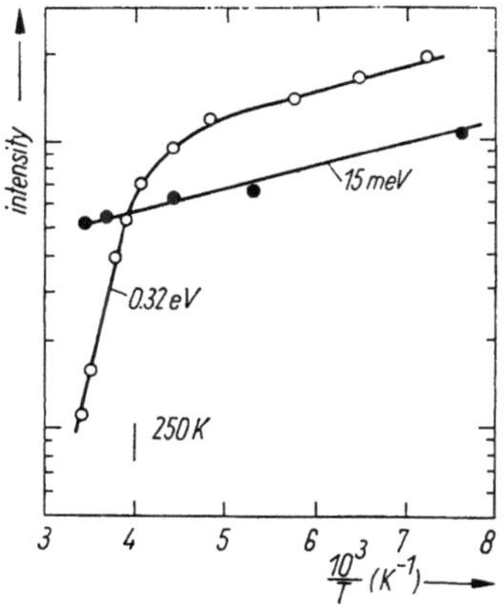

Fig. 8

Fig. 7. Photoluminescence of a) A- (——— 77 K, - - - 100 K) and b) B-peaks (——— 4.2, - - - 128, — · — 298 K)

Fig. 8. Temperature dependence of A (●) and B (○) emission peaks

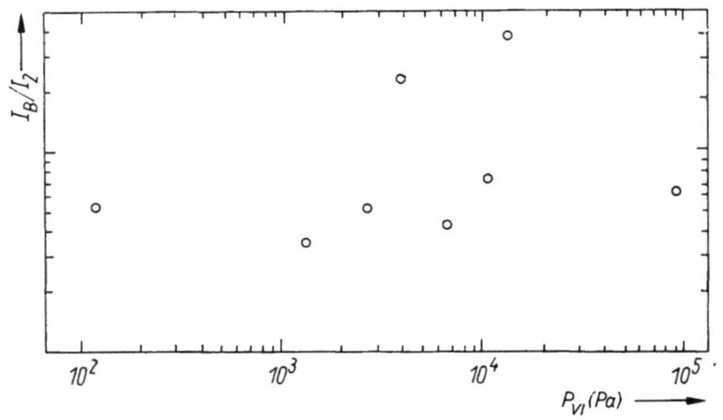

Fig. 9. Emission intensity ratio of B- and I_2-peaks as a function of P_{VI}

A-peak shifts to the low-energy side with increasing temperature, but the B-peak shifts to the high-energy side.

Fig. 9 shows the emission intensity ratio of B- and I_2-peaks (I_B/I_{I_2}) as a function of P_{VI}. No clear dependence was observed.

4. Discussion

The dependence observed in Fig. 2 is the same one found in II–VI binary system [10, 11] and can be explained by assuming that the diffusion process in the vapour phase is rate determining. The calculated transport rate (solid line) was obtained by analogy with binary systems without introducing any adjustable parameters. The good agreement between calculated and observed results suggests that the solid–vapour equilibrium in the growth tube is nearly established through the growth process and the deviation from stoichiometry of the grown crystals is controlled effectively.

From the comparison of the photoluminescence with the reflection spectra in Fig. 3b, it is known that E_X-peak is the free excitonic emission due to the radiative annihilation of free excitons. The I_2-peak is the emission due to the radiative annihilation of excitons bound to a neutral donor. The energy difference between E_X- and I_2-peaks is 4.8 meV and this corresponds to that of ZnSe [12].

Small two peaks observed close to band edge emission in Fig. 3a are due to I_1^D and I_1^D − 1 LO peak, respectively. I_1^D-peak is known as the emission due to the radiative annihilation of excitons bound to a neutral acceptor [13]. The assignment of I_1^D-peak was made by the composition dependence of the emission energy in ZnS_xSe_{1-x} close to ZnSe [14].

A broad peak at about 430 nm in Fig. 4a appears instead of I_1^D and I_1^D − 1 LO, but the origin is not clear.

The reflection spectra become unclear with increasing P_{VI} and this might be caused by the introduction of defects due to the deviation from stoichiometry.

As seen in Fig. 5, the emission intensity of I_2-peak becomes to decrease with increasing P_{VI} whereas that of the A-peak increases with increasing P_{VI}. This fact suggests that the native defect such as a zinc vacancy V_{Zn} which increases with increasing P_{VI}, is closely related with the origin of A-peak [15]. On the other hand, the behaviour of I_2-peak can be explained by considering that the donor species responsible for I_2-emission is compensated by the acceptor V_{Zn}, forming D–A pair (donor–acceptor pair) or so-called S–A emission [16] and the compensated donor does not contribute to the I_2-emission.

By assuming that the A-peak is due to D–A pair emission related with V_{Zn}, the emission energy was estimated by the following equation:

$$E_{D\text{-}A} = E_g - (E_D + E_A)\,. \tag{2}$$

Here, E_g, E_D, and E_A are the band gap energy of $ZnS_{0.25}Se_{0.75}$, the ionization energy of a donor and that of an acceptor. The band gap energy of $ZnS_{0.25}Se_{0.75}$ can be estimated by assuming that the energy difference between band gap and free excitonic emission is the same for ZnSe and $ZnS_{0.25}Se_{0.75}$. Therefore, the band gap energy of $ZnS_{0.25}Se_{0.75}$ is the sum of the binding energy of free exciton of ZnSe (18 meV) [12] and the emission energy due to free exciton of $ZnS_{0.25}Se_{0.75}$ (2.9649 eV). A donor ionization energy such as Al is (26.3 ± 0.6) meV [17, 18]. The acceptor level due to V_{Zn} is estimated from the following empirical equation [19]:

$$E_A = \frac{E_{x_1} - E_1^D}{0.1}\,. \tag{3}$$

Here, E_{x_1} and E_1^D are the emission energy due to the annihilation of free excitons and that of excitons bound to a neutral V_{Zn}, respectively. By using these values, the emission energy $E_{D\text{-}A}$ was estimated to 2.553 eV and this corresponds to the emission energy of the A-peak.

From this calculation, it is probable that the A-peak is the D–A pair emission due to a shallow donor and an acceptor V_{Zn}. As seen in Fig. 7, the peak A shifts to the low-energy side with increasing temperature and the activation energy estimated from Fig. 8 is about 15 meV. This result is compatible with the origin of D–A pair emission, because the D–A pair emission shifts to the low-energy side with increasing temperature [19].

The B-peak shifts to the high-energy side with increasing temperature and this shows that the B-peak is due to the so-called S–A emission [16, 20].

The quenching behaviour of B-peak in Fig. 8 is the same as in the case of ZnSe [20].

As seen in Fig. 9, the relative emission intensity ($I_{S\text{-}A}/I_{I_2}$) does not have clear dependence and this means that the V_{Zn} introduced into the crystal does not form the S–A pair. This result supports that the V_{Zn} forms the D–A pair rather than the associated pair.

Fig. 6 shows the dark and photoconductivity at room temperature as a function of P_{VI}. The dark conductivity increases with increasing P_{VI} by two orders of magnitude, but the photoconductivity shows an opposite dependence. The obvious change of the dark conductivity must be caused by the deviation from stoichiometry and also corresponds to the increase of V_{Zn} with increasing P_{VI}. Therefore, the behaviour of the dark conductivity can be explained by considering $V_{Zn}^{\cdot}$ which acts as an acceptor.

Since it is known that the crystal with stoichiometric composition has an equal and minimum concentration of cation- and anion-type native defects compensated electrically, it is expected that the dark conductivity of the crystal becomes lowest. For this reason, the composition close to stoichiometry should be realized for the crystal grown under lowest P_{VI}, i.e., P_{min}.

It is expected also that the crystal with stoichiometric composition has the lowest concentration of trapping or recombination centres, which affects the photoconductivity by reducing the lifetime of free electrons and holes created by light irradiation.

Therefore, it seems reasonable that the highest photoconductivity was observed for the crystal grown under P_{min} condition. The decrease of photoconductivity with increasing P_{VI} can be explained by the increase of a defect acting as a trapping and a recombination centre.

From the results of photoluminescence and electrical conductivity, it is expected that one part of V_{Zn} compensates the shallow donor contributing to the A-peak as a D–A pair, and the other part of V_{Zn} is an isolated defect contributing to the increase of dark conductivity with increasing P_{VI}.

5. Conclusions

The stoichiometry of $Zn_{1+y}(S_{0.25}Se_{0.75})$ single crystals was controlled through the vapour growth process by regulating the sum of chalcogen partial pressure $P_{VI}(= P_{S_2} + P_{Se_2})$, and by keeping the pressure ratio $P_r(= P_{S_2}/P_{Se_2})$ constant in each vapour growth experiment.

With increasing P_{VI}, the photoluminescence intensity of the broad peak at 485.5 nm increased, but the intensities of edge emissions, I_2, and free excitonic peaks, decreased.

The dark electrical conductivity at room temperature of the grown crystals increased from 10^{-14} to 10^{-12} $(\Omega cm)^{-1}$ with increasing P_{VI}.

From these results, it was known that the deviation from stoichiometry of $ZnS_{0.25}Se_{0.75}$ is controlled effectively by this method and the results can be explained by the increase of native defects such as zinc vacancies V_{Zn}.

Acknowledgement

The authors are grateful to Prof. K. Igaki for his kind supports and advices in this study.

References

[1] K. Mochizuki and K. Igaki, J. Crystal Growth **45**, 218 (1978).
[2] X. W. Fan and J. Woods, IEEE Trans. Electron Devices **28**, 428 (1981).
[3] H. J. Łożykowski, J. Sekulski, and B. Konstańska-Sekulska, J. Lum. **20**, 83 (1979).
[4] C. Lawther and J. Woods, J. Lum. **18**, 724 (1979).
[5] H. Kukimoto, S. Oda, and H. Katayama, J. Lum. **12**, 923 (1976).
[6] T. Yokogawa, T. Taguchi, S. Fujita, and M. Satoh, IEEE Trans. Electron Devices **30**, 271 (1983).
[7] E. T. Handelman and W. Kaiser, J. appl. Phys. **35**, 3519 (1964).
[8] K. C. Mills, in: Thermodynamic Data for Inorganic Sulfides, Selenides, and Tellurides, Butterworths, London 1974 (pp. 686, 690).
[9] T. Kiyosawa, K. Igaki, and N. Ohashi, Trans. Japan. Inst. Metals **13**, 248 (1972).
[10] K. Igaki and K. Mochizuki, J. Crystal Growth **24/25**, 162 (1974).
[11] K. Mochizuki and K. Igaki, Japan. J. appl. Phys. **18**, 1447 (1979).
[12] S. Satoh and K. Igaki, Japan. J. appl. Phys. **19**, 1953 (1980).
[13] S. Huang, Y. Nozue, and K. Igaki, Japan. J. appl. Phys. **22**, L420 (1983).
[14] K. Mochizuki, J. Crystal Growth **58**, 87 (1982).
[15] H. Röppischer, J. Jacobs, and B. Novikov, phys. stat. sol. (a) **27**, 123 (1975).
[16] T. Koda and S. Shionoya, Phys. Rev. **136**, A541 (1964).
[17] J. L. Merz, H. Kukimoto, K. Nassau, and J. W. Shiever, Phys. Rev. B **6**, 545 (1972).
[18] J. L. Merz, K. Nassau, and J. W. Shiever, Phys. Rev. B **8**, 1444 (1973).
[19] R. E. Halsted and M. Aven, Phys. Rev. Letters **14**, 64 (1965).
[20] S. Iida, J. Phys. Soc. Japan **25**, 177 (1968).

(Received May 2, 1984)

phys. stat. sol. (a) 85, 257 (1984)

Subject classification: 19; 22.9

Sektion Physik der Karl-Marx-Universität Leipzig[1]) *(a),*
Institute of Chemical Kinetics and Combustion, Academy of Sciences of the USSR, Sibirian Branch, Novosibirsk (b),
and Zentralinstitut für Festkörperphysik und Werkstofforschung der Akademie der Wissenschaften der DDR, Dresden[2]) *(c)*

Time-Resolved EPR on Polyacetylene

By

G. Völkel (a), S. A. Dzuba (b), A. Bartl (c), W. Brunner (a), and J. Fröhner (c)

Time-resolved EPR is applied to distinguish between different types of paramagnetic centres in polyacetylene. The electron spin echo (ESE) registration of EPR spectra allows to study the role of oxygen in the appearance of trapped paramagnetic centres in cis- and trans-rich polyacetylene. Using pulsed saturation transfer experiments considerably different spectral diffusion within the spectra of the trapped centres may be detected for pure and oxidized samples. The existence of two types of paramagnetic centres in the EPR spectra of polyacetylene is confirmed by EPR line-shape simulations.

Mit Hilfe der zeitaufgelösten EPR wird versucht, die im Polyacetylen vorhandenen unterschiedlichen Arten von paramagnetischen Zentren einzeln zu charakterisieren. Die Elektronen-Spin-Echo (ESE)-Messungen erlauben es, den Einfluß des Luftsauerstoffs auf die Beweglichkeit der paramagnetischen Zentren im cis- und trans-Polyacetylen zu untersuchen. Saturation-Transfer-Experimente zeigen beträchtliche Unterschiede der Spektraldiffusion in den EPR-Spektren für reine und oxidierte Proben. Die Existenz von zwei Typen paramagnetischer Zentren im Polyacetylen wird durch Linienformanalyse der EPR-Spektren bestätigt.

1. Introduction

Time-resolved EPR is a powerful tool in studying motional effects of paramagnetic centres. Recently, it has successfully been applied to study paramagnetic centres in polyacetylene [1 to 3]. It has been shown that measurements of phase relaxation allow a direct calculation of the intrachain soliton diffusion rate [1, 2].

Moreover, pulse measurements in connection with cw-EPR allow also two different types of centres to be distinguished, a mobile one with a narrow EPR line and a trapped one with a much broader line [3]. The correlation time for the trapped centre was estimated earlier from dynamic nuclear polarization experiments to be at least longer than 10^{-7} s [4, 5].

The diffusional motion of paramagnetic centres should result in a spectral diffusion throughout the EPR spectrum. Such spectral diffusion can be studied by the pulsed saturation transfer method developed in [6, 7]. Using a magnetic field jump for the investigation of the saturation transfer to different parts of the EPR spectrum, it is possible to study very slow spectral diffusion. This technique seems to be interesting for studying the trapped paramagnetic centre in polyacetylene.

As is known from the literature [8 to 10] the presence of oxygen plays an important role for trapping of paramagnetic centres. It will be shown that the ESE technique is

[1]) Linnéstr. 5, DDR-7010 Leipzig, GDR.
[2]) Postfach, DDR-8027 Dresden, GDR.

a useful tool for studying the role of oxygen in the appearance of trapped paramagnetic centres in cis- and trans-rich polyacetylene as well.

2. Experimental

The samples were prepared at 220 K according to the procedure of Shirakawa and Ikeda [11]. After the preparation samples 1, 2, and 6 (see Table 1) were evacuated

Table 1

Summary of the experimental data for the investigated polyacetylene samples

sample	presence of oxygen	heat treat-ment	concen-tration (10^{18} cm^{-3})	T (K)	EPR linewidth*) (mT)	ESE linewidth**) (mT)	T_1 (µs)	τ_c (µs)
1	no	no	1	300	0.56	0.83	70	40
				77	0.73	0.96	500	300
2	no	yes	5	300	0.07	no echo signal		
3	yes	no	3	300	0.65	1.11	100	20
				77	0.79	1.13	220	30
				27			490	30
4	yes	yes	2.5	300	0.45	0.99	70	30
				77	0.74	1.08	260	50
				27			660	70
5	yes	yes	2	300	0.22	0.90	90	40
				77	0.36	1.11	260	40
				27			600	50
6	after heat treatment	yes	5	300	0.22	—	—	—
				77	0.33	—	—	—

The symbol (—) means that measurement was not possible (no echo).
*) Peak to peak.
**) At half height.

immediately without contamination by oxygen, whereas samples 3, 4, and 5 were stored in air for 15 min and only then evacuated. The heat treatment was generally done at +150 °C for 30 min except for sample 5 which was previously heat-treated at 100 °C for 30 min followed by a second treatment at 150 °C for 1 h.

The electron spin echo (ESE) experiment was performed in the X-band with a pulse heterodyne spectrometer described earlier [12]. The pulse width was varied from 30 to 100 ns for a $\pi/2$ pulse. The dead time of the spectrometer amounts to about 250 ns. The device was equipped with a cavity containing modulation coils. Applying a current pulse up to 20 A yielded a magnetic field jump ΔB up to 1.6 mT. This arrangement allowed us to saturate one part of the EPR line at the field position B_0 and then immediately observe changes of the longitudinal magnetization in all other parts of the line using the pulse sequence shown in Fig. 1. All technical details have been described in [6, 7]. Usually the experiments were performed using a constant time interval t and scanning the magnetic field jump ΔB. In this manner pulse saturation transfer experiments can be done.

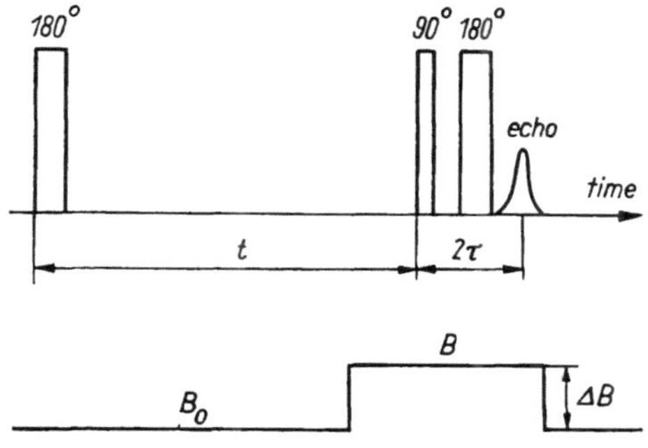

Fig. 1. Time scheme of microwave pulses (top) and magnetic field jump (bottom). The first pulse is the saturating pulse whereas the echo is formed by the second and third pulses

The EPR experiments were performed with a VARIAN E 112 spectrometer. The measurements were made at room temperature, at liquid nitrogen temperature, $T = 77$ K, and at liquid neon temperature, $T = 27$ K.

3. ESE-Detected EPR Spectra

Magnetic field dependent registration of the echo amplitude results in ESE-detected EPR spectra. As compared with the conventional EPR technique, there exist two sources of additional broadening in ESE measurements. The first is related with the fact that microwave pulses excite spins, the resonant fields of which lie in an interval with a width approximately equal to B_1.

Another source of broadening appears due to the mechanism of instantaneous diffusion [13]. This mechanism results in a more rapid phase relaxation at the maximum of the spectrum as compared with its wings. So the shape of the spectrum depends on the time interval τ between the microwave pulses. However, this relaxation has a simple exponential form at each spectral position [13] and the shape at $\tau = 0$ can be extrapolated by comparing results at different τ values.

Fig. 2 shows ESE-detected EPR spectra taken at equidistant time intervals τ. The amplification was adjusted for each spectrum to obtain the same signal amplitude at the point of the maximum. By comparing the spectra from $\tau = 0.28$ to 0.84 μs it is possible to extrapolate the spectrum to $\tau = 0$ (dashed curve of Fig. 2).

In the above experiments we used pules sequences $\pi/2$–π with pulse widths equal to 100 and 200 ns, respectively. The calculated value of B_1 is of the order of 0.1 mT. The additional broadening of EPR spectra is about 0.2 mT in this case, as can be seen from the spectrum of irradiated quartz (p–p EPR linewidth 0.08 mT) taken under the same experimental conditions (Fig. 2).

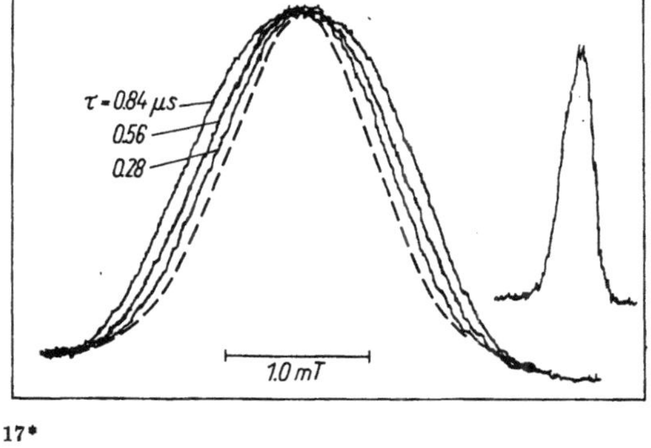

Fig. 2. ESE-detected EPR signals of sample 3 at 77 K for different τ values. The amplification has been adjusted for each spectrum to obtain the same signal amplitude at the line centre. On the right side is shown the ESE-detected spectrum of a γ-irradiated quartz sample with a peak-to-peak EPR linewidth of 0.08 mT taken under the same experimental conditions

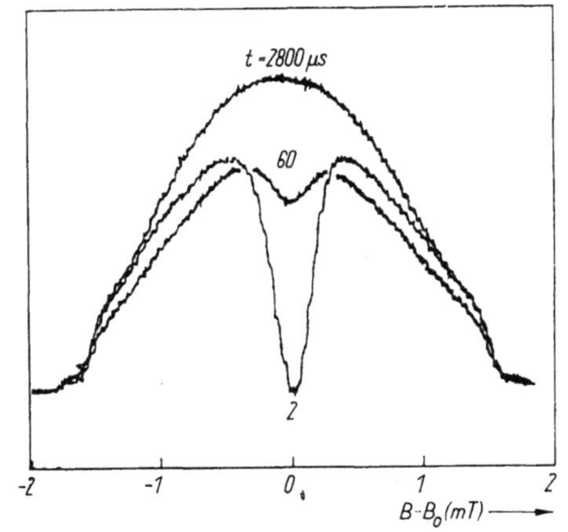

Fig. 3. Field jump scanned ESE-detected spectrum taken at different times after application of a saturating pulse. Sample 3 at 77 K. Some distortions in the wings of the line appear due to the nonrectangular shape of the magnetic field jump for bigger ΔB

For ESE-detected EPR spectra it is more convenient to determine the linewidth at half maximum. These linewidths for different samples are given in Table 1. It is interesting to note that they do not differ very much from each other in contrast to the EPR spectra.

4. Pulsed Saturation Transfer Experiments

A typical example of such an experiment using a magnetic field jump is shown in Fig. 3. The pulse width was 40 and 80 ns for the $\pi/2$ and π pulse, respectively. As can be seen immediately from Fig. 3 the saturation localized previously at the centre of the spectrum spreads later on throughout the spectrum. The important feature of the saturation transfer data of Fig. 3 is that the saturation appears in every part of the spectrum at the same time, independently of the distance from the centre of the hole. So spectral diffusion has a strongly uncorrelated character.

For uncorrelated spectral diffusion it is very easy to obtain a quantitative relation for the longitudinal magnetization behaviour [14]. Let us divide the whole spectrum into N parts. The magnetization of each part will be designated as $M_i(t)$. At the time $t = 0$ the set $M_i(0)$ is determined by the shape of the hole originally burnt into the spectrum. Let the equilibrium magnetization of each part be $M_{0i} = M_i(\infty)$, and without loss of generality one is allowed to set $\sum_{i=1}^{N} M_{0i} = 1$. If τ_c is the mean lifetime at the i-th spectral position, the kinetic equations for $M_i(t)$ take the form [14]

$$\frac{\mathrm{d}M_i(t)}{\mathrm{d}t} = \frac{M_{0i} - M_i(t)}{T_1} - \frac{M_i(t)}{\tau_c} + \frac{M_{0i} \sum_{j=1}^{N} M_j(t)}{\tau_c}; \qquad i = 1, 2, \dots, N. \tag{1}$$

The last term in (1) is characteristic of an uncorrelated spectral diffusion. The probability that the magnetization appears at the i-th spectral position is proportional only to the statistical weight of this position, $M_{0i}/\sum_{j=1}^{N} M_{0j}$.

Equation (1) can be easily solved independently of the value of N. The solutions have the form

$$-\ln\left[1 - \frac{M_i(t)}{M_{0i}}\right] = \frac{t}{T_1} - \ln\left[1 - \sum_{j=1}^{N} M_j(0) + \left\{\sum_{j=1}^{N} M_j(0) - \frac{M_i(0)}{M_{0i}}\right\} \mathrm{e}^{-t/\tau_c}\right]. \tag{2}$$

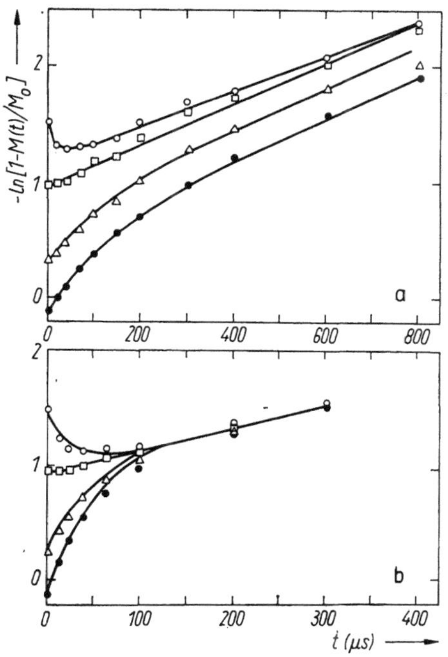

Fig. 4. Time kinetics of the longitudinal magnetization. The experimental points are taken at different distances from the centre of the hole burnt by the saturating pulse as indicated in the figure: ● 0, △ 0.15, □ 0.30, ○ 0.45 mT; a) sample 1 at 77 K, b) sample 3 at 27 K. The curves drawn in 4b correspond to (2) (see text)

Fig. 4 presents experimental data for different spectral positions, plotted in coordinates convenient for comparing with (2).

Note two properties of (2). Firstly, if spectral diffusion does not take place, there follows $\tau_c \to \infty$, so the curves should be straight lines, parallel to one another. Such a situation occurs in the case of Fig. 4a for long times $t > 300$ μs. Secondly, for times $t \gg \tau_c$ all curves should collapse into a single straight line. This situation happens in the case of Fig. 4b. The curves of Fig. 4b correspond to (2) with appropriate choice of τ_c and $M_i(0)$. The parameters T_1 and $\sum_{j=1}^{N} M_j(0)$ have been obtained from the asymptotic straight line. The agreement between theory and experiment is good, this means that spectral diffusion really has an uncorrelated character. The experimental points of Fig. 4b are best fitted with τ_c equal to 30 μs. In the same manner the τ_c values at other temperatures and for other samples were obtained. The data are presented in Table 1.

The sample without oxygen does not exhibit such a remarkable spectral diffusion (Fig. 4a). Some tendency for collapsing in short times could be attributed to a small admixture of oxygen. For a rather long time (Fig. 4a) the curves change to parallel straight lines. So the experimental points of Fig. 4a can not be fitted by (2) for any choice of τ_c. It is possible to note only that τ_c should not be less than 300 μs according to our estimate.

5. Discussion

The observed spin echo linewidth is strikingly larger than the EPR linewidth. This confirms the conclusion that at least two types of centres may exist in polyacetylene. On centre has a narrow line with a transverse relaxation time in the order of $T_M \leqq 100$ ns [1] and is difficult to observe by the ESE method. This centre has been attributed to the highly mobile spins in polyacetylene. The other type of paramagnetic centre has a much longer transverse relaxation time in the order of 350 ns and a wide EPR line.

As compared with literature data [3] the linewidth in our experiment (Table 1) is slightly narrower. The reason for this difference seems to be due to the fact that in [3] the two sources of additional broadening mentioned above have not been taken into account. The importance of both of these mechanisms increases with increasing B_1. According to the experimental conditions in [3] B_1 can be estimated to be in the order of 0.2 mT. This is twice the value used in our experiment. Consequently B_1 should have a remarkable influence on the ESE-detected EPR linewidth. With big B_1 and long τ values we also observe a line doubling. The separation between the two lines depends strongly on the B_1 and τ values. We conclude that this effect can be interpreted as a consequence of instantaneous diffusion [13]. As indicated above this mechanism results in a more rapid phase relaxation for the line centre as compared with the wings. In this case a much lower ESE intensity appears at the centre of the spectrum than in the wings resulting in a doubling of the line. From this point of view the suggestion about an additional localized chain defect offered in [3] requires more careful reinvestigation.

Furthermore we note that no broad ESE line could be observed in the case of sample 2 originally prepared without oxygen and then heat-treated. Even after admission of oxygen (sample 6) no broad ESE signal appeared whereas the EPR line broadened to 0.22 mT. In the case of non-oxidized polyacetylene the broad ESE line disappeared after heat treatment. The reason may be a strong interaction of these centres with mobile centres appearing after heat treatment which shortens the phase relaxation. It is interesting to note that supplementary to previous EPR data reported in the literature [10] different sequences of sample treatment result in a different nature of paramagnetic species as can be seen from comparing EPR and ESE data for samples 5 and 6 (cf. Table 1).

From dynamic nuclear polarization experiments [5] it is known that in cis-rich polyacetylene the correlation time of motion τ_c for the unpaired electron is at least larger than 10^{-7} s. Our pulsed saturation transfer experiments allow this boundary to be extended to a value of $\tau_c \geqq 3 \times 10^{-4}$ s.

In contrast to this situation the sample prepared in the presence of oxygen (sample 3 in Table 1) exhibits considerably faster spectral diffusion. This spectral diffusion has a noncorrelated character and shows a very weak temperature dependence. Till now we have found no clear explanation for this behaviour. Possibly this faster motion along or between chains may be related to some special structural properties of trans-sequences isomerized due to the presence of oxygen [15]. Furthermore it should be noted that after heat treatment (samples 4 and 5) the centres with the broad EPR line show no remarkable change in motion.

In order to confirm the existence of the two types of paramagnetic centres we compared the tails of EPR lines of the different samples before and after heat treatment. It appeared that the form of wings 1.5 mT outside the line centre was perfectly the same for samples 1 and 2 on the one hand and for 3, 4, and 5 on the other. A fitting procedure of the EPR spectrum of polyacetylene, assuming the single line to be a super-

Table 2

Analysis of polyacetylene spectra in Gaussian and Lorentzian portions

sample	Gaussian linewidth*) (mT)	Lorentzian linewidth*) (mT)	portion of Gaussian intensity (%)
2	1.38	0.065	7
3	1.32	0.643	31
4	1.36	0.457	22
5	1.19	0.258	22

*) Peak to peak.

position of a small Lorentzian line and a broad Gaussian line also shows the existence of two types of paramagnetic centres (Table 2). The linewidth of the Gaussian line is about 1.3 mT for all samples. This part of the spectrum represents low mobile paramagnetic centres in oxygen-trapped states and in cis-regions of the polymer. The linewidth of the Lorentzian part changes between 0.6 and 0.06 mT depending on the degree of isomerization because with increasing trans-content these paramagnetic centres become more mobile.

Acknowledgements

The ESE investigations were performed in the Wissenschaftsbereich Struktur der Materie der Sektion Physik der Karl-Marx-Universität and the authors are indebted to Prof. W. Windsch for his encouragement and G. Reinhard for technical assistance. The authors wish also to acknowledge promotive discussions with Prof. G. Lehmann, Prof. L. Wuckel, and Dr. B. Pietraß.

References

[1] N. S. Shiren, Y. Tomkiewicz, T. G. Kazyaka, A. R. Taranko, H. Thomann, L. Dalton, and T. C. Clarke, Solid State Commun. **44**, 1157 (1982).
[2] N. S. Shiren, Y. Tomkiewicz, H. Thomann, L. Dalton, and T. C. Clarke, J. Physique Suppl. 6, **44**, C3-223 (1983).
[3] M. Mehring, H. Seidel, H. Weber, and G. Wegner, J. Physique Suppl. 6, **44**, C3-217 (1983).
[4] M. Nechtschein, F. Devreux, F. Genoud, M. Guglielmi, and K. Holczer, J. Physique Suppl. 6, **44**, C3-209 (1983).
[5] M. Nechtschein, F. Devreux, R. L. Greene, T. C. Clarke, and G. B. Street, Phys. Rev. Letters **44**, 356 (1980).
[6] S. A. Dzuba and Yu. D. Tsvetkov, Khim. Fiz. **1**, 1197 (1982).
[7] S. A. Dzuba, A. G. Maryasov, K. M. Salikhov, and Yu. D. Tsvetkov, J. magnetic Resonance **58**, 95 (1984).
[8] K. Holczer, J. P. Boucher, F. Devreux, and M. Nechtschein, Phys. Rev. B **23**, 1051 (1981).
[9] M. Nechtschein, F. Devreux, F. Genoud, M. Guglielmi, and K. Holczer, Phys. Rev. B **27**, 61 (1983).
[10] P. Bernier, M. Rolland, C. Linaya, M. Aldissi, J. Sledz, F. Schue, J. M. Fabre, and L. Giral, Polymer J. **13**, 201 (1981).
[11] H. Shirakawa and S. Ikeda, Polymer J. **2**, 231 (1971).
[12] W. Brunner and F. Gentzsch, Exper. Tech. Phys. **19**, 39 (1971).
[13] J. R. Klauder and P. W. Anderson, Phys. Rev. **125**, 912 (1962).
[14] S. A. Dzuba, K. M. Salikhov, and Yu. D. Tsvetkov, Chem. Phys. Letters **79**, 568 (1981).
[15] J. C. W. Chien and X. Yang, J. Polymer Sci. Polymer Letters Ed. **20**, 97 (1982).

(Received June 21, 1984)

phys. stat. sol. (a) 85, 265 (1984)

Subject classification: 10.2; 20.1; 20.3; 22.5.4

Department of Physics, Nizam College, Osmania University, Hyderabad[1])

Two Types of F-Centres and Thermoluminescence in BaFCl Crystals

By

K. SOMAIAH and V. HARI BABU

Thermoluminescence and optical absorption studies are employed to idetify two types of F-centres in BaFCl crystals, being formed at fluorine and chlorine ion vacancies. Four glow peaks at 345, 365, 385, and 410 K and two prominent optical absorption bands at 440 and 550 nm are obtained. The assignment of the 440 and 550 nm absorption bands to $F(\overline{F})$ and $F(\overline{Cl})$ centres, respectively, is confirmed by the thermal stability of the absorption bands and glow curves. High-temperature irradiation glow curves and absorption bands suggest that the glow peaks at 365 and 385 K correspond to the absorption band at 550 nm and are caused by $F(\overline{Cl})$ centres, while the high-temperature 410 K peak correspond to 440 nm absorption band caused by $F(\overline{F})$ centres. The $F(\overline{Cl})$ centres are found to be less stable than the $F(\overline{F})$ centres. Thermoluminescence emission spectra at all near glow peak temperatures show only one emission band at 390 nm.

Thermolumineszenzuntersuchungen und Messungen der optischen Absorption werden benutzt, um zwei Arten von F-Zentren in BaFCl-Kristallen zu untersuchen, die an Fluor- und Chlorionenlücken gebildet werden. Vier Glowmaxima bei 345, 365, 385 und 410 K und zwei herausragende optische Absorptionsbanden bei 440 und 550 nm werden gefunden. Die Zuordnung der 440 und 550 nm-Absorptionsbanden zu $F(\overline{F})$- bzw. $F(\overline{Cl})$-Zentren wird durch die thermische Stabilität der Absorptionsbanden und Glowkurven bestätigt. Glowkurven und Absorptionsbanden nach Hochtemperaturbestrahlung weisen darauf hin, daß die Glowmaxima bei 365 und 385 K mit der Absorptionsbande bei 550 nm korrespondieren und durch $F(\overline{Cl})$-Zentren verursacht werden, während das Hochtemperaturmaximum bei 410 K mit der 440 nm-Absorptionsbande korrespondiert und durch $F(\overline{F})$-Zentren verursacht wird. Es wird gefunden, daß die $F(\overline{Cl})$-Zentren nicht so stabil sind wie die $F(\overline{F})$-Zentren. Die Thermolumineszenzspektren in der Nähe der Glowmaximatemperaturen zeigen nur eine Emissionsbande bei 390 nm.

1. Introduction

Although single crystals of BaFCl have been grown [1] as early as 1964, scarce attention has been paid to their optical and electrical properties. BaFCl, SrFCl, and BaFBr have been used as host lattices for rare-earth ions to study the effect of 4f states [2]. In these mixed crystals the anion vacancies are of two types, fluoride and chloride. Since BaFCl crystallizes in the tetragonal system (D_{4h}^7) the fluorine and chlorine vacancies possess tetragonal symmetry with the fourfold axis parallel to the optical *c*-axis of these crystals [3]. Two types of F-centres could be created in such crystals [4, 5], the $F(\overline{F})$ and $F(\overline{Cl})$ centres, corresponding to trapped electrons in fluorine and chlorine vacancies, respectively. The F-centre at the fluorine site $F(\overline{F})$ has symmetry D_2d and $F(\overline{Cl})$ has C_4-symmetry. Nicklaus and Fischer [4] identified these two types of F-centres by means of optical absorption (OA) bands and also concluded that additive coloration predominantly produces fluorine-type F-centres. Later the assignment of optical

[1]) O.U.B.F.-38, Hyderabad 500001, India.

absorption bands to $F(\bar{F})$ and $F(\overline{Cl})$ centres as proposed by Nicklaus and Fischer was proved to be exactly opposite [5]. This was supported by ESR, magnetic circular dichroism (MCD), stress-induced linear dichroism (SLD), ENDOR, and theoretical calculations as attempted by several workers [5, 6]. Agreeing with the proposals as suggested by the above experimental results in assigning the optical absorption bands to two types of F-centres, we also attempted with thermoluminescence (TL) to establish the identity of two types of F-centres and present the results of undoped BaFCl in this paper.

2. Experimental

Well-transparent crystals of flux grown [7] BaFCl having optimum dimensions have been irradiated by either X-rays or γ-rays. The optical absorption spectra of the coloured crystals have been recorded by a Cary 14(R) spectrophotometer. A strip heater in combination with a Philips temperature controller is used for high-temperature irradiation and optical absorption measurements. A conventional apparatus as described elsewhere [8] has been used for thermoluminescence and emission measurements with the EMI 6256S photomultiplier tube and a 0.25 m Jarrel-Ash monochromator.

3. Results

3.1 Glow curves

In our earlier paper [8] on BaFCl, four glow peaks at 345, 365, 385, and 415 K have been reported. Among these peaks the high-temperature peak at 415 K has been attributed to $F(\bar{F})$ centres in view of its stability as shown by TL and OA results, while the 385 K peak was tentatively attributed to $F(\overline{Cl})$ centres which are considered to be less stable as compared to $F(\bar{F})$ centres. In order to confirm the lower stability of $F(\overline{Cl})$ centres, the authors have undertaken a systematic study of the effect of room temperature annealing, high-temperature irradiation, and optical bleaching at different temperatures, on the thermoluminescence of undoped BaFCl crystals. The results have suggested the existence of two types of F-centres, i.e. $F(\bar{F})$ and $F(\overline{Cl})$ centres, and $F(\bar{F})$ centres are more stable than $F(\overline{Cl})$ centres.

As BaFCl crystals exhibited phosphorescence decay at room temperature, thermoluminescence has been studied after different times of annealing at room temperature. Fig. 1 shows the glow curves of 15 min X-irradiated BaFCl crystals. In crystals annealed for 3 and 5 min, two overlapping peaks situated at 345 and 365 K are obtained. In curve 1 which is obtained after annealing for 3 min, the intensity of the 345 K peak is higher than that of the 365 K peak. In curve 2 which is obtained after annealing for 5 min, the 345 K peak is smaller than the 365 K peak. On annealing for 10 min the 345 K peak just appears as a shoulder to the 365 K peak and the 365 K increases considerably. Further annealing eliminates the first peak totally and the intensity of the second peak also starts decreasing. For still longer annealing times (curves 4 to 6) the intensity of the second peak rapidly decreases and the peak also shifts towards high temperatures, i.e. from 365 to 380 K.

In order to isolate the low-temperature peak from its immediate high-temperature peak the crystal has been irradiated at various temperatures. In Fig. 2, curve 1 which was recorded at RT, the irradiation time being 15 min, shows two broad peaks positioned at 335 and 355 K. The peak at 335 K totally disappears and the other peak at 355 K shifts to 365 K, when the crystal is irradiated at 345 K (curve 2). Only one peak at 365 K with lower intensity is obtained as the irradiation temperature is increased to 360 K (curve 3). Curve 4 represents the glow curve obtained for a sample irradiated

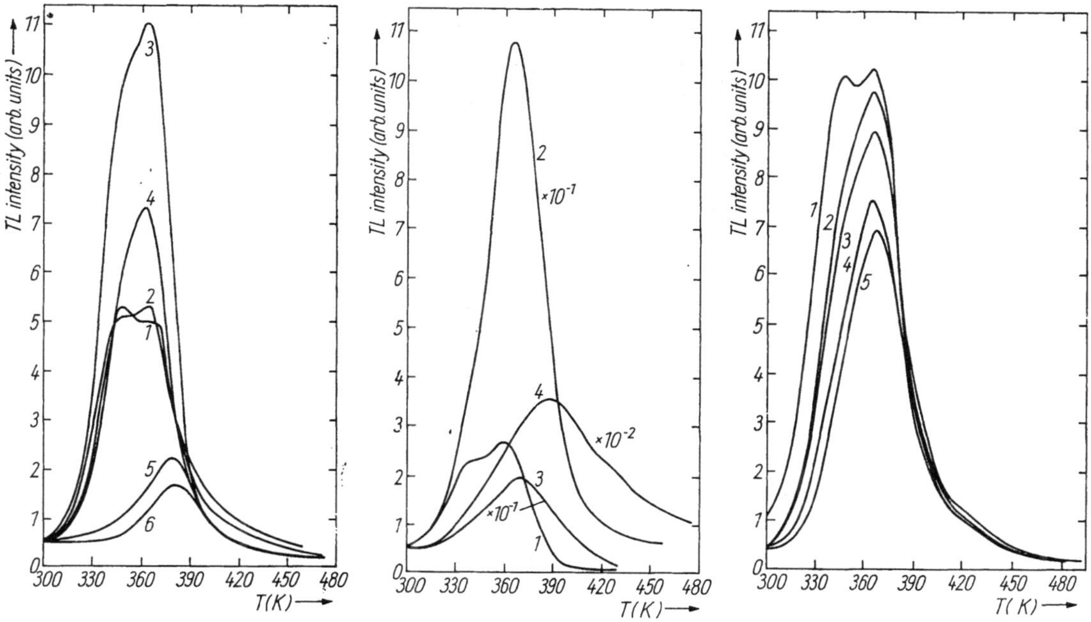

Fig. 1 Fig. 2 Fig. 3

Fig. 1. Effect of room temperature annealing on TL glow curves of BaFCl. t_a = 3 (1), 5 (2), 10 (3), 20 min (4), 17 (5), 20 h (6)

Fig. 2. Effect of high-temperature irradiation on TL glow curves of BaFCl (explanation of curves see text)

Fig. 3. Effect of bleaching on TL glow curves of BaFCl with 440 nm unpolarized light. (1) without bleaching, (2) to (5) bleached for 5 (2), 10 (3), 30 (4), 60 min (5)

for 1 h at 395 K. In this, a peak at 380 K and shoulder at 410 K may be seen. From these curves we can see that the 355 K peak as obtained at room-temperature irradiation, shifts to high temperatures due to high-temperature irradiation.

The optical absorption spectra of X-irradiated BaFCl at room temperature consist of two prominent bands at 440 and 550 nm. Hence, a systematic study of the bleaching effect on TL glow curves has been made. For this a BaFCl crystal has been X-irradiated every time for 10 min and subsequently bleached for 5, 10, 30, and 60 min with 440 nm unpolarized light. In Fig. 3, curve 1 represents the glow curve pattern of the unbleached crystal which consists of two clear peaks at 345 and 360 K and two shoulders at about 380 and 415 K. On bleaching for 5 and 10 min the first peak at 345 K becomes a shoulder and the intensity of the 360 K peak gradually decreases (curves 2 and 3). Bleaching for 30 and 60 min resulted in almost total elimination of the first peak at 345 K and a further reduction in the intensity of the 360 K peak.

Bleaching studies with 550 nm unpolarized light are similar to those presented in Fig. 3.

3.2 Optical absorption

Fig. 4a shows the optical absorption spectra of BaFCl crystals X-irradiated for different times 5, 10, 60, and 120 min. Two absorption bands at 440 and 550 nm and small bulgings at 310 and 700 nm are seen in all curves. Initially the absorption bands are broad and they become sharp as the time of irradiation increases. Closer examination

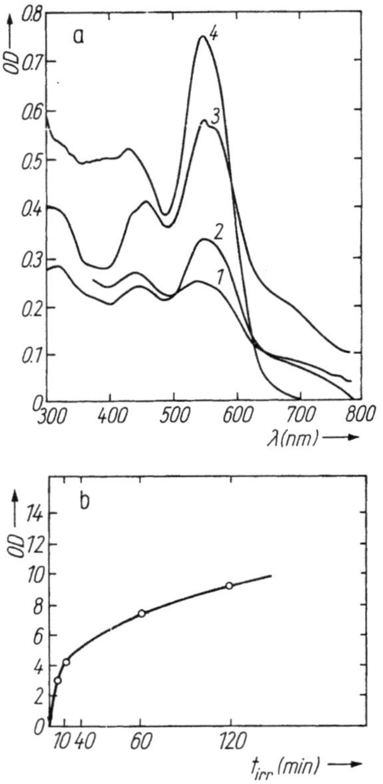

Fig. 4. a) Effect of irradiation time on optical absorption bands of BaFCl. $t_{irr} = 5$ (1), 10 (2), 60 (3), 120 min (4); b) $F(\overline{Cl})$ centre growth curve for BaFCl

reveals that the 550 nm band has a double structure for all irradiation times, while it is seen in the 440 nm band after irradiating for 60 min only. Fig. 4b shows the variation of optical density of the 550 nm band with time of irradiation. From this figure one may observe that the growth is rapid initially and saturates for longer times of irradiation.

In order to see the effect of temperature on the absorption bands the BaFCl crystals were X-irradiated for 15 min at room temperature and the spectra have been recorded at elevated temperatures. The results are shown in Fig. 5a and from this figure the following observations are made:

(i) The intensity of the four bands decreases with heating. The rate of decrease in intensity of the 550 nm band is larger than for other bands.

(ii) In the spectra taken at 378 K and above, the 550 nm band is completely absent whereas the 440 and 310 nm bands still persist.

(iii) The peak positions shift to higher wavelengths with increase in temperature and the bands also broaden.

To investigate the nature of absorbing centers and their stability with temperature, BaFCl crystals were X-irradiated for 10 min at different temperatures and optical absorption spectra were recorded at room temperature. The results so obtained are shown in Fig. 5b from which one can observe the following features:

(i) Irradiation at room temperature gives two prominent absorption bands at 440 and 550 nm the latter being more pronounced than the former.

(ii) As the temperature of irradiation is increased to 343 K both 440 and 550 nm bands decrease first. On further increase of temperature to 383 K, both bands at 440

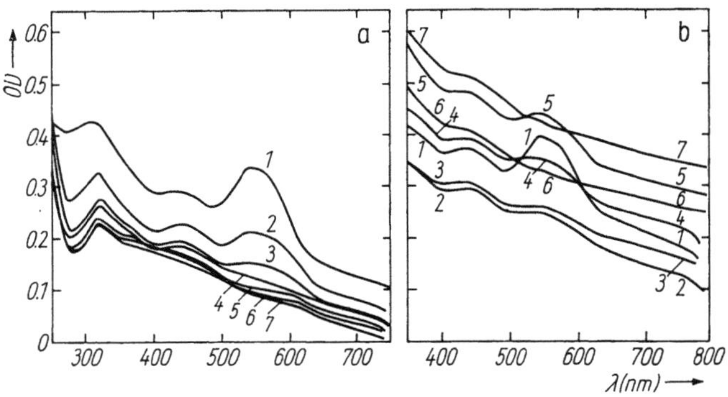

Fig. 5. a) Effect of temperature on optical absorption bands of BaFCl. RT (1), $T = 358$ (2), 368 (3), 378 (4), 398 (5), 408 K (6). b) Effect of high-temperature irradiation on optical absorption bands of BaFCl. RT (4), $T_{irr} = 343$ (2), 353 (3), 363 (4), 383 (5), 393 (6), 403 K (7)

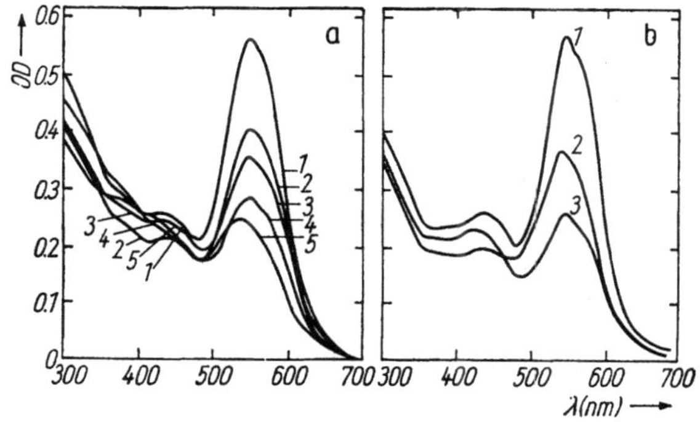

Fig. 6. a) Effect of bleaching on optical absorption bands of BaFCl with 440 nm unpolarized light. (1) without bleaching, (2) to (5) bleached for 5 (2), 10 (3), 15 (4), 20 min (5). b) Effect of bleaching on optical absorption bands of BaFCl with 550 nm unpolarized light. (1) without bleaching, (2), (3) bleached for 5 (2), 10 min (3)

and 550 nm increase in intensity considerably as compared to that recorded at room temperature (curve 3).

(iii) Irradiation at 393 and 403 K results in complete elimination of the 550 nm band. Although the height of the 440 nm band decreases it is still present as a borad band.

In order to see whether any additional bands develop on bleaching, systematic bleaching studies with 440 and 550 nm unpolarized light have been undertaken and the results are shown in Fig. 6 a and b, the irradiation time being 15 min for all runs. From Fig. 6a it is observed that, although the intensity of both the absorption bands decreases with bleaching time, the decrease in intensity of the 550 nm band is more pronounced than that of the 440 nm band. Bleaching with 550 nm light shows that the 550 nm band decreases at a faster rate than the 440 nm band.

3.3 Thermoluminescence emission

Emission spectra corresponding to different glow peaks have been recorded at elevated temperatures. Fig. 7 shows that the emission spectra of the 10 min γ-irradiated BaFCl at RT are very weak and a broad band peaking around 390 nm (3.18 eV) is obtained. The emission obtained at 323 K is represented by curve 1. It is observed that the emission is enhanced with temperature (curves 2 to 4) up to 450 K and then decreases. No additional bands have been noticed due to increase in temperature.

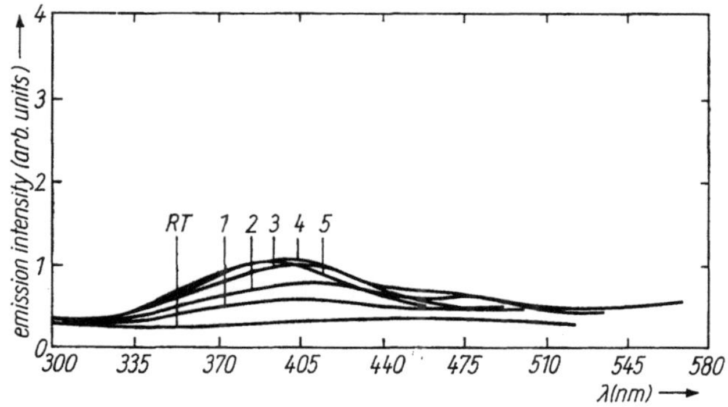

Fig. 7. Emission spectra of BaFCl (γ-irradiated) at (1) 323, (2) 348, (3) 393, (4) 450, (5) 475 K

4. Discussion

As mentioned earlier [8] in the present investigation also X- or γ-irradiation produced four glow peaks at 345, 365, 385, and 410 K. Room temperature annealing of γ-irradiated crystals showed only one TL glow curve at 415 K and absorption band at 440 nm. It is suggested that the 415 K peak corresponds to the 440 nm absorption band which is caused by $F(\bar{F})$ centres. As seen from Fig. 1 the low-temperature peak at 345 K is very unstable. The easy elimination of the 345 K peak due to annealing suggests that it might be formed by the emptying of electrons from shallow impurity traps and subsequent recombination with holes. Thus the low-temperature glow peak at 345 K may be attributed to impurities. Increase in intensity and shifting of the 365 K peak with annealing time may be understood as follows. For shorter periods of annealing the electrons released from comparatively deeper traps may be retrapped at shallow traps and this probability increases the thermoluminescence output. However, the shifting of peaks with annealing periods may have two significances (i) either there is a very broad distribution of trapping levels in depth or (ii) there is a very narrow distribution of traps in depth but due to high density of electrons in the traps mutual interaction may take place. Hence two factors are responsible for the release of electrons from the traps as they may have two kinds of energies. One kind of energy is the result of the thermal vibrations of the lattice and this is the only energy that is considered to be responsible for the ejection of electrons from traps. The other type of energy may arise due to the interaction taking place between the electrons trapped in the same depth. It is quite likely that the contribution of this energy to the thermal energy increases the total energy of the electron, thereby enhancing the probability of its release from the trap. In such a case, when the level is very densely populated, a thermoluminescence curve will give a lower level of trap depth as compared to the situation when the level is thinly populated. Hence the shifting of the 365 K thermoluminescence peak with annealing time can be explained in either of the two ways.

The second peak at 365 K is found to grow at a faster rate than the other peaks [8]. In the present investigation optical absorption spectra (Fig. 4a) obtained under similar conditions show that the 550 nm band increases with time of irradiation at a faster rate than the 440 nm band. High-temperature irradiation at 343 K produces the 365 K peak which is well resolved (Fig. 2). However, its intensity rapidly decreases as the temperature of irradiation is further increased. Later the other two peaks at 385 and 410 K appear. The optical absorption spectra obtained at high temperature or taken at room temperature after irradiation at high temperatures (Fig. 5 a and b) show that the 550 nm band bleaches faster than the 440 nm band.

The effect of optical bleaching on the 365 K TL peak has shown that its intensity decreases on bleaching with either 440 or 550 nm light. But the amount of decrease in intensity of the 365 K peak by 550 nm light is larger than that by bleaching with 440 nm light. Optical absorption spectra taken under similar conditions of bleaching have shown that though both bands 550 and 440 nm decrease on bleaching with 550 and 440 nm light, the decrease of the 550 nm band on 550 nm light bleaching is larger than for the 440 nm band. These results and those at high temperature confirm that the 365 K peak arises due to thermal ionization of $F(\overline{Cl})$ centres, and $F(\overline{Cl})$ centres are less stable than $F(\bar{F})$ centres.

Thermoluminescence studies have shown that the third peak at 385 K is generally observed as a shoulder on the high-temperature side of the second peak. However, it is clearly seen in crystals after prolonged irradiation and quenching (results not shown). These results suggest that the third glow peak at 385 K might have been formed due to thermal ionization of $F(\overline{Cl})$ centres having two different thermal stabilities, an "old" idea proposed by Jain and Mahendru [9] and Mahendru and Radhakrishna [10] in order to explain F-centers formed in two stages.

Thermoluminescence emission spectra obtained on γ-irradiated BaFCl crystals have shown only one strong emission band around 390 nm (3.18 eV). This suggests that this emission band is common for all glow peaks. Hageseth [11] has observed a narrow TL emission band at 3.08 eV in KBr and proposed that the F-centre electron after being released thermally to the conduction band is captured by a V-centre. This process has been fitted to an energy level diagram where the band gap for KBr was taken to be 7.13 eV. In these BaFCl crystals also V-centres causing absorption at about 340 nm (3.65 eV) at LNT have been reported by Yuste et al. [5]. These centres are also found to be stable at room temperature to some extent. Hence the recombination of the thermally excited F-centre electron with a V-centre is possible, because the sum of the energies of absorption and emission which is about 6.83 eV, is approximately equal to the band gap of ionic crystals.

As the glow emission spectra are characteristic of the impurity ions (either initially present or developed after γ-irradiation) it is evident that these ions act as luminescence centres and, since similar emission spectra are observed for different glow peaks, they are probably all due to the same luminescence centres.

However, from the thermal cleaning behaviour of TL and OA, it is clear that TL is observed, when holes are thermally released from different anion vacancies created on γ-irradiation. The possible mechanism is the recombination of released holes with electrons resulting in the emission of TL glow. The mechanism might involve the recombination of holes with the electrons trapped at the anion vacancy or a lattice defect, and the non-radiative transfer of recombination energy to the nearest impurity ion raising it to the excited level.

Acknowledgements

The authors are thankful to the Head, Department of Physics, Osmania University, for providing experimental facilities, K.S. wish to thank University Grants Commission, New Delhi, for financial assistance through a minor research project.

References

[1] F. K. Fong and P. N. Yocom, J. chem. Phys. **41**, 1383 (1964).
[2] J. F. Scott, J. chem. Phys. **49**, 2766 (1968).
[3] ASTM, Card Index 3-0304.
[4] E. Nicklaus and F. Fischer, phys. stat. sol. (b) **52**, 453 (1972).
[5] M. Yuste, L. Taurel, M. Rahman, and D. Lemoyne, J. Phys. Chem. Solids **37**, 1961 (1976).
[6] S. Lefrant and H. H. Harkar, Solid State Commun. **19**, 853 (1976).
[7] K. Somaiah and V. Hari Babu, Indian J. pure appl. Phys. **14**, 702 (1976).
[8] K. Somaiah, P. Veeresham, K. L. N. Prasad, and V. Hari Babu, phys. stat. sol. (a) **56**, 737 (1979).
[9] S. C. Jain and P. C. Mahendru, Phys. Letters (Netherlands) **5**, 186 (1963).
[10] P. C. Mahendru and S. Radhakrishna, J. Phys. C **12**, 796 (1979).
[11] G. T. Hageseth, Phys. Rev. B **5**, 4060 (1972).

(Received May 23, 1984)

phys. stat. sol. (a) 85, 273 (1984)

Subject classification: 13.4; 19; 20.3; 22.1.1; 22.1.2

Department of Physics, Moscow State University[1])

Energy Transfer between Excited Adsorbed Dye Molecules and Charged Defects in Insulator–Semiconductor Structures

By

S. V. VINTSENTS, V. F. KISELEV, and G. S. PLOTNIKOV

Two competition channels of energy migration of electronically excited dye molecules adsorbed on the germanium and silicon oxidized surfaces are discovered. In the low adsorbed dye-molecules concentration range the excitation energy is mainly consumed for discharging of negatively and positively charged oxide layer traps. Energy distribution of the traps is qualitatively determined. Their charging is accompanied with red shift of the dye molecules fluorescence maximum. Excitation energy dissipation in adsorption phase dominated if the adsorbed molecules concentration is high enough.

Zwei konkurrierende Kanäle der Energiemigration elektronisch angeregter Farbstoffmoleküle, die auf oxydierten Oberflächen von Germanium und Silizium adsorbiert sind, werden beobachtet. Im Gebiet geringer Konzentration adsorbierter Moleküle wird die Anregungsenergie im wesentlichen für die Entladung negativ und positiv geladener Haftstellen in der Oxidschicht verwendet. Die Energieverteilung der Haftstellen wird von einer Rotverschiebung begleitet. Bei hoher Konzentration adsorbierter Moleküle überwiegt die Dissipation der Anregungsenergie in de adsorbierten Phase.

1. Introduction

The study of the mechanism of electron excitation energy migration of adsorbed molecules has two aspects of special interest. One of these stems from the generally acknowledged role of excited molecular states in the elementary acts of adsorption and catalysis [1]; the other is associated with elucidating the effect of spectral sensitization of photoconduction in semiconductors by adsorbed dye molecules. In experiments with photoconduction it is often impossible to distinguish whether the sensitization is due to energy transfer or to charge transfer from the dye molecule to the solid [2]. The electron excitation of adsorbed molecules may arise both in the act of adsorption itself, and under the action of photons (photoadsorption, photocatalysis) or charged particles (electron- or ion-stimulated desorption and catalysis) incident on the surface. Vast information regarding these processes may be obtained by studying the luminescence of adsorbed molecules. On dielectric surfaces the main channel of excitation relaxation is the energy transfer to adjacent adsorbed molecules (channel 1), resulting in concentration quenching of fluorescence [3, 4]. In semiconductors we have discovered an alternative channel of energy migration (channel 2): dissipation into the solid, resulting in the discharging of superficial charged defects [5].

A convenient model object for investigations of this kind is furnished by an insulator–semiconductor (IS) structure with adsorbed dye molecules [5]; the typical band diagram of this structure is shown in Fig. 1a. Extensive knowledge concerning the paths of excitation energy migration can be gained from simultaneous measurements of adsorbed dye luminescence and surface charge. In view of the close relationship between the adsorption activity and the electronic subsystem of the semiconductor

[1]) 117234 Moscow, USSR.

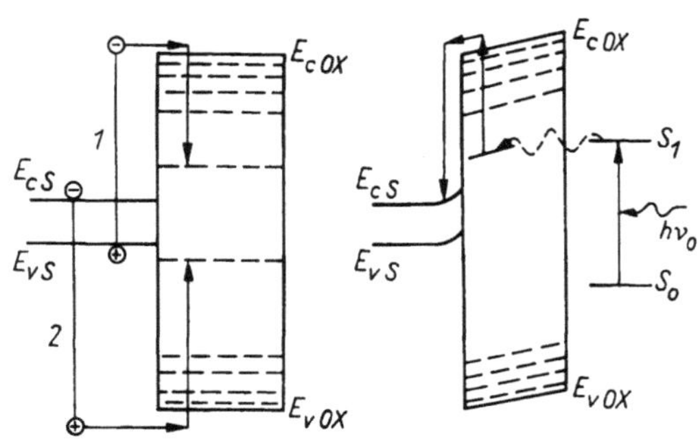

Fig. 1. a) Band diagram of the IS structure. Transitions 1 and 2 correspond to injection of electrons and holes from the semiconductor, respectively. b) Scheme of discharging of negatively charged insulator traps when adsorbed dye molecules are photoexcited from the ground state S_0 to the singlet excited state S_1. E_{cs}, $E_{c\,ox}$, E_{vs}, E_{vox} are the edges of conduction and valence bands of semiconductor and oxide, respectively

[1], the IS-dye system may be expected to exhibit the following features: (i) luminescence quenching due to excitation energy transfer from dye molecules to charged insulator defects resulting in the discharge of the latter (reduction of surface charge); (ii) shift of the luminescence maximum due to the change in the charge of adsorption sites or to the action of microfields produced by adjacent defects having captured charge carriers from the semiconductor, and (iii) ignition of luminescence due to the transfer of energy released in the course of capture of charge carriers by adsorption sites or adjacent defects, to adsorption complexes. In the present paper we confine our analysis to the first two effects.

2. Sample Preparation and Investigative Techniques

We studied the fluorescence of a number of organic dye molecules, adsorbed on oxidized ⟨111⟩ faces of Ge and Si single crystals. The samples were prepared by thermal oxidation of n-type high-resistance Ge ($\varrho = 30\ \Omega$ cm) and Si ($\varrho = 15$ kΩ cm). The thickness of the oxide layer, as deduced from ellipsometric measurements, varied from 3 to 30 nm. The following dyes were used: erythrosin (E), coumarin (C), rhodamine B (RB), aminobenzoton (AB), methylene blue (MB), manganese naphthalcyanine (NcMn), adsorbed from alcohol solutions with subsequent vacuum treatment to remove the solvent. The concentration of the adsorbed dye molecules was assessed with the aid of an adsorption piezoresonance balance. The luminescence spectra were recorded by a Hitachi MPF-4 spectrofluorimeter.[2]) Negative charging of insulator traps (IT) in IS structures was accomplished by exposing the samples to monochromatic light of a xenon lamp DKSŠ-1000 [6]. The measurements of photocharging of IT were based on the field effect technique with a high-amplitude sine-wave signal [5, 6]. Luminescence spectra were taken between two successive exposures required to charge the IT.

3. Results and Discussion

As with insulators [7], the adsorption of E and C on uncharged Ge and Si samples shifted the fluorescence maxima λ_{max} towards larger wavelengths ($\Delta\lambda_0 \approx 20$ nm). Photocharging of IT, as demonstrated elsewhere [8, 9], gave rise to new effects: an extra shift of λ_{max}, and considerable quenching of fluorescence. The effect of quenching of fluorescence of tetracene adsorbed on the surface of Si and GaAs was reported also in [10].

Let us begin our discussion with the quenching of fluorescence. Illumination of negatively charged Ge and Si samples in the absorption band of dyes (E and C, respectively) caused a considerable reduction in the negative charge Q_s (Fig. 2a). It is observed

[2]) The authors are grateful to Prof. L. V. Levshin for his generous assistance.

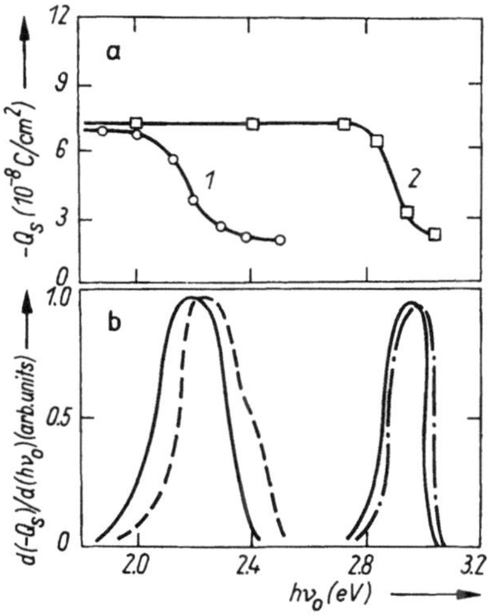

Fig. 2. a) Negative excess charge $-Q_s$ in deep insulator traps IT as a function of the energy of light quanta $h\nu_0$ for E-GeO_2-Ge (1) and C-SiO_2-Si (2). $N_E = N_C \approx \approx 2 \times 10^{13}$ cm^{-2}; $d_{GeO_2} \approx d_{SiO_2} \approx 5$ nm. b) Respective derivatives of charge drain spectra (solid lines) and absorption bands of erythrosin (dashed line) and coumarin (dash-dotted line) in adsorbed state

that the energy of fluorescence-exciting light quanta $h\nu_0$ (≈ 2.3 eV for E, and ≈ 3.0 eV for C) was strictly below the threshold W_1 of negative charging of IT, corresponding to the transition of photoexcited charge carriers of the semiconductor into the band of delocalized oxide states (transition 1 in Fig. 1a), and thus the illumination could not be responsible for extra charging of IT.[3]) Control experiments indicate that the adsorption of E and C molecules itself does not affect the concentration of IT and charge Q_s. The fall of Q_s cannot be attributed to photo-devastation of IT by photons emitted by fluorescent dye molecules. The exposure of IS structures without dye molecules to light quanta $h\nu_1$ of the same intensity and energy did not alter the charge

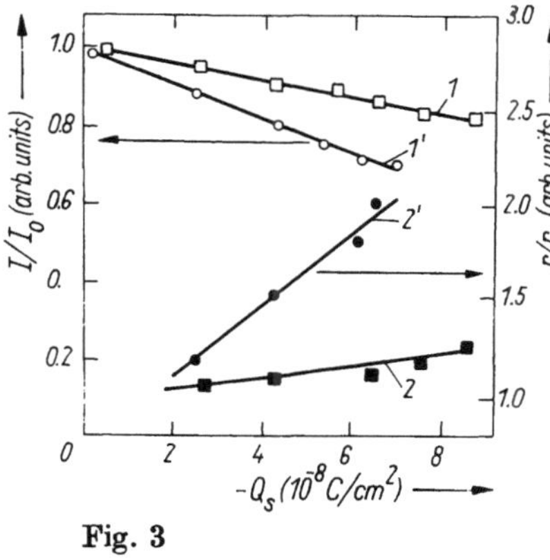

Fig. 3

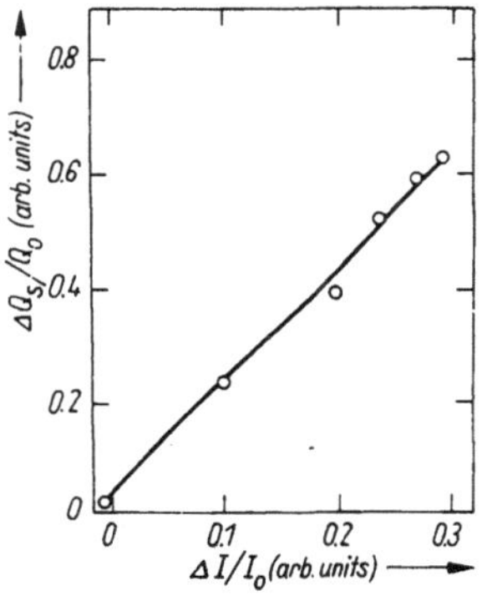

Fig. 4

Fig. 3. Relative quantum yield I/I_0 and coefficient of fluorescence anisotropy r/r_0 of E on GeO_2-Ge ($d_{GeO_2} = 5$ nm) (curves 1′, 2′) and C on SiO_2-Si ($d_{SiO_2} = 5$ nm) (curves 1, 2) against charge $-Q_s$. $N_E = N_C \approx 2 \times 10^{13}$ cm^{-2}

Fig. 4. Rejected charge $-\Delta Q_s$ against fluorescence quenching $\Delta I/I_0$ in the E-GeO_2-Ge system. $N_E = 2 \times 10^{13}$ cm^{-2}; $d_{GeO_2} = 5$ nm

[3]) With the employed intensities of exciting light ($h\nu_0 < W_1$), extra charging due to transitions of electrons via localized fluctuative levels of the amorphous oxide layer [11] was negligibly small (dashed lines in Fig. 1).

of IT, because of the small interaction cross section between photons and electrons localized in IT.

All this implies that the observed reduction of IT charge Q_s (Fig. 2a) can only be explained by the existence of an efficient channel (channel 2) of nonradiative transfer of the energy of electron excitation of E and C molecules to charged IT, knocking the electrons from these traps into the conduction band of GeO_2 (SiO_2), with subsequent sink of charge into the semiconductor. This is shown schematically in Fig. 1b. As follows from Fig. 2b, the derivative of the discharging spectrum of IT lies reasonably close to the absorption band of the respective dye in adsorbed state.

The dissipation of excitation energy of molecules in the dielectric layer was accompanied by quenching of fluorescence I/I_0 (I_0 and I being the intensities at the dye emission maximum before and after charging, respectively) (Fig. 3). The efficiency of channel 2 is the higher, the higher the density of charged insulator traps in the IS structure is. Indeed, from Fig. 3 it is clear that the quantum yield of luminescence I/I_0 declines with increasing Q_s: the number of charges ejected from IT ($-\Delta Q_s$) is found to be proportional to the number of energy quanta released in the course of luminescence decay ($\Delta I/I_0$). An example is given in Fig. 4 for the structure E–Ge–GeO_2.

The energy of luminescence quanta $h\nu_l$ (2.2 eV for E; 2.5 eV for C) equals or slightly exceeds the depth of electron IT E_t, measured from the bottom of the band of delocalized states of GeO_2 (SiO_2). For Ge–GeO_2 structure the value of E_t, as estimated from measurements of thermally stimulated emission, is 1.7 to 1.9 eV [11, 12]; for SiO_2 E_t = 2.4 to 2.5 eV (on the basis of data on photodischarging of traps) [13].

In amorphous insulators (GeO_2, SiO_2) the traps are spread over a certain range of energies [12]. By using dyes with different energies of the quanta at the maximum of fluorescence band (NcMn: 1.5 eV, MB: 1.75 eV, AB: 1.88 eV, RB: 2.07 eV, E: 2.2 eV), we were able to accomplish controlled ejection of electrons from traps of varying depth in Ge–GeO_2 structure. Fig. 5a shows the value of the excess charge $-Q_s$ in IT as a function of the energy of fluorescence-exciting quanta $h\nu_0$. The efficiency of photodischarging $\varphi = \Delta Q_s/Q_0$ against the energy of luminescence quanta

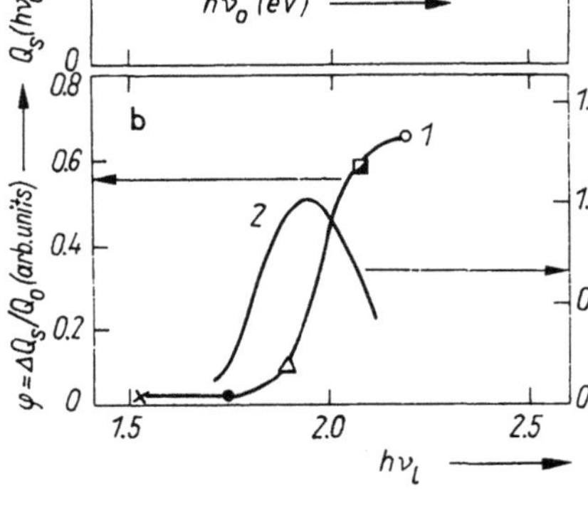

Fig. 5. a) Normalized spectra of photodischarging of deep electron traps of a GeO_2 layer in: (1) E–GeO_2–Ge; (2) RB–GeO_2–Ge; (3) AB–GeO_2–Ge; (4) MB–GeO_2–Ge; (5) NcMn–GeO_2–Ge; d_{GeO_2} = = 5 nm; $N_{dye} = 2 \times 10^{13}$ cm^{-2}; Q_0 is the amount of charge in IT prior to photodischarging. b) Charge rejected from IT $\varphi = -\Delta Q_s/Q_0$ as a function of quantum energy of luminescence $h\nu_l$ (1); energy distribution of deep electron traps in a GeO_2 layer (2)

$h\nu_l$ is plotted in Fig. 5b. The derivative $d\varphi/dh\nu_l$, characterizing the distribution of traps over energies, deviates little from the curve of thermally stimulated emission, being slightly shifted towards higher energies [11, 12], which is typical of photodischarging curves [13].

The value of φ was found to be practically independent of temperature T in the range from 170 to 300 K. This implies that the transport of ejected electrons proceeds mainly via the band of delocalized states, since the hopping mechanism of transfer over localized fluctuative states is temperature-dependent (dashed line in Fig. 1) [11].

An additional proof of existence of channel 2 was obtained from fluorescence polarization measurements. As follows from Fig. 3, the anisotropy of fluorescence r/r_0 (r_0 and r being the values of anisotropy for uncharged and charged samples, respectively) increases with increasing charge $-Q_s$ in IT. In the case of energy migration between acceptors with the same energy, the value of r/r_0 should go down with increasing density of acceptors [14]. In our case (see Fig. 3), the concentration of adsorbed molecules was low ($N_{E,C} \approx 2$ to 3×10^{13} cm^{-2}); then the channel of excitation energy migration between like adsorbed molecules (channel 1) is inefficient (see below). The observed behaviour of r/r_0 directly points to the transfer of energy between highly dissimilar species: adsorbed molecules and insulator traps.

Let us now turn to the second effect, the shift of the luminescence spectral maximum λ_{max} associated with IT charging. Negative charging of IT was observed to shift λ_{max} towards larger wavelengths in accordance with the concentration of charged IT (Fig. 6). After the drain of charge λ_{max} resumed its initial value. The observed shifts of λ_{max} point to the direct influence of the captured charge carriers of the semiconductor on the properties of the adsorptive complex, which substantiates our concept of a donor–acceptor mechanism of adsorption and charging of the semiconductor surface [1, 15]. Apparently, Coulomb microfields associated with charged IT, cause an additional polarization of adsorption complexes and an extra shift of λ_{max}.

Further on, we investigated the effects of the thickness d of the oxide layer in the IS structure on the shift of λ_{max} and the quenching of luminescence. An example is given in Fig. 7 for the E–Ge–GeO_2 system. According to the data on photocharging

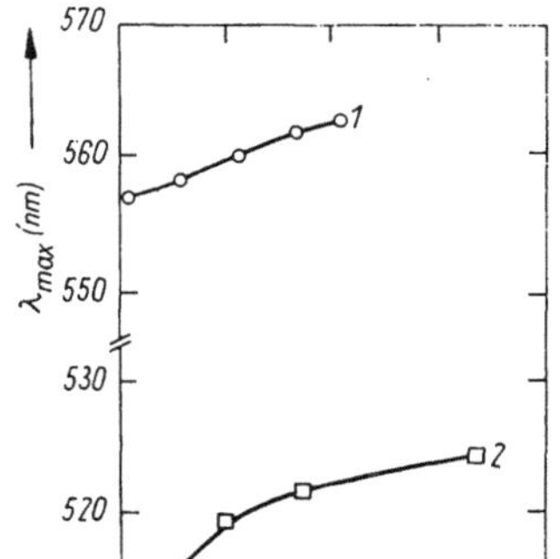

Fig. 6

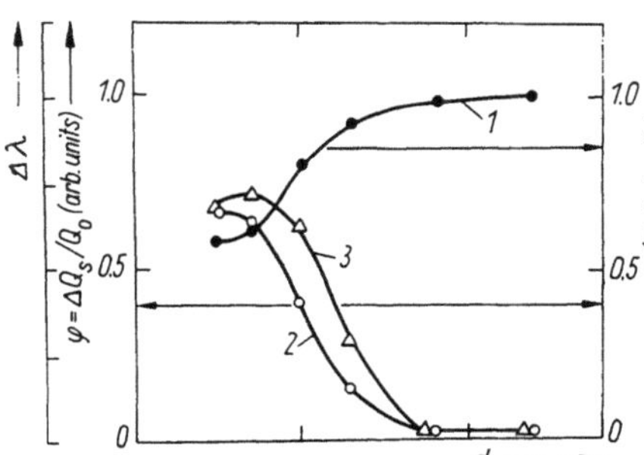

Fig. 7

Fig. 6. Fluorescence spectral maximum of E (1) and C (2) adsorbed, respectively, on Ge (d_{GeO_2} = = 5 nm) and Si (d_{SiO_2} = 5 nm) against charge $-Q_s$. $N_E = N_C = 2 \times 10^{13}$ cm^{-2}

Fig. 7. Fluorescence intensity I/I_0 (1), photodischarging efficiency φ (2), spectral shift $\Delta\lambda$ (3) as functions of the thickness of oxide layer d in the E–GeO_2–Ge system. $N_E = 2 \times 10^{13}$ cm^{-2}

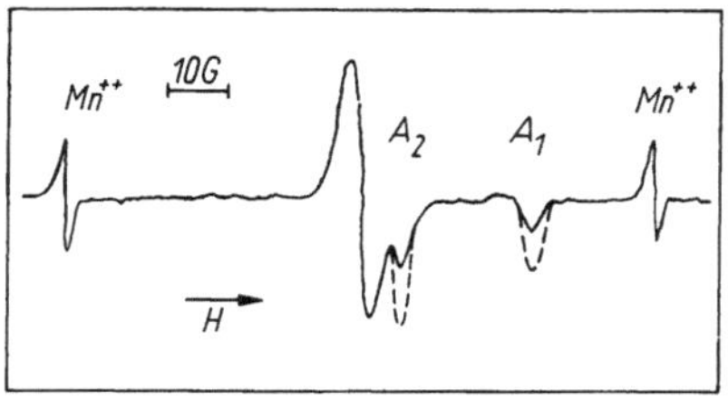

Fig. 8. EPR spectra of a light-exposed Ge–GeO_2 structure (solid line) and Ge–GeO_2–C (dashed line)

[12], the centre of gravity of negatively charged IT in this system is located at $d_0 = 5$ to 6 nm. In Fig. 7 the increase in d beyond 5 nm is accompanied by the rise in fluorescence intensity (curve 1) and the drop in the efficiency of photodischarging φ (curve 2). A similar behaviour of I/I_0 as a function of d was observed also with the C–SiO_2–Si system. The minimum of I/I_0 and the centre of gravity of negatively charged traps here fell with $d_0 \approx 7$ to 10 nm. The magnitude of $\Delta\lambda$ for Ge (as well as for Si) drops abruptly with d increasing beyond 5 nm (curve 3, Fig. 7).

So far we have been speaking of the recharging of deep negatively charged electron traps in the dielectric layer. However, the oxide GeO_2 (SiO_2) layers contain also hole traps. The sites capable of being positively charged in oxide layers are represented by oxygen vacancies: E_1' and E_2' sites which are photocharged by harder radiation [12]. This process is associated with transfer of holes to vacancy-type defects (transition of electrons from the depth of the oxide valence band into the semiconductor conduction band with the capture of excited holes by E_1' and E_2' sites), transition 2 in Fig. 1a. Then the above-mentioned "ionized" sites assume paramagnetic properties, responsible for EPR signals A_1 and A_2 [16]. The threshold of positive charging is $W_2 = 3.3$ eV. The authors of [17] investigated the EPR spectra of C–GeO_2–Ge structures ($d_{GeO_2} = 20$ nm). Exposure of samples to light in the absorption band of coumarin ($h\nu < W_2$) raised the strength of the EPR signals A_1 and A_2 (see Fig. 8) with the following parameters: A_1: $\Delta H = 4$ G, $g = 1.993 \pm 0.001$, concentration of spin centres $N_1 = 3$ to 4×10^{11} cm^{-2} (without dye, $N_1 = 1$ to 2×10^{11} cm^{-2}); A_2: $\Delta H = 2$ to 3 G, $g_2 = 2.001 \pm 0.001$, $N_2 = 3 \times 10^{11}$ cm^{-2} (without dye, $N_2 = 2 \times 10^{11}$ cm^{-2}). Exposure of pure GeO_2 oxides with adsorbed coumarin did not give rise to any EPR signals. Irradiation of dye-containing structures increases the number of ionized E_1' and E_2' sites owing to migration of molecular excitation energy towards these sites.

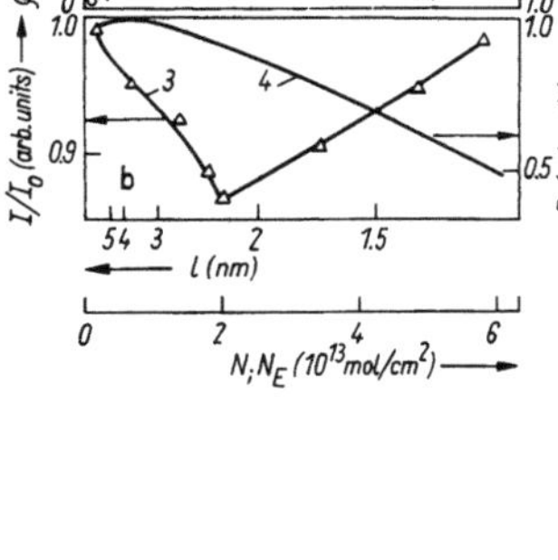

Fig. 9. Dependences of $-\Delta Q_s/Q_0$ (1), r/r_0 (2), I/I_0 (3), I_0 (4) on N_E in E–GeO_2–Ge system; $-\Delta Q_s/Q_0$ (1') against N_{E+C} in (E + C)–GeO_2–Ge system; $-\Delta Q_s/Q_0$ (1'') against N_{E+RB} in the (E+RB)–GeO_2–Ge system; $Q_0 = 4.5 \times 10^{-8}$ C cm^{-2}

Consider now the dissipation of energy of excited molecules via the competing channel 1 into the adsorptive phase. Fig. 9 shows the concentration dependence of $\Delta Q_s/Q_0$, I/I_0, and r/r_0 for the E–GeO_2–Ge system. Apart from the concentration of adsorbed erythrosin molecules N_E, the x-axis is graduated in terms of the mean separation between adsorbed molecules l, calculated for a uniform distribution of molecules, each covering the area $\omega_E = 1.2\ nm^2$. As ought to be expected, the dissipation via channel 2 prevails at the initial stage of adsorption (as in Fig. 2, 3). The quantitaties $-\Delta Q_s/Q_0$ (curve 1), r/r_0 (curve 2) rise with increasing density of dye molecules, while the relative intensity of fluorescence I/I_0 (curve 3) falls. The three curves (1, 2, 3) in Fig. 9 have an extremum at $N_E = 2 \times 10^{13}\ cm^{-2}$. With further increase in N_E the proportion of charge removed from IT decreases, and the values of I and r approach I_0 and r_0, respectively (i.e., the values for uncharged samples). This indicates that at $N_E > 2 \times 10^{13}\ cm^{-2}$ channel 1 becomes dominant over channel 2.

The drop in the intensity of luminescence of adsorbed molecules I_0 on the surface of an uncharged specimen in the high-occupancy range (small l) (curve 4 in Fig. 9) correlates in general with the data available on the luminescence of dyes adsorbed on a dielectric surface (SiO_2; see [3] and references cited in [10]), and, according to [4, 18], may be associated with nonradiative energy exchange between adsorbed molecules via an induction-resonance mechanism [19]. Then, as N_E increases, the probability of transitions of this kind will rise, and ultimately will dominate over the probability of energy transfer from adsorbed molecules to IT (channel 2). If in the adsorptive phase the induction-resonance mechanism of energy exchange is the predominant one, the probability of energy transfer from a donor molecule to an acceptor molecule will strongly depend on the overlapping of emission and absorption spectra of the molecules in question.

To corroborate these conclusions, we performed additional experiments on adsorption of two pairs of luminescent dyes on the GeO_2–Ge structure: E + RB, and E + C. The second dye (RB or C) was deposited from alcohol solution on germanium surface, which already contained erythrosin at $N_E = 2 \times 10^{13}\ cm^{-2}$. The upper abscissa in Fig. 9 represents the absolute concentrations of additionally adsorbed molecules of RB (N_{RB}) and C (N_C).

In the system E + C the probability of energy exchange between E and C is practically zero, because the emission band of E does not overlap with the absorption band of C. Indeed, additional adsorption of C on the uncharged germanium sample has no effect on the relative strength of luminescence of E $(I_E/I_E^*)_C$ (curve 1 in Fig. 10) in the entire range of N_C (here I_E^* and I_E are the luminescence intensities at the emission band maximum of E before and after adsorption of C, respectively). The proportion of charge $\Delta Q_s/Q_0$, ejected from IT (curve 1′ in Fig. 9) is not affected by the increase

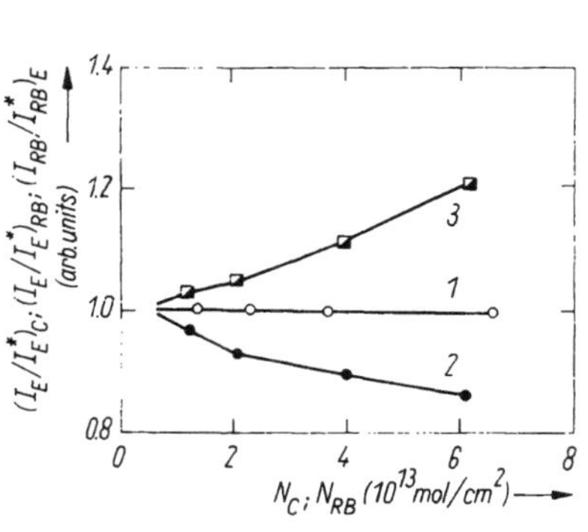

Fig. 10. $(I_E/I_E^*)_C$ against N_C (1) in the (E + C)–GeO_2–Ge system ($N_E = 2 \times 10^{13}\ cm^{-2}$); $(I_E/I_E^*)_{RB}$ and $(I_{RB}/I_{RB}^*)_E$ against N_{RB} in the (E + RB)–GeO_2–Ge system (2, 3)

in N_C, because additional adsorption of C does not improve the efficiency of the alternative path of energy migration (channel 1).

A different situation is encountered with additional adsorption of RB. The emission band of E exhibits considerable overlapping with the absorption band of RB; therefore, when fluorescence is being excited in the absorption band of E (2.3 eV), it is effectively quenched with simultaneous stimulation of fluorescence of RB at 2.06 eV. The latter is a consequence of nonradiative energy transfer from E to RB. Recall that the direct excitation of RB molecules by this light is negligibly small, its absorption band is shifted towards lower energies ($\approx$ 2.2 eV) with respect to that of E. From Fig. 10 it is clear that as N_{RB} increases, the relative strength of luminescence of E $(I_E/I_E^*)_{RB}$ (curve 2) goes down, and $(I_{RB}/I_{RB}^*)_E$ (curve 3) goes up (here I_{RB}^* and I_{RB} are the intensities at the emission maximum of RB before and after adsorption, respectively).

Since the overlap integral of the emission band of energy donor E with the absorption band of energy acceptor RB approximately equals the overlapping of the respective bands of E molecules, it ought to be expected that the probability of energy exchange via an induction-resonance mechanism from E to RB will be the same as that between E molecules, as long as their surface density N is the same. In other words, the efficiency of energy transfer via channel 1 will remain the same if we replace some of the E molecules by the same number of RB molecules. The contribution of channel 1 to the dissipation of energy of excited molecules is the same, as long as the concentration N_E is the same as N_{E+RB} ($N = 2 \times 10^{13}$ cm^{-2} + N_{RB}). Indeed, from Fig. 9 it is clear that the efficiency of photodischarging goes down with increasing N_E (curve 1) and N_{E+RB} (curve 1'') at approximately the same rate. This gives an additional proof of the existence of an efficient channel 2 of discharging of IT, which operates independently of channel 1 and transfers the energy from dye molecules into the solid.

As in the case of insulators [3, 4], the increase in the density of dye molecules on the surface of uncharged semiconductor (Ge) was accompanied by the shift of the luminescence maximum towards longer wavelengths (Fig. 11). The authors of [3, 4] attributed this to the effect of aggregation at large N_E. Apparently, the same mechanism works in our case. However, in a semiconductor the location of the maximum is additionally affected by the field produced by negatively charged IT. As follows from Fig. 11, the negative charging of IT leads to a red-shift of the emission band of erythrosin.

The ratio between luminescent E molecules and their total number on the germanium surface in our experiments is estimated as 10^{-3}. In the investigated range of N_E the

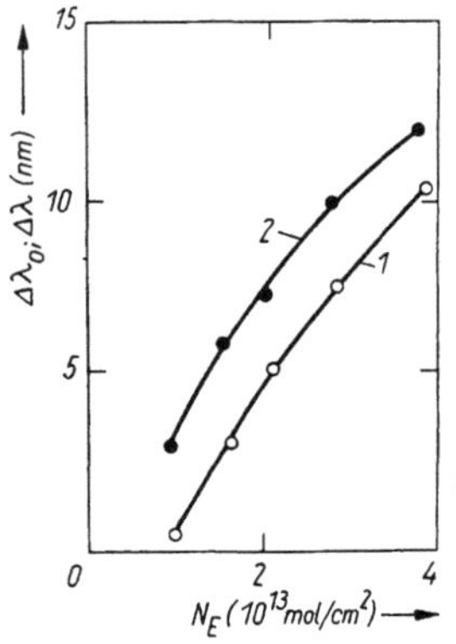

Fig. 11. Shifts $\Delta\lambda$ and $\Delta\lambda_0$ against N_E in the E–GeO_2–Ge system for $Q_s = 0$ (1) and $Q_s = 1.2 \times 10^{-9}$ C cm^{-2} (2). $\lambda_{max} = 556$ nm at $N_E = 10^{13}$ cm^{-2}

concentration of excited molecules ($\approx 10^{10}\,cm^{-2}$) is about an order of magnitude below the number of negatively charged IT ($\approx 10^{11}\,cm^{-2}$). The additional red-shift of the fluorescence band in Fig. 11 is most probably associated with an additional polarization of excited molecules in the fields of charged IT.

4. Conclusion

The discovered effects of electron-excited, adsorbed molecules on the charged traps in IS structures, and the reciprocal influence of the charge in these traps on the parameters of fluorescence, open new opportunities for studying the adsorptive and electronic processes taking place in these structures. The next stage of these investigations, aimed at elucidating the mechanism of energy migration between dye molecules and charged defects in IS structures, is the study of the kinetics of fluorescence quenching and the ejection of electrons from insulator traps, which we have now started in our laboratory.

References

[1] V. F. Kiselev and O. V. Krylov, Adsorbtsionnye protsessy na poverkhnosti poluprovodnikov i dielektrikov, Izd. Nauka, Moskva 1978.
[2] I. A. Akimov, Yu. A. Cherkasov, and M. I. Cherkashin, Sensibilizirovannyi fotoeffekt, Izd. Nauka, Moskva 1980.
[3] S. Garoff, R. B. Stephens, C. D. Hanson, and G. K. Sorensen, J. Lum. **24/25**, 773 (1981).
[4] S. Garoff, R. B. Stephens, C. D. Hanson, and G. K. Sorensen, Optics Commun. **41**, 257 (1982).
[5] V. F. Kiselev, S. N. Kozlov, and G. S. Plotnikov, Zh. tekh. Fiz., Pisma **7**, 982 (1981).
[6] P. K. Kashkarov, V. F. Kiselev, and S. N. Kozlov, Surface Sci. **75**, 231 (1978).
[7] D. Oelkrug, M. Plaushinat, and R. W. Kessler, J. Lum. **18/19**, 434 (1979).
[8] G. S. Plotnikov, A. M. Saletskii, and S. V. Vintsents, Poverkhnost: Fiz., Khim., Mekh. **11**, 71 (1982).
[9] S. V. Vintsents, P. K. Kashkarov, V. F. Kiselev, and G. S. Plotnikov, Dokl. Akad. Nauk SSSR **268**, 373 (1983).
[10] T. Hayashi, T. G. Castner, and R. W. Boyd, Chem. Phys. Letters **94**, 461 (1983).
[11] P. K. Kashkarov, S. N. Kozlov, and A. V. Petrov, Izv. vuzov SSSR, Fizika **5**, 36 (1982).
[12] P. K. Kashkarov and S. N. Kozlov, Izv. vuzov SSSR, Fizika **2**, 79 (1976).
[13] D. J. Di Maria, The Physics of SiO_2 and Its Interfaces, Pergamon Press, 1978 (p. 164).
[14] V. L. Ermolaev, E. N. Bodunov, E. B. Sveshnikov, and T. A. Shakhverdova, Bezyzluchatelnyi perenos energii elektronnogo vozbuzhdeniya, Izd. Nauka, Leningrad 1977 (p. 283).
[15] V. F. Kiselev, Dokl. Akad. Nauk SSSR **176**, 124 (1967).
[16] S. N. Karyagin, P. K. Kashkarov, V. F. Kiselev, and S. N. Kozlov, phys. stat. sol. (a) **37**, 17 (1976).
[17] S. V. Vintsents, S. N. Karyagin, and G. S. Plotnikov, Vestnik MGU **24**, 32 (1983).
[18] R. Rosetti and L. E. Brus, J. chem. Phys. **73**, 572 (1980).
[19] T. Forster, Disc. Faraday Soc. **27**, 7 (1959).

(Received June 6, 1984)

phys. stat. sol. (a) 85, 283 (1984)

Subject classification: 14.3; 22.9

Department of Chemistry, University of Nottingham[1])

AC Conductivity in Single Crystals and Compactions of TCNQ Complex Salts

By

I. Diaconu[2]), D. D. Eley, and M. R. Willis

AC conductivity in the range 10 to 10^5 Hz of N,N'-diethyl- and N,N'-dibenzyl-4,4'-bipyridylium $(TCNQ)_4$ consists of a frequency independent part, identified with dc conductivity, and a frequency dependent part resulting from hopping of carriers between localised sites. The density of hopping carriers is a sensitive indicator of the degree of perfection of the crystal and is significantly increased by grinding and compression.

AC conductivité dans l'échelle 10 à 10^5 Hz de N,N'-diethyl- et N,N'-dibenzyl-4,4'-bipyridylium $(TCNQ)_4$ est composée d'une partie qui est indépendante de la fréquence, identifiée avec conductivité dc, et une partie qui est dépendante de la fréquence occasionée par le saut de porteurs entre des emplacements localisés. La densité de porteurs sautants est un indicateur qui est sensible au degré de perfection du cristal et qui est augmentée considérablement à force de broyer et de comprimer.

1. Introduction

Recent studies on the temperature dependence of the electrical conductivity of TCNQ complex salts [1] have revealed departures from simple Arrhenius behaviour, which have been interpreted as resulting from the existence of states in the band gap. Using the theory of Roberts and Schmidlin [2, 3], a good fit to the experimental data was achieved. However, there is little direct experimental evidence for the existence of such states other than the observation of Curie law [4] magnetic behaviour at low temperatures.

AC conductivity measurements have been widely assumed to reflect the existence of localised sites between which electrons can hop in response to an alternating field. Whereas band conduction is frequency independent, the hopping conduction is frequency dependent (as $\sigma \propto \omega^n$) with $0.5 \leqq n \leqq 1$) and tends to dominate at low temperatures and high frequencies [5].

A good model substance for the investigation of localised states is N,N'-diethyl-4-4'-bipyridyl $(TCNQ)_4$. The crystal structure has been reported [6] and shows no significant signs of disorder. The electrical conduction of the single crystals appears to be intrinsic with an anisotropy consistent with the crystal structure. Furthermore, compactions show a temperature variation of thermopower in accord with band theory. Making reasonable assumptions, both electron and hole mobilities and their temperature dependencies have been deduced. The corresponding dibenzyl compound [7] yields well formed single crystals which enable the anisotropy of ac conductivity to be studied and provides an opportunity to compare and behaviour of single crystals and compactions, to assess the extent to which the grinding and compression necessary for the preparation of compactions influence the electrical behaviour.

[1]) University Park, Nottingham NG7 2RD, Great Britain.

[2]) Permanent address: P. Poni Institute of Macromolecular Chemistry, Iassy, Romania.

2. Experimental

The materials were prepared as already described [6, 7]. Crystals of the dibenzyl compound were grown by slow cooling of acetonitrile solutions. Compactions were formed by compressing a finely powdered sample in an evacuated die at 70 MN m^{-2} for five minutes.

The specimens were mounted in the cell (Fig. 1) in a sandwich configuration using high conductivity silver paint electrodes and fine copper leads. DC conductivity measurements were made using a Keithley 640 electrometer, and ac capacitance and conductivity measurements were made using a General Radio 1621 Precision Capacitance Measurement System. Because of the high conductivity of the materials the electrode areas were kept as small as possible ($\approx$ 1 mm^2) so that the conductances fell within the measuring range of the bridge. To eliminate the impedances of the leads the bridge was used in the three terminal mode.

The cell shown in Fig. 1 is built of stainless steel and is evacuable. Lead connections are via BNC connections giving an empty cell resistance greater than 10^{16} Ω. An electrical heater coupled with a Pye Oven Programmer and platinum resistance thermometer provides accurate temperature control. For measurements below room temperature the cell was mounted in a large Dewar vessel containing liquid nitrogen

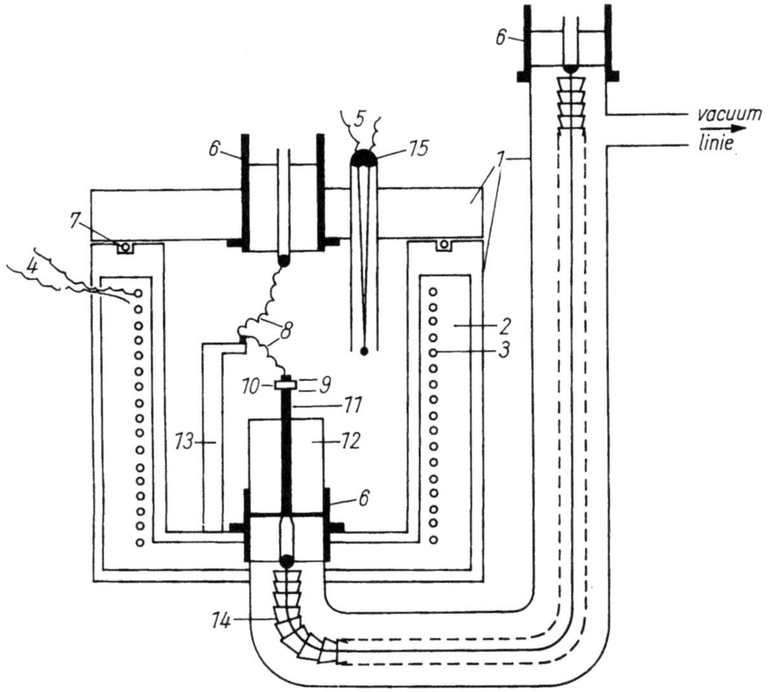

Fig. 1. Electrical measurement cell; 1 stainless steel, 2 alumina, 3 heater, 4 heater leads to temperature controller, 5 platinum resistance thermometer, 6 BNC 50 Ω connection, 7 rubber 0-ring, 8 thin copper wires, 9 high-conductivity silver paint, 10 sample, 11 copper rod, 12 teflon, 13 plastic support, 14 ceramic beads, 15 epoxy resin seal

3. Results

3.1 On compactions of N,N'-diethyl-4,4'-bipyridylium $(TCNQ)_4$ or DEBP $(TCNQ)_4$

Fig. 2 shows the variation of the conductivity, as measured by ac (σ), with temperature and frequency. Also shown is the dc value. It is clear that the frequency independent portion apparent at higher temperatures and low frequencies is the dc conductivity. Thus, the measured conductivity may be written as the sum of the frequency independent (dc) conductivity (σ') and a frequency dependent component $(\sigma(\omega))$

$$\sigma = \sigma' + \sigma(\omega) . \tag{1}$$

Fig. 3 shows $\sigma(\omega)$, obtained by subtracting σ', as a function of the frequency and temperature. The frequency variation follows,

$$\sigma(\omega) \propto \omega^n \tag{2}$$

with n in the range 0.5 to 1, but with n increasing slightly with frequency, particularly at lower temperatures where a value of unity is reached in the decade 10^4 to 10^5 Hz. It should be noted, however, that the experimental error is greater at low frequencies where σ and σ' approach one another. Fig. 4 shows σ' (measured by dc) as a function of temperature and $\sigma(\omega)$, as a function of temperature and frequency. The dc conductivity varies with temperature as

$$\sigma' = \sigma_0' \, e^{-E_a/kT} \tag{3}$$

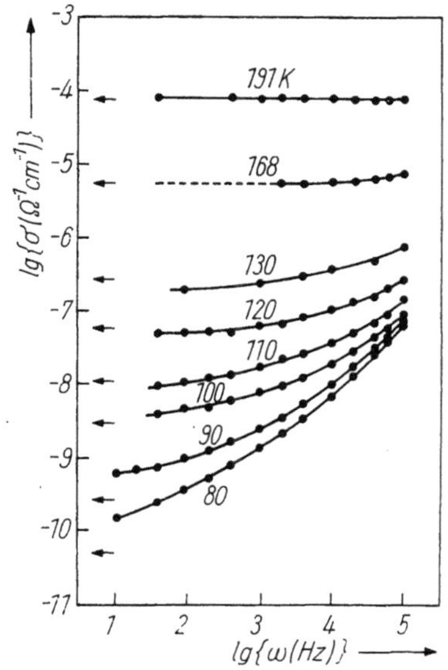

Fig. 2

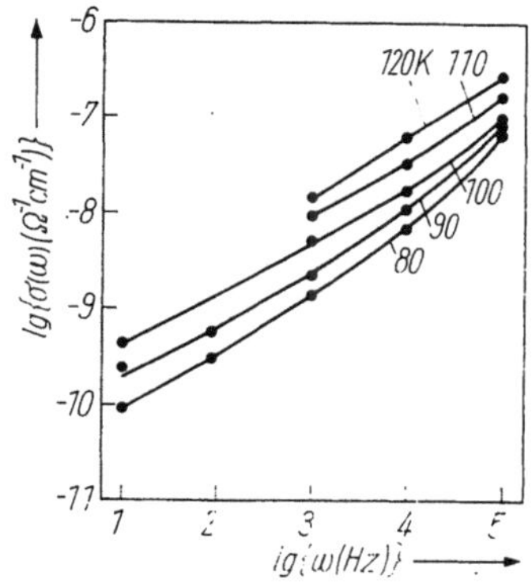

Fig. 3

Fig. 2. Frequency dependence of conductivity (σ) of DEBP$(TCNQ)_4$ at different temperatures. The corresponding dc values are indicated by arrows

Fig. 3. Frequency dependence of frequency dependent component of conductivity ($\sigma(\omega)$) at different temperatures for DEBP $(TCNQ)_4$

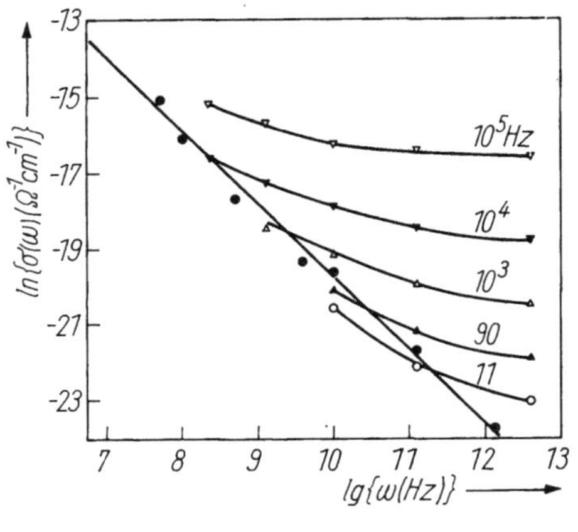

Fig. 4. Temperature dependence of the frequency dependent component of conductivity ($\sigma(\omega)$) for DEBP $(TCNQ)_4$. Also shown is the dc conductivity (σ'), indicated by ●

with $E_a = 0.162$ eV, in good agreement with the published value over this temperature range [6]. The apparent activation energy of $\sigma(\omega)$ is frequency dependent, approaching zero at high frequencies and low temperatures.

Fig. 5 shows the variation of capacitance as a function of frequency. The capacitance is strongly temperature dependent at low frequencies, but tends to a common value at high frequencies. Now when (2) is obeyed, the Kramers-Kronig relation may be used to obtain the frequency dependence of the capacitance

$$C = C_\infty + A\omega^{n-1}\,, \tag{4}$$

where C_∞ is the capacitance at high frequencies, C the total capacitance and A a constant. From Fig. 5, C_∞ is taken as the value at 10^5 Hz, i.e. 5.95×10^{-13} F. From Fig. 6 it can be seen that (4) is closely obeyed. Furthermore, the value of n (Table 1) obtained from the slopes of Fig. 3 and Fig. 6 are in reasonable agreement, particularly at lower temperatures. Thus the material shows the expected consistency between conductivity and capacity behaviour, and there is no indication of any change of mechanism towards either end of the frequency range studied.

The deduction of carrier mobility from space charge limited currents requires a value of the relative permittivity. A low frequency (e.g. 10^3 Hz), room temperature

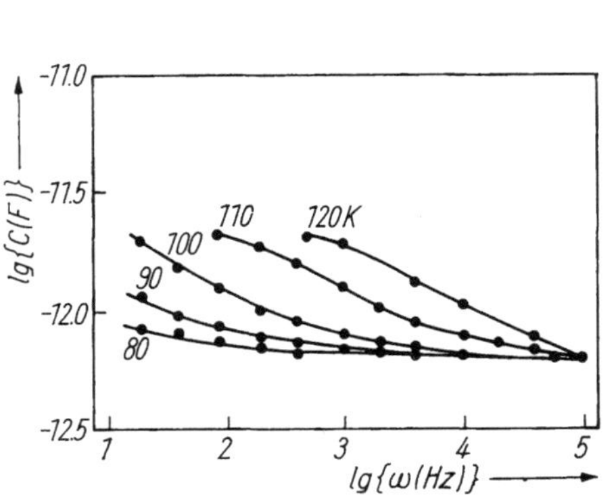

Fig. 5

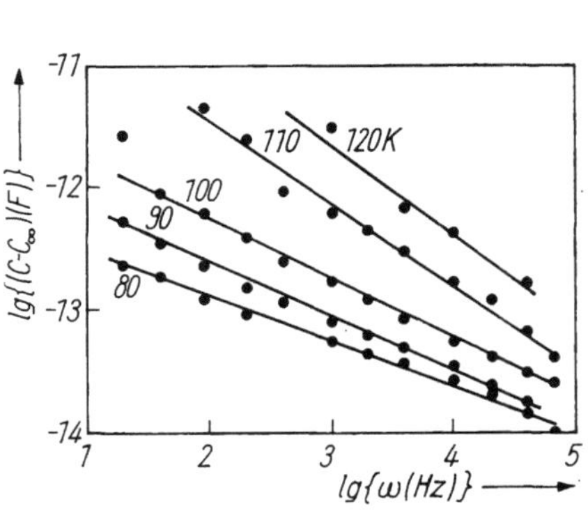

Fig. 6

Fig. 5. Frequency dependence of capacitance at various temperatures for DEBP $(TCNQ)_4$

Fig. 6. Dependence of lg ($C - C_\infty$) on frequency, at various temperatures, for DEBP $(TCNQ)_4$

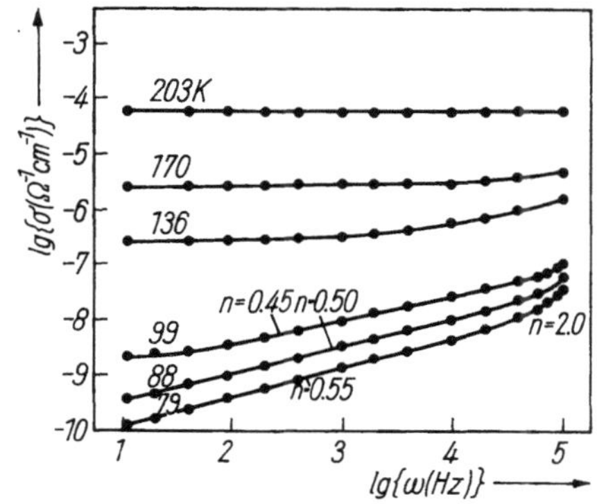

Fig. 7. Frequency dependence of conductivity (σ) for polycrystalline DBBP (TCNQ) at various temperatures

measurement is usually assumed to be a good approximation to the static value [8]. The strong temperature and frequency dependence of the capacitance suggests that this assumption is not valid for TCNQ complex salts, and measurements at much lower frequencies are required.

Table 1

Values of n at different temperatures

T (K)	n (Fig. 3)	n (Fig. 6)
80	0.6 to 1	0.625
90	0.53 to 1	0.54
100	0.50 to 0.83	0.50
110	0.52 to 0.675	0.35
120	0.6	0.3

3.2 On compactions and single crystals of N,N'-dibenzyl-4,4'-bipyridylium (TCNQ)$_4$ or DEBP (TCNQ)$_4$

This material was studied to provide a comparison between single crystals and compactions. The behaviour of the compaction as a function of frequency and temperature (Fig. 7) follows a similar pattern to that of the diethyl compound and is reported in less detail. The conductivity of single crystals as measured by ac along the three crystal axes is shown in Fig. 8. On the basis of previous conductivity studies, graphs (i), (ii), and (iii) correspond to the a, c, and b axes, respectively [9].

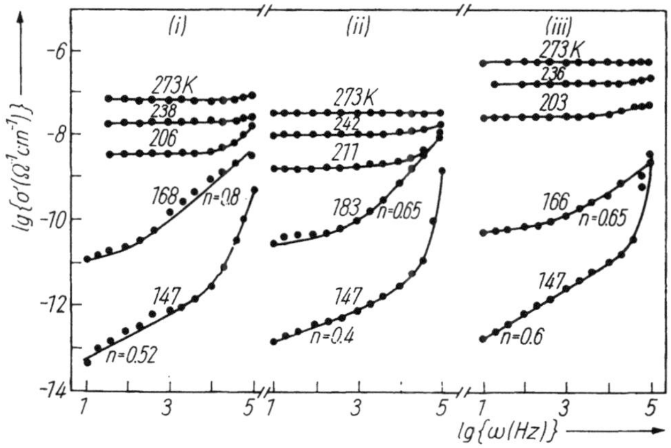

Fig. 8. Frequency dependence of conductivity of single crystal DEBP $(TCNQ)_4$ at various temperatures. (i), (ii), and (iii) correspond to the a, c, and b axes, respectively

4. Discussion

4.1 Polycrystalline N,N′-diethyl-4-4′-bipyridylium (TCNQ)₄

The results shown here are typical of the behaviour of materials in which hopping of carriers occurs between localised sites [10]. It is of particular interest to discover the nature, origin, and density of these sites. It is unlikely that they are sites in the ideal structure as hopping between identical pairs of sites gives,

$$\sigma(\omega) \propto \frac{\omega^2}{\omega^2 + \Omega^2}, \tag{5}$$

where Ω is the natural frequency of hopping between the centres. In a crystal of a TCNQ complex salt there are three crystallographically non-equivalent chemical species, and so the number of different hopping processes is small and insufficient to give the distribution necessary to observe $\sigma(\omega) \sim \omega^n$ behaviour over a wide frequency range. It is assumed such a distribution must arise from defects or impurities, although Jonscher has discussed other possibilities [11].

The results reported here show a striking similarity to those reported for the related di-propyl compound [5]. Now the two materials were separate preparations with different starting materials and hence the impurities are unlikely to have the same nature and distribution. It is therefore proposed that the sites arise from crystal imperfections. Since the crystallisation conditions are the same it is likely that these imperfections are at equilibrium concentration at the temperature of crystal growth and are frozen into the crystal. In addition it is to be expected that the process of compaction has some influence on defect concentration.

Webb and Williams [12] discussing somewhat similar data for thin films of tetrathiotetracene, have applied the theory of Mott and Davis [13] for hopping at the Fermi level, which gives,

$$\sigma(\omega) = \frac{4\pi}{3} (\ln 2)\, e^2 kTN(E_F)^2 \alpha^{-5} \left[\ln \frac{\nu_{ph}}{\omega}\right]^4 \omega\,, \tag{6}$$

where α is the fall-off of the electronic wave function with distance from the defect centre, and ν_{ph} is the phonon frequency. Putting α as the reciprocal of the intermolecular spacing, and $\nu_{ph} = 10^{12}$ Hz, it is possible to estimate the density of states at the Fermi level $(N(E_F))$ as $1.3 \times 10^{20}\ eV^{-1}\ cm^{-3}$. Assuming a band width of 0.1 eV this corresponds to $1.3 \times 10^{25}\ m^{-3}$, which provides an approximate indication of the concentration of imperfections This compares with $1.3 \times 10^{25}\ m^{-3}$ from the size of the Curie law component of magnetic susceptibility [13] in polycrystalline samples of related compounds.

4.2 N,N′-dibenzyl-4,4′-bipyridyl (TCNQ)₄

The measurements on single crystals shown in Fig. 8 show a qualitative similarity to those of the compaction. The frequency independent part reflects the anisotropy of conductivity with the highest conductivity in the direction TCNQ stacks. The similarity in behaviour of the frequency dependent components along the three crystal axes indicates that the hopping sites are uniformly distributed throughout the structure and not predominantly along one conduction path.

Calculations of the density of states at the Fermi level, as in Section 4.1, yield values of $1.2 \times 10^{24}\ m^{-3}$ for the single crystal and $1.5 \times 10^{25}\ m^{-3}$ for the compaction. These calculations involve a number of assumptions, and hence the values are only approximate. Nevertheless, the relative magnitudes should be much more accurate. Thus, it can be concluded that the process of grinding increases the density of imperfections at the Fermi level by an order of magnitude. It follows therefore, that the magnitude of the frequency dependent contribution to conductivity is a sensitive method of comparing the crystal perfection of different samples of the same material. It should prove a very useful way of comparing the quality of single crystals of these materials grown by different techniques. Only when crystals of much higher perfection can be grown will accurate determination of mobility by time-of-flight measurements be possible.

5. Conclusions

a) Ac conductivity of single crystals of TCNQ complex salts is frequency dependent indicating the existence of localised sites, which may affect or dominate electrical properties.

b) The concentration of localised sites at the Fermi level is increased tenfold by the grinding and compression involved in formation of compactions.

c) Ac conductivity represents a sensitive indication of the degree of perfection of crystals grown for electrical measurements.

d) The assumption that permittivity values from low frequency measurements are approximately equal to the static value is likely to give rise to large errors.

Acknowledgements

The authors' thanks are due to the British Council for the award of a scholarship to I.D., and to K. Mistry and C. Raven for some additional measurements.

References

[1] N. Drew, D. D. Eley, W. Waclawek and M. R. Willis, Mol. Cryst. liquid Cryst. **86**, 1963 (1982).
[2] G. G. Roberts and F. W. Schmidlin, Phys. Rev. **180**, 785 (1969).
[3] F. W. Schmidlin and G. G. Roberts, Phys. Rev. B **9**, 1578 (1974).
[4] G. J. A. Ashwell, D. D. Eley, S. C. Wallwork, M. R. Willis, and J. Woodward, Ann. New York Acad. Sci. **313**, 417 (1978).
[5] G. J. Ashwell, I. Diaconu, D. D. Eley, S. C. Wallwork, and M. R. Willis, Z. Naturforsch. **34a**, 1 (1979).
[6] G. J. Ashwell, D. D. Eley, S. C. Wallwork, and M. R. Willis, Proc. Roy. Soc. **A343**, 461 (1975).
[7] T. Sundaresan and S. C. Wallwork, Acta cryst. **B28**, 2472 (1972).
[8] J. P. Farges, A. Brau, and F. Gutman, J. Phys. Chem. Solids **33**, 1723 (1972).
[9] G. J. Ashwell, I. Diaconu, D. D. Eley, S. C. Wallwork, and M. R. Willis, 7th Molecular Crystal Symp., Tokyo 1975.
[10] M. Pollak and T. H. Geballe, Phys. Rev. **122**, 1742 (1961).
[11] F. Argall and A. K. Jonscher, Thin Solid Films **2**, 185 (1968).
[12] J. B. Webb and D. F. Williams, J. Phys. C. **12**, 3173 (1979).
[13] N. F. Mott and E. A. Davis, Electronic Processes in Non-Crystalline Materials, Oxford University Press, 1971.
[14] J. Woodward, Ph.D. Thesis, Nottingham 1976.

(Received May 25, 1984)

phys. stat. sol. (a) 85, 291 (1984)

Subject classification: 14.3; 22.9

Laboratoire d'Optique de la Matière Condensée, Equipe de recherche associée au C.N.R.S., Université Pierre et Marie Curie, Paris VI[1])

Conductivity in One-Dimensional Disordered Cs_2TCNQ_3 Salts

By

M. MESSAOUDI and J. DURAN

Room temperature, low frequency conductivity measurements of two isomers (I and II) of the Cs_2TCNQ_3 compounds are made. The conductivity–voltage characteristic display informative data concerning the interaction between disorder and conductivity in these materials. The isomer I clearly exhibits a disorder-dominated conductivity as can be seen from deviations to the Ohmic law. The isomer II, on the contrary, shows up various features connected with the existence of stopping barriers. In particular, it is shown that the interdimensional crossover which takes place at low frequency plays a significant role in these experiments. Moreover the conductivity–voltage characteristic measured along the **a** axis (perpendicular to the stacking axis) displays a very interesting behavior whereby the disorder properties interplay in an unusual manner.

Cet article concerne un certain nombre de mesures de conductivité effectuées à basse fréquence et à température ordinaire sur deux composés isomères du Cs_2TCNQ_3 (I et II). Les caractéristiques courant–tension de ces cristaux contiennent un ensemble d'informations intéressantes sur les interactions entre le désordre et les propriétés de transport. Ainsi les cristaux de type I montrent de manière évidente un processus de conduction quasi-unidimensionnel limité par des barrières de potentiel aléatoires. Les isomères de type II révèlent une conduction particulière limitée par des barrières de potential infranchissables qui met en lumière le rôle essentiel joué par les transferts entre colonnes adjacentes. De plus, dans ces derniers échantillons les expériences effectuées perpendiculairement à l'axe d'empilement mettent en évidence une caractéristique courant–tension qui décroit en fonction du champ contrairement aux observations usuelles.

1. Introduction

Transport properties in one-dimensional organic semiconductors have recently attracted considerable attention [1, 2] since they are expected to display a number of unusual or anomalous properties related to the intrinsic interplay of the electronic diffusion with the disordered nature of the materials.

Basically, it is well known that at low field, low frequency, and low temperature, purely one-dimensional conductors cannot exist, because any potential barrier will definitely switch off the conduction. Under these circumstances the charge transfer salts have been shown to exhibit pseudo-one-dimensional characteristic whereby the excitation migrates preferentially along stacking axes but is able to jump from one column to the next leading to what is known as an interdimensional cross-over [3]. Consequently the conductivity of these materials will reflect the "forced random walk" nature of the electronic trajectories. Therefore, they are expected to display anomalous properties as soon as frequency, temperature, or field dependences are concerned because all these parameters, at least under favorable circumstances, are likely to induce some change in the paths followed by the excitation.

It is a particularly interesting feature of the hopping conduction to relate directly to local disorder; among the numerous materials which can be relevant to the same

[1]) Tour 13, 4 place Jussieu, F-75230 Paris Cedex 05, France.

theoretical treatment, the one-dimensional semiconductors turn out, for obvious reasons, to be the more sensitive to disorder and, ultimately, the more demonstrative.

So far, most of the published work in the field have dealt with the temperature or frequency dependence of the conductivity of these materials. However, we feel that, in a number of cases, the field dependence of the conductivity may, as well; bring interesting information about the process involved. Roughly stated, and depending on the nature of the impurities or defects, the increasing electric field will force the more or less two- or three-dimensional electronic path to turn back to the one-dimensional one, at the expense of some loss in the overall conduction. Such an effect has been anticipated theoretically by Alexander et al. [3]. Although it has not yet been experimentally demonstrated, it is the purpose of this paper to show that under certain circumstances, typical behaviors of this sort can be effectively observed.

This study has been triggered by the tentative synthesis in this laboratory of "perturbed" Cs_2TCNQ_3 crystals. Although various attempts have been made in this field leading to rather promising results, we have focussed our attention on a particular isomeric form of the regular Cs_2TCNQ_3 crystals which we have shown elsewhere [4] to grow along an unusual axis and to possess most of the structural properties of the well-known crystal, but differs from the standpoint of conductivity properties. In the following we will label by the index II this new crystal whereas the usual material will be referred to by the index I.

In order to introduce the following conductivity measurements concerning both type I and II Cs_2TCNQ_3 crystals, it is worthwhile to recall here various relevant features which we have discussed in another paper [4]. The experiments dealing with optics (visible and infrared), EPR and X-rays allowed us to prove the similarity between the two synthesized crystals except with the following properties:

(i) Crystals of type I grow along the $\boldsymbol{b}$ stacking axis. The lattice parameter along this axis is 1.0 ± 0.01 nm. Type II crystals grow along the $\boldsymbol{a}$ axis nearly perpendicular to the stacking axis $\boldsymbol{b}$. The lattice parameter along $\boldsymbol{b}$ is somewhat larger than in crystals of type I (1.08 ± 0.01 nm). All the other lattice parameters are identical within experimental error.

(ii) Low-field, room-temperature, low-frequency resistivity measurements for these two compounds led to noticeable differences, i.e.:

$$Cs_2TCNQ_3\ (I) \qquad \varrho_b = 2.2 \times 10^3\ \Omega\ \text{cm} \qquad \varrho_a = 5 \times 10^4\ \Omega\ \text{cm}$$

$$Cs_2TCNQ_3\ (\text{II}) \qquad \varrho_b = 2.8 \times 10^3\ \Omega\ \text{cm} \qquad \varrho_a = 3 \times 10^4\ \Omega\ \text{cm}\ .$$

Briefly stated, type II crystals display a smaller anisotropy and a slightly larger resistivity along the principal axis. In turn, this increase in longitudinal resistivity has been related to a measurable shift of the near IR charge-transfer band towards higher energy indicating a higher repulsive energy and a larger resistivity [4].

2. Conductivity Measurements

Since our main concern here deals with the voltage dependence of the conductivities we restricted our experiments to dc or low-frequency (< 5 kHz) ac measurements. The experimental setup involved the usual dc bridges or lock-in when dealing with ac currents. The best crystals were carefully selected and properly washed and dried before any experiment. Various kinds of silver paints have been used during the course of this study leading, in most cases, to reproducible results. Several experiments performed at relatively high voltages involved a non-negligible thermal dissipation of the Joule effect calories, which might have changed the crystal temperatures thereby giv-

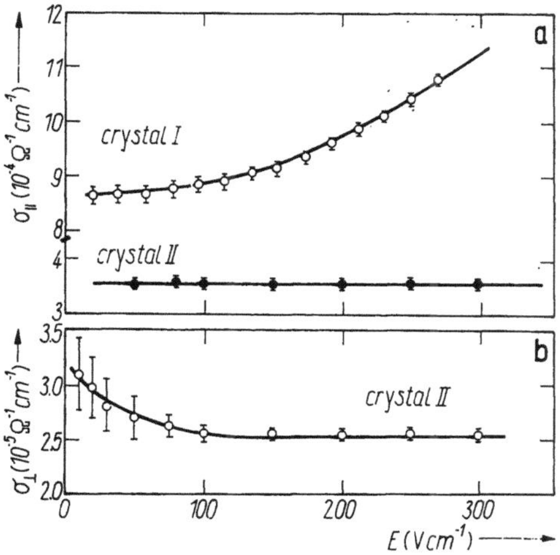

Fig. 1. Electric field dependence of the conductivity of Cs_2TCNQ_3 a) type I (open circles) and II (full points) crystals measured along the principal ***b*** axis (point experimental, lines theory) and b) type II crystals measured along the ***a*** axis both at room temperature and low frequency (500 Hz)

ing misleading results. We took much care about this problem and repeated several experiments by immersing the samples in oil or other non-conductive liquids.

The experimental data resulting from a repeated set of experiments are reported in Fig. 1 a and b.

We will first deal with the voltage dependence of the conductivity measured along the longitudinal axis ***b*** in type I crystals. The curve reported in Fig. 1a exhibits a curvature analogous to the well-known VDR (voltage-dependent resistor) characteristic. Basically the involved process deals with the interplay between conductivity and disorder. The problem has first been elucidated by Bernasconi and Schneider [5], Derrida and Orbach [6], and Grüner [2]. In both cases it involves phonon-assisted hopping conduction where two processes take place. The first one is a back-and-forth random hopping along the chains where the probability to progress is dependent on the applied electric field (so-called drift velocity). This process implies the existence of randomly distributed barriers lying at a distance l from each other. The mean value of the conductance (or the hopping rate) is introduced via the nearest neighbor transition rate W. So the drift velocity v is written as [6]

$$v = 2\left\langle \frac{1}{W} \right\rangle^{-1} \sinh\left(\frac{Eel}{k_B T}\right),$$

where E stands for the low-frequency external electric field, it follows that

$$\sigma(E) = \sigma(0)\left(\frac{k_B T}{Eel}\right) \sinh\left(\frac{Eel}{k_B T}\right)$$

and the zero-field conductivity $\sigma(0)$ is related to

$$\sigma(0) = \frac{2e^2 l}{k_B T}\left\langle \frac{1}{W} \right\rangle^{-1}.$$

These equations turn out to provide a satisfactory fit to our experimental results (Fig. 1) and allow us to determine

$$l = (1.2 \pm 0.3)\ \mu m\,,$$

$$\left\langle \frac{1}{W} \right\rangle^{-1} = 2 \times 10^{-3}\ cm^2/s\,.$$

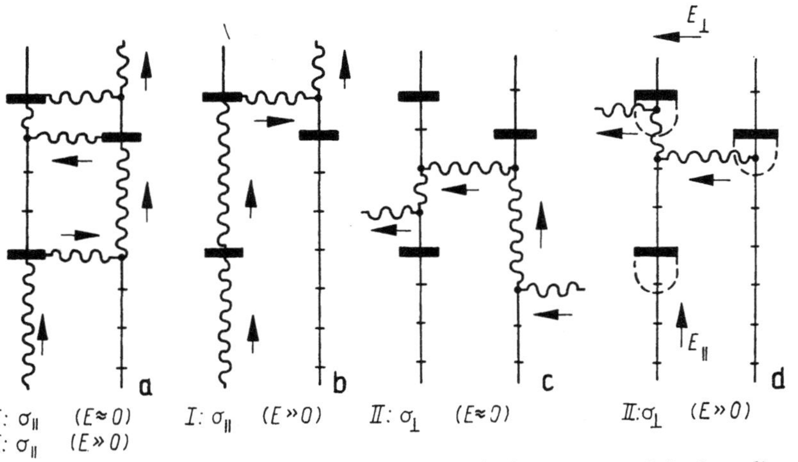

Fig. 2. Sketch of the various processes showing the importance of the interdimensional cross-over effect in the case of high barriers for I and II crystals. The potential barriers are symbolized by horizontal segments whereas the trajectory of the charge transfer is represented by a zig-zag line

An important point which must be noted here and which is relevant to the following considerations concerns the existence of interdimensional cross-over as explicitly mentioned by Alexander et al. [3]. Since we are dealing with dc or low-frequency ac measurements, the probability that the charge transfer occurs with neighboring chains is non-zero. Actually, and except in the parallel conductivity for type I crystals, this process turns out to play a major role for the conductivity.

Fig. 2 reports, as a sketch, the various processes which we discuss in the following. Firstly, we are concerned with the small-barrier conductivity displayed by type I crystals. As shown on Fig. 2 a and b, the electric field applied along the stacking axis tends to favorize the one-dimensional character of the conduction. However, it can be easily understood that, if the potential barriers (due to defects or impurities) can be overcome, the interdimensional cross-over will not play a significant role in the conductivity as reflected in the preceding equations. Actually, the anisotropy of the transition rates is a function of the ratio of the conductivities parallel and perpendicular to the $\boldsymbol{b}$ axis, according to [3]

$$r = \frac{\langle X_{\parallel}^2 \rangle}{\langle X_{\perp}^2 \rangle} = \frac{\sigma_{\parallel}}{\sigma_{\perp}} \gg 1 ,$$

where $X_{\parallel,\perp}$ stand for the hopping probabilities along and perpendicular to the principal axis.

Our measurements show that the typical barrier–barrier distances lie in the micron range, so that the charge transfer will hop normally along the chain at least a thousand times before reaching a potential barrier. Since we know that the ratio r is ranging between 20 and 50 in our experiments, we deduce that even if the system prefers to avoid all the potential barriers by jumping to the next column, this effect will only contribute 1/20 to the overall conductivity, so that the preceding calculation need not take this effect into consideration explicitly. However, it must be emphasized that this effect does take place and that the preceding equations analyse the process which allow the electronic excitation to jump over the various barriers as a function of the electric field.

We consider now the experimental results obtained for a type II Cs_2TCNQ_3 crystal along the principal axis as reported in Fig. 1. As can be realized, the field dependence of the conductivity turns out to be almost zero, so that we propose that at low frequency and for sufficiently high barriers the system will meet another cross-over to a situation where the conductance would still be provided by the principal axis trips but where the high potential barriers would be avoided by interdimensional cross-over. Be T_{1d} the characteristic time inverse of the transition probability to jump from one column to the next, and be l the distance between two high barriers. Then if v is the velocity of the excitation along the principal chain, the ratio R given by

$$R = \frac{l}{vT_{1d}}$$

depicts the number of times an excitation will jump to another column during its trip between two barriers. Here $vT_{1d} = l_{1d}$ stands for the mean distance travelled over a single chain.

So that if $R \ll 1$ then Alexander et al.'s calculation is valid and the conductivity will turn to zero. Now, if $R \gg 1$ we get essentially an interdimensional cross-over effect and the overall conductivity will be slightly decreased as a consequence of the perpendicular trips. Keeping in mind the fact that type II crystals exhibit a larger perpendicular conductivity than the type I samples, which corresponds to an increase by a factor of 2.8 of the interchain transition probability, we may tentatively estimate the effect and try to determine l_{1d} starting from our experimental results.

Following the preceding equation, the total resistivity will be given by

$$\varrho = \frac{\varrho_{\parallel} l_{1d} + \varrho_{\perp} L}{l_{1d} + L},$$

where $\varrho_{\parallel}$ and $\varrho_{\perp}$ stand for the resistivities parallel and perpendicular to the stacking axis, respectively, and L for the total distance perpendicular to the chain followed by the excitation. Since we deal with nearly one-dimensional systems ϱ is written as

$$\varrho = \varrho_{\parallel} + \frac{L}{l_{1d}} \varrho_{\perp} .$$

Our experimental data lead to

$$L \approx 0.02 l_{1d} .$$

Since a normal trip would necessarily associate one l_{1d} trip to one jump to the next column and approximating the two lattice parameters parallel and perpendicular to the principal axis to be equal, we would get

$$L \approx 0.001 l$$

if the one-dimensional trip was determined by the barriers.

Finally we can estimate the R ratio to be roughly equal to 20 which indicates that, at least at very small fields, the trip is not influenced by the barriers.

Actually, we feel that the preceding estimate suffers a number of approximations which cast a serious doubt on the R value. Nevertheless, the fact that the zero-field conductivities are approximately identical is a clear indication that the barriers do not play any role at zero field. On the contrary, at higher field they force the system to jump from one column to the next one but, since no overstepping is possible, this will not change the Ohmic law for conductivity as seen in Fig. 1a (crystal II).

We now turn to the consideration of the informative results obtained on type II crystals when the electric field was parallel to the $\boldsymbol{a}$ axis.

As seen on Fig. 1b the field dependence of the conductivity exhibits a rather unusual behavior whereby, at least at low field, the conductivity decreases when the electric field is increased. In the absence of a detailed theory of this process, this feature may well be understood in the framework of the preceding considerations concerning type II crystal.

As recognized earlier, type II crystals exhibit a large number of very high potential barriers which force the electronic excitation to by-pass these defects by interdimensional cross-over. In turn, as can be seen on Fig. 1b, at low perpendicular field, the barriers should not play any role since, as we have shown previously the "forced" interdimensional cross-over would still reinforce our conclusions.

At higher field, and as soon as the longitudinal component of the electric field pushes the excitation against the potential barriers, it is very likely that this will induce a decrease in the conductivity by confining electronic excitations in imperfect regions of the crystals. Although this last explanation need be confirmed through a detailed theoretical analysis of the problem, we feel that it is coherent with our preceding conclusions.

3. Conclusion

We have reported various experimental results concerning the field dependence of the conductivity parallel and perpendicular to the principal axis of two different Cs_2TCNQ_3 crystals synthesized under different experimental conditions. Although these two samples exhibit a number of common properties, they differ with respect to conduction properties. Actually, this particular feature allowed us to draw some conclusions concerning the interplay between defects and the deviations from the classical Ohmic law. In turn, this study puts in evidence the possibility to modify, at will, the field–conduction characteristics which might turn out to be quite useful in the future.

Acknowledgements

This paper has benefitted from enlightening discussions with S. Alexander and P. Evesque. Also, the skilful help of M. Lemal and M. Lanuza has been appreciated.

References

[1] J. S. BLAKEMORE, J. E. LANE, and D. A. WOODBURY, Phys. Rev. B **18**, 6797 (1978).
[2] G. GRÜNER, Proc. Internat. Conf. Low-Dimensional Synthetic Metals, Denmark 1980, in: Chem. Scripta **17** (1980).
[3] S. ALEXANDER, J. BERNASCONI, W. R. SCHNEIDER, and R. ORBACH, in: Physics in One Dimension, Springer Series in Solid State Sciences, Vol. 23, Ed. J. BERNASCONI and T. SCHNEIDER, Springer-Verlag, Berlin 1981 (p. 277).
[4] M. MESSAOUDI, A. KUHN, J. TOULOUSE, and J. DURAN, to be published.
[5] J. BERNASCONI and W. R. SCHNEIDER, J. Phys. A **15**, 2729 (1983).
[6] B. DERRIDA and R. ORBACH, Phys. Rev. B **27**, 4694 (1983).

(Received April 27, 1984)

phys. stat. sol. (a) 85, 297 (1984)

Subject classification: 16; 2; 22.1.2

Department of Physics, Indian Institute of Technology, Kanpur[1])

Intensity and Temperature Dependence of Steady-State Photoconductivity Down to 8 K and DOS Distribution Obtained from These Measurements in a-Si:H

By

D. S. MISRA, A. KUMAR[2]), and S. C. AGARWAL

Steady state photoconductivity (σ_{ph}) measurements are made on thin films of a-Si:H prepared by dc glow discharge of 3% silane mixed with argon as a function of temperatures (8 K $\leqq T \leqq$ 330 K) and intensities ($10^{-2}F_{|\sigma|\to 0} \leqq F \leqq F_{|\sigma|\to 0}$, $F_{|\sigma|\to 0} \approx 30$ mW/cm²) in heat dried (A) and light soaked (B) states. σ_{ph} decreases continuously, without any evidence of thermal quenching effects, with temperature and becomes constant in the low temperature range ($T \approx 25$ K) irrespective of the state of the sample. Also, σ_{ph} in state B is found to be larger than in state A below $T \approx 125$ K. For all temperatures $\sigma_{ph} \propto F^{\gamma}$ where γ (300 K) is 0.72 and 0.80 for states A and B, respectively. γ decreases with a decrease in temperature, increases again, and becomes constant in the region where σ_{ph} is independent of temperature ($T \approx 25$ K). The results are explained with the help of recombination model and thermalization of carriers. DOS are calculated from the intensity dependence of σ_{ph} following Huang et al.

Messungen der stationären Photoleitung (σ_{ph}) werden an dünnen, durch Gleichstromentladung von Argon mit 3% Silan hergestellten Schichten von a-Si:H als Funktion der Temperatur (8 K $\leqq T \leqq$ 300 K) und Intensität ($10^{-2}F_0 \leqq F \leqq F_0$, $F_0 \approx 30$ mW/cm²) im wärmegetrockneten (Al) und lichtgesinterten (B) Zustand durchgeführt. σ_{ph} nimmt ohne jeden Hinweis auf thermische Tilgungseffekte mit der Temperatur ab und wird im Tieftemperaturbereich ($T \approx 25$ K) konstant unabhängig vom Zustand der Probe. Es wird gefunden, daß σ_{ph} im Zustand B größer ist als im Zustand A unterhalb $T \approx 125$ K. Für alle Temperaturen gibt $\sigma_{ph} \propto F^{\gamma}$, wobei γ (300 K) gleich 0,72 oder 0,80 für den Zustand A bzw. Zustand B ist. γ nimmt mit sinkender Temperatur ab, nimmt wieder zu und wird konstant in dem Bereich, wo σ_{ph} unabhängig von der Temperatur ist ($T \approx 25$ K). Die Ergebnisse werden mit einem Rekombinationsmodell und Thermalisierung der Träger erklärt. DOS wird aus der Intensitätsabhängigkeit von σ_{ph} nach Huang et al. berechnet.

1. Introduction

Photoconductivity measurements give information about the recombination kinetics in semiconductors. Both steady-state and transient photoconductivity (SPC and TPC) measurements have been done on a-Si:H (for a review see [1], see also [2]). The present investigation concentrates only on SPC measurements.

Spear et al. [3] reported the spectral (λ), intensity (F), and temperature (T) dependence of photoconductivity (σ_{ph}) in a-Si:H. In the spectral dependence, the samples show maximum photoresponse for photon energies between 1.8 and 2.0 eV. σ_{ph} is found to be proportional to γ, i.e., $\sigma_{ph} \sim F^{\gamma}$ with γ (slope of the $\log \sigma_{ph}$ versus $\log F$

---

[1]) Kanpur 208016, India.

[2]) Present address: Physics Department, H.B.T.I., Kanpur.

curve) ranging between 0.6 and 0.9 depending upon F and T. The dependence of σ_{ph} on T is divided in three regions (I: $300\ K \leqq T \leqq 450$ K, II: $250\ K \leqq T \leqq 300$ K, and III: $T < 250$ K). In regions I and II, the activation energy ($\Delta E_{\sigma ph}$) of σ_{ph} is found to be the same and in region III it has a lower value. The transition from region II to III takes place around $T = 250$ K. On the basis of these results Spear and LeComber [4] proposed a model and concluded that the values of $\Delta E_{\sigma ph}$ can be explained by considering the recombination of electrons near E_A to the holes trapped near E_Y (Fig. 1). Further, it is argued that $\Delta E_{\sigma ph}$ is lower at low T (region III) because the dominant conduction mechanism in this temperature range is hopping of electrons in localized states near E_A. On the other hand, the results from other laboratories [5, 6] are in agreement only qualitatively and the interpretations differ from those by Spear and LeComber [4]. For instance, Wronski and Daniel [5] who observe that $\sigma_{ph} \sim F^{\gamma}$ report that $\Delta E_{\sigma ph}$ in their samples is greater in region III than in region II. They explain their results by a model given by Rose [7] and find no evidence for a peak in DOS near E_x (i.e., 0.4 eV below $E_{c'}$, Fig. 1). Fuhs et al. [6] on the basis of the detailed measurements of SPC and TPC conclude that there is no evidence for a change from band conduction to hopping conduction in the temperature range ($100\ K \leqq T \leqq$ $\leqq 500$ K) investigated.

Vanier et al. [8] and Persans and Fritzsche [9] have observed some new features in σ_{ph} of a-Si:H films. These are ascribed to what is known as thermal and infrared quenching which have been observed in crystalline semiconductors [10, 11]. Vanier et al. [8] have found that changing the position of E_F towards one of the band edges by doping with PH_3, B_2H_6, O_2, N_2, and air results in the elimination of these effects. This is in agreement with the conclusions of Persans and Fritzsche [9] also. Thus, the quenching effects are observed only when the dark Fermi level (E_F) lies near midgap.

Huang et al. [12] have reported SPC and TPC measurements at different temperatures and have deduced the DOS in the upper half of the band gap from these data.

All these authors have done measurements only down to liquid nitrogen temperature (≈ 77 K). Recently, Hoheisel et al. [13] have reported measurements of σ_{ph} down to 4 K. They report that σ_{ph} becomes constant and γ approaches unity for $T \approx$ ≈ 50 K.

In the present investigation, SPC measurements are done for the intensities $10^{-2}F_0 \leqq$ $\leqq F \leqq F_0$ ($F_0 \approx 30$ mW/cm²) and temperatures $8\ K \leqq T \leqq 330$ K in the heat dried (A) as well as the light soaked (S-W effect, B) states. Section 2 describes the experimental details and the results are reported in Section 3.

In Section 4 the interpretations of the results are discussed and the DOS distribution is calculated following Huang et al. [12].

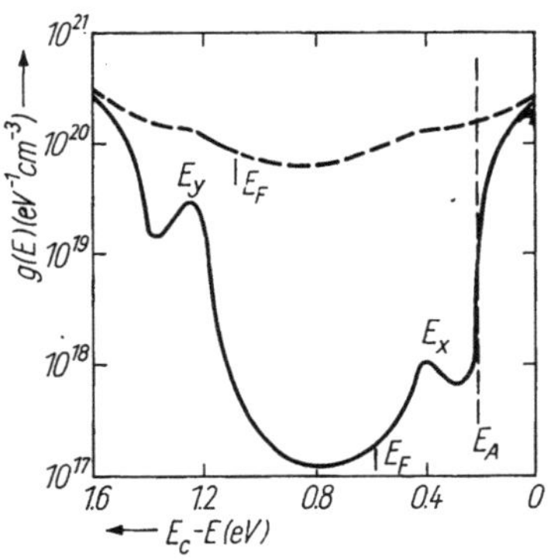

Fig. 1. Density-of-states distribution (DOS) for a-Si specimens obtained by Spear and LeComber [17] using the field effect technique. The solid curve is DOS of an a-Si:H sample prepared at $T_s \approx 520$ K and the dashed curve is DOS for an evaporated Si sample. In the evaporated Si simple the measurements could only be done in a small range of energy (solid portion of the dashed curve). DOS of a-Si:H sample shows two peaks; one at E_x (≈ 0.4 eV below E_c) and the other at E_y (1.2 eV below E_c)

2. Experimental

Thin films of a-Si:H of 1 μm thickness are prepared by the dc glow discharge of 3% silane in argon at a substrate temperature of 580 K. These films are characterized by electron microscopy and IR spectroscopy. The results are reported elsewhere [14].

Undoped samples of a-Si:H with nichrome electrodes in a coplanar configuration are used for the photoconductivity measurements. For the measurements of σ_{ph} at low temperatures, the samples are mounted on the cold-finger of a closed cycle He refrigerator (Lakeshore Cryotronics, USA). The high-temperature σ_{ph} measurements are done in another cryostat where temperature can be varied up to 500 K. σ_{ph} is measured for the band-gap light ($\lambda \approx 670$ nm) from an Oriel filter monochromator. The intensity of light is varied using neutral density filters. All the measurements are done, in a vacuum $\approx 10^{-6}$ Torr, in state A (annealed in dark at 150 °C for 2 h) and in state B (exposed to white light from a 100 W tungsten halogen lamp for 2 h in vacuum). The samples exhibit ohmic characteristics in the presence of light up to the electric field $\approx 10^3$ V/cm in states A and B for temperatures ranging from 330 to 8 K.

3. Results

Variation of σ_{ph} with temperature for a typical sample of a-Si:H is shown in Fig. 2 in the states A and B. σ_{ph} in states A and B decreases monotonically with decreasing T, in agreement with others [3, 5, 6, 13]. σ_{ph} for all the samples, irrespective of whether they are in state A or B, becomes constant below $T \approx 25$ K. In contrast with Wronski and Daniel [5], and in agreement with Spear et al. [3], $\Delta E_{\sigma_{ph}}$ in our case is highest (≈ 0.12 eV) in temperature range II (250 K $\leqq T \leqq$ 330 K) (Fig. 2 and 3) and then decreases continuously in state A. $\Delta E_{\sigma_{ph}}$ and σ_{ph} in state B are found to be smaller for 250 K $\leqq T \leqq$ 330 K. However, σ_{ph} in state B is greater than that in state A for all samples for $T < 125$ K. It may be mentioned that the thermal quenching effects [8, 9] are not visible in the entire temperature range.

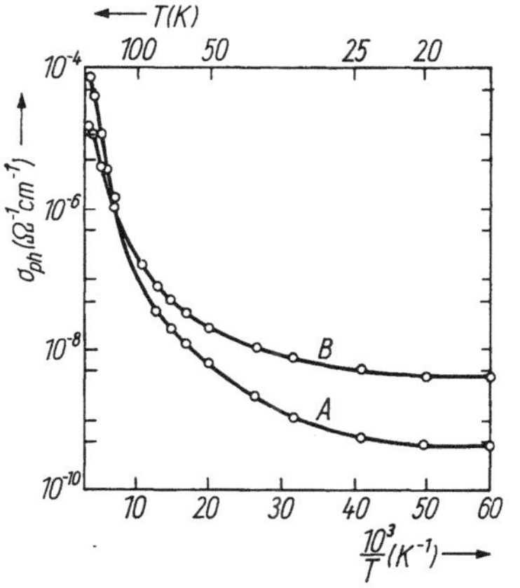

Fig. 2

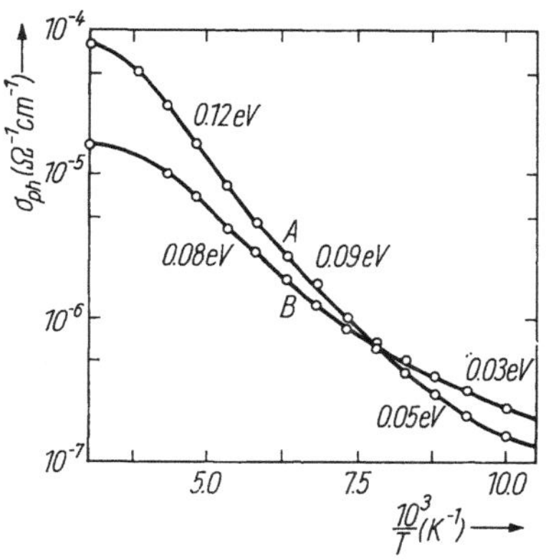

Fig. 3

Fig. 2. Plots of ln σ_{ph} vs. $10^3/T$ of a typical a-Si:H sample in heat dried (A) and light soaked (B) states

Fig. 3. Plots of ln σ_{ph} vs. $10^3/T$ for an a-Si:H sample in heat dried (A) and light soaked (B) states. These curves are for the same sample as in Fig. 2 and are plotted on an extended temperature scale for the reasons of clarity. The slopes of straight lines fitted in different temperature ranges give activation energy of photoconductivity ($\Delta E_{\sigma_{bh}}$), as indicated

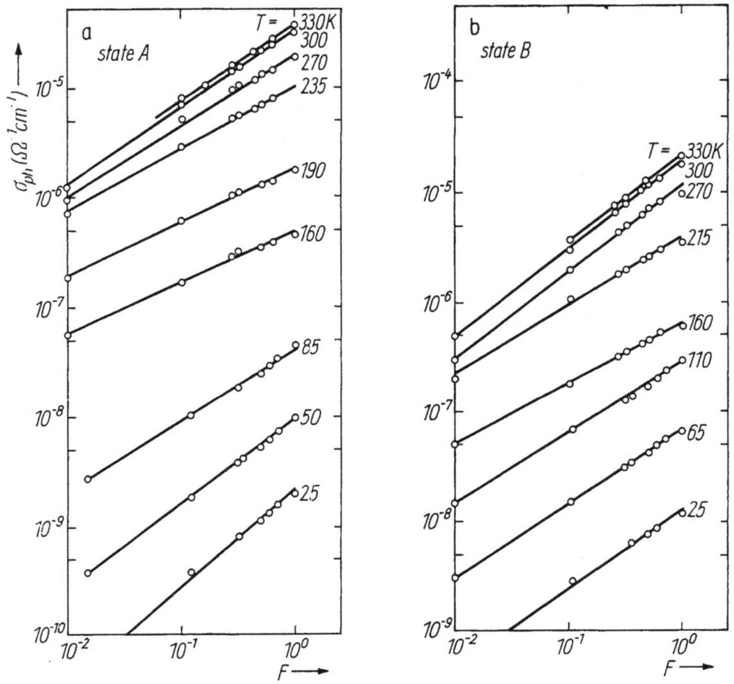

Fig. 4. Intensity (F) dependence of σ_{ph} in a-Si:H sample a) in heat dried (A) state and b) in light soaked (B) state, at different temperatures

The intensity (F) dependence of a typical sample at different temperatures for intensities ranging from $10^{-2}F_0$ to F_0 is shown in Fig. 4 for states A and B, respectively. The temperature dependence of γ is shown in Fig. 5. $\gamma \approx 0.72$ and 0.80 for states A and B, respectively, near 300 K. It first decreases with a decrease in temperature, then increases, and becomes constant (0.85 for state A and 0.7 for state B) in the region ($T \approx 25$ K) where σ_{ph} is independent of temperature. These results are in agreement with others [12, 13].

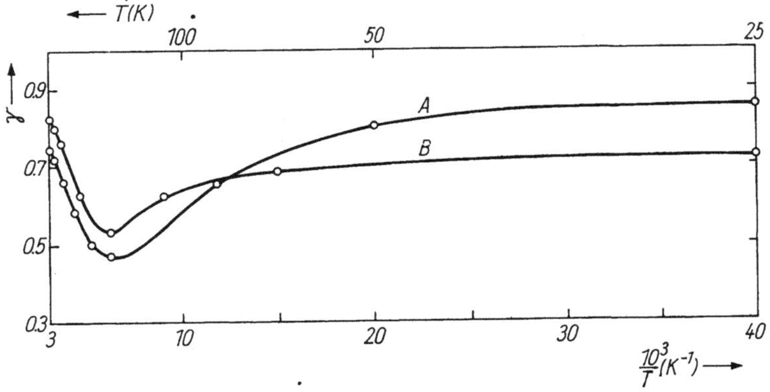

Fig. 5. Temperature dependence of γ (slope of σ_{ph} vs. F curves in Fig. 4) in a-Si:H in heat dried (A) and light soaked (B) states

4. Discussion

Our results of temperature and intensity dependence of σ_{ph} can be explained qualitatively by Spear and LeComber's model [4] in a limited temperature range (150 K $\leqq T \leqq$ 330 K). $\Delta E_{\sigma_{ph}}$ in region II (250 K $\leqq T \leqq$ 330 K) is ≈ 0.12 eV (see Fig. 3) which should be compared with that predicted by the model (≈ 0.20 eV). Similarly the model predicts a square-root intensity dependence (i.e., $\gamma = 0.50$) while we find that $\gamma = 0.60$ in temperature range II. Paul and Anderson [2] have, however, argued that the recombination path suggested by Spear and LeComber [4] need not be a unique explanation. Other paths (i.e., E_C to $E_{Y'}$, E_A to E_F, see Fig. 1) may match the experimentally observed values of $\Delta E_{\sigma_{ph}}$ and γ, as well.

However, the results of σ_{ph} at low temperatures are more difficult to explain. Spear and LeComber [4] suggest that for all $T < 250$ K, σ_{ph} is controlled by hopping of electrons near E_A and, therefore, $\Delta E_{\sigma_{ph}} \approx 0.08$ eV. Although in the temperature range 150 K $\leqq T \leqq$ 250 K, $\Delta E_{\sigma_{ph}}$ in the present case is ≈ 0.09 eV (see Fig. 3), for $T < 150$ K it is much smaller and at $T \approx 25$ K becomes almost zero.

Hoheisel et al. [13] have explained their σ_{ph} results at low temperatures (4 K $\leqq T \leqq$ 500 K) by suggesting that below 30 K, σ_{ph} arises from the thermalization of photoexcited carriers in the states above the mobility edges before they are localized in tail states and, therefore, the thermalization time (τ_{th}) rather than the trapping or recombination time determines the value of σ_{ph}. Since τ_{th} is independent of the carrier density, σ_{ph} becomes constant below 30 K and γ approaches unity. Furthermore, τ_{th} does not depend upon the density of defects but is an intrinsic property of the amorphous silicon network, being determined by the density of states near mobility edges. In the present case also the value of γ is found to be approaching unity (see Fig. 5) at low temperatures (where σ_{ph} becomes constant) irrespective of whether the sample is in state A or B. Thus, it seems likely that the thermalization of carriers is responsible for the observed behaviour of σ_{ph} at low temperatures.

The increase in σ_{ph} below 125 K in state B (S-W effect) shown in Fig. 2 and 3 may be related to the possibility that after the S-W effect, the Fermi level has moved towards the midgap and is in such a position that the thermal quenching [8, 9] effects start becoming visible. This argument is strengthened by the fact that the temperature (≈ 125 K) where σ_{ph} in state B starts becoming higher than that in A (Fig. 2 and 3) is about the same where a broad peak due to thermal quenching in $\sigma_{ph}(T)$ is observed by others [8, 9]. Alternatively, it may be that the S-W effect has created more states in the gap in such a way that the recombination paths, in this range of temperature, are different in states A and B. Another possibility may be that the thermalization time of carriers is changed in state B, since the light-induced effects may increase the DOS near mobility edges. However, such an increase in DOS near mobility edges would decrease τ_{th} and, therefore, a decrease in σ_{ph}, not an increase, is expected.

4.1 DOS distribution from SPC measurements

Even though we find that the Spear and LeComber model [4] can qualitatively explain our results in a limited temperature range, it does not necessarily imply that the DOS distribution in our case has a structure in the form of peaks near E_x or E_y. This argument is favoured when we find that in agreement with Huang et al. [12] the intensity dependence of σ_{ph} at different temperatures can be fitted to the DOS distribution which is the sum of two exponentials without any structure. The two exponentials are fitted in different regions of the mobility gap, i.e., (i) near the dark Fermi level and (ii) near conduction band edge (E_c).

It may be noted that the exponential DOS is only one of the several trap distributions for which $\sigma_{ph} \sim F^{\gamma}$. The others are Gaussian, a sum of two exponentials, etc. [15].

If we take

$$g(E) = g_0 \exp\left(-\frac{E_c - E}{kT_0}\right), \tag{1}$$

where T_0 is related to γ as

$$\gamma = \frac{T_0}{T + T_0}, \tag{2}$$

then T_0 determines the trap distribution in the vicinity of the quasi-Fermi level for traps (E_{ft}). This is close to the quasi-Fermi level for free electrons (E_{fn}) in a-Si:H [12]. Following Huang et al. [12] the position of E_{fn} can be determined using

$$E_c - E_{fn}(F, T) = E_c - E_F(T) - kT \ln \frac{\sigma_{ph}(F, T)}{\sigma_{dc}}, \tag{3}$$

where $E_{fn}(F, T)$ and $\sigma_{ph}(F, T)$ are dependent on intensity and temperature and $E_F(T)$ is the position of the dark Fermi level corrected for the temperature dependence of the mobility gap [12].

Equations (2) and (3) are used to obtain the values of T_0 for different positions of $E_{fn}(T)$ from the data of σ_{ph} versus F up to $T \approx 150$ K in states A and B. Results of T_0 versus energy are shown in Fig. 6. The values of T_0, in states A and B, are 850 and 1300 K for $|E_c - E| > 0.4$ eV, respectively. For $|E_c - E| < 0.3$ eV, the T_0 values lie around 150 and 190 K for states A and B, respectively. These values of T_0 (state A) are fitted to the DOS distribution which is the sum of two exponentials. Results are shown in Fig. 7. It can be observed that the DOS distribution is steeper near the band

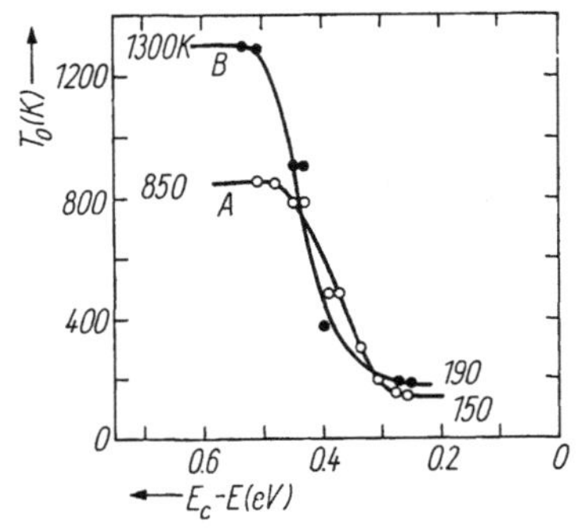

Fig. 6

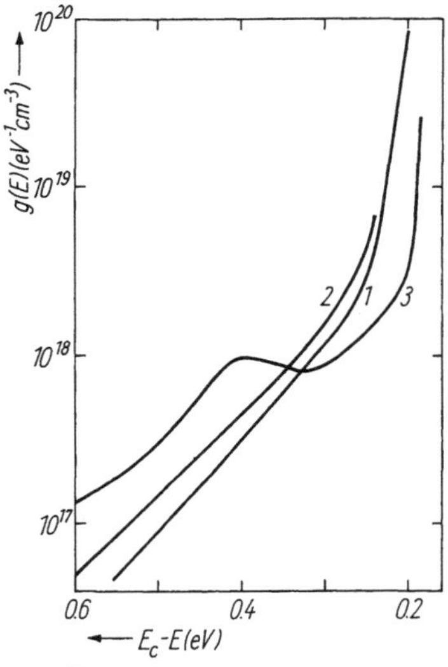

Fig. 7

Fig. 6. Plots of T_0 vs. $E_c - E$ in a-Si:H in heat dried (A) and light soaked (B) states

Fig. 7. DOS obtained by fitting T_0 vs. $E_c - E$ data (Fig. 6) to exponential distribution of states in different portions of the mobility gap (i) near dark Fermi level (E_F) and (ii) near conduction band edge (E_c) for state A; (1) present results, (2) field effect [1], (3) field effect [17]

edge. The higher value of T_0 in state B for $|E_c - E| > 0.4$ eV suggests that the DOS distribution in this region after S-W effect has become less steeper as compared to that in state A. The value of T_0 is higher in the state B (190 as compared to 150 in state A) for $|E_c - E| < 0.3$ eV also. This suggests that the DOS distribution near band edge is flattened due to light soaking. However, the two values of T_0 are too close to allow us to draw any definite conclusion. For the calculation of DOS (Fig. 7), g_0 (2) is obtained by using DOS at $E_F(\approx 5 \times 10^{16}\ \text{eV}^{-1}\ \text{cm}^{-3})$ obtained by C–V, C–ω, and C–T measurements on Schottky diodes [16].

5. Conclusions

We thus conclude that in our case the dependence of σ_{ph} on F and T appears to be governed by two processes in two different temperature ranges: (i) the recombination model suggested by Spear and LeComber [4] for higher $T(T > 150$ K) and (ii) thermalization of carriers suggested by Hoheisel et al. [13] for low $T(T < 150$ K). From the temperature dependence of γ, the transition from a recombination limited process to a thermalization process seems to be taking place around $T \approx 130$ K where γ starts increasing with temperature.

We have suggested two possible explanations for a larger σ_{ph} for $T < 125$ K in state B as compared to that in state A. These are (i) a shift of Fermi level towards midgap after the S-W effect and (ii) a change in recombination path due to the S-W effect in this range of T. However, at present it is not possible to point out specifically which of these is responsible for the observed behaviour.

Finally, the DOS distribution has been fitted to an exponential distribution. For comparison we have also plotted the DOS distribution obtained using the field effect technique by others [1, 17] (Fig. 7). However, it has been noted that a Gaussian and a sum of exponentials [15] also predict that $\sigma_{ph} \sim F^\gamma$. Thus, a distinction in the type of distribution on the basis of intensity and temperature dependences of σ_{ph} may not be possible. It can, however, be said that the assumption of presence of a structure in the form of a peak near 0.4 eV below E_c does not appear to be necessary for fitting the data.

References

[1] H. Fritzsche, Solar Energy Mater. **3**, 447 (1980).
[2] W. Paul and D. A. Anderson, Solar Energy Mater. **5**, 229 (1981).
[3] W. E. Spear, R. J. Loveland, and A. Al-Sharbaty, J. non-crystall Solids **15**, 410 (1974).
[4] W. E. Spear and P. G. LeComber, in: Photoconductivity and Related Phenomenon, Chap. 5, Ed. J. Mort and D. M. Pai, Elsevier, New York 1976.
[5] C. R. Wronski and R. E. Daniel, Phys. Rev. **23**, 794 (1981).
[6] W. Fuhs, M. Milleville, and J. Stuke, phys. stat. sol. (b) **89**, 495 (1978).
[7] A. Rose, Concepts in Photoconductivity and Allied Problems, Kreiger, New York 1978.
[8] P. E. Vanier, A. E. Delahoy, and R. W. Griffith, J. appl. Phys. **52**, 5235 (1981).
[9] P. Persans and H. Fritzsche, J. Physique **42**, C4-697 (1981); Solid State Commun. **36**, 851 (1980); Phil. Mag. **46**, 435 (1982).
[10] R. H. Bube, J. Phys. Chem. Solids **1**, 234 (1957).
[11] R. Newman, H. H. Woodbury, and W. W. Tyler, Phys. Rev. **102**, 613 (1956).
[12] C. Y. Huang, S. Guha, and S. J. Hudgens, Phys. Rev. B **27**, 7460 (1983).
[13] M. Hoheisel, R. Carius, and W. Fuhs, J. non-crystall. Solids **59/60**, 457 (1983).
[14] D. S. Misra, P. N. Dixit, and S. C. Agarwal, Bull. Mater. Sci. **3**, 347 1981.
[15] E. Bhattacharya and K. L. Narasimhan, Preprint, T.I.F.R. Bombay 1983.
[16] D. S. Misra, Ph. D. Thesis, I.I.T., Kanpur 1984 (unpublished).
[17] W. E. Spear and P. G. LeComber, Phil. Mag. **33**, 935 (1976).

(Received June 19, 1984)

See Short Note by A. D. POGREBNYAK et al.

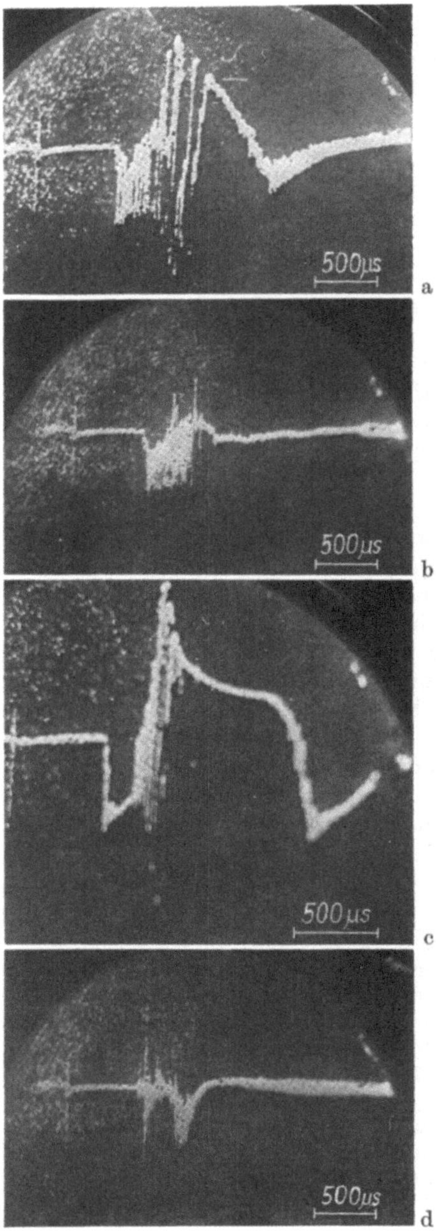

Fig. 2

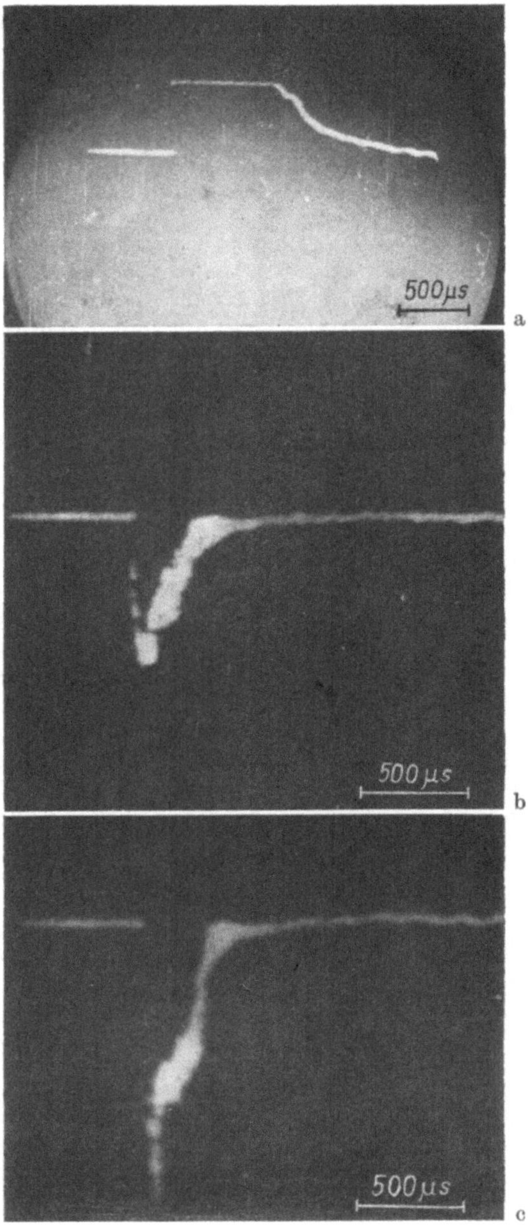

Fig. 3

Fig. 2. Oscillograms of rf emission for a), b) Si and c,) d) GaAs at different radiation power density W (Wm^{-2}): a) 5.6×10^9, b) 3.7×10^7, c) 6.3×10^9, d) 2.9×10^7. The laser operates in a free generation regime

Fig. 3. Oscillograms of rf emission for a Si crystal under laser irradiation with single pulses at different power density W (Wm^{-2}): a) 3.7×10^7, b) 5.7×10^9, c) 8.9×10^9

phys. stat. sol. (a) 85, K1 (1984)
Subject classification: 1.6 and 11; 21

Sektion Physik der Friedrich-Schiller-Universität Jena[1)]

Formation of Pt Silicides by a Millisecond Laser Pulse

By

P. MICHEL, B. WEBER, and G. GÖTZ

Recently, the generation of metal silicides by short temperature pulses became highly interesting. The object of this note is the formation of Pt silicides by irradiation with a millisecond laser pulse.

Thin Pt layers (124 nm) sputter deposited on (100) Si wafers with a rate of 3 nm s^{-1} at a pressure of 0.67 Pa were irradiated in air by one pulse of a free running Nd-glass laser with a pulse length of 1 ms. Three regimes for laser irradiation were realized: irradiation of the frontside of the sample, irradiation of the backside, and irradiation from the frontside through an evaporated antireflection Si layer.

The samples were investigated by means of the RBS technique using 1.4 MeV He^+ ions to determine the concentration profiles /1/. The phase identification is provided by glancing-angle X-ray diffraction using a Read camera /2/.

In the case of frontside irradiation it was seen from RBS spectra that the silicide reaction starts at an energy density of about 90 Jcm^{-2}, but already local melting spots are seen on the surface. At an energy density of about 100 Jcm^{-2} the whole surface is molten. Because of the high reflectivity of the Pt layer (R_{Pt} = 0.8) the threshold energy starting the silicide reaction is very high. The melting spots on the surface promote a laterally inhomogeneous melting process because the energy absorption is higher in the molten spots. X-ray diffraction patterns show that the samples contain the silicide phases PtSi and $Pt_{12}Si_5$ which are surrounded by pure Si. An interesting fact is related to the intensity of diffraction lines at various laser energies, which is a measure for the portion of the phase in the layer. The intensity of the PtSi lines changes with energy, whereas the intensity of the $Pt_{12}Si_5$ lines remains equal. This

1) Max-Wien-Platz 1, DDR-6900 Jena, GDR.

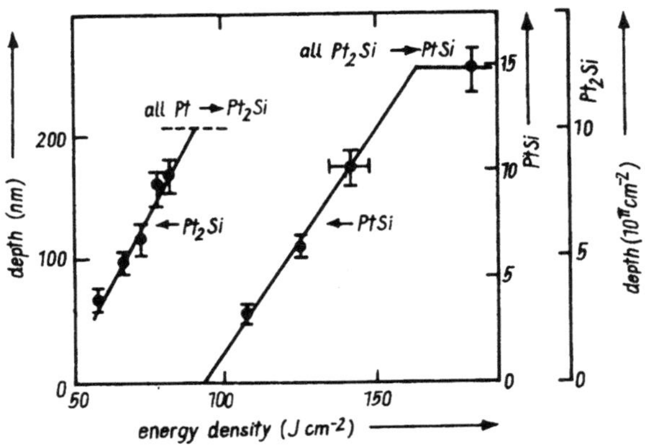

Fig. 1. Growth of Pt_2Si and PtSi layers in dependence on laser energy

gives the idea that the $Pt_{12}Si_5$ formation occurs already at low laser energy densities as the first phase. Summarizing, the formation of a single phase Pt silicide was not possible by frontside irradiation with a ms laser pulse. It is known from the literature (see for instance /3/) that the irradiation with ns pulses always gives rise to multiphase systems. It was shown in /4/ for Pd silicide formation with ms pulses by calculation of the temperature in the silicide layer that a temperature peak exists which affects the nucleation conditions and either leads to a very narrow energy window for a solid phase process or prevents the solid phase process.

In the case of backside irradiation the energy absorption in the relatively thick Si layer ($\approx$ 0.25 mm) results in heating of the whole sample similar to furnace annealing with the except of very short times. The experimental RBS results are shown in Fig. 1, the phase sequence of Pt silicide formation is the following:

- The Pt_2Si phase grows at energy densities between 50 and 90 Jcm^{-2}.
- After conversion of the whole Pt layer into Pt_2Si (at about 100 Jcm^{-2}) the PtSi layer growth starts at the Pt_2Si-Si interface.
- The phase sequence is similar to furnace annealing /3/. The main difference consists in a significantly sharper Pt_2Si-PtSi interface for laser irradiation.
- The X-ray diffraction patterns show that in dependence on laser energy density single phase Pt_2Si and PtSi layers where obtained.

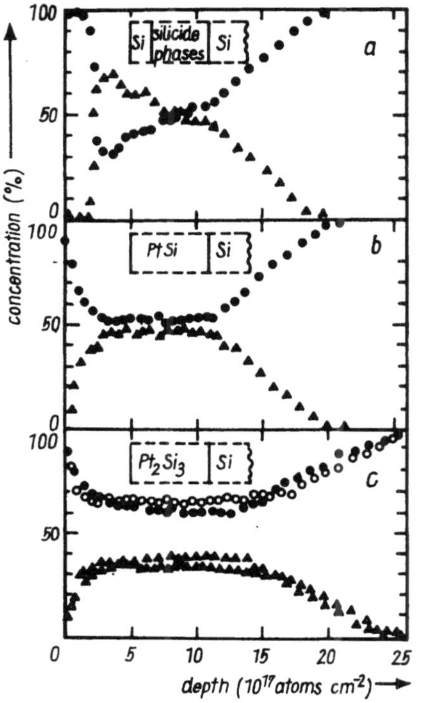

Fig. 2. Depth profiles of Si-Pt-Si systems a) after evaporation of the antireflection Si layer and b), c) after laser irradiation; points: Si, triangles: Pt; b) 56 Jcm^{-2}, c) ● , ▲ 75 J/cm^{-2}, o, △ 98 Jcm^{-2}

To realize the formation of single phase Pt silicides by frontside irradiation, a 40 nm thick Si layer was deposited in front of the Pt layer by e-beam evaporation under UHV conditions whereby the reflection of the layer system with respect to the laser energy is minimized.

The concentration profiles calculated from RBS measurements of the evaporated (a))and laser irradiated (b) 56 Jcm^{-2}, c) 75 and 98 Jcm^{-2}) samples are shown in Fig. 2. It is seen in Fig. 2a that Pt_2Si and PtSi was formed already before evaporation of Si.

The reason for this is a technical one. To get the ultra-high vacuum for evaporation of the Si layer, it is necessary to heat the whole vacuum chamber to a temperature of about 500 K. Thereby the sample is heated, too.

After laser irradiation the following phase sequence was observed:

- At an energy density of about 40 Jcm^{-2} the PtSi layer is completely formed. The PtSi phase is stable up to about 60 Jcm^{-2} (Fig. 2b).

- At about 75 Jcm^{-2} a new phase is produced. The phase is stable up to 100 Jcm^{-2} (Fig. 2c).

For the new phase we found a mean composition of $PtSi_2$ (75 Jcm^{-2}) or Pt_2Si_3 (98 Jcm^{-2}). However, with respect to the equilibrium phase diagram a $PtSi_2$ phase does not exist. Shibata et al. /5/ have found a phase with a mean composition of $PtSi_2$ by irradiation of a Pt-Si layer system with a scanning cw Ar laser in a multimode regime, too (scanning velocity: 12.0 cms^{-1}, focus diameter: 50 μm). They suppose the formation of the superconducting metastable phase Pt_2Si_3. Our X-ray diffraction patterns show diffraction lines which

Table 1

line number	1	2	3	4	5	6
measured distance of lattice planes (nm)	0.332	0.296	0.197	0.167	0.164	0.138
calculated distance of lattice planes (nm)	0.333	0.298	0.194	0.166	0.165	0.136

do not correspond to lines of known silicide phases. The Pt_2Si_3 phase has a hexagonal structure. A calculation of distances of lattice planes yields values corresponding to the lines of diffraction patterns, which are shown in Table 1. Thus, we suppose the formation of the metastable Pt_2Si_3 phase, too.

Summarizing, three possibilities of reacting Pt silicide layers with ms laser pulses were proved. Whereas frontside irradiation leads to multiphase systems, backside irradiation gives single phase silicides in a sequence similarly to furnace annealing. An evaporated Si layer in front of the Pt surface enables the formation of homogeneous stable Pt silicide phases at low energies. At higher energies the metastable Pt_2Si_3 phase was formed.

References

/1/ W.K. CHU, J.W. MAYER, and M-A. NICOLET, Backscattering Spectrometry, Academic Press, New York 1978.

/2/ S.S. LAU, W.K. CHU, J.W. MAYER, and K.N. TU, Thin Solid Films 23, 203 (1974).

/3/ H. MUTA and D. SHINODA, J. appl. Phys. 43, 2913 (1972).

/4/ H.-D. GEILER, F. THRUM, and G. GÖTZ, phys. stat. sol. (a) 70, K159 (1982).

/5/ T. SHIBATA, J.F. GIBBONS, and T.W. SIGMON, Appl. Phys. Letters 36, 566 (1980).

(Received July 6, 1984)

phys. stat. sol. (a) 85, K5 (1984)
Subject classification: 1 and 2; 22.5.2

Department of Chemistry, Faculty of Science, Yamagata University[1)]

The Neutron Diffraction of Molten TlCl, TlBr, and TlI

By
T. SATOW, O. UEMURA, K. HOSHINO, and T. WATANABE

Molten alkali-metal halides are well known to be typical ionic liquids and existing diffraction data have indicated that the short-range arrangement of constituent ions is very similar to that in their solid state /1/. The structural function of these liquids is characterized by a strong Coulomb peak in the unlike ion pair distribution function in r-space and a rapid convergence in the total interference function in K-space. On the other hand, in the case of molten CuCl in which a covalent character is contained in the bonding of unlike atoms a different structural behaviour is found, i.e., the absence of structure in the cation-cation correlation /2/.

Crystalline thallium monohalide (TlX) maintains a body-centred cubic lattice even at high temperatures near the melting point. It is noteworthy to investigate what local order is present in molten TlX. The authors have carried out neutron diffraction measurements on liquid TlCl, TlBr, and TlI, and have attempted to estimate the three partial Tl-Tl, Tl-X, and X-X functions applying the isomorphous substitution approximation.

Measurements were made using an ISSP-ND diffractometer at 10 MW JRR-3 reactor (Japan Atomic Energy Research Institute, Tokai-mura, Japan). The TlX powder was sealed in a thin-walled cylindrical quartz tube of 13 mm inner diameter after evacuating it at 130 °C for several hours. The measuring conditions and the analysis of the intensity data were almost identical to those used in the case of liquid Sb_2Se_3 previously reported elsewhere /3/.

Fig. 1 gives total structure factors S(K) for liquid TlCl, TlBr, and TlI at respective temperatures as a function of K $(=4\pi\sin\Theta/\lambda)$, where b_i, 2Θ, and λ are the neutron scattering lengths of the i-species, the scattering angle, and the wavelengths of the i-species, respectively. It may be supposed from the functional form of these curves that these liquids are rather ionic; for example, the first peak of these curves is considerably higher than the next two

1) Yamagata 990, Japan.

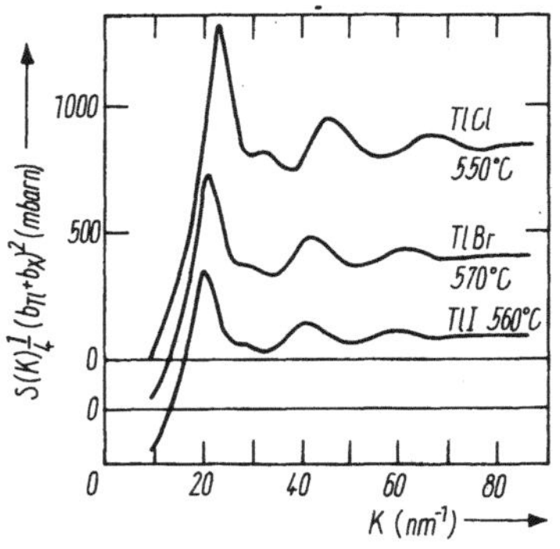

Fig. 1. Total structure factors S(K) for liquid TlCl (550 °C), TlBr (570 °C), and TlI (560 °C) as a function of K ($=4\pi\sin\Theta/\lambda$)

peaks and the S(K) curve becomes damped very quickly at the high angle side in comparison with metallic or covalent liquids. Atomic radial distribution functions (RDF) derived by Fourier transformation from S(K) are shown in Fig. 2. Every RDF has the first maximum at r = 0.313, 0.343, and 0.352 nm, respectively. These values all are slightly smaller than those of their solid state counterparts. The peak maxima in RDF are too broad to estimate the coordination number through the peak area.

Fig. 3 shows normalized structure factors $S(K\ r_1)$ of three molten thallium monohalides, where r_1 represents the first peak position in RDF's as mentioned above. The position of the corresponding peak maxima and minima agrees well between the three modified curves. Then, we have attempted to derive the three partial structure factors, i.e., $S_{Tl-Tl}(K)$, $S_{Tl-Br}(K)$, and $S_{Br-Br}(K)$ using three normalized $S(K\ r_1/r_{1TlBr})$, where r_{1TlBr} is the interatomic distance of molten TlBr. Calculated results of the three partial functions for the molten TlBr are given in Fig. 4a, b. The cation-anion pair correlation function, $g_{TlBr}(r)$, has a sharp first peak completely isolated from the second peak, indicating that a strong attractive force is exerted between unlike ions, whereas the other two are roughly in phase. The ratio of the first peak position for like and unlike ion pair correlations, r_{Tl-Tl}/r_{Tl-Br} (or r_{Br-Br}/r_{Tl-Br}) is nearly 1.4, which is near the value, 1.5, reported in some sodium-chloride type molten salts /1/. This suggests that TlX has a tendency not to retain cesium-chloride type local order, but to take sodium-chloride type order when melting.

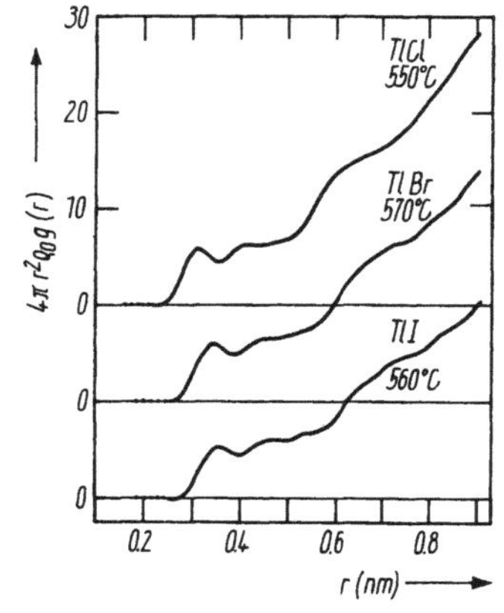

Fig. 2. Atomic radial distribution functions $4\pi r^2 \varrho_0\, g(r)$ for liquid TlCl, TlBr, and TlI

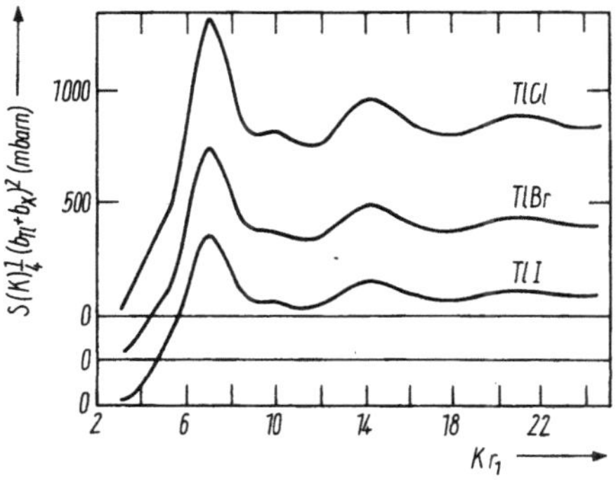

Fig. 3. Normalized structure factors $S(K\,r_1)$ of liquid TlCl, TlBr, and TlI, where r_1 represents the first peak position in the RDF

However, the application of the isomorphous substitution method for the derivation of partial functions has not yet been established in the field of ionic liquids. It may be expected, therefore, that the diffraction study of molten TlX by the isotopic substitution method which can obtain the partial functions without any assumptions will be performed in the near future.

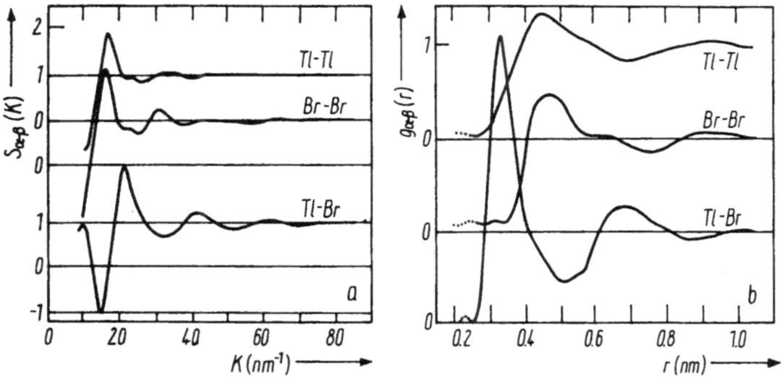

Fig. 4. a) Partial structure factors $S_{Tl-Tl}(K)$, $S_{Tl-Br}(K)$, and $S_{Br-Br}(K)$ and b) pair correlation functions $g_{Tl-Tl}(r)$, $g_{Tl-Br}(r)$, and $g_{Br-Br}(r)$ for liquid TlBr

References

/1/ J.E. ENDERBY and G.W. NEILSON, Adv. Phys. 29, 323 (1980).

/2/ D.I. PAGE and K. MIKA, J. Phys. C 4, 3034 (1971).

/3/ T. SATOW, O. UEMURA, S. AKAIKE, and S. TAMAKI, J. non-crystall. Solids 29, 215 (1978).

(Received April 27, 1984)

phys. stat. sol. (a) 85, K9 (1984)

Subject classification: 1.6; 2; 17; 22.6.1

Central Institute of Inorganic Chemistry (a) and
Central Institute of Electron Physics (b),
Academy of Sciences of the GDR, Berlin[1])

In Vitro Characterization of Bioactivity of Glassy or Glass-Crystalline Implant Materials Using Auger Electron Spectroscopy

By

G. BERGER (a) and V. ATZRODT (b)

Several investigations have shown the ability of special glasses and glass-ceramics to bond directly to living bone without encapsulating by connective tissue /1 to 6/. The formation of a calcium phosphate-rich film is believed to be an indication of the development of a stable bonding between implant materials and bone /7/. The conditions of establishing these films are particularly satisfied in the case of materials which develop an open structure of glass network by aqueous attack of body fluid and have a high corrosion rate. The kinetics of this corrosion or development of the Ca- and P-rich films have been investigated using Auger electron spectroscopy (AES) and other methods /8 to 10/.

The purpose of this short note is to report on compositional modifications of such glasses and glass-ceramics by additions of Al_2O_3 or ZrO_2 which are known to protect the surface corrosion producing a special surface layer.

Composition data of some glasses and glass-ceramics used in this experiments are shown in Table 1. Bulk samples were broken and exposed to 0.2 m tris(hydroxymethyl)aminomethane and hydrochloric acid solution buffered to pH 7.4 (tris buffer solution) at 37 $^{\circ}C$ for 1 h. The ratio of surface area to volume of solution was about 0.1 cm^{-1}. During the exposure period the samples were moved three times per second. After the reaction with the solution, samples were dried and then covered by about 40 nm of carbon to prevent charging up during Auger measurements.

The Auger analysis was carried out using a scanning Auger microprobe. Low beam energy and currents were used (2 keV and 200 to 300 nA, respectively) to prevent radiation damages and to minimize charging up of the sample surfaces. Depth profiling was realized using 1 keV argon ions in raster mode

1) Rudower Chaussee 5, DDR-1199 Berlin, GDR.

T a b l e 1

Compositional range and selected properties of some investigated glasses and glass-ceramics

	compositional range (wt%)						bonding ability to bone (EMA)	determination of the resistance of glass (glass-ceramic) to water attack at 98 °C*)	
	P_2O_5	CaO	SiO_2	MgO, alkali fluoride	Al_2O_3	ZrO_2		ml(0, 01 in HCl)	RC**)
glass-ceramic 1	10 to 15	30 to 35	40 to 50	less than 10 wt%	-	-	+++	2.68	5
glass-ceramic 2					1	-	+++	1.43	4
glass-ceramic 3					3	-	+++	0.68	3
glass-ceramic 4					5	-	++	0.55	3
glass-ceramic 5					-	5	+++	0.75	3
glass 6	12 to 25	50 to 52	23 to 37	-	-	-	+++	2.16	5
glass-ceramic 7				-	-	5	+++	0.78	3
glass 8	71.5	7	-	-	21.5	-	-	0.05	1

*) grain tetration method; (GDR standard TGL 14 809)

**) resistance classification (GDR standard TGL 14809)

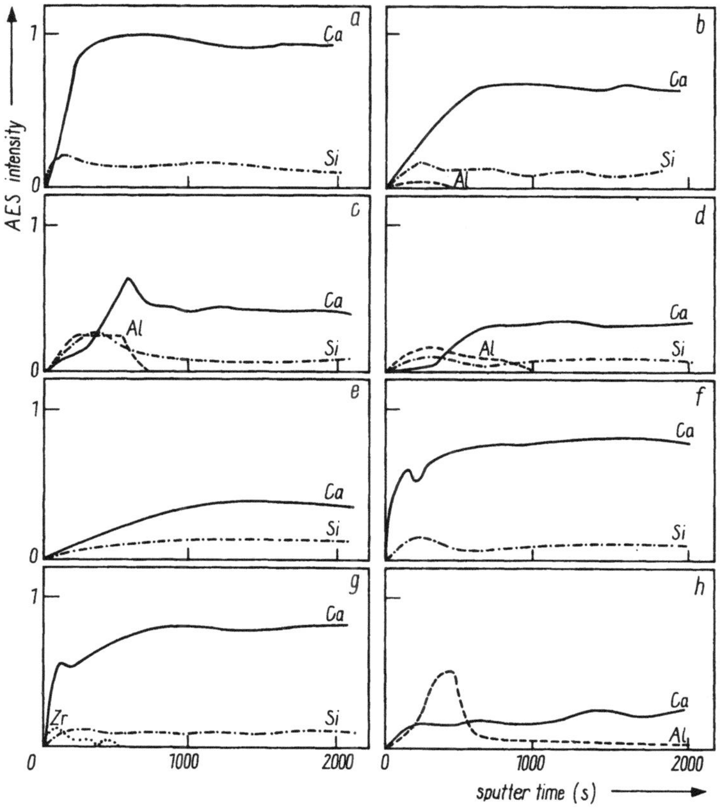

Fig. 1. In-depth profiles of samples (see Table 1) after reaction in tris buffer solution (without carbon over-layers). a) glass-ceramic 1, b) glass-ceramic 2, c) glass-ceramic 3, d) glass-ceramic 4, e) glass-ceramic 5, f) glass 6, g) glass-ceramic 7, h) glass 8

(current density about 45 $\mu A/cm^2$). Compositional profiles from the surface into the bulk of the material were obtained which are shown in Fig. 1. Because it is not necessary to get more than semi-quantitative information, the peak heights have been normalized using the same assumption as suggested by Pantano, Jr., et al. /8/. Furthermore, the results were compared with those of electron microprobe analysis (EMA) which were found by in vivo studies of the bonding ability to bone /5, 11/.

As can be seen from Fig. 1, only glass-ceramics of type 1, 2, and 3 (up to 3 wt% Al_2O_3) had no Al-rich film on the Ca-rich layer which forms the majority of the corrosion zone. In contrast, glass-ceramic 4 and especially glass

8 build up an Al-rich protective layer after treatment with aqueous solution. The peak of the Al-rich film is highest for glass 8 (see Fig. 1), which contains 21.5 wt% Al_2O_3. The bulk compositions were obtained after about 10^3 s sputtering time for glass-ceramics 3, 4, and 5 and even after 700 s for glass 8. This means that Al_2O_3 and ZrO_2 additions lead to smaller corrosion zone thicknesses than expected in comparison with determinations of the resistance of the materials to water attack (see Table 1). Although the water resistance for glass-ceramics 5 and 7 (see Fig. 1) containing 5 wt% ZrO_2 was also higher than for the corresponding glass-ceramic 1 and glass 6 (both without additions of Al_2O_3 or ZrO_2), these samples, as well as glass-ceramics 1, 2, and 3 and glass 6, have shown bonding to living bone.

ZrO_2 is incorporated in the surface layer as well as very small additions of Al_2O_3 (about 1 wt%) in glass-ceramic 2. We believe that this fact is responsible for the high bonding ability of these materials in contrast to the glass-ceramic 4 which shows a smaller degree of direct contact to bone. Glass 8 is a non-bioactive material and there is no direct contact to bone. Determining the bonding ability to living bone afer 12 weeks in vivo tests we have found similar composition profiles at the implant-bone interface but the reaction zones had larger depths (dependent on the chemical compositions: 10 to 60 μm).

Thus, it can be supposed that this method will be possibly expanded to an efficient in vitro test for bioactivity when the samples in the examined compositional range of the system P_2O_5-CaO-SiO_2 get a treatment of only 1 h in stimulated physiological body fluid.

The authors wish to express their gratitude to A. Spitzer and Dipl.-Min. I. Kosche for sample preparation and Dr. H. Lange for helpful discussion.

References

/1/ L.L. HENCH, R.J. SPLINTER, T.C. ALLEN, and T.K. GREENLEE, Jr., J. Biomed. Mater. Res. (Symp.) 2, 117 (1971).

/2/ B. BLENCKE, H. BRÖMER, K. DEUTSCHER, and E. PFEIL, Glaskeramik für Osteoplastik und Osteosynthese, Bundesministerium für Forschung und Technologie, Forschungsbericht T 77-91.

/3/ H.H. KÄS, DT-OS 2, 349 859, C 03 c 3/22.

/4/ G. BERGER, H. MARX, P.-O. KUNTH, R. SAUER, G. SOLOW, ST. KÖHLER, K. RETEMEYER, V. THIEME, and H. HOFMANN, DD-WP A 61 F/25 497 4/3.

/5/ F.G. WIHSMANN, G. BERGER, G. BOCHYNEK, V. THIEME, H. HOFMANN, ST. KÖHLER, and K. RETEMEYER, Wiss. Z. Friedrich-Schiller-Univ. Jena, Math.-Naturwiss. R. 32, 553 (1983).

/6/ W. HÖLAND, K. NAUMANN, W. VOGEL, and J. GUMMEL, Wiss. Z. Friedrich-Schiller-Univ. Jena, Math.-Naturwiss. R. 32, 571 (1983).

/7/ M. OGINO and L.L. HENCH, J. non-crystall. Solids 38/39, 673 (1980).

/8/ C.G. PANTAO, Jr., D.B. DOVE, and G.Y. ONODA, Jr., Appl. Phys. Letters 26, 601 (1975).

/9/ C.G. PANTANO, Jr., D.B. DOVE, and G.Y. ONODA, Jr., J. non-crystall. Solids 19, 41 (1975).

/10/ A.E. CLARK, Jr., C.G. PANTANO, Jr., and L.L. HENCH, J. Amer. Ceram. Soc. 59, 37 (1976).

/11/ G. BERGER, G. BOCHYNEK, and H. POPP, to be published.

(Received July 4, 1984)

phys. stat. sol. (a) 85, K15 (1984)

Subject classification: 1.6; 9; 11; 17; 22

Zentralinstitut für Elektronenphysik
der Akademie der Wissenschaften der DDR, Berlin[1)]

DC Plasma Nitridation of Thin Aluminium Films

By

T. JUNG and L. ROTH

In recent years a variety of methods for deposition of thin dielectric films at low temperatures has been developed, for example plasma enhanced chemical vapour deposition (PECVD) and various kinds of sputter deposition techniques. As an alternative to these techniques conversion of bulk or thin film material into a dielectric by solid/plasma reaction has been studied extensively in selected cases (e.g. oxidation of silicon /1, 2/, aluminium /3/, GaAs /1, 4/). Regarding reactions of solid surfaces with plasma activated nitrogen the present knowledge is much smaller /5 to 9/.

In this note we report on studies of the nitridation of thin aluminium films in a nitrogen dc discharge. Special interest has been directed towards the temperature dependence of the process in the range between 50 and 550 $^{\circ}$C.

The substrates used were n-doped Si ([100], 4Ωcm) and Cr-doped SI-GaAs ([100], > 10^6 Ωcm). No special treatment has been performed prior to inserting into the reaction chamber.

Our experiments were carried out in a stainless steel vacuum chamber which has been pumped down by a turbomolecular pump in combination with a liquid nitrogen trap to a background pressure of 2.5×10^{-6} Pa. After pumping down the system an Al film of 100 nm thickness was deposited by thermal evaporation from a tungsten wire. The thickness was measured by a quartz oscillator. Without breaking the vacuum the chamber then was backfilled by 5N pure nitrogen to a pressure of 130 Pa (1 Torr) and the sample was heated to the desired temperature. Subsequently a dc voltage of 600 V was applied to the sample (cathode) and to the ring-shaped Al anode, which resulted in a gas discharge of some mA. The current density was 0.4 mAcm^{-2} on the average and showed only little variation during the experiments. The nitridation time for all experiments reported here was the same (30 min).

---

1) Rudower Chaussee 5, DDR-1199 Berlin, GDR.

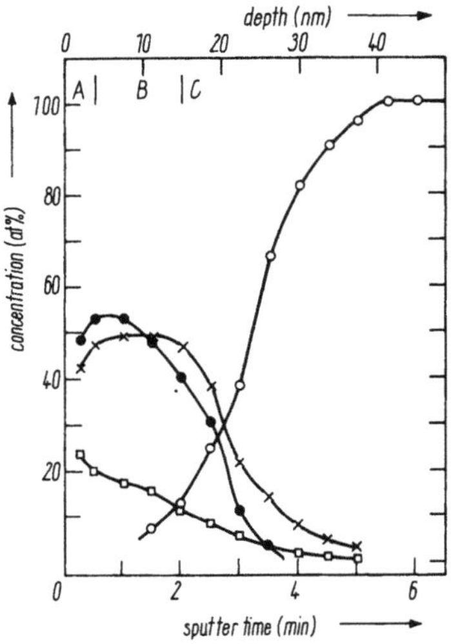

Fig. 1. Auger depth profile of an Al/GaAs sample nitrided at 325 °C for 30 min. ● Al bonded to N, o Al bonded to Al, × nitrogen, □ oxygen

After preparation the samples were analysed by Auger depth profiling in a PHI 590 scanning Auger microprobe. The main analysis parameters were: electron beam voltage 3 kV, current density 40 Am^{-2}, energy resolution 0.6%, modulation voltage 2 V, Ar ion energy 2 kV, beam density 50 μAcm^{-2}. The ion beam was rastered over 2.8x2.8 mm^2. For quantitative evaluation the Al_{LVV} Auger line was used which appears at 68 and 57 eV for unreacted and nitrided aluminium, respectively. Sensitivity factors were taken from /10/. Assuming the elemental sensitivity factors of N, O, Al to be applicable to the given matrix the sensitivity factor of Al in AlN was found to be 0.13 (at 3 keV electron beam voltage). However, a precise evaluation of the Al 57 eV line intensity appeared to be difficult because of interaction first with the 51 eV line (oxide) and later in the profile with the 68 eV line (pure Al) and its satellites. Therefore, the more sensitive 68 eV Auger line has been used for thickness considerations.

A typical Auger depth profile is shown in Fig. 1. We may distinguish between three regions: Region A contains a relatively large amount of oxygen (compared to nitrogen), i.e. it exhibits an oxynitride. Region B appears to be a stoichiometric aluminium nitride with relatively low oxygen content, and region C is characterized by a slow decrease of the nitrogen, oxygen, and reacted aluminium lines. This regions are followed by one of pure Al and finally by the substrate material (not shown here).

In Fig. 2 we have plotted the total film thickness as a function of preparation temperature. Whereas in the region up to 400 °C the film growth rate is rising very slowly, it shows a strong increase at higher temperatures. The depth scales applied in both figures may contain some uncertainty due to differences in the sputter yield of Al and the nitride. From sputtering experiments of pure Al films, however, we have found a nearly equal sputter yield of both materials in the given form.

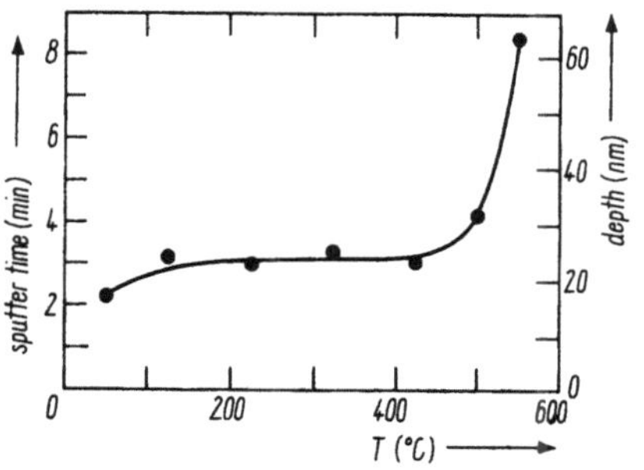

Fig. 2. AlN film thickness of samples prepared for 30 min at different temperatures. Thickness was determined by the half maximum value of the Al 68 eV line in the depth profile

The mechanism of film growth is assumed to be a combination of a low energy implantation of nitrogen ions (note that a nitrogen plasma usually contains almost only positively charged ions) and thermally stimulated diffusion. Field assisted migration we take out of consideration because the nitrogen ions should become neutralized soon after incorporation into the solid film. The strong increase of the film growth rate above 500 °C cannot be understood by simple diffusion theory and should become subject of further studies.

The oxygen content in the films is assumed to be due to the water and oxygen partial pressure of the atmosphere during preparation, but it is not clear whether it originates from the residual gas of the vacuum chamber or from contaminations of the supplied nitrogen gas.

References

/1/ T. SUGANO, F. KOSHIGA, K. YAMASAKI, Q.V. HO, and Y. HIRAYAMA, J. Fac. Engineering, Univ. Tokyo (B), XXXV, 553 (1980).

/2/ A.K. RAY and A. REISMAN, J. Electrochem. Soc. 128, 2460, 2466 (1981).

/3/ L.J. MILES and P.H. SMITH, J. Electrochem. Soc. 110, 1240 (1963).

/4/ R.P.H. CHANG, Thin Solid Films 56, 89 (1979).

/5/ R.P.H. CHANG, C.C. CHANG, and S. DARAK, Appl. Phys. Letters 36, 999 (1980).

/6/ J.I. PANKOVE, J.E. BERKEYHEISER, S.J. KILPATRICK, and C.W. MAGES, J. electronic Mater. 12, 359 (1983).

/7/ L. LÁSKA and J. KODYMOVÁ, Czech. J. Phys. B27, 1219 (1977).

/8/ A.G. SCHROTT and S.C. FAIN, Jr., Surface Sci. 123, 204 (1982).

/9/ A.S. KORHONEN, E.H. SIRVIO, and M.S. SULONEN, Thin Solid Films 107, 387 (1983).

/10/ L.E. DAVIS, N.C. MACDONALD, P.W. PALMBERG, G.E. RIACH, and R.E. WEBER, Handbook of Auger Electron Spectroscopy, 2nd ed., Physical Electronics Industries, 1976.

(Received June 1, 1984)

phys. stat. sol. (a) 85, K19 (1984)
Subject classification: 1.5 and 13.3; 17; 21.1.1

Sektion Physik der Technischen Universität Dresden,
WB Oberflächen- und Elektronenphysik, Dresden[1])

Influence of Work Function Change Due to Oxygen Chemisorption on the Secondary-Ion Emission Probability

By

K. UHLMANN and M. SCHMIDT

In some recent papers (e.g. /1, 2/) it has been shown experimentally that it is possible to describe the dependence of the ionization probability $P^{\pm}$ (for positive and negative secondary ions, respectively) upon the substrate work function Φ by the exponential laws of Nørskov and Lundquist /3/

$$P^{+} \sim \exp\left[-(I - \Phi)/c\, v\right],$$

$$P^{-} \sim \exp\left[-(\Phi - A)/c\, v\right] \qquad (1)$$

(I ionization potential, A affinity of the departing atom, v outward velocity of the sputtered atom, c constant).

We checked the validity of formulas (1) in case of oxygen adsorption on Fe(111) at room temperature. The detailed results of the adsorption studies are shown elsewhere /4, 5/.

Our experiments were carried out in two different ultra-high vacuum systems, one of them equipped with a cylindrical mirror analyzer for Auger electron spectrometry (AES) and a quadrupole mass spectrometer for secondary-ion mass spectrometry (SIMS), the other with a four-grid retarding field analyzer which enables low-energy electron diffraction (LEED), AES, and work function change ($\Delta\Phi$) measurements. The experimental conditions and the results coincide for both devices as shown by comparison of the AES measurements. $\Delta\Phi$ measurements during oxygen chemisorption on Fe(111) have been carried out by Arabczyk and Müssig /6/.

The mass spectrometer used for SIMS was a Q 156 (Riber) without energy filter. For excitation an Ar^{+} ion beam of an energy of 3 keV and with a current density of about 1 nA/cm^2 was applied.

Adsorption experiments were performed by flooding the chamber with oxygen ($p = 1.3 \times 10^{-6}$ Pa) with the ion pump running. Secondary-ion intensities

---

1) Mommsenstr. 13, DDR-8027 Dresden, GDR.

of Fe^+, Fe_2^+, FeO^+, and Fe_2O^+ were recorded separately during oxygen composition. The intensity of the O^- line was only measured after each half Langmuir ($1\ L = 1.33x10^{-4}$ Pa s) of oxygen exposition at a somewhat higher primary current density. The maximum exposition was in all cases 5 L. Because of the relatively low count rates, the statistical error of the measured secondary-ion intensities was about 15% (except for Fe^+ which was 3%). The intensity values used for further calculations were got by smoothing the measured intensity versus dose curves.

Since no absolute values, neither of the ionization probabilities nor of the work function can be found by our methods, equations (1) have to be reduced to

$$\frac{P^+}{P_o^+} = \exp\left(\frac{\Delta\Phi}{c\,v}\right)$$

and

$$\frac{P^-}{P_o^-} = \exp\left(-\frac{\Delta\Phi}{c\,v}\right)\ , \qquad (2)$$

respectively, where $P_o^{\pm}$ denotes the ionization probability on the uncovered surface.

Unlike the experiments of Yu /1/ the coverage could not be limited to a few percent of a monolayer in our case. That means that the influence of the adatoms has to be taken into consideration. In addition some problems arise from the fact that after preparation of the surface prior to adsorption experiments a remaining oxygen coverage Θ_o (in our case ≈ 0.1) and remaining intensities of oxygen containing secondary ions $I_o^{\pm}$ can be observed. However, this oxygen cannot be regarded as being part of the chemisorption layer. It can be expected that after the final heating of the iron sample up to about 600 K only tight-bond oxides exist on the surface. Thus, we found that the residual coverage and residual intensities of oxygen containing ions can be simply subtracted. In this way we got for Fe^+ and Fe_2^+

$$\frac{P^+}{P_o^+} = \frac{I^+}{I_o^+(1 - \Theta_c)} \qquad (3)$$

and for O^-, FeO^+ and Fe_2O^+

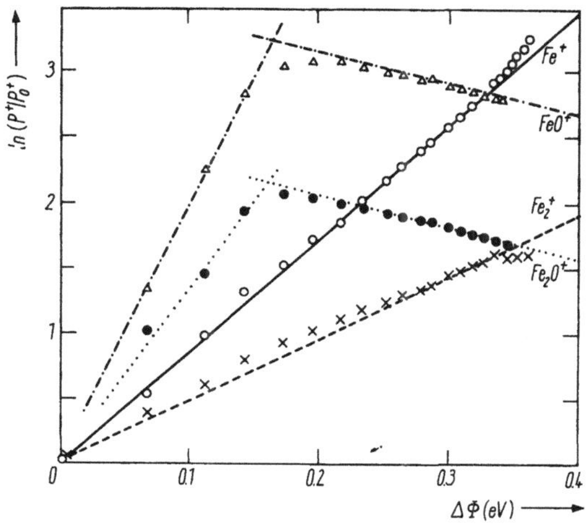

Fig. 1. Plot of $\ln(P^+/P_o^+)$ versus work function change $\Delta\Phi$ due to oxygen chemisorption on Fe(111)

$$\frac{P_-^{\pm}}{P_o^{\pm}} = \alpha \frac{I_-^{\pm} - I_o^{\pm}}{\Theta_c} \quad , \tag{4}$$

where $\Theta_c = \Theta - \Theta_o$ is the corrected coverage, $I_-^{\pm}$ are the measured intensities, and α is a parameter which has to be chosen in such a manner that $P^{\pm}/P_o^{\pm}$ tends to one for $\Theta_c \to 0$. The coverage Θ was calculated from Auger peak-to-peak heights of the O (KLL) peak at 510 eV and the Fe (L_3VV) peak at 703 eV which have been calibrated by the help of a LEED p(1x1) superstructure.

The results of the measurements are shown in Fig. 1. We find for Fe^+ and Fe_2^+ the dependence expected from (2) up to a work function change of 0.33 eV ($\Theta_c \approx 0.75$). The ionization probabilities of the oxygen containing ions, however, seem to follow (2) only up to $\Delta\Phi \approx 0.15$ eV ($\Theta_c \approx 0.35$). With higher coverages a straight line with a negative slope can be found for FeO^+ and Fe_2O^+.

The results for O^- are not plotted in Fig. 1 because of the lack of measured values at low coverages. With higher coverage a behaviour can be found which corresponds to that of FeO^+ and Fe_2O^+.

To sum up, it can be stated that only up to an oxygen coverage of less than half a monolayer the ionization probability changes due to the change of the work function. With higher coverages changes in the electronic structure due to oxide formation already become dominant. Thus, the plot of Fig. 1 can be used as a tool for adsorption investigations. It shows the beginning of the oxidation sooner than other methods. In our case with AES the oxidation could be detected only beginning with a monolayer coverage (regarding the $M_{23}VV$ peak).

References

/1/ M.L. YU, J. Vacuum Sci. Technol. A 1, 500 (1983).

/2/ W. SPECKMANN, S. PRIGGE, and E. BAUER, in: Secondary Ion Mass Spectrometry SIMS III, Springer-Verlag, Berlin/Heidelberg/New York 1982 (p. 128).

/3/ J.K. NØRSKOV and B.I. LUNDQUIST, Phys. Rev. B 19, 5661 (1979).

/4/ M. SCHMIDT, W. KÄTZEL, K. BERNDT, and F. STORBECK, Proc. 8th IVC, 4th ICSS, and 3rd ECOSS, Cannes 1980, Vide, Couches Minces 35, Suppl. 370 (1980).

/5/ M. SCHMIDT, Dissertation A, TU Dresden, 1983.

/6/ W. ARABCZYK and H.-J. MÜSSIG, Thin Solid Films 34, 103 (1976); Dissertation A and B, respectively, TU Dresden, 1977.

(Received July 30, 1984)

phys. stat. sol. (a) 85, K23 (1984)

Subject classification: 1.6

A.N. Sevchenko Scientific Research Institute
for Applied Physics Problems,
V.I. Lenin Belorussian State University, Minsk[1)]

On X-Ray Surface Diffraction

By

S.A. MAKSIMENKO

It has been shown by Baryshevskii /1, 2/ that besides the well known Laue and Bragg geometries in dynamical diffraction there is another one when the initial beam of radiation incidents on a crystal at a grazing angle. The total external reflection of the incident and diffracted rays occurs when the reflected planes are normal to the crystal surface. As a result two waves, besides the initial one, are propagated outside the crystal (Fig. 1) and the reflected radiation intensity is redistributed between these two waves. This phenomenon has been observed by Marra et al. /3/ and is being actively investigated at present /4 to 7/.

The special case, when the Bragg angle Θ_B is about $\pi/2$, has not been considered in previous investigations of the Baryshevskii diffraction scheme. In the ordinary situation ($\gamma = \pi/2 - \Theta_B \approx 1$) the total external reflection condition for both reflected waves can be satisfied only at small angles Ψ between the diffraction plane (DP) and the crystal surface. (These angles are of the same order as the mirror reflection one: 10^{-2} to 10^{-3} rad. DP is formed by the wave vectors of the incident and diffracted waves $\vec{k}_0$ and $\vec{k}_1 = \vec{k}_0 + \vec{\tau}$. The coefficient 2π is included to define the reciprocal lattice vector $\vec{\tau}$.) On the contrary, at $\gamma \ll 1$ this condition is satisfied at arbitrary position of DP relative to the crystal surface (Fig. 2).

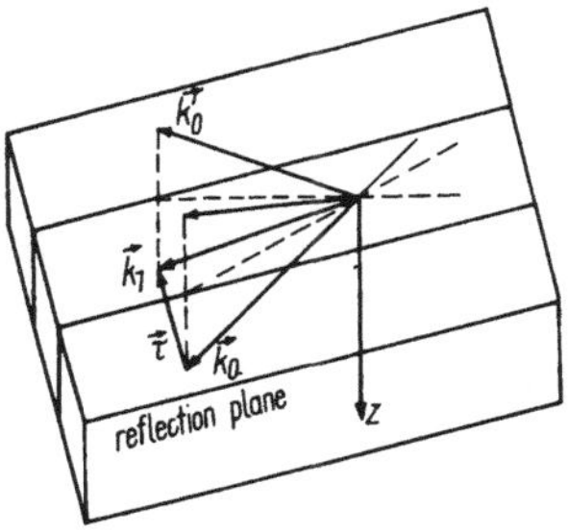

The realization of the proposed diffraction scheme permits us to weaken essentially the dependence of the deviation parameter from the exact Bragg condition $\alpha = \vec{\tau}(2\vec{k}_0 + \vec{\tau})/k_0^2$ upon the crystal

Fig. 1. Surface diffraction scheme

1) 220080 Minsk 80, USSR.

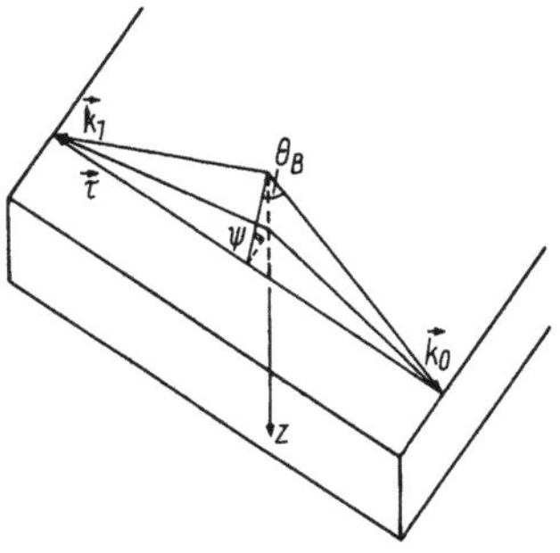

Fig. 2. Diffraction scheme at $\Theta_B \approx \pi/2$

orientation in the surface plane. As follows from simple geometric constructions, at small crystal rotation $\delta\varphi$ around the z-axis (the normal to the surface) from the position when the exact Bragg condition is satisfied, we obtain in second order in $\delta\varphi$:

$$\alpha = 2\left[(\delta\varphi)^2 - \delta\varphi \cos\Psi \sin 2\gamma\right]. \tag{1}$$

When $\sin\gamma \approx 1 \gg \delta\varphi$ formula (1) takes the ordinary form $\alpha = -2\sin 2\Theta_B \Delta\Theta_B$. But the first term in (1) cannot be neglected for $\gamma \ll 1$. Moreover, this one becomes basic when DP is normal to the surface. In this case α takes a square-law dependence on the rotation angle around the z-axis. It is known /5/ that the diffracted-reflected wave intensity curve is kept completely within the range 0.1' to 0.5'. The square-law dependence on $\delta\varphi$ allows to extend this range to the first to third order and to avoid, in such a way, considerable experimental difficulties connected with the extreme angular narrowness of the diffraction peak.

The diffracted-reflected intensity for different values of γ as a function of $\delta\varphi$ is represented in Fig. 3. The calculations were carried out for Si with [440] reflection plane and incidence angle 10'.

The formula obtained in /1, 2/ for the amplitude of the diffracted-reflected wave was used:

$$B = \frac{2(k_{1z} - k_{2z})c_1 c_2}{(k_{0z} + k_{2z})(k'_{0z} + k_{1z})c_1 - (k_{0z} + k_{1z})(k'_{0z} + k_{2z})c_2},$$

where

$$c_{1,2} = \frac{2g_{21}}{p \pm \sqrt{p^2 + 4g_{12}g_{21}}}, \quad p = g_{11} - g_{22} + \alpha.$$

The normal components of wave vectors inside a crystal can be found from the dispersion equation and have the following form:

$$k_{1(2)z} = k_0\left(\frac{k_{0z}^2}{k_0^2} + \frac{g_{11} + g_{22} - \alpha}{2} \pm \sqrt{p^2 + 4g_{12}g_{21}}\right)^{1/2}.$$

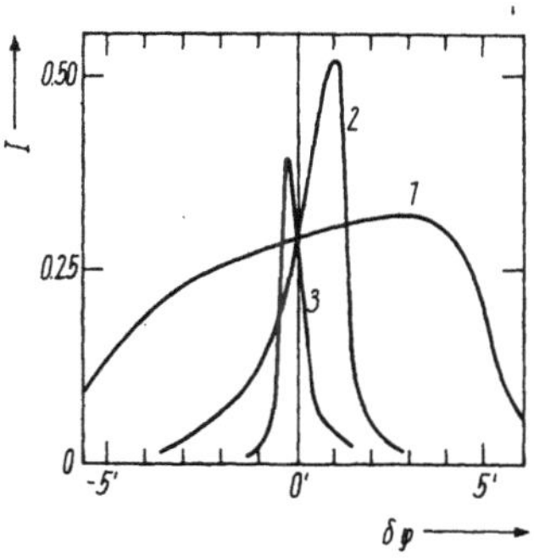

Fig. 3. Plots of the diffracted-reflected intensity for various $\Delta k/\tau$ = (1) 10^{-5}, (2) 10^{-4}, (3) 10^{-3}

The expression for the wave vector normal component of the diffracted-reflected wave follows from the energy conservation law for elastic scattering:

$$k'_{0z} = \sqrt{k_0^2 - k'^2_{0t}} = \sqrt{k_{0z}^2 + \alpha k_0^2}, \quad \vec{k}'_{0t} = \vec{k}_{0t} + \vec{\tau},$$

k_{0z}, $\vec{k}_{0t}$ are the normal and tangential components of $\vec{k}_0$, g_{ij} are the known coefficients of two-wave diffraction.

For the case of equality of g_{ij} the above formulas can be found in /5/, for example. The possibility of precise α measurements of the diffraction peak following from the connection $k'_{0z} = k'_{0z}(\alpha)$ has been discussed in the same paper. The diffraction scheme proposed in the present note is an alternative way for its realization.

It should be noted in conclusion that the condition $\gamma \ll 1$ gives a rigid restriction on the incident wave vector value: $k_0 = \tau/2 + \Delta k$, $\Delta k/\tau \ll 1$. The latter creates some difficulties connected with the selection of the X-ray source which should have a sufficient spectral density in a definite region.

References

/1/ V.G. BARYSHEVSKII, Zh. eksper. teor. Fiz., Pisma 2, 112 (1976).

/2/ V.G. BARYSHEVSKII, Nuclear Optics of Polarized Media, Belorussian State University, Minsk 1976 (in Russian).

/3/ W.C. MARRA, P. EISENBERGER, and A.Y. CHO, J. appl. Phys. 50, 6927 (1979).

/4/ H. VINEYARD, Phys. Rev. B 26, 4146 (1983).

/5/ P.A. ALEKSANDROV, A.M. AFANASIEV, M.K. MELKONYAN, and S.A. STEPANOV, phys. stat. sol. (a) 81, 47 (1984).

/6/ R.M. IMAMOV and A.L. GOLOVIN, phys. stat. sol. (a) 77, K91 (1983).

/7/ M.A. ANDREEVA and R.N. KUZMIN, Solid State Commun. 49, 743 (1984).

(Received July 13, 1984)

phys. stat. sol. (a) 85, K 27 (1984).
Subject classification: 12.2 and 20; 22.5.4

Department of Physics, University College of Science, Osmania University, Hyderabad[1]

Strain-Optical Constant (p_{44}) of Mixed Crystals of KCl-KBr of Equimolar Concentration

By
G.S. KUMAR, S. SUDHAKAR, R. ETHIRAJ, and V.G. KRISHNA MURTHY

Recently, Kumar et al. had determined the piezo-optic constants of mixed crystals of KCl-KBr /1/ and KBr-KI /2/. A theory of piezo-optic birefringence characterised by strain-optical constants (p_{11} - p_{12}) for mixed crystals of equimolar concentration of NaCl structure was developed by Kumar et al. /3/. This theory is now extended to the study of birefringence characterised by the constant p_{44} for mixed crystals of KCl-KBr of equimolar concentration and is valid for any mixed crystal having the same structure and composition.

A crystal with its edges parallel to [$\bar{1}$10], [001], and [110] (to be called x', y', and z', respectively) is stressed along z' and observation is made along y'. The strain-optical constant p_{44} and induced refractive index changes are related to strain ε' through the following expression based on Pockel's theory /4/:

$$n_{z'} - n_{x'} = -p_{44} \frac{n^3}{2} \varepsilon' \qquad (1)$$

where n is the refractive index in the unstrained condition, $n_{x'}$ and $n_{z'}$ are the altered refractive indices for light vibrating in x' and z' directions.

Attributing the changes in refractive indices to

i) change in density of the crystal,

ii) Lorentz-Lorenz anisotropy due to material continuum,

iii) Lorentz-Lorenz anisotropy due to dipoles within the assumed cavity,

iv) Coulomb anisotropy due to charges within the cavity, and following the same procedure as Kumar et al. /3/, an expression for the birefringence was developed and is given by

1) Hyderabad 500 007, India.

$$dn_{z'} - dn_{x'} = \left[-\frac{n^2-1}{5n} - \frac{1+\sigma_{x'}}{n}\left(1+\frac{4\pi}{f}\right)\left\{0.1067\,Nr^3(n^2+2)^2u^2 + \right.\right.$$

$$\left.\left. + 0.3201(n^2+2)(n^2-1)u + \frac{0.0844(n^2-1)^2}{Nr^3}\right\}\right] +$$

$$+ \frac{N(n^2+2)^2}{72\pi nr^3}\left[\frac{3}{\sqrt{2}}\left(1+\frac{4\pi}{f}\right)\left\{\alpha_{Cl}\alpha_{Br} - \frac{\alpha_{Cl}^2+\alpha_{Br}^2}{2}\right\} + \right.$$

$$\left. + \frac{4\sqrt{2}\pi\alpha_k r}{f}(s_1\alpha_{Cl} - s_3\alpha_{Br})\right] +$$

$$+ \frac{N(n^2+2)^2}{18\,Nr^3}\left[\left(\frac{\alpha_k r}{f}\,\frac{3}{\sqrt{2}}\sigma_{x'} - \frac{1}{\sqrt{2}}\right)\left(s_1\alpha_{Cl} - s_3\alpha_{Br}\right)\right]\varepsilon', \qquad (2)$$

where N number of molecules per unit volume, r interionic distance, $u = \alpha^+(\alpha^-)/r^3$, f oscillator strength, s_1, s_2, and s_3 are the displacements of excess electric charges associated with Cl^-, Br^-, and K^+ ions, α polarizability.

From (2) it may be observed that the birefringence ($dn_{z'} - dn_{x'}$) does not become zero in the undeformed state ($\varepsilon' = 0$), but is equal to

$$\frac{N(n^2+2)^2}{72\pi nr^3}\left[\frac{3}{\sqrt{2}}\left(1+\frac{4\pi}{f}\right)\left\{\alpha_{Cl}\alpha_{Br} - \frac{\alpha_{Cl}^2+\alpha_{Br}^2}{2}\right\} + \right.$$

$$\left. + \frac{4\sqrt{2}\pi\alpha_k r}{f}(s_1\alpha_{Cl} - s_3\alpha_{Br})\right]. \qquad (3)$$

However, if we consider an adjacent unit cell to the one considered, the birefringence in the undeformed state works out to be opposite in sign but equal in magnitude. Thus the net birefringence (3) for the crystal as a whole turns out to be zero in the undeformed state. Using (1) and (2), the strain-optic constant p_{44} is given by

$$p_{44} = -\frac{2C_{44}}{n^4}\,\frac{1+\sigma_{x'}}{E'}\left[-\frac{(n^2-1)^2}{5} - \left\{0.1067\,Nr^3(n^2+2)^2u^2 + \right.\right.$$

$$\left.\left. + 0.3201(n^2+2)(n^2+1)u + \frac{0.0844}{Nr^3}(n^2-1)^2\right\}(1+\sigma_{x'})\left(1+\frac{4\pi}{f}\right)\right], \qquad (4)$$

Table 1

Variation of polarizabilities with wavelength

wave-length (nm)	refrac-tive index n	Brewster constant $-C_{110}$ (10^{-12} m^2/N)	p_{44}	polariza-bility of K^+ α^+ (10^{-30} m^3)	effective polariza-bility of the negative ions α^-_{eff} (10^{-30} m^3)
600	1.510	7.15	0.023	0.75	4.17
560	1.512	7.10	0.023	0.75	4.19
520	1.516	7.04	0.023	0.75	4.22
480	1.520	6.94	0.022	0.75	4.25
450	1.525	6.77	0.021	0.75	4.30
400	1.535	6.45	0.020	0.75	4.37
360	1,547	6.16	0.019	0.75	4.47
320	1.566	5.85	0.017	0.75	4.61
270	1.609	5.38	0.014	0.77	4.92

where

$$E' = \frac{4(C_{11} - C_{12})(C_{11} + 2C_{12})C_{44}}{C} ,$$

$$\sigma_{x'} = \frac{2C_{11}C_{44} - (C_{11} - C_{12})(C_{11} + 2\,C_{12})}{C} ,$$

and

$$C = 2\,C_{11}C_{44} + (C_{11} - C_{12})(C_{11} + 2\,C_{12}) .$$

C_{11}, C_{12}, and C_{44} are the elastic constants of the crystals.

Using the values of p_{44} calculated from the data of C_{110} by Kumar et al. /2/ and elastic constants C_{11}, C_{12}, and C_{44} by Slagle and Mc Kinstry /5/, the polarizability of positive and effective polarizability of negative ions are evaluated at different wavelengths and are given in Table 1.

The effective polarizability of the negative ions increases with decrease of wavelength as was observed in the case of pure crystals by Ethiraj et al. /6/. The variation of the polarizability of the positive ion is, however, small. The polarizabilities of Cl^- and Br^- ions in KCl and KBr for sodium light are 3.413

and 4.662, respectively, as reported by Ethiraj and Bansigir /7/. The effective polarizability of the negative ion obtained in the present investigation is comparable with them.

Our thanks are due to Prof. K.G. Bansigir, Jiwaji University, Gwalior, India for useful discussions.

References

/1/ G.S. KUMAR, R. ETHIRAJ, and V.G. KRISHNA MURTY, Acta cryst. A35, 857 (1979).

/2/ G.S. KUMAR, S. SUDHAKAR, R. ETHIRAJ, and V.G. KRISHNA MURTY, J. Mater. Sci. Letters, in the press.

/3/ G.S. KUMAR, R. ETHIRAJ, and V.G. KRISHNA MURTY, Acta cryst. A36, 530 (1980).

/4/ F. POCKELS, Lehrbuch der Kristalloptik, Teubner, Berlin 1906.

/5/ O.D. SLAGLE and H.A. MCKINSTRY, J. appl. Phys. 38, 446 (1967).

/6/ R. ETHIRAJ, V.G. KRISHNA MURTY, and K.G. BANSIGIR, Acta cryst. A34, 318 (1978).

/7/ R. ETHIRAJ and K.G. BANSIGIR, Acta cryst. A29, 157 (1973).

(Received June 27, 1984)

phys. stat. sol. (a) 85, K31 (1984)
Subject classification: 11 and 20; 22.1.2; 22.2.1

Institute of Nuclear Physics, Tomsk Polytechnical Institute[1]

Observation of RF Pulses from Solids during Laser Irradiation

By

A.D. POGREBNYAK, SH.R. MASTOV, M.F. KUZNETSOV, K.P. AREFIEV, S.V. RAKITIN, and S.A. VOROBIEV

The present note demonstrates a new effect of electromagnetic radiation emission in the radio-frequency range, while irradiating solids by powerful pulses of a laser radiation. In the experiments we used a pulsing ruby laser of $\lambda = 694.3$ nm wavelength. Semiconductors, such as p-type Si (BKBD-500) and n-type GaAs:Te (the Te concentration was about 2.3×10^{16} cm^{-3}), with polished surfaces were used as samples. Their dimensions were 20x20x3 mm^3. All the experiments have been carried out under normal conditions. Laser irradiation of the samples has been performed both by single pulses of $\tau = 30$ ns duration and by pulse trains with the duration of t = 350 μs. In the last case the duration of every pulse was $\tau \approx 2.5$ μs and the number of pulses was from 12 to 15. The laser beam was focused on the surface of the investigated materials using a lens of F = 10 cm focal distance, that permitted adjusting of radiation power density.

The scheme of the experimental set-up is shown in Fig. 1. The recording channel for electromagnetic pulses consisted of a variable-capacitance transducer - a ring serial joined up by coaxial cable RK-50 to an amplifier output. The amplifier had a bandwidth from 20 Hz to 1.0 MHz.

The signal from the amplifier output was applied to the oscilloscope S8-1 being in storing operation. The sensitivity of the recording channel was about

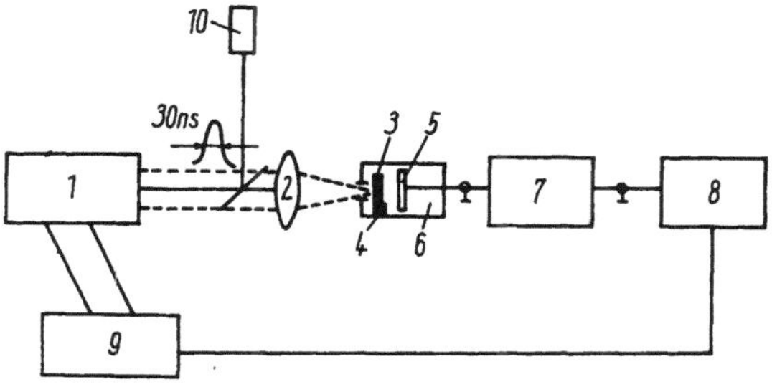

Fig. 1. Scheme of the experimental set-up: (1) ruby laser, (2) lens, (3) sample, (4) insulator, (5) antenna, (6) screen, (7) amplifier, (8) oscilloscope, (9) multiple-channel generator, (10) calorimeter

1) P.O. Box 25, 634050 Tomsk, USSR.

10 μV. The beam energy of the laser radiation has been measured by calorimeter JKT-2M, the range of which is from 0.01 to 150 J. The stability of the laser pulse energy was not poorer than ±5%. Synchronization of the pump lamp and electrooptical laser gate (1) with the recording oscilloscope (8) has been done by means of a multiple-channel generator with controlled time delays (9). To exclude the effect of external fields on the results of measurements, the capacitance pick-up and samples under study were placed into a grounded metal cell with a window for transmission of the laser beam. The laser has been placed in a double metal screen.

Fig. 2 (see on the photo pages before the Short Notes part) shows the oscillograms of rf emission from Si (a), b)) and GaAs (c), d)) crystals being observed after exposure of the laser radiation pulse train to crystals for different radiation power densities $W(Wm^{-2})$. Here $W = E/st$, E is the energy of the laser beam, s is the area of the irradiated surface, t is the laser irradiation duration. Fig. 3 (see on the photo pages before the Short Notes part) represents the oscillograms of rf emission pulses when exposing a single laser pulse with $\tau = 30$ ns to a GaAs crystal for different radiation power densities.

From a comparison of rf emission data for laser irradiation with the pulse train with single pulse irradiation of the samples one can discuss the obtained oscillograms in the following way. In the case of small values of the laser power density W, corresponding to near-threshold values of rf emission (oscillograms b), d) in Fig. 2) a group of short rf pulses of negative polarity is observed. At high W values the observed rf pulses are more complicated (oscillograms a), c) in Fig. 2). At the first moment of pulse train irradiation, i.e., while affecting the first part of the pulse train, an effect being similar to the case of small W is observed. Then, when supplying the pulse energy into the starting process, transient effects are observed, i.e., the superposition of rf emission from a few single pulses of laser radiation. Further, under laser pulse train exposure, an rf emission signal will be detected being typical for every material under study. The given rf signal is, apparently, connected with the kinetic processes of cracking in the materials under study. Similar effects have appeared with mechanical failure of brittle materials /1/.

In the case of exposure of short laser pulses ($\tau = 30$ ns) at small near-threshold radiation power density W, an rf pulse of negative polarity is observed (oscillogram a) in Fig. 3). As the power density increases, a pulse of positive polarity appears, which increases with the rise of W (oscillograms b), c) in Fig. 3). The shape and the amplitude of the observed rf pulse is strongly different for each of the investigated materials. The characteristics of the observed rf emission depend also on the surface strength of materials.

References

/1/ YU. P. MALYSHKOV, V.F. GORDEEV, V.P. DMITRIEV, V.A. SMIRNOV, T.V. FURSA, and V.I. ULCHENKO, Zh. tekh. Fiz. 54, 336 (1984).

(Received July 17, 1984)

phys. stat. sol. (a) 85, K35 (1984)

Subject classification: 10.2 and 11; 22.1.2

Zentralinstitut für Elektronenphysik
der Akademie der Wissenschaften der DDR, Berlin[1] (a),
Zentralinstitut für Kernforschung
der Akademie der Wissenschaften der DDR, Dresden (b), and
Central Research Institute, Hungarian Academy of Sciences, Budapest (c)

Damage and Trapping Behaviour of Crystalline Silicon at Low Energy Deuterium Implantation

By

D. HILDEBRANDT (a), H. STRUSNY (a), R. GROETZSCHEL (b), E. KOTAI (c), and F. PASZTI (c)

Introduction Effects which occur during irradiation of solids with light ions at low energies (< 1 keV) have attracted considerable interest in the fields of fusion research and semiconductor devices. In plasma-wall interaction they can influence processes which play an important role in the boundary layer of Tokamak devices.

Furthermore such radiation effects can be used to determine plasma properties, if special probes are applied. For this purpose mainly carbon and silicon single crystal samples are exposed in the plasma boundary and thereupon investigated by different surface analysis methods.

The trapping and damage behaviour of silicon has been used for energy and flux measurements of plasma particles /1 to 5/. Therefore a detailed knowledge of these effects at low energy bombardment of deuterium is necessary.

Trapping of deuterium implanted into silicon with fluences from 3×10^{15} to 3×10^{18} D/cm^2 was studied by Staudenmaier et al. /1/ in the energy range from 50 to 1000 eV and Doyle et al. /6/ at 1500 eV.

Some lattice damage investigations of silicon at similar irradiation parameters were carried out by Withrow et al. /3, 4/. The present work was undertaken to complete the existing data of damage and to compare the trapping and damage behaviour of low energy deuterium-implanted silicon crystals.

Experiment Si⟨111⟩ samples of high purity (10 Ωcm, n-type) were irradiated at approximately normal incidence with D_2^+ ions at energies from 25 to 2500 eV per nucleon and at fluences from 5×10^{15} to 5×10^{17} D/cm^2.

---

1) Hausvogteiplatz 5-7, DDR-1086 Berlin, GDR.

The implantation was carried out with a Varian ion source without mass separation in an UHV chamber at a deuterium pressure of 10^{-3} Pa. Background pressures lower than 10^{-6} Pa have been obtained using a turbomolecular pump. The contribution of other ionized species than D_2^+ in the ion beam should be negligible as checked by a quadrupole mass spectrometer. Current measurements were carried out with a Faraday cup. To achieve considerable fluences at energies lower than 100 eV we used a post-deceleration method. In order to check the influence of the energy distribution of the primary ions and neutrals the ion beam was post-decelerated to zero energy. The damage effect at this zero energy was negligible.

The lattice damage of the implanted Si samples was investigated by 1 MeV $^4He^+$ Rutherford backscattering in channeling geometry with a depth resolution of 4.5 and 20 nm at a scattering angle of 97° and 135° using the Van de Graaff accelerators inthe KFKI Budapest and in the ZfK Rossendorf.

The number of displaced atoms per unit volume $N_d(x)$ in the depth x was calculated using the procedure according to Eisen /7/. The integral of $N_d(x)$ over the depth of the disorder concentration gives the amount of displaced atoms projected per unit area of the sample surface. This number was also calculated by the surface approximation method /7/. Both procedures give the same results for our experiment.

The amount of trapped deuterium in the samples was determined by elastic recoil detection (ERD) with 3 MeV $^4He^+$ ions /8/.

Results The results of our measurements of the amount of retained deuterium as a function of the implantation fluence at different incident ion energies are illustrated in Fig. 1. Here are also shown results of Staudenmaier et al. /1/ and Doyle et al. /6/.

Our results are very similar to those of Staudenmaier et al., whereas the measurements of Doyle et al. yielded a somewhat higher eetention.

At high fluences the amount of retained deuterium satturates. A similar saturation behaviour was also found for the damage induced by the deuterium bombardment in the same samples. Fig. 2 shows the dependence of the amount of damage on the implantation fluence together with curves deduced from measurements of Withrow et al. /4/. In a first approximation the results are consistent.

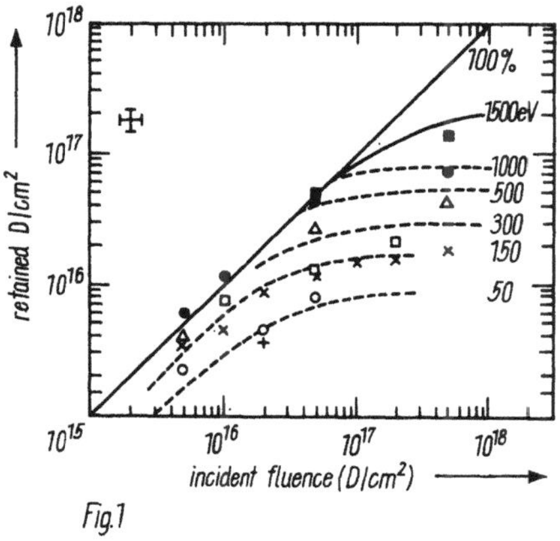

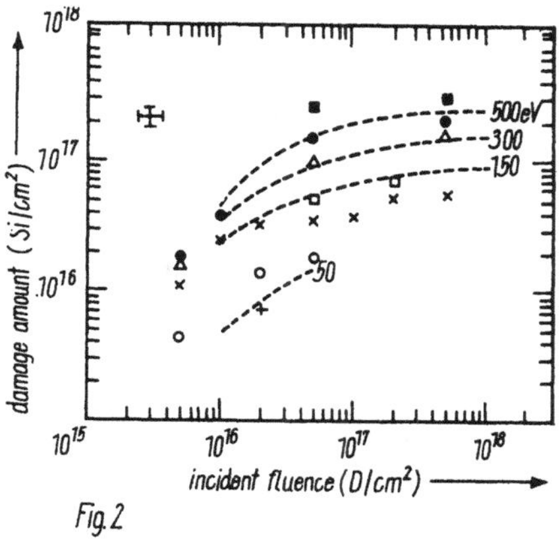

Fig. 1. The amount of retained deuterium as a function of the implantation fluence at different incident ion energies; this work: + 25, ○ 50, × 100, □ 200, △ 400, ● 1000, ■ 2500 eV; ---- /1/ —— /6/

Fig. 2. The dependence of the amount of damage on the implantation fluence at different incident deuterium energies; this work: + 25, ○ 50, × 100, □ 200, △ 400, ● 1000, ■ 2500 eV; ---/4/

If we compare the amount of damage and the retained deuterium in the same samples we get a mean ratio of 3.3 displaced Si atoms per retained deuterium atom. This value is nearly independent of the incident energy. Assuming the same locations for deuterium and damage in Si and a complete disordered layer /9/ we can conclude from the value 3.3 displaced Si atoms per retained deuterium atom a 30% saturation concentration of deuterium in Si.

The strong correlation of damage and trapping suggests that the induced damage is not only determined by energy deposition in the collision cascade. Thus the amount of energy deposited by elastic collisions should decrease with decreasing incident energy from 2500 to 25 eV. But the damage yield per trapped deuterium atom is nearly constant.

We assume that the effective displacement energy is lowered by the incorporation of deuterium in the Si lattice, which also leads to a pronounced volume expansion /10/. The known displacement energy of silicon of about 14 eV /11/ can also not explain damage creation at the lowest incident energies, because the maximum energy transferred by a deuterium atom to a Si atom is about 1/4 of the impact energy. Thus the effective displacement energy should be lower than 10 eV. In order to examine this hypothesis further investigations with hydrogen are planned.

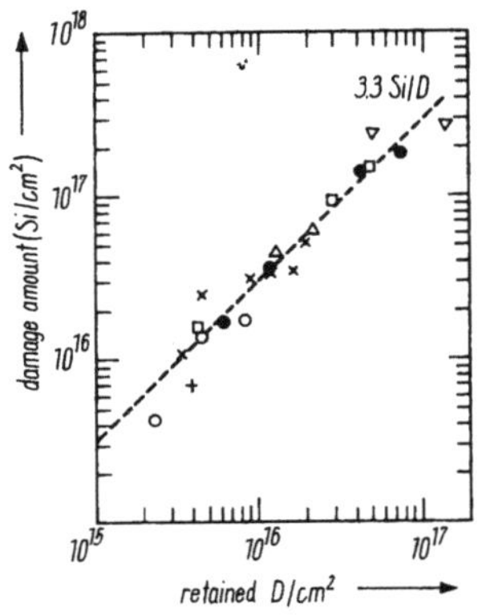

Fig. 3. Comparison of the amount of damage and retained deuterium; + 25, o 50, × 100, Δ 200, □ 400, ● 1000, ∇ 2500 eV

The similar saturation behaviour of trapping and damage in Si can be useful if hydrogen fluxes of low energies are investigated. In this case hydrogen damage investigations can give more accurate results than hydrogen trapping measurements because adsorption layers of hydrogen at the surface can cause serious problems.

The authors are indebted to Dr. H. Wolff for a critical reading of the manuscript and valuable comments.

References

/1/ G. STAUDENMAIER, J. ROTH, R. BEHRISCH, J. BODANSKY, G. ECKSTEIN, P. STAIB, S. MATTISON, and S.K. GRENTS, J. nuclear Mater. 84, 149 (1979).

/2/ S.A. COHEN and G.M. MC CRACKEN, J. nuclear Mater. 84, 157 (1979).

/3/ S.P. WITHROW, R.A. ZUHR, J.B. ROBERTO, R.B. APPLETON, and M.T. ROBINSON, J. nuclear Mater. 98, 231 (1981).

/4/ S.P. WITHROW, R.A. ZUHR, and J. ROTH, J. Vacuum Sci. Technol. 20, 1323 (1982).

/5/ H. GROTE, D. HILDEBRANDT, H. STRUSNY, N. FRIED, F. PASZTI, and G. VIZKELETHY, ZIE-Preprint 83-5, Berlin 1983.

/6/ B.L. DOYLE, W.R. WENDLER, B.K. BRICE, and S.T. PITRAUX, J. nuclear Mater. 93/94, 551 (1980).

/7/ W.K. CHU, J.W. MAYER, and M-A. NICOLET, Backscattering Spectrometry, Academic Press, New York 1978.

/8/ H. GROTE, D. HILDEBRANT, H. STRUSNY, E. KOTAI, A. MANUABA, and F. PASZTI, ZIE-Preprint 83-4, Berlin 1983.

/9/ D. HILDEBRANDT, to be published.

/10/K. WITTMAACK and G. STAUDENMAIER, J. nuclear Mater. 93/94, 581 (1980).

/11/ J.J. LAFERSKI and P. RAPPAPORT, J. appl. Phys. 30, 1296 (1959).

(Received July 24, 1984)

phys. stat. sol. (a) 85, K39 (1984)

Subject classification: 10.2 and 11; 22.1.2

Institute of Semiconductor Physics, Academy of Sciences of the USSR, Siberian Branch, Novosibirsk[1] (a)
Institute of Nuclear Physics, Moscow State University (b), and
Moscow State University (c)

Defect Structure Study with Planar Channeling in Pulse-Annealed Ion-Implanted Silicon

By

A.V. DVURECHENSKII (a), B.P. KASHNIKOV (a), G.P. POKHIL (b), V.P. POPOV (a), A.V. TULINOV (b), and A.A. TURINGE (c)

The present paper offers a study of the impurity distribution function in the cell of a highly doped silicon crystal. We explored samples implanted with arsenic with an energy E_{As} = 100 keV up to a concentration of ≈ 1 at%. Recrystallization of the implanted layer was performed with electron pulse annealing. The electron energy was 10 keV, pulse duration 500 ns, and energy density 2J/cm^2 /1/. Besides, in order to eliminate the defects due to quenching we thermally annealed the crystal during 2.5 s up to 1100 °C.

In calculations of the impurity distribution function we used the method based on measurements of the backscattering yield on impurity atoms under planar channeling in a thin crystal layer with thickness of the order of half of a wavelength of the vibrational motions of a particle in the channel /2/. The basis of the method consists in the search of an approximate solution of the following equation by the method of reduction /3/:

$$Y(\Theta, x) = \hat{K} \hat{M} f(y) + \nu \qquad (1)$$

which relates the function of interest f(y) with the experimentally measured backscattered yield $Y(\Theta, x)$ as a function of particle incidence angle on the crystal (Θ) and depth x at which the particle is scattered on the impurity atom. Here $\hat{M}$ is obtained from a physical model that provides the probability of particle large angle scattering at depth x when the particle flow falls at an incidence angle Θ to the planar channel; $\hat{K}$ is the apparatus distortion of the scattering probability measured with finite energy and angular resolution; ν is the statistical noise which is supposed to have a normal distribution.

1) prospekt Akademika Lavrenteva 13, 630090 Novosibirsk, USSR.

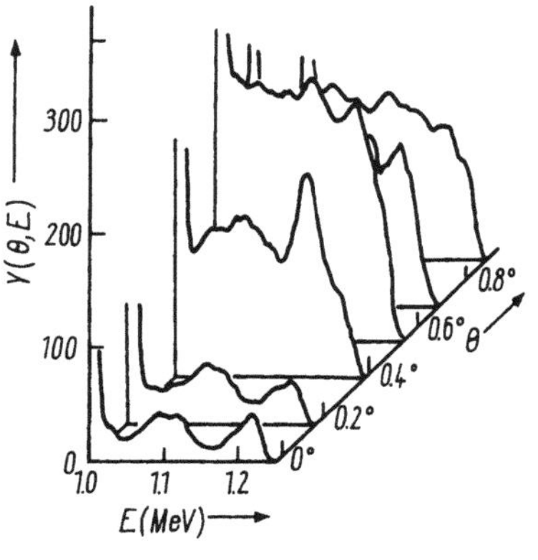

Fig. 1. Dependence of the backscattering yield (TS) of He particles scattered on As atoms at the angle (Θ) between the (110) plane and the He ions beam

Unlike the conventional measurement of the impurity position with channeling /4/, the discussed method /2/ enables one to obtain a detailed distribution function of the impurity along the channel cross-section.

We measured the backscattering of 1.4 MeV He^+ ions on our samples. The ion beam with an angular divergence $\leq 0.02^o$ is incident normally on the crystal whose planar channel is close to the beam direction. A detector with energy resolution E ≈ 11 keV was mounted at a glancing angle of ≈ 8.0^o to the surface to improve the depth resolution Δx, which was ≈ 4 nm. The (110) plane was slightly tilted to the scattering plane. During measurements of the angular dependence of the yield of scattered particles the crystal was tilted around the axis formed by intersection of the crystal surface with the scattering plane. Fig. 1 shows the measured energy spectra of particles scattered on atoms at different angles of the beam incidence on the crystal for the electron-annealed sample. Similar spectra have been obtained for particles scattered on As and Si for samples after electron annealing and after additional thermal annealing.

The experimentally measured backscattering yield Y(Θ, E) was converted according to a linear law to Y(Θ, x), then an approximate solution of (1) by the algorithm received in /2/ was found. We have used an apparatus function as measured at the same target and in the same geometry. In our case the method of reduction enabled us to obtain an impurity distribution function it would have when measured in a device with spatial resolution of ≈ 0.005 nm.

Fig. 2a shows the distribution functions of As and Si atoms over the (110) channel cross-section after electron annealing, in Fig. 2b after additional thermal annealing. The thermal annealing results in a redistribution of atoms over the channel cross-section. Practically all Si atoms are incorporated in the channel walls. The As atoms form several groups. One group of ≈ 10% is located in the centre of the channel, another of ≈ 90% near the channel wall. Even without special treatment one can see that it is divided into subgroups that

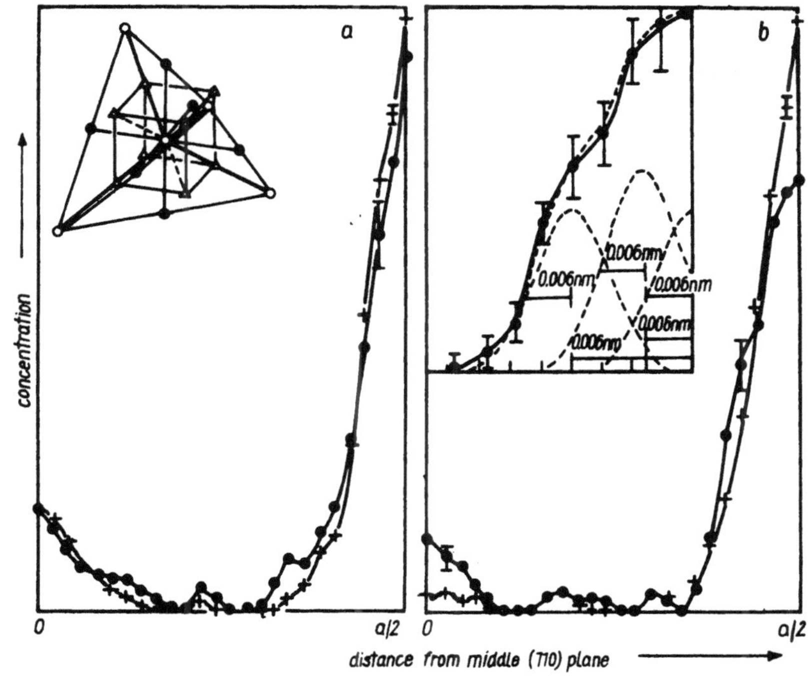

Fig. 2. Distribution functions of As (●) and Si (+) atoms over the cross-section of the (110) flat channel: a) after electron annealing; the insert shows possible positions of atoms corresponding to the peak of the distribution function in zero (channel centre), o atoms on the lattice sites, ● atoms displaced from lattice by a/2 along ⟨100⟩, Δ atoms displaced from lattice by ($\sqrt{3}/8$)a along ⟨111⟩ ; b) after additional heating; the insert shows decomposition of the distribution function near the channel wall in three Gaussians by the least squares method

have different small shifts with respect to the channel wall. Taking into account the cubic symmetry of the crystal we can obtain possible positions of the impurity in a unit cell as points of intersection of three groups of planes, the (110) wall and mid-planes of the planar channel. The insert in Fig. 2a shows possible positions of the impurity and host atoms near an arbitrary lattice point obtained by the above method. The distribution function of the impurity near the channel wall was represented as a superposition of three Gaussian functions with identical width. The fitting was made with the least squares method. The heights, widths, and the position of the centres of two shifted groups were varied. Fig. 2b presents the result of this fitting. As and Si atoms in electron-annealed samples are distributed over the (110) cross-section of the channel almost identically. The central peaks (at zero) appear to be due to point defects produced during rapid crystallization. They may be two types. One is the Si or As atom, shifted by a/4 along ⟨100⟩ (di-interstitial configuration). Another defect type is the same atom, displaced by $(\sqrt{3}/8)a$ from the substitutional site along ⟨111⟩ (⟨111⟩ bond-centered atom or shifted in the antibonding direction).

Additional annealing eliminates quenching defects in silicon and results in decomposition of a supersaturated As solution /5/. The peak in the centre of the As distribution is related to precipitations observed in silicon and germanium by the TEM method between the (110) and (111) planes /5, 6/. However, the dominating type of defects is As atoms slightly shifted from the substitutional sites. Expansion of the distribution function near the channel wall into three Gaussians is most likely. In this case the width of each Gaussian is ≈ 0.006 nm, which corresponds to the amplitude of thermal vibrations of impurity atoms. From an analysis of the Gaussian height and the crystal symmetry one can conclude that two Gaussians, in the wall and at a distance of 0.016 nm, correspond to the atoms shifted by 0.022 nm along the ⟨100⟩ direction. There are no atoms in the sites and it is difficult to choose a symmetrical defect for the third Gaussian (Fig. 2b). It may be As atoms, shifted by 0.008 nm along ⟨110⟩. Apparently in our case several types of defects are responsible for electrically inactive arsenic. Their structure could be identified by measuring the As distribution function over cross-sections of other channels.

References

/1/ A.V. DVURECHENSKII, G.A. KACHURIN, E.V. NIDAEV, and L.S. SMIRNOV, Pulsed Annealing of Semiconductor Materials, Nauka, Moscow 1982.

/2/ G.P. POKHIL et al., Abstr. XIV. All-Union Conf. ICPSC, Moscow 1984 (p. 94).

/3/ YU.P. PYTEV, in: Problems of Reduction in Exper., Vol. 120, No. 2 (1983).

/4/ R.B. ALEXANDER et al., in: Ion Implantation in Semiconductors and Other Materials, Ed. B.L. CROWDER, Plenum Press, New York 1973 (p. 477).

/5/ B.P. KASHNIKOV, A.I. POPOV, and V.P. POPOV, Abstr. Internat. Conf. Energy Pulse Modification of Semicond. and Related Mater., Dresden 1984.

/6/ V.N. ROZHANKII and N.D. ZAKHAROV, Fiz. tverd. Tela 13, 440 (1971).

/7/ V.D. OSVENSKII, L.M. MORGULIS, S.P. GRISHINA, and N.M. KLIMOVA, Fiz. tverd. Tela 11, 1314 (1969).

(Received July 26, 1984)

phys. stat. sol. (a) 85, K43 (1984)
Subject classification: 10.2 and 11; 22.1.2

Department of Physics, Belorussian State University, Minsk[1)]

Damage Profiles in Ion Implanted Silicon

By

V.D. TKACHEV, G. HÖLZER, and A.R. CHELYADINSKII

X-ray rocking curves are sensitive to lattice distortions near the surface, resulting from processes such as impurity diffusion, ion implantation, high-energy irradiation. They may be used for reconstruction of the strain and the defect profiles. In the present note we study the distortion profiles in ion implanted silicon as a function of the ion fluence and the character of their changes during annealing.

Damage distributions cannot be obtained directly from rocking curves and, therefore, the opposite problem is solved. From a chosen profile of strain and defects the diffraction curve is calculated and compared with the experimental result. Variations of the profiles by hand /1/ or by computer /2, 3/ over all possible values of strains and damage in every point of the distribution require a great expenditure of computer time. In this note the problem is solved by a functional description of the defect and the strain profiles and optimization of the function parameters by computer to obtain satisfactory agreement between calculated and measured rocking curves.

Distributions with depth t of the strains $\frac{\Delta d}{d}(t)$ and the damages W(t) are given by

$$\frac{\Delta d}{d}(t) = \begin{cases} x_2 \exp\left[-\left(\frac{t - x_1}{\sqrt{2}\, x_3}\right)^{x_6}\right] & \text{when } t < x_1 \,, \\ x_2 \exp\left[-\left(\frac{t - x_1}{\sqrt{2}\, x_4}\right)^{x_7}\right] & \text{when } t \geq x_1 \,, \end{cases} \tag{1}$$

$$W(t) = 1 - x_5 \frac{\Delta d}{d}(t) \,.$$

We assume the parameters x_6, x_7 to be usually equal to 2 (Gaussian distribution), but the optimization procedure permits their variation in the vicinity of this value. Optimization of the parameters $x_1, \ldots, x_j$ of the profile is carried out by the least-squares-fit method with minimizing of the function

1) Universitetskii gorodok, 220080 Minsk, USSR.

$$F(x_1, \ldots x_j, \Delta\Theta) = \sum_{i=1}^{k} \varrho_k \left[\frac{J_{exp}(\Delta\Theta_k) - J_{calc}(x_1, \ldots, x_j, \Delta\Theta_k)}{J_{exp}(\Delta\Theta_k)}\right]^2 .$$

This is the coincidence criterion of the calculated diffraction curves $J_{calc}(x_1, \ldots x_j, \Delta\Theta_k)$ and the measured ones $J_{exp}(\Delta\Theta_k)$ in the points k . ϱ_k is the weight function, $\Delta\Theta_k = \Theta_k - \Theta_B$, Θ_B is the Bragg angle.

From the kinematical theory /4/ the amplitude reflected by N laminae containing n individual planes with account being taken of the phase difference due to the strain $(\Delta d/d)_j$ and the damage W_j is given by

$$R_N = \sum_{j=1}^{N} |q_j| \exp\left\{i\left(\frac{n\delta_j}{2} + \Phi_j\right)\right\} \frac{\sin\frac{n\delta_j}{2}}{\frac{\delta_j}{2}} ,$$

where $q_j = -i \frac{d\lambda}{V \sin\Theta_B} \frac{e^2}{m c^2} \left|F^j_{hkl}\right|$ is the amplitude reflected by one plane of the laminae j for unit incident amplitude, e^2/mc^2 is the classical electron radius, λ the X-ray wavelength, V the volume of the unit cell, d the interplanar spacing. $\delta_j = (4\pi/\lambda)d \cos\Theta_B \Delta\Theta + 2\pi(\Delta d/d)_j$ is the phase deviation in neighbouring planes of the laminae:

$$\Phi_j = \sum_{i=1}^{j-1} n_i \delta_i ; \quad \Phi_1 = 0 ;$$

Φ_j is the sum of the phase displacement in the (j - 1)-th laminae.

$$F^j_{hkl} = \exp(-M_j) F^o_{hkl} = W_j F^o_{hkl} ,$$

F^o_{hkl} and F^j_{hkl} are the structure factors for reflection (hkl) in perfect and damaged crystals, respectively,

$$M_j = \frac{8\pi^2}{\lambda^2} \sin\Theta_B \overline{U^2_j} ;$$

$\overline{U^2_j}$ is the average squared displacement of atoms from lattice nodes due to the structural defects in laminae j. The degree of damage W(t) changes from 1 for the perfect crystal to 0 for the completely disturbed structure.

By the dynamical diffraction theory /5/ the amplitude in the "tails" of the main peak is

$$R_o(\Delta\Theta) = \frac{1}{2p}\left(1 + \frac{1}{4p^2}\right) ; \quad p = -\frac{\sin 2\Theta_B}{e^2/mc^2} \frac{\pi V}{\lambda^2 \left|F^o_{hkl}\right|} \Delta\Theta .$$

The total amplitude is $R(\Delta\Theta) = R_N(\Delta\Theta) + R_o(\Delta\Theta)$ and the intensity

$$J(\Delta\Theta) = R_N(\Delta\Theta)R_N^*(\Delta\Theta) + R_o^2(\Delta\Theta) + 2\,R_o(\Delta\Theta)\,\mathrm{Re}\left[R_N(\Delta\Theta)\right].$$

The measurements of the rocking curves were carried out with the double crystal diffractometer in (n, -n) arrangement with CuKα(333) reflection. Silicon n-type crystals (ρ_o = 0.3 Ωcm) irradiated by Si^+ ions with energy of 200 keV and fluences from 6×10^{13} to 6×10^{14} cm^{-2} were investigated. Isochronal 15 min annealing of the implanted samples was performed in evacuated quartz tubes. Temperature deviations did not exceed ±2 °C.

In Fig. 1 the strain profiles of silicon implanted by Si^+ ions with different fluences are shown. The positions of maxima of the strain distributions and the distributions of the energy loss by atom collision are in good agreement. The strain profiles are not symmetric. The "tails" reach the surface corresponding to the distribution of radiation defects in the damaged layer caused by implantation /7/. For comparison the defect distribution D for the sample irradiated by Si^+ with fluence 6×10^{14} cm^{-2} is shown in Fig. 1 (curve 5). The defect distribution was obtained by the method given in /8/ based on the dependence of the anodic oxidation rate on the defect concentration. The agreement between the defect distribution and the strain profile is quite satisfactory.

As seen from Fig. 1, the width of the damaged layer increases with the ion dose. In the maximum of the distribution less defects than proportional to the dose (sublinear) are accumulated, and in the "tails" of the distribution near substrate the dependence is superlinear. For example, at a depth of 300 nm Δd/d increases by a factor 7 by increasing the ion dose twice. The sublinear character of the defect accumulation in the maximum of their distribution is caused, in our opinion, by annihilation of newly created primary defects with earlier accumulated

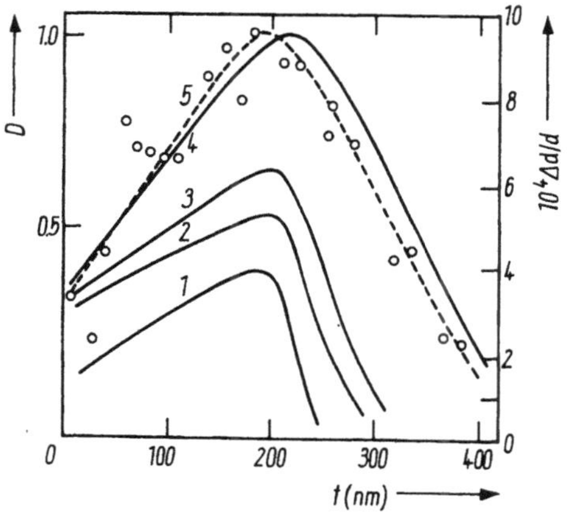

Fig. 1. Profiles of strain (curves 1 to 4) and defects (curve 5) in Si^+ implanted silicon. (1) 1×10^{14}, (2) 2×10^{14}, (3) 3×10^{14}, (4), (5) 6×10^{14} cm^{-2}

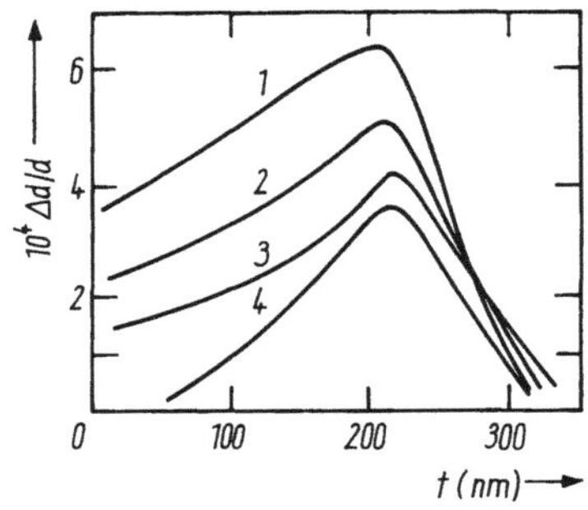

Fig. 2. Change of strain profiles of implanted silicon during annealing. (1) 20, (2) 170, (3) 340, (4) 385 °C

stable defects. Moreover, annihilation is more probable in the region of maximum concentration. The superlinear accumulation of the defects in the "tails" of the distribution is possibly caused by a decrease of the primary defect flow from the implanted layer due to a decrease of their mobility in the elastically deformed region near the boundary between the perfect substrate and the distorted layer which has a different lattice period. The decrease of the primary defect mobility increases their momentary concentration during irradiation and consequently the probability of complex formation.

The change of the radiation damage profiles of the ion implanted layers during isochronal 15 min annealing is illustrated by Fig. 2. It is seen that defects anneal faster near the surface during heat treatment. An analogous behavior has been seen earlier in silicon irradiated with fast neutrons /9/.

References

/1/ R.N. KYUTT, P.V. PETRASHEN, and L.M. SOROKIN, phys. stat. sol. (a) 60, 381 (1980).

/2/ V.S. SPERIOSU, H.L. GLASS, and T. KOBAYSHI, Appl. Phys. Letters 34, 539 (1979).

/3/ B.C. LARSON and F. BARHORST, J. appl. Phys. 51, 3181 (1980).

/4/ R.W. JAMES, The Optical Principles of the Diffraction of X-Rays, Cornell University Press, Ithaca (New York) 1965.

/5/ R.W. JAMES, Solid State Phys. 15, 53 (1963).

/6/ D.K. BRICE, Radiat. Eff. 6, 77 (1970).

/7/ L.S. SMIRNOV (Ed.), Fizicheskie protsessi v obluchennikh poluprovodnikakh, Nauka, Novosibirsk 1977 (p. 165).

/8/ M. KISIELEWICZ, Radiat. Eff. Letters 58, 109 (1981).

/9/ V.F. STELMAKH and A.R. CHELYADINSKII, Elektronnaya Tekh., Ser. Mater. 7, 77 (1980).

(Received April 26, 1984)

phys. stat. sol. (a) 85, K47 (1984)
Subject classification: 10.1; 21

Dipartimento di Fisica and Gruppo Nazionale di Struttura della Materia del C.N.R., Università di Bologna[1)]

Influence of Dislocation Core Density on Overheating Absence

By
R. TOGNATO

In a previous paper /1/ the author explained the absence of overheating in single crystal f.c.c. metals, having free outer surfaces corresponding to "smooth" planes, with the presence, already below the thermodynamical melting temperature T_m, of 19-atom liquid clusters. The 19-atom liquid clusters were supposed to be distributed on the interface between the dislocation cores and the surrounding bulk material. The dislocation core density was the one normally present in as-grown single crystals (i.e. $\varrho \approx 10^{12}$ m/m^3).

The aim of the present short note is to prove that this model may also be applied to single crystals with lower dislocation densities.

The following considerations have been checked for lead (T_m = 600 K) at T = 599 K with reference to the volume occupied by a gram atom, $V_{g\,atom}$.

If the core-cluster atom system is treated as a solid solution which follows Henry's rule, the random distribution of the f_{19} clusters on M nearly flat catalytic surfaces induces relative changes in free energy given by

$$\Delta G/MkT = \left[D(19)/kT\right](f_{19}/M) + \ln\left[1 - (f_{19}/M)\right] - (f_{19}/M)\ln\left[(M/f_{19}) - 1\right], \quad (1)$$

where $D = 2\mu^c(v_l - v_c)^2(3v_l)^{-1} \approx 7.6\times10^{-23}$ J/atom is the elastic free energy due to a liquid atom supported by core atoms, $\mu^c \approx 4.5\times10^9$ N/m^2 is the rigidity modulus of dislocation cores, $v_l = 3.25\times10^{-29}$ m^3 and $v_c = 3.15\times10^{-29}$ m^3 are the volumes at the disposal of a lead atom in the liquid and in the core phase, respectively.

The trend of (1) is shown in Fig. 1. The minimum value of $\Delta G/MkT$ corresponds to $f_{19}/M \approx 0.475$.

The number, M, of catalytic surfaces may be written /1/ as $M = l_{tot}/6d'_{\{111\}} = \varrho v_c N_0/6d'_{\{111\}} \approx \varrho V_{g\,atom}/6d'_{\{111\}}$, where l_{tot} is the total length of dislocation cores in the volume occupied by a gram atom, ϱ is a dislocation density

1) via Irnerio 46, 40126 Bologna, Italy.

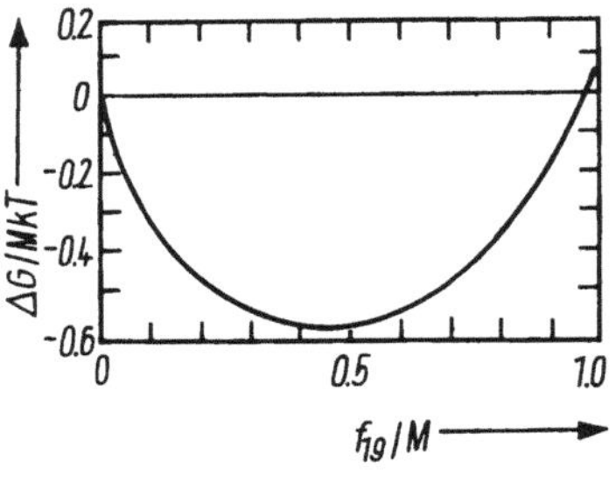

Fig. 1. Relative free energy changes, $\Delta G/MkT$, versus catalytic position occupation, f_{19}/M

(m/m^3), N_o the Avogadro number, and $d'_{\{111\}} \approx 3.09 \times 10^{-10}$ m is the interplanar distance between neighbouring planes $\{111\}$ in the cores.

From the previous considerations and to avoid misunderstanding, from now onwards it is preferable to treat the matter in terms of total length of dislocation cores in the volume occupied by a gram atom instead of the usual dislocation density. It is evident that the total length decreases of the dislocations correspond to decreases of M and consequently to the number of clusters f_{19}. There is a linear dependence between f_{19} and l_{tot} (i.e. ϱ).

The mean distance between clusters $\bar{d}_{cl} \approx 10 d'_{\{111\}}$ is not influenced by the f_{19} decreases. The mean distance is given by $\bar{d}_{cl} \approx [f_{19}/v_c N_c]^{-1/3}$ where $N_c = n(l_{tot}/d'_{\{111\}}) \approx n\varrho v_c N_o/d'_{\{111\}} \approx n\varrho V_{g\,atom}/d'_{\{111\}}$ is the total number of atoms contained in the dislocation cores (n = 61 /1/).

It is recalled that $\bar{d}_{cl}$ is the distance between the centres of two clusters. Considering that the cluster radius is $\approx 1.5 d'_{\{111\}}$ the real distance between clusters is $\approx 7 d'_{\{111\}}$.

Fig. 2 shows the variations of f_{19} and N_c with l_{tot} at the temperature given. The lower value of total length of the dislocation cores corresponds to a single dislocation going from one face to its opposite of a cube whose volume corresponds to $V_{g\,atom}$. For $V_{g\,atom} \approx 18.89 \times 10^{-6}$ m^3 there results $(V_{g\,atom})^{1/3} = l_{tot} \approx 2.65 \times 10^{-2}$ m (corresponding to $\varrho \approx 10^3$ m/m^3).

Considering that at $T = T_m$ the clusters become embryos which, growing at the expence of the unclustered atoms and coalescing between themselves, become the critical nuclei f^*_{300} of the solid-liquid phase transition, the attention may be focused on the clusters f'_{19}, among the total f_{19}, that have a finite coalescing probability. Since our interest is fixed upon clusters with a distance $d_{cl} < \bar{d}_{cl}$ we propose a distribution function of the type

$$f(d_{cl}/d'_{\{111\}}) = \left[1/(2\pi)^{1/2}\lambda\right] \exp\left[-(1/2)((d_{cl} - \bar{d}_{cl})/d'_{\{111\}}\lambda)^2\right] .$$

The condition of non-interaction between clusters (Henry's rule) makes it pos-

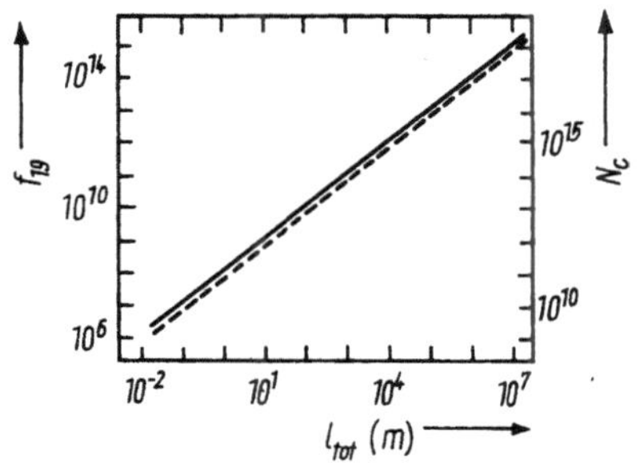

Fig.2

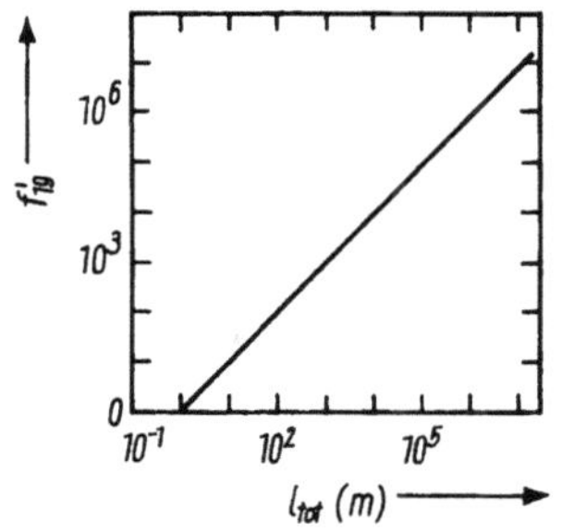

Fig.3

Fig. 2. Dependence of the cluster number, f_{19}, (continuous line) and of the number of atoms contained in the dislocation cores, N_c, (dashedline) on the total length, l_{tot}, of dislocation cores

Fig. 3. Number of clusters, f'_{19}, susceptible of coalescing, versus the total length, l_{tot}, of the dislocation cores in the volume occupied by a g atom

sible to write $(d_{cl})_{min} \approx 5d'_{\{111\}}$ (i.e. $(d_{cl} - \bar{d}_{cl})/d'_{\{111\}} \approx -5$) which corresponds to a real distance of $\approx 2d'_{\{111\}}$, a value that ensures also that the two clusters belong to different catalytic surfaces. By symmetry $(d_{cl})_{max} \approx 15\, d'_{\{111\}}$ (i.e. $(d_{cl} - \bar{d}_{cl})/d'_{\{111\}} \approx 5$).

The growth and coalescence of two embryos such as $d_{cl} \approx 5d'_{\{111\}}$ give an embryo with a $(6d'_{\{111\}} + 2d'_{\{111\}}) \approx 2.47\times10^{-9}$ m diameter corresponding to that of a critical nucleus containing 300 liquid atoms.

Following the previous considerations, for a variance $\lambda^2 \approx 0.5$, there results

$$\int_{-5}^{5} f(d_{cl}/d'_{\{111\}})d(d_{cl}/d'_{\{111\}}) = 1$$

and

$$\int_{-10}^{5} f(d_{cl}/d'_{\{111\}})d(d_{cl}/d'_{\{111\}}) = \int_{5}^{10} f(d_{cl}/d'_{\{111\}})d(d_{cl}/d'_{\{111\}}) = 0 .$$

It may be seen that for $(d_{cl} - \bar{d}_{cl})/d'_{\{111\}} = -10$ it turns out that $d_{cl} = 0$, a physical situation that has no meaning.

The number of clusters f'_{19} will be given by

$$f'_{19} = \left[\int_{-5}^{-4} f(d_{cl}/d'_{\{111\}})d(d_{cl}/d'_{\{111\}})\right] f_{19} = \left[5.5\times10^{-9}\right] f_{19} .$$

The integrals have been solved numerically with the usual methods. Fig. 3 shows the f'_{19} dependence on l_{tot}.

Recalling that the occurrence of only one critical nucleus in dislocation cores gives a nucleation rate $J = (kT/h)f^{*}_{300}$ of the order of $2x10^{13}$ $(V_{g\,atom})^{-1}s^{-1}$ /1/, it is possible, on the basis of the present considerations, to assert that overheating may be absent also for monocrystals with $l_{tot} \approx 1$ m (corresponding to $\varrho \approx 10^{5}$ m/m^{3}).

This paper neglects completely possible increases of dislocation density, i.e. of the total length of the dislocation cores due to unbalanced stresses between clusters and core atoms /2/.

References

/1/ R. TOGNATO, phys. stat. sol. (a) 80, 147 (1983).
/2/ R. TOGNATO, phys. stat. sol. (a) 73, 237 (1982).

(Received July 12, 1984)

phys. stat. sol. (a) 85, K51 (1984)

Subject classification: 11; 19; 22.1.2

Institute of Physics of Solids and Semiconductors, Academy of Sciences of the Belorussian SSR, Minsk[1)]

Relaxation Time and Electron Scattering Cross Section in Irradiated p-Silicon at Temperatures of Liquid Helium from Cyclotron Resonance Data

By

V.I. GATALSKAYA, E.M. GOLOLOBOV, and L.A. KUROCHKIN

The knowledge of the characteristics of kinetic processes in semiconductors is an important source of information on the mechanism of charge carrier (CC) scattering in a crystal, since these processes depend strongly on the type and distribution of impurities, change of temperature, electric field strength, and other physical parameters.

The SHF methods are most preferable in treating the kinetic parameters and allow to carry out contactless measurements, important for studying high resistivity samples at temperatures of liquid helium. The cyclotron resonance (CR) method enables a very accurate and independent determination of two most important values to be carried out: relaxation time of the signal of CC τ and effective mass m* (separately for electrons from different valleys, light and heavy holes). It seems to be ideal for investigating these kinetic parameters and has been employed by us for the first time to irradiated germanium (/1/ and references therein) and n-type silicon /2/.

The aim of the present note is the study of the CR spectra both in initial and irradiated p-type silicon single crystals, the separation of partial contributions to the total scattering of different groups of scattering centers, and the quantitative estimation of the kinetic parameters of the centers of radiation origin (relaxation time τ and scattering cross section σ). On this basis conclusions are drawn regarding possible mechanisms of scattering of CC. Such data on irradiated p-silicon are not available in literature.

We chose pure oxygen-free p-type single crystals ($\rho_0 = 2\times10^3\,\Omega$cm) grown by zone melting and doped with boron up to 5×10^{12} cm^{-3}. The specimens for CR and Hall measurements were cut from a Si plate. Then the specimens with identical

---

1) Podlesnaya 17, 220726 Minsk, USSR.

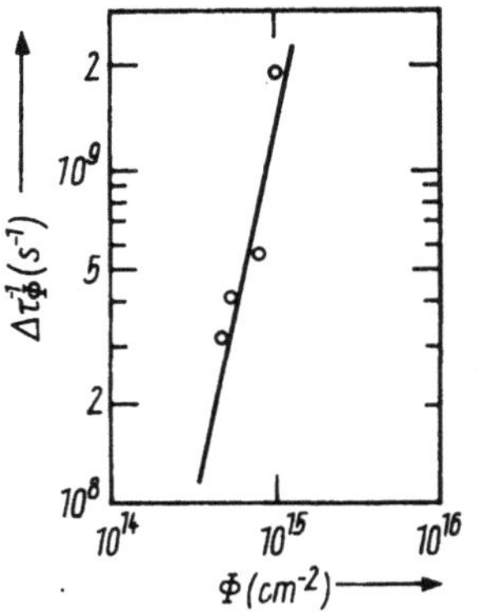

Fig. 1. Dose dependence of $\Delta\tau_{\Phi}^{-1}$ for irradiated p-Si at T = 4.2 K

parameters were irradiated by fast electrons (25 MeV) at room temperature in the dose range from 10^{13} to 10^{15} cm^{-2}.

An autodyne radiospectrometer CR, operating at 36 GHz over the temperature range from 1.8 to 4.2 K was designed and described in /3, 4/. The amount of CC necessary for observing CR at helium temperature was excited by interband light, the CR signal was recorded at a modulation frequency of 2.3 kHz. The Hall measurements were carried out under dc in the temperature range from 77 to 300 K by the compensation method, stabilizing the current through the sample and its temperature to an accuracy of 1% and 0.1 K, respectively.

From the CR absorption spectra measurements of the initial samples it was established that in p-Si with $\rho_0 = 2\text{x}10^3\,\Omega\text{cm}$ at 1.8 to 4.2 K the dominant scattering is on transverse acoustic lattice oscillations described as

$$\tau^{-1} = 3.3\text{x}10^8\,T^{3/2}\;,$$

where τ^{-1} is the inverse electron momentum relaxation time in the initial crystals. The scattering on neutral and ionized centers can be neglected, as the boron concentration is small and the recombination processes of excess charge carriers, when excited by interband light at helium temperature, proceed slowlier than their capture by ionized centers.

By irradiation of p-Si samples the intensity of the CR signal decreases and the linewidth broadens. The dose dependence (Fig. 1) of the partial contribution to the total scattering obtained by subtracting the value τ^{-1} of the initial sample from that of τ_{Φ}^{-1} of the irradiated sample ($\Delta\tau_{\Phi}^{-1} = \tau_{\Phi}^{-1} - \tau^{-1}$) is approximated by the function

$$\Delta\tau_{\Phi}^{-1} = A\Phi^n\;,$$

where n = 2. For all exposure doses the $\Delta\tau_{\Phi}^{-1}$ value in the range from 1.8 to 4.2 K depends weakly on temperature, decreasing with decreasing temperature (Fig. 2).

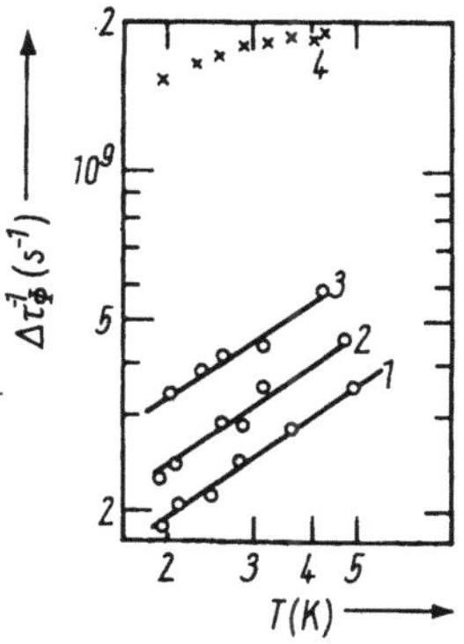

Fig. 2. Temperature dependence of $\Delta\tau_{\Phi}^{-1}$ for irradiated p-Si: Φ = (1) 5×10^{14}, (2) 6×10^{14}, (3) 8×10^{14}, (4) 1×10^{15} cm^{-2}

To determine the energetic position of the radiation defects (RD) and their concentration in the irradiated crystals the temperature dependence of the Hall effect in a wide radiation dose range was measured; the experimental and the calculated curves were compared with each other. For each particular case the electroneutrality equations were solved by a computer. As follows from the analysis of the data, the irradiation of Si crystals with doses up to 5×10^{13} cm^{-2} causes the appearance of RD with E_V + 0.2 eV, but at higher doses with E_V + 0.28 eV, the concentration of which increases with radiation dose. Beginning from $\Phi = 5\times10^{14}$ cm^{-2} it exceeds the concentration of the doping boron impurity and reaches the value of 4×10^{13} cm^{-3} at $\Phi = 1\times10^{15}$ cm^{-2}. From literature the RD with E_V + 0.29 eV can be attributed to divacancies /5, 6/: for the RD with E_V + 0.2 eV there are no reliable literature data, probably it is a complex which includes boron.

The concentration of divacancies N_W practically exceeding RD by an order of magnitude with E_V + 0.2 eV for radiation doses between 10^{14} and 10^{15} cm^{-2} depends on the dose in the following way:

$$N_W = C\Phi^m ,$$

where m = 2.2. This is in accordance with a relationship close to quadratic; the process of divacancy formation proceeds according to the kinetic scheme proposed in /7/.

Thus the principal scattering centers in irradiated p-Si with radiation doses of 10^{14} to 10^{15} cm^{-2} are divacancies and the value of additional scattering on RD, as follows from the CR and galvanomagnetic measurements, is proportional to the RD concentration and is due to the scattering on divacancies.

Knowing the partial momentum relaxation time connected with scattering on divacancies, one can determine the effective scattering cross section on these defects σ_W:

$$\Delta\tau_{\Phi}^{-1} = N_W\langle v\rangle\sigma_W ,$$

where $\langle v\rangle$ is the average thermal velocity of electrons.

Assuming additivity of different scattering mechanisms, the scattering cross section of electrons on divacancies, σ_w, appears to be equal to 1.6×10^{-11} cm^2, which exceeds by an order of magnitude that of scattering on the singly ionized center.

The obtained value of scattering cross sections for irradiated p-Si cannot be accounted for in terms of common scattering mechanisms. The interpretation of this phenomenon of a gigantic scattering cross section in terms of an elastic deformation of the crystal lattice connected with a rearrangement of atoms around RD used for the first time for explaining galvanomagnetic measurement /8/ and CR data /1/ for irradiated germanium seems interesting.

On the basis of the theory of elasticity the interaction of the electron with the static defect /9/ is described by $U(r) = C(a/r)^3$, where a is the radius of the field effect. The estimation of σ_w for electron scattering on divacancies as on the range of elastic deformation around the RD was carried out in the Born scattering potential approximation:

$$\sigma_w = 2\pi\int_0^{\pi}|f(\Theta)|^2 \sin\Theta\, d\Theta = 4\pi\left(\frac{2m^*}{\hbar^2}\right)^2\left|\int U(r)\,\frac{\sin qr}{qr}\,r^2\,dr\right|^2 =$$

$$= 4\pi\left(\frac{2m^*}{\hbar^2}\right)^2\left|\int U(r)\,r^2 dr\right|^2 \approx 4\pi(2a)^2\left(\frac{C}{E_0}\right)^2$$

$$\left(ka \ll 1,\quad \sin qr \approx qr;\qquad q = 2k\sin\frac{\Theta}{2},\quad E_0 = \frac{\hbar^2}{ma^2}\right).$$

In the present case the effective cross section σ_w reaches 1×10^{-11} cm^2 which is in good agreement with our independent estimations from CR measurements. The weak temperature depencence of σ_w between 1.8 and 4.2 K is also confirmed by our CR investigations.

References

/1/ N.N. SIROTA, V.I. LUKASHOVA, and L.A. KUROCHKIN, phys. stat. sol. (a) 47, 677 (1978).

/2/ L.A. KUROCHKIN and V.I. GATALSKAYA, Dokl. Akad. Nauk Beloruss. SSR 26, 128 (1982).

/3/ L.A. KUROCHKIN and V.I. LUKASHOVA, Izv. Akad. Nauk Beloruss. SSR, Ser. fiz.-mat. Nauk 5, 113 (1975).

/4/ L.A. KUROCHKIN and V.I. LUKASHOVA, Dokl. Akad. Nauk Beloruss. SSR 21, 1085 (1977).

/5/ N.I. AKULOVICH, V.V. PETROV, and V.D. TKACHEV, Fiz. Tekh. Poluprov. 10, 1038 (1976).
/6/ V.V. PETROV and V.D. TKACHEV, Fiz. Tekh. Poluprov. 10, 966 (1976).
/7/ M. HIRATA and M. FUKUMOTO, Japan. J. appl. Phys. 13, 8 (1974).
/8/ V.V. EMTSEV, M.I. KLINGER, and T.V. MASHOVETS, Zh. eksper. teor. Fiz., Pisma 19, 575 (1974).
/9/ J.D. ESHELBY, Solid State Phys. 3, 79 (1956).

(Received June 5, 1984)

phys. stat. sol. (a) 85, K57 (1984)

Subject classification: 11 and 13.4; 22.2.1

A.N. Sevchenko Scientific Research Institute for Applied Physics Problems, V.I. Lenin Belorussian State University, Minsk[1)]

Temperature Dependence of the Charge-Carrier Removal Rate in n-GaAs Containing Defect Clusters

By

V.M. LOMAKO and P.YA. STAROSTIN

It is known that upon irradiation of semiconductors with high-energy particles complex structure damages, i.e. defect clusters (DC) are introduced. Their first simulations are given in /1/.

The charge-carrier removal in DC containing silicon has been analyzed in /2, 3/.

Theoretical studies of the functional dependences of the charge-carrier removal rate on the doping level in DC containing GaAs have been performed and an estimation of some parameters by comparison of the results obtained with the experimental data has been given in /4, 5/.

The present note reports on the results of a theoretical calculation of the temperature dependence of the charge-carrier removal rate in the interval from 77 to 300 K for DC containing n-GaAs.

The spatial and energetic schemes of the DC model are given in Fig. 1. It is suggested that beyond the DC core only doping impurity atoms are present, and within the core ($r \leqq r_1$), besides these, there are radiation defects of high concentrations. The coordinate function of the potential barrier height $\varphi(r)$ and the electric charge density $\varrho(\varphi)$ are related through the Poisson equation by a self-consistent expression:

$$\nabla^2 \varphi = -\frac{4\pi}{\varepsilon_a} \varrho(\varphi) , \qquad (1)$$

where

$$\varrho(\varphi) = e\left[n(\varphi) - p(\varphi) + N_a^-(\varphi) + M_a^-(\varphi) - N_d^+(\varphi) - M_d^+(\varphi)\right] .$$

Concentrations of free carriers $n(\varphi)$, $p(\varphi)$ and electrically charged defects and impurities are expressed with account of their changes with the potential barrier

1) 220106 Minsk, USSR.

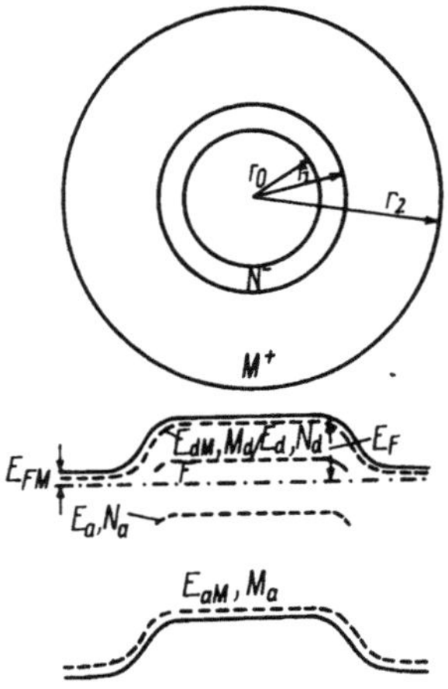

Fig. 1. Spatial and energetic schemes of DC; r_2, r_1, r_o are the radii of the space charge, the core, and the neutral part of the core, respectively; N_d, N_a, E_d, E_a are the concentrations and the energy levels for donor- and acceptor-type radiation defects; M_d, M_a, E_{dM}, E_{aM} are the concentrations and the energy levels for donor- and acceptor-type impurity atoms; F is the Fermi-level position; E_{FM}, E_F are the Fermi levels in DC matrix and core; N_a, $N_d \gg M_d$; $M_d > M_a$

coordinate function /5/. Integration of the Poisson equation is carried out under the following limiting conditions:

$$\varphi(r)\Big|_{r \geqq r_2} = 0, \; \varphi(r)\Big|_{r \leqq r_o} = \varphi_o; \quad \frac{d\varphi}{dr}\Big|_{r \geqq r_2} = \frac{d\varphi}{dr}\Big|_{r \leqq r_o} = 0 \tag{2}$$

and with the DC charge neutrality condition

$$\int_0^{r_2} r^2 \varrho(\varphi) d r = 0 \; . \tag{3}$$

The initial charge-carrier removal rate was obtained using the expressions given in /5/. The Fermi-level position inside the DC and the value of the potential barrier height were determined from the electroneutrality equations by numerical methods.

The results of calculations of the temperature dependence of the initial charge-carrier removal rate R(T) in n-GaAs are given in Fig. 2. It is seen that the R(T) functions weakly increase with decreasing temperature what is realized for cases where radiation defect levels are in the upper half of the gap (curves 3 and 4) or near its centre (curve 1). The weakly decreasing functions are realized for the cases where defect levels are near the resolved intrinsic bands (curve 2).

In /5/ it is shown that the energy region of the radiation defect level positions in the upper half of the gap (the acceptor-type defect levels should be higher than levels with $E_a \approx E_v + 1.0$ eV, and donor-type defect levels - higher than level $E_d \approx E_c - 0.2$ eV) is a region, where the theoretical dependence of the initial

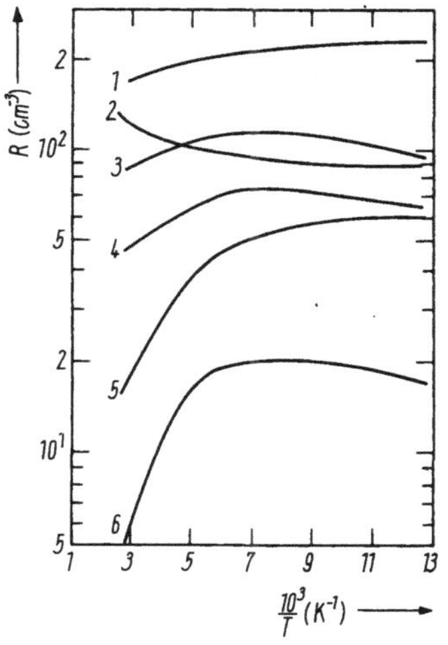

Fig. 2. Calculated temperature dependence of the initial charge-carrier removal rate in n-GaAs containing DC. r_1 = 30 nm, $M_d = 10^{15}$ cm^{-3}, $M_a = 10^{14}$ cm^{-3}; $N_a = N_d = 10^{20}$ cm^{-3};
(1) $E_a = E_v + 0.73$ eV, $E_d = E_c - 0.73$ eV;
(2) $E_a = E_v + 0.3$ eV, $E_d = E_c - 0.3$ eV;
(3) $E_a = E_v + 0.9$ eV, $E_d = E_c - 0.3$ eV;
(4) $E_a = E_v + 1.0$ eV, $E_d = E_c - 0.2$ eV;
(5) $E_a = E_v + 1.2$ eV, $E_d = E_c - 0.2$ eV;
(6) $E_a = E_v + 1.4$ eV, $E_d = E_c - 0.1$ eV.
$E_{dM} = E_c - 0.005$ eV; $E_{aM} = E_v + 0.034$ eV.
The concentration of the uncompensated charge in the core $N = 10^{19}$ cm^{-3}

charge-carrier removal rate $R(n_0)$ on the doping level gives the best fit to the experimental data in fast neutron-irradiated n-GaAs /6/.

The calculated values of the carrier removal rates for the above-stated region of defect ionization energies vary weakly with temperature, in agreement with the experimental data on the carrier removal in neutron-irradiated GaAs /7/. The temperature dependence of the carrier removal rate increases as defect levels are closer to the conduction band. Since at the same time the discrepancy with the experimental data grows, the energy region of defect level positions near the conduction band should be excluded from the region of forecasting of radiation defect level energy positions in DC.

Thus, the temperature dependence of the carrier removal rate in GaAs containing DC agrees with the experimental data for definite values of defect ionization energies in the core. These dependences supplement the concentration dependences of the rate of carrier removal by DC in n-GaAs upon the determination of defect level depths in DC cores. If at the same time the comparison of the doping level dependence of the initial charge-carrier removal rate shows that the radiation defect levels in the DC core are in the upper half of the gap, then the temperature dependences obtained from the energy region mentioned in /5/ exclude defect states near the conduction band. This increases the accuracy in the determination of the energy positions of defect levels forming the DC core.

References

/1/ B.R. GOSSIK, J. appl. Phys. 30, 1214 (1959).

/2/ R.R. HOLMES, IEEE Trans. nuclear Sci. 17, 137 (1970).

/3/ N.G. GOLEMINOV, E.A. KRAMER-AGEEV, and YU.A. MIRONOV, phys. stat. sol. (a) 28, 371 (1975).

/4/ V.M. LOMAKO and P.YA. STAROSTIN, phys. stat. sol. (a) 63, 585 (1981).

/5/ V.M. LOMAKO and P.YA. STAROSTIN, phys. stat. sol. (a) 65, 39 (1981).

/6/ T.I. KOLCHENKO and V.M. LOMAKO, phys. stat. sol. (a) 48, 263 (1978).

/7/ H.J. STEIN, J. appl. Phys. 40, 5300 (1969).

(Received June 5, 1984)

phys. stat. sol. (a) 85, K61 (1984)
Subject classification: 18.2; 21.1

Department of Chemistry, University of California, San Diego[1)]

Structure and Magnetic Properties of $Nd_2Fe_{14}BH_{2.7}$[2)]

By
K. OESTERREICHER and H. OESTERREICHER

We report here on the structural and magnetic properties of $Nd_2Fe_{14}BH_{2.7}$ and compare them in Table 1 with the ones of $Nd_2Fe_{14}B$ as given in /3/. $Nd_2Fe_{14}B$ was shown /1 to 3/ to have an axial anisotropy field of order 76 kOe at room temperature but to exhibit complex spin reorientation behavior /2, 3/ at lower temperature. As this and related materials have technological potential (hydrogen storage, magnetic recording or permanent magnet materials) it became of interest to study the hydriding properties in order to attempt to influence both macroscopic (pulverization) and microscopic properties of this compound.

$Nd_2Fe_{14}B$ as prepared from the melt forms with a tetragonal structure /1 to 5/. It readily takes up hydrogen at room temperature under 50 atm H_2 to a composition $Nd_2Fe_{14}BH_{2.7}$. The material pulverizes in the process although the phenomenon is much less pronounced compared to $LaNi_5$. No noticeable

Table 1

Structural and magnetic parameters of $Nd_2Fe_{14}B$ and $Nd_2Fe_{14}BH_{2.7}$

	a_o (nm)	c_o (nm)	c_o/a_o	M*) (μ_B/FU)	T_C(K)	H_A(kOe)
$Nd_2Fe_{14}B$	0.8792	1.2174	1.385	28.8	603	76
$Nd_2Fe_{14}BH_{2.7}$	0.8930	1.2360	1.384	31.8	.	20 (300 K) 54 (77 K)

*) For $Nd_2Fe_{14}BH_{2.7}$ the formula unit was taken to be $Nd_2Fe_{14}B$. Values are for 300 K.

---

1) La Jolla, California 92093, USA.

2) Work partly supported by University of California Energy Research Group.

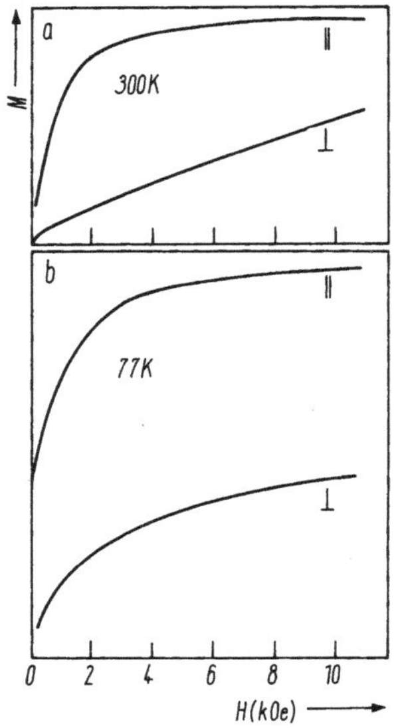

Fig. 1. Magnetization M (arbitrary units) versus applied field H (kOe) for $Nd_2Fe_{24}BH_{2.7}$ at a) 300 and b) 77 K, respectively

change in H composition was observed on reducing the pressure to 1 atm H_2. The hydride appears to be rather stable at least down to 0.01 atm H_2. An X-ray pattern was obtained of the hydride with CuK_α radiation and showed the structure of $Nd_2Fe_{14}B$ with enlarged lattice parameters (a_o = 0.8930 and c_o = 1.236 nm, c/a = 1.384). The X-ray diffraction pattern remains of excellent clarity and roughly similar intensities to $Nd_2Fe_{14}B$. X-ray diffraction on pulverized materials oriented in a field of 10 kOe shows that the easy axis of magnetization is along the c axis.

$Nd_2Fe_{14}BH_{2.7}$ looses H around 450 K and we therefore did not determine an ordering temperature. The room temperature moment of a powder at 10.6 kOe, which was oriented parallel to the field at 10.6 kOe, is 31.8 μ_B per formula unit $Nd_2Fe_{14}B$.

Fig. 1a, b shows magnetization versus applied field at room temperature and 77 K, respectively, of powders oriented parallel and perpendicular to the field direction. Accordingly an anisotropy field H_A = 20 kOe is extrapolated at 300 K.

Fig. 2 shows the temperature dependence of magnetization of powders aligned parallel and perpendicular to the field (at room temperature). While "parallel powders" show only a minor decrease in magnetization with temperature, the perpendicular powders exhibit a broad minimum. The upturn in magnetization at the lowest temperatures reminds one of the spin reorientation of $Nd_2Fe_{14}B$.

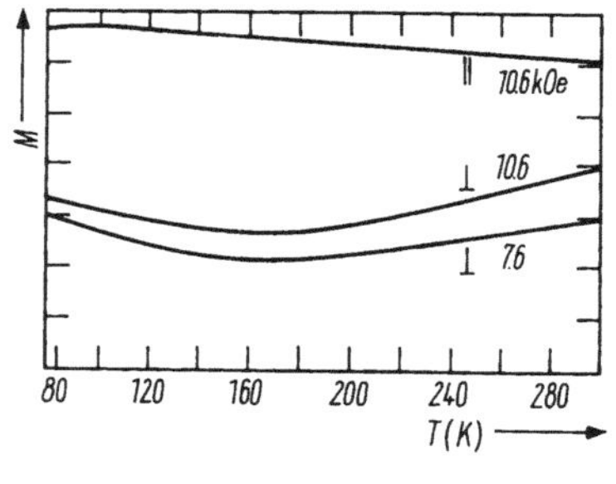

Fig. 2. Magnetization M (arbitrary units) versus temperature T(K) of $Nd_2Fe_{14}BH_{2.7}$ oriented parallel (∥) and perpendicular (⊥) to the field direction

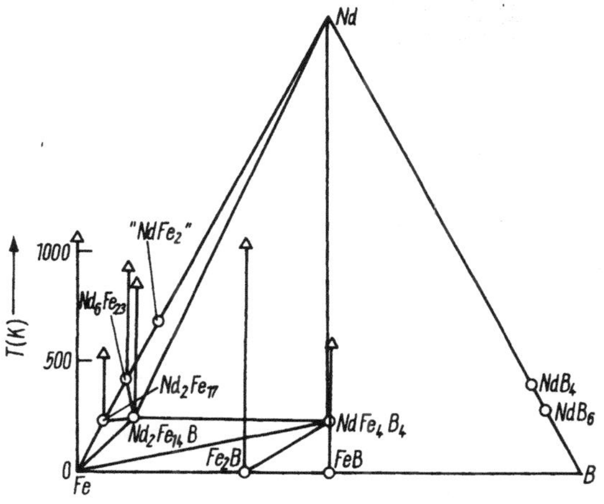

Fig. 3. Curie temperature T_C and phase relations, in Nd-Fe-B (data are from this laboratory), T_C should increase towards a line extending from Fe_2B to a fictitious "$NdFe_2$" (see text). An increase in T_C upon hydride formation of $Nd_2Fe_{14}B$ can be anticipated as a result of a shift towards this line in terms of electronic configuration

It is interesting to note that hydride formation of $Nd_2Fe_{14}B$ increases the saturation magnetization M_s. This appears to be a result of the increased volume of the unit cell. We have also previously shown /6, 7/ that hydrides of $Y_2Mn_{23-x}Fe_x$ and related systems mimic the magnetic behavior of electron-richer compositions (higher x). In this band filling model, high values of M_s and T_C should be obtained approximately along the line from "$NdFe_2$" to Fe_2B (Fig. 3) in Nd-Fe-B assuming both Nd and B to contribute three electrons. H uptake results in a situation which corresponds to a shift towards this line.

Concerning H_A we note that hydriding of $Nd_2Fe_{14}B$ has resulted in a change to approximately the value of $Y_2Fe_{14}B$ (H_A = 20 kOe at 300 K). We ascribe this change to the reduction in Nd-Fe exchange upon hydriding. This diminishes, at room temperature, the influence of Nd anisotropy so that we observe mainly the Fe anisotropy in the hydride. This in turn further corroborates the idea that H_A of $Nd_2Fe_{14}B$ originates from both Nd and Fe components. However, Nd-Fe exchange appears to be weak and there exists a peculiar lack of strong temperature dependence of H_A in $Nd_2Fe_{14}B$.

The situation at lower temperature is more complex. H_A obviously increases on going to lower temperatures (Fig. 2). However, beyond the upturn in M of "perpendicular powders", H_A values obtained from straightforward extrapolation obviously do not retain their classical meaning.

References

/1/ M. SAGAWA, S. FUJIMURA, N. TOGAWA, H. YAMAMOTO, and Y. MATSURA, J. appl. Phys. 55, 6 (1984).

/2/ F. SPADA, C. ABACHE, and H. OESTERREICHER, J. less-common Metals 99, L21 (1984).

/3/ H. OESTERREICHER, F. SPADA, and C. ABACHE, Mater. Res. Bull. (August 1984).

/4/ J.F. HERBST, J.J. CROAT, F.E. PINKERTON, and W.B. YELON, Phys. Rev. B 29, 4176 (1984).

/5/ D. GIVORD, H.S. LI, and J.M. MOREAU, Solid State Commun. 50, 497 (1984).

/6/ H. OESTERREICHER and H. BITTNER, phys. stat. sol. 41, K101 (1977)

/7/ H. OESTERREICHER and H. BITTNER, J. Magnetism magnetic Mater. 15/18, 1264 (1980).

(Received July 23, 1984)

phys. stat. sol. (a) 85, K65 (1984)

Subject classification: 7 and 18; 19; 22.8.2

Grupo de Física Molecular e Magnetismo,
Departamento de Física, Universidade de Brasília[1)]

The Usefulness of the Photoacoustic Cell for Magnetic Measurements[2)]

By

N.M.L. KÖCHE, P.C. MORAIS, and K. SKEFF NETO

In the conventional photoacoustic spectroscopy (PAS) a gas-microphone cell has been used to detect the photoacoustic signal. Basically, the PAS signal is detected as a periodic gas pressure fluctuation in an enclosed chamber, as described elsewhere /1/.

In the present letter we report magnetic measurements, namely, ferromagnetic resonance and magnetization curve, taken with a gas-microphone cell, similar to that used in the PAS detection. In Fig. 1 we show the block diagram used in our experiments. The spacial configuration of the modulation coils (A) is similar to that used by Foner /2/. The waveguide (Fig. 1) is attached to an X-band FMR spectrometer. In Fig. 2a we show a detail of the end of the waveguide and in Fig. 2b a detailed description of the gas-cell is made. To illustrate the versatility of the system described above we have used a commercial ferrite. The magnetization measurements are made by the microphone attached to the PAS cell (Fig. 2b) and the FMR measurements are made using the microphone attached to the end of the waveguide (Fig. 2a). The screw in Fig. 2b is used to prevent vibrations of the elastic membrane when FMR measurements are made.

For magnetization measurements we observe that the presence of a dc magnetic field modulated by an ac magnetic field (modulation coils) moves the sample around an equilibrium position. The pressure fluctuation inside the gas cell is detected by the capacitive

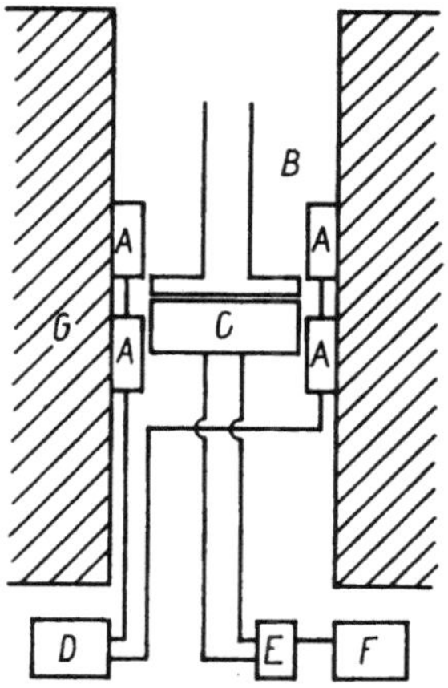

Fig. 1. Block diagram: (A) modulation coils, (B) waveguide, (C) PAS cell, (D) signal generator and audio amplifier, (E) pre-amplifier, (F) lock-in amplifier, (G) magnetic poles

1) 70910 - Brasília, D.F., Brazil.

2) Work partially supported by CNPq.

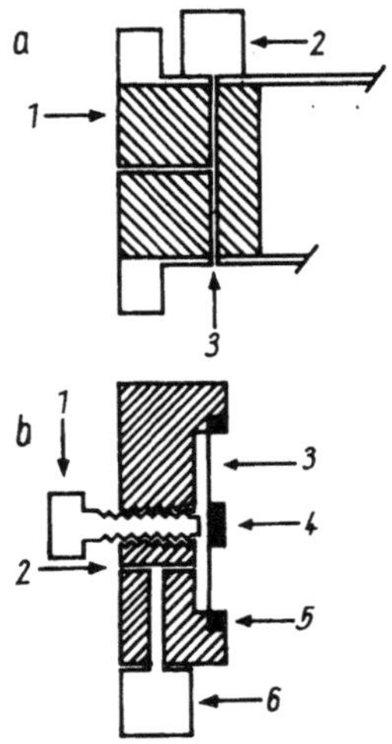

Fig. 2. a) 1 teflon, 2 microphone, 3 duct for pressure adjust; b) 1 screw, 2 duct for pressure adjust, 3 elastic membrane, 4 sample, 5 teflon O-ring, 6 microphone

microphone. The amplitude of this pressure fluctuation is taken as proportional to the magnetization of the sample. From this we obtain the magnetization curve as shown in Fig. 3, curve a. The linearity of the system is displayed in the inset of Fig. 3, where we have in curve c the output of the lock-in amplifier versus milligrams of iron(Mössbauer iron foil) at 7 kG. In Fig. 3, curve d we show the residual signal due to the metallic foil of the capacitive microphone. At this point we note that the FET of the capacitive microphone has been pulled out and connected directly with the preamplifier. Furthermore, we note that the motion of the elastic membrane must be harmonic, for linearity requirements. In Fig. 3, curve b we have the FMR curve detected by photoacoustic spectroscopy at 9.1 GHz. For a more detailed description of this measurement see /3, 4/.

In conclusion, we have pointed out that similar experimental arrangements can be used successfully in a variety of experimental measurements. The inset in Fig. 3 shows clearly the sensibility and the linearity of the method.

The authors wish to express their sincere thanks to the members of the resonance group of the UFMG - Brazil, specially to Prof. R. Gazzinelli, A.S. Chaves, and G.M. Ribeiro for their hospitality and interest.

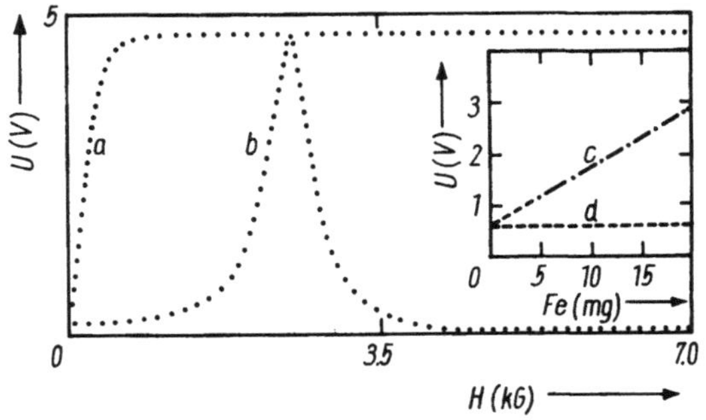

Fig. 3. Experimental data of the ferrite at room temperature;(a) magnetization curve, (b) ferromagnetic resonance curve, (c) output of the lock-in amplifier versus milligrams of iron at 7 kG, (d) residual signal of the metallic foil of the microphone

References

/1/ A. ROSENCWAIG, in: Optoacoustic Spectroscopy and Detection, Ed. Y.H. PAO, Academic Press, London 1977 (p. 193); in: Advances in Electronics and Electron Physics, Vol. 46, Ed. L. MARTON, Academic Press, New York 1978 (p. 208); in: Photoacoustics and Photoacoustic Spectroscopy, Chem. Analysis, Vol. 57, Ed. P.F. ELVING, F.D. WINEFORDNER, I.M. KOLTHOFF, and A. WITLEY, Interscience Publishers, 1980.

/2/ S. FONER, Rev. sci. Instrum. 27, 548 (1956); 30, 548 (1959); 45, 1181 (1974); 46, 1425 (1975).

/3/ O.A.C. NUNES, M.A. AMATO, and K. SKEFF NETO, phys. stat. sol. (a) 68, K25 (1981).

/4/ A.L. TRONCONI, K. SKEFF NETO, and M.A. AMATO, phys. stat. sol. (a) 76, K209 (1983).

(Received June 14, 1984)

phys. stat. sol. (a) 85, K69 (1984)
Subject classification: 14.1 and 18.2; 22.8

Institute of Physics of Solids and Semiconductors, Academy of Sciences of the Belorussian SSR, Minsk[1)]

Crystallographic, Magnetic, and Electrical Properties of Nickel-Substituted Chromium Telluride

By

G.I. MAKOVETSKII and V.P. DYMONT

In /1/ it is shown that the substitution of chromium atoms by vanadium and manganese ones in the $(Cr_{1-x}M_x)_{1-\delta}Te$ (M = Mn, V) solid solution system results in a non-monotonic change of the Curie temperature as a function of composition. The non-monotonic change of the Curie temperature as a function of composition is also observed in the $Cr_{1-x}Ti_xTe$ system /2/. A small rise of the Curie temperature, when compared to that of chromium telluride, is observed for some compositions of the above systems. In this connection it is of interest to study magnetic and other physical properties of the $Cr_{1-x}Ni_xTe$ system.

The samples were prepared by direct alloying of chromium (99.9%), tellurium (99.99%), and nickel (99.8%). The appropriate amounts of the starting materials were carefully mixed, placed in quartz tubes evacuated to 10^{-2} Pa. The synthesis was performed according to the following procedure. The mixture was slowly heated up to 1220 K with a 24 h exposure at 1020 K. The samples obtained were ground to powders, pressed to pellets, annealed during 100 h at 1220 K, and then quenched in cold water.

The X-ray diffraction study was performed on powder samples with CuK_α radiation at room temperature. All reflections on the X-ray diffraction patterns can be indexed through the hexagonal NiAs-type structure. However, for the $Cr_{0.85}Ni_{0.15}Te$ sample, a considerable broadening of the reflections (102) and (103) was observed, which can be attributed to the monoclinic distortions of the NiAs structure.

Fig. 1 presents the lattice parameters a and c versus the nickel concentration. The parameters a and c are seen to decrease with increasing Ni concentration.

---

1) Podlesnaya 17, 220726 Minsk, USSR.

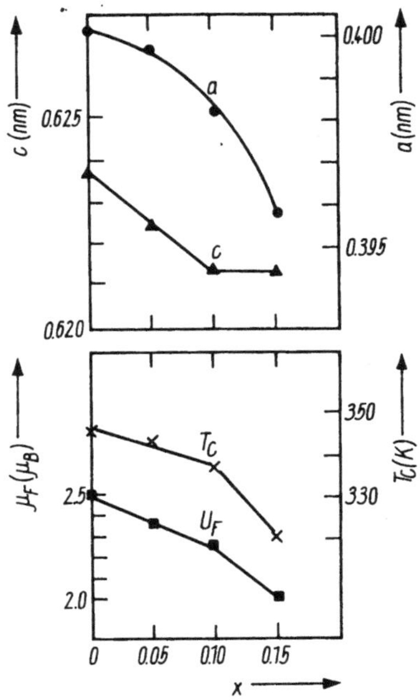

Fig. 1. a) Unit cell parameters a and c; b) Curie temperature and effective magnetic moment as a function of nickel concentration x

All synthesized samples were ferromagnets. Their saturation magnetization was measured in the temperature range 100 to 400 K. Measurements were made by the Faraday method in a 4.5 kOe field. The transition temperature was determined by extrapolating the linear part of the σ^2(T) curve to the temperature axis. In Fig. 1b, the dependence of the Curie temperature on composition is given. Increase of the Ni concentration leads to a decreasing transition temperature. In this figure, the effective magnetic moment as a function of composition is also presented. While substituting Cr atoms by Ni ones, the magnetic moment decreases from 2.5 μ_B for CrTe to 2.0 μ_B for $Cr_{0.85}Ni_{0.15}Te$.

The temperature dependence of the electrical resistivity for samples with different Ni concentration was investigated. Measurements were made by a two-probe technique in the temperature range 100 to 900 K. The curves of the temperature-dependent electrical resistivity are given in Fig. 2. All compositions exhibit metal-type conduction from 100 to 600 K. The curves of the ϱ(T) dependence show breaks typical of ferromagnetic materials due to the destruction of the ferromagnetic ordering from 300 to 350 K. The analysis of the (T) curve for the $Cr_{0.85}Ni_{0.15}Te$ sample supports that this composition possesses the distored lattice of the NiAs type. For samples of this composition, a considerable increase of the temperature coefficient of the electrical resistivity in the temperature range below the temperature of the magnetic phase transition is observed. As is shown in /3/, the ϱ(T) curves of quenched and annealed samples of Cr_5Te_6 composition (the quenched sample has the NiAs-type structure, the annealed one the monoclinic one) differ just in the temperature coefficient of the resistance. A sharp increase of the sample resistance in the temperature range 600 to 700 K is, probably, due to the

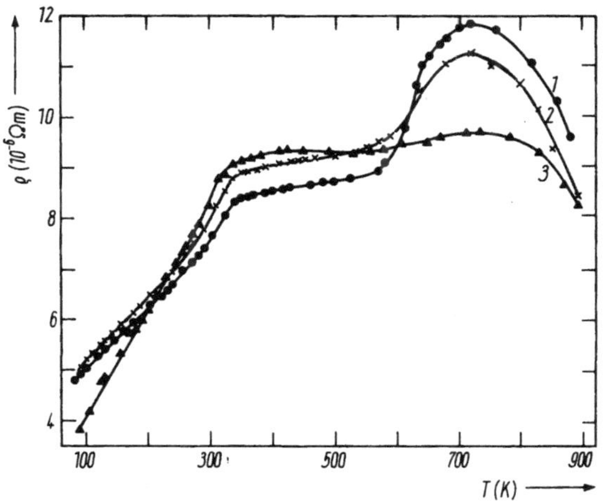

Fig. 2. Electrical resistivity as a function of temperature; (1) $Cr_{0.95}Ni_{0.05}Te$, (2) $Cr_{0.9}Ni_{0.1}Te$, (3) $Cr_{0.85}Ni_{0.15}Te$

phase change at heating of quenched samples. Thus, solid solutions of the $Cr_{1-x}Ni_xTe$ system with the NiAs structure exist in the concentration range $0 < x \leq 0.1$. The substitution of chromium atoms by nickel ones results both in decreasing temperature of transition from the ferromagnetic to the paramagnetic state and in reducing mean magnetic moment. All samples investigated exhibit (temperature-dependent) metal-type conduction.

References

/1/ E.B. BLANKOVA, B.E. STARICHENKO, M.A. UIMIN, N.I. PEREPECHAEVA, and A.P. UNRAU, Dep. VINITI N 1520-77.

/2/ G.I. MAKOVETSKII and V.P. DYMONT, Dokl. Akad. Nauk Beloruss. SSR 6, 521 (1984).

/3/ N.P. GRAZHDANKINA, Fiz. i Tekh. vyssok. Dav. 1, 43 (1980).

(Received July 19, 1984)

phys. stat. sol. (a) 85, K73 (1984)
Subject classification: 13.3 and 20.3; 22.1.1; 22.6

Department of Physics, Moscow State University[1)]

Fluorescence Decay of Adsorbed Dye Molecules on Solid Surfaces

By

V. A. BESPALOV, V.F. KISELEV, G.S. PLOTNIKOV, and A.M. SALETSKII

It was shown in /1/ that for dielectrics or semiconductors with thick (>20 nm) oxide layer the main channel of adsorbed molecule excitation relaxation was the energy transfer to adjacent adsorbed molecules (channel 1). An alternative channel 2, i.e. the dissipation into the solid, becomes competitive in semiconductors with a thin (< 10 nm) oxide film. In this case the energy acceptors are the valence band electrons (channel 2A), electrons captured by insulator traps (IT) (channel 2B), or interface traps (channel 2C) /1, 2/.

The study of fluorescence (F) decay kinetics of adsorbed molecules and lifetimes of the excited states τ gives us direct information about the relative efficiency of the energy transfer through these channels. Lifetime measurements /3, 4/ of Rhodamin B (RB) molecules on the surface of molecular crystals, semiconductors, and quartz indicate that $\tau \approx 35$ to 50 ps, which is about 3 orders smaller than τ_0 for isolated molecules in diluted solutions (1.5 to 5 ns) /4, 5/. There are no data about changes of electrophysical parameters of the surface accompanying the F-decay in these works and it is impossible to say anything about the prevailing excitation relaxation channels and their contribution to τ.

We studied the F-decay kinetics of RB molecules adsorbed in concentration $N = 2x10^{13}$ mol cm^{-2} on the polished and etched surfaces of quartz and on thermally oxidized ⟨111⟩ faces of Ge n-type single crystals (ρ = 30 Ωcm) with charged and uncharged IT's with the help of a second fluorimeter /6/.[2)] The thickness of the GeO_2 layer was 5 nm. Discharging of a charged IT at the surface was most effective for samples with this thickness of GeO_2 /1/. The concentration dependence of the F-decay and negative charge discharging had a maximum at a chosen concentration N (see Fig. 9 in /1/). The measurements of the surface IT charge were based on the field effect technique. The techniques of IT photocharging, RB adsorption, and N determination were the same as in /1/.

---

1) Leninskie Gory, 119899 Moscow, USSR.

2) The authors are grateful to Dr. V.Z. Pashenko for his generous assistance.

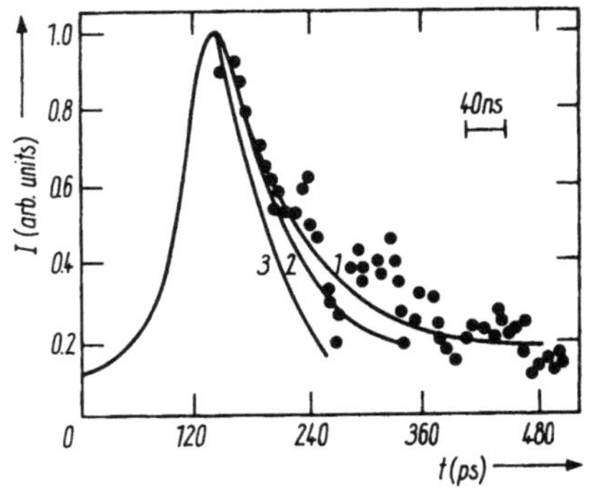

Fig. 1. Experimental curves of the fluorescence decay of Rhodamine B molecules (1) on the quartz surface, (2) on the Ge surface with uncharged insulators traps (IT), and (3) on the Ge surface with charged IT

Fig. 1 shows the experimental curves of the F-decay I(t). The information from the spectrograph (points on Fig. 1 for quartz sample) was processed by Eniac and represented as smooth curves (we do not give points for the other curves). Information about the lifetime τ was read out simultaneously (first three lines of Table 1). I(t) was an exponential curve with τ = 56 ps for quartz, where excitation relaxation was realized through channel 1 only. The statistical weight of the other exponential curves in Fig. 1 was less than 0.01.

In the case of a semiconductor we have supplementary channels of energy migration to the acceptors in the solid (channels 2A, 2B, 2C) resulting in energy relaxation acceleration (Fig. 1 and Table 1).

Elementary exponential curves of energy relaxation over channels 2A + 2C (Ge with uncharged IT) and through channel 2B (to the charged IT) in Fig. 2 were calculated by Eniac using the experimental curves 2 and 3 on Fig. 1 and the fact that the energy relaxation through channel 1 was close to an exponential. Respective time constants are given in Table 1.

Table 1

channels	τ(ps)	I(t)	G_m (n=6)	R_o (nm)	R (nm)	K (10^{10} s^{-1})	comments
1	56	expon.	0.001	5	2.5	1.3	exper.
1+2A+2B	42	non-expon.	-	-	-	-	exper.
1+2A+2B+2C	36	non-expon.	-	-	-	-	exper.
2A+2B	168	expon.	0.012	8	5	0.3	calc.
2C	46	expon.	0.005	5	2.5 to 3.5	0.8 to 1.5	calc.

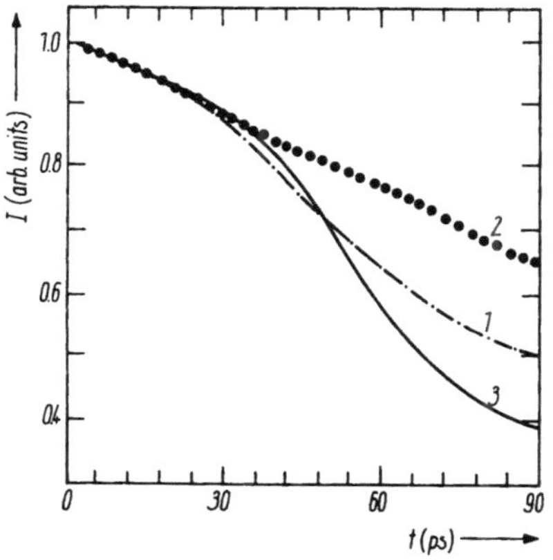

Fig. 2. Exponential curves of excitation relaxation through (1) channel 1, (2) channels 2A+2B, (3) channel 2C

If the concentration of energy acceptors in a condensed phase is not very high, the energy relaxation of excited molecules is explained by the Dexter-Förster induction-resonance mechanism of energy exchange /2, 7/ with a rate constant

$$K_{DF} = \frac{1}{\tau_0}\left(\frac{R_0}{R}\right)^n . \qquad (1)$$

Here τ_0 is the lifetime of an isolated excited molecule, n = 6, 8, and 10 for dipole, the dipole-quadrupole, and quadrupole interaction, respectively; R_0 is the critical radius, R is the distance between donors and acceptors. We fitted the exponential curves on Fig. 2 to a theoretical decay function $\Phi(t)$ /8/ to identify the interaction law in the case of energy migration through various channels,

$$\Phi(t) = \exp\left[-\frac{t}{\tau_0} - \frac{2}{\sqrt{\pi}}\Gamma\left(1-\frac{3}{n}\right)\gamma_A\left(\frac{t}{\tau_0}\right)^{3/n}\right], \qquad (2)$$

where n is the power index in formula (1), $\gamma_A = (\sqrt{\pi}/2)(C_A/C_A^0)$; here C_A and C_A^0 are the concentration and critical concentration of the acceptor A, respectively. In our case $\gamma_A \approx 0.8$.

It happens to be that all curves in Fig. 2 are closest to the theoretical function $\Phi(t)$, when n = 6 with a very small root-mean-square deviation G_m (see Table 1).

Thus the energy relaxation of photoexcited molecules is well explained by a dipole induction-resonance mechanism /2, 7/ - both in the adsorption phase and in a semiconductor.

We have estimated in /1/ that in our case only $\approx 10^{10}$ mol cm^{-2} are excited at the surface during illumination, with is nearly an order of magnitude less than the IT number (discharged via energy migration through channel 2B, $N_{IT} \approx 10^{11}$ electrons cm^{-2}). Under these conditions, when I(t) is close to exponential the relation $I/I_0 = \tau/\tau_0$ must be satisfied. Here τ and τ_0 are the mean lifetimes of the excited states, I and I_0 are F-yields in presence and absence of quenching, respectively /9/. In our case this ratio is equal to 0.92.

We have estimated K_{DF} (see Table 1) using reasonable values of R and R_0 for quartz-RB and Ge-GeO_2 + RB systems and supposing τ_0 = 5 ns. Values of K_{DF} appeared to be 4 to 5 orders of magnitude higher than the electron transfer ratio constant K_e. This constant determines the process of electron transfer from the higher singlet level of an excited molecule to the conduction band of Ge with simultaneous regeneration of a cation-radical via electron tunneling from IT to the lower singlet level. The rate constant K_e of this electron exchange is limited by the second process and may be estimated using the well-known formula for tunneling through a rectangular barrier. Thus in our case the energy is the main deactivation process.

References

/1/ S.V. VINTSENTS, V.F. KISELEV, and G.S. PLOTNIKOV, phys. stat. sol. (a) 85, No. 1 (1984).

/2/ D.L. DEXTER, J. Lum. 18/19, 779 (1979).

/3/ N. NAKASHIMA, K. YOSHIHARA, and F. WILLING, J. chem. Phys. 73, 3553 (1980).

/4/ Y. LIANG, A.M. PONTE GONSALVES, and D.R. NEGUS, J. phys. Chem. 87, L1 (1983).

/5/ M.J. SHARE, F.E. TREOLOAR, K.P. CHIGGINO, and P.J. THISTLETAWAITE, J. Photochem. 18, 335 (1982).

/6/ V.Z. PASHENKO, S.P. PROTASOV, A.B. RUBIN et al., Biochem. Biophys. Acta 408, 143 (1976).

/7/ TH. FÖRSTER, Disc. Faraday Soc. 27, 7 (1956).

/8/ M. INOKUTI and F. HIRAJAMA, J. chem. Phys. 43, 1978 (1968).

/9/ C. BOJARSKI, Acta Phys. Polon. A63, 251 (1983).

(Received August 7, 1984)

phys. stat. sol. (a) 85, K77 (1984)
Subject classification: 20.3; 22.2.4

Institute of Semiconductor Physics and Optics of the Technical University Braunschweig[1)]

Cathodoluminescence of InGaAsP

By

F.R. KESSLER, U. BARKOW, and K. JÜRGENS

The quarternary system InGaAsP has an energy gap of 0.75 to 1.4 eV corresponding to luminescence light emission in the wavelength region between 0.9 to 1.7 μm. In this spectral region the transmission of light through glass fibres is extremely high. The energy gap and the lattice constant can be adjusted independently by choosing the appropriate composition ratio. Thus the preparation of high-quality epitaxial layers is relatively simple. For this reason InGaAsP is becoming increasingly interesting as a material for transmitters in optical signal communication, and therefore its luminescence properties are of special interest /1/.

In /2/ the spectral distribution of luminescence radiation from $In_{0.73}Ga_{0.27}As_{0.63}P_{0.37}$ and InP was measured under cathode-ray excitation: This had the advantage that luminescence could be attained without electrical contacts being required as in the case of LED technology and without the disturbing effects by scattering of excitation radiation as in the case of photoluminescence. A comparison between these results and spectra calculated by means of the principle of detailed balance for band-to-band transitions /3/ was made.

The sample was prepared as an epitaxial layer grown from the liquid phase onto 100-oriented InP substrates. Excitation was made with pulsed electron beams focused onto the sample in the form of a line. Thus, it was possible to obtain optimum illumination of the entrance slit of the spectrometer. All measurements were made at room temperature with acceleration voltages of about 9 kV.

The luminescence radiation is generated inside the sample, so the measured spectra have to be corrected for the effects of self-absorption in order to enable a comparison with calculated spectra. The absorption coefficients needed for these corrections and for calculating spectra using the principle of detailed

1) Postfach 3329, D-3300 Braunschweig, FRG.

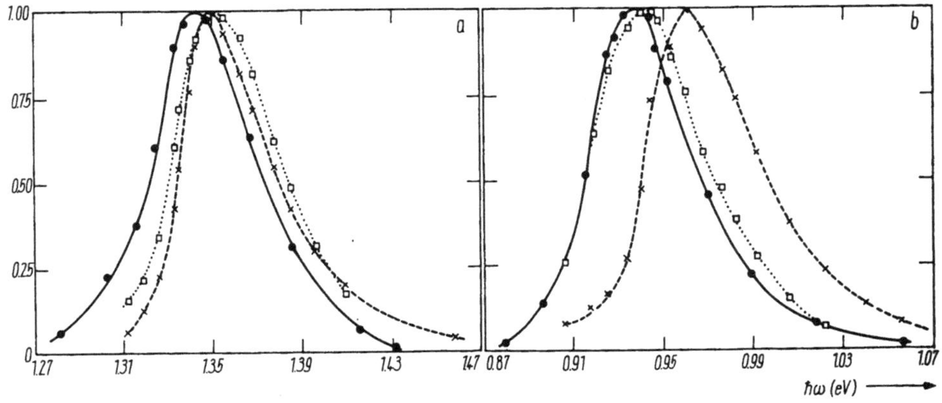

Fig. 1. Luminescence spectra of a) InP, b) $In_{0.73}Ga_{0.27}As_{0.63}P_{0.37}$; ——— experimental spectra, • • • • • experimental spectra corrected for self-absorption, - - - theoretical spectra for interband transitions

balance are taken from Kowalsky et al. /4/ for InGaAsP and from Li Sheng /5/ and Seraphin and Bennett /6/ for InP. Results are shown in Fig. 1, intensities have been drawn in arbitrary units.

The maximum intensity of luminescence radiation for InP, corrected for self-absorption, corresponds to the calculated value. Therefore, the luminescence must be considered as being dominated by band-gap emission, since the calculation was made with respect to band-to-band transitions. In contrast to the result for InP, the maximum intensity for InGaAsP, corrected for self-absorption, was shifted to a lower energy region by 15 meV in comparison to the calculated spectrum. One could thus conclude that in the case of quarternary layers, the band-to-band recombination is not dominating. Yamazoe et al. /7/ arrived at the same conclusion from measurements of the photoluminescence of InGaAsP. Since, however, in the case of electron-beam excitation as in our experiment, the penetration depth of the excitation was reduced, our results are more valid for the surface region.

A calculation including exciton states in the recombination process gives an incorrect result, because the corresponding shift of about 15 meV is much greater than the exciton binding energy of about 5 meV. The measured spectra can be explained, if the luminescence is interpreted as being due to transitions from the conduction band to activator states. The energetic distance between the activator level and the upper edge of the valance band has to be 15 meV in this case. Further experiments with specially prepared epitaxial layers would be helpful to allow further investigation into the dominant process of recombination in this quarternary semiconductor, and to analyse the nature of recombination centers, which might possibly be surface states.

We thank the Institut für Hochfrequenztechnik, Technische Universität Braunschweig, for preparing the InGaAsP sample and Prof. Dr. A. Schlachetzki for his cooperation.

References

/1/ A. SCHLACHETZKI, Phys. Bl. 37, 337 (1981).

/2/ K. JÜRGENS, Diplomarbeit, TU Braunschweig, 1983.

/3/ W. VAN ROOSBROECK and W. SHOCKLEY, Phys. Rev. 94, 1158 (1954).

/4/ W. KOWALKY, A. SCHLACHETZKI, and F. FIEDLER, phys. stat. sol. (a) 68, 153 (1981).

/5/ S. LI SHENG, Appl. Phys. Letters 29, 126 (1976).

/6/ B.O. SERAPHIN and H.E. BENNETT, in: Semiconductors and Semimetals, Vol. 3, Ed. A.C. BEER and R.K. WILLARDSON, Academic Press, New York 1967 (p. 530).

/7/ Y. YAMAZOE, T. NISHINO, Y. HAMAKAWA, and T. KARIYA, Japan. J. appl. Phys. 19, 1473 (1980).

(Received July 31, 1984)

phys. stat. sol. (a) 85, K81 (1984)
Subject classification: 7 and 16; 14.3; 22.4.3
Institut für angewandte Physik, Universität Karlsruhe[1)]

Bipolar Photoconductivity and Hall Effect of the Acoustoelectric Current in CdTe

By
M. ADOLF, J. ROSENZWEIG, and U. BIRKHOLZ

In a bipolar semiconductor the investigation of the electrical conductivity

$$\sigma = \sigma_p + \sigma_n \tag{1}$$

and of the Hall mobility

$$\mu_H = \frac{E_H \sigma}{jB} = A \frac{\mu_p \sigma_p - \mu_n \sigma_n}{\sigma} \tag{2}$$

(σ total conductivity, σ_p, σ_n hole and electron conductivities, μ_H, μ_p, μ_n Hall mobility and mobilities of holes and electrons, A scattering factor) is not sufficient to determine densities and mobilities of the charge carriers. The measurement of the Hall effect for the acoustoelectric current (HEAC), first reported by Kietis et al. /1, 2/, can give an additional relation. Using the expression for the acoustoelectric current j_{AE} /3/

$$j_{AE} = \frac{\mu_p \sigma_p - \mu_n \sigma_n}{\sigma} \alpha \Phi \tag{3}$$

(α absorption coefficient, Φ energy density of the acoustic wave) a coefficient μ_{HEAC} /1/ is defined in analogy to the Hall effect of the drift current

$$\mu_{HEAC} = \frac{E_{HEAC} \sigma}{j_{AE} B} = A \frac{\mu_p^2 \sigma_p + \mu_n^2 \sigma_n}{\mu_p \sigma_p - \mu_n \sigma_n} \tag{4}$$

independent of absorption and energy density of the acoustic wave. The CdTe samples used in these experiments were p-type ($\mu_p \sigma_p \gg \mu_n \sigma_n$) in the dark. The Hall mobility in the dark $\mu_{H,d}$ therefore corresponds to the hole mobility

$$\mu_{H,d} = A \mu_p \, . \tag{5}$$

The hole mobility is supposed to be independent of illumination. Using σ, μ_H,

---

1) Kaiserstr. 12, D-7500 Karlsruhe, FRG.

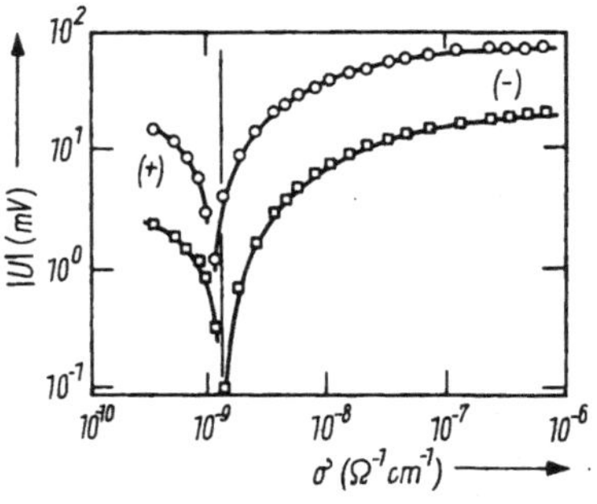

Fig. 1. Hall voltage U_H (□) and acousto-electric voltage U_{AE} (○) as functions of the photoconductivity

μ_{HEAC}, and $\mu_{H,d}$ the densities and mobilities of holes and electrons can be determined. Calculations were carried out with A = 1.

Photosensitive CdTe single crystals were grown from the melt with excess Te using the Bridgman method. Compensation was achieved by doping with $CdCl_2$ as described by Höschl et al. /4/. The dark resistivity of the p-type crystals was $3x10^9$ Ωcm at 300 K. The conductivity type was inverted to n-type by increasing illumination with band gap light of 854 nm. Measurements were performed at room temperature. The crystals with dimensions 1x2x6 mm^3 were orientated in the [111] direction. For the acoustoelectric investigations longitudinal acoustic waves of 23 MHz were used.

The Hall voltage U_H and the open-circuit acoustoelectric voltage U_{AE} as functions of the photoconductivity are presented in Fig. 1. The corresponding coefficients μ_H and μ_{HEAC} are shown in Fig. 2. In contradiction with formulae (2), (3), and (4) the acoustoelectric quantities U_{AE} and μ_{HEAC} change the sign at lower photoconductivities than the Hall effect. A vertical line indicates the inversion point $\mu_H = 0$ in all figures. The electron mobility and the electron density calculated from these data (Fig. 3 and 4) show a non-physical singularity in the vicinity of the Hall voltage inversion. In order to explain the different inversion points of μ_H and μ_{HEAC} a trapping of the charge carriers is assumed. The trapping can be described by a factor f, which is the ratio of the free bunched space charge to the total bunched space charge generated by the acoustic wave /5/.

According to Greebe /5/ for an n-type semiconductor the acoustoelectric current is

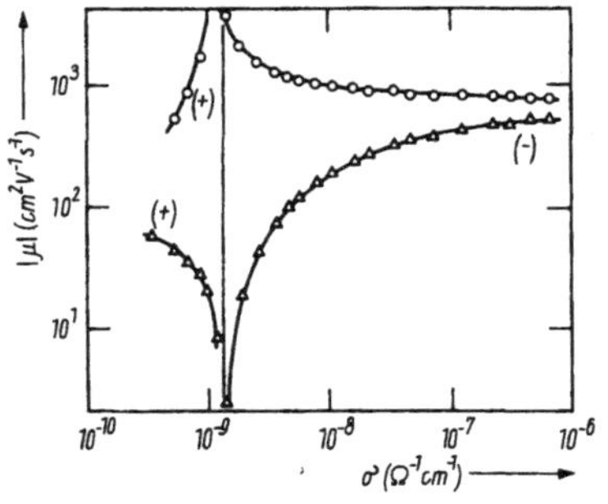

Fig. 2. Dependence of the Hall coefficients μ_H(Δ) and μ_{HEAC} (o) on photoconductivity

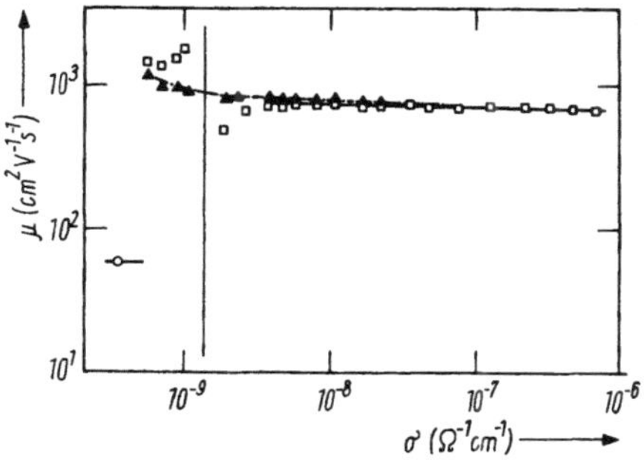

Fig. 3. Dependence of the electron mobility μ_n on photoconductivity calculated with (▲) and without (□) the trapping factor $f_r = 0.79$. Hole mobility μ_p (○) measured in the dark is assumed to be independent of photoconductivity

$$j_{AE} = -\mathrm{Re}(f)\mu_n \alpha \Phi . \tag{6}$$

If the acoustic frequency is sufficiently low, diffusion processes can be neglected and only the real part of f contributes to j_{AE} /6/.

In the case of bipolar conduction with trapping, equations (3) and (4) have to be modified:

$$j_{AE} = \frac{f_p \mu_p \sigma_p - f_n \mu_n \sigma_n}{\sigma} \alpha \Phi , \tag{7}$$

$$\mu_{HEAC} = \frac{f_r \mu_p^2 \sigma_p + \mu_n^2 \sigma_n}{f_r \mu_p \sigma_p - \mu_n \sigma_n} \tag{8}$$

(f_p, f_n trapping factors of holes and electrons, $f_r = f_p/f_n$).

Choosing $f_r = 0.79$ the electron mobility as function of photoconductivity becomes continuous and the singularity in the calculated electron density is removed (Fig. 3 and 4). Hole trapping dominates if $f_r < 1$. Neglecting electron trapping ($f_n \approx 1$) the energy range of the hole trap levels can be estimated according to /5/. Probably the relevant trap levels are situated between 0.2 and 0.4 eV above the valence band.

The electron density increases proportional to the generation rate G (Fig. 4). The number of

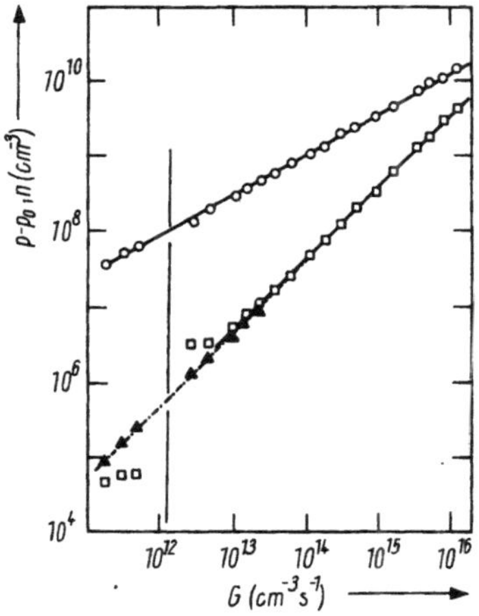

Fig. 4. Generated hole density ($p - p_0$) (○) and electron density n as functions of generation rate G; n calculated with (▲) and without (□) the trapping factor $f_r = 0.79$; p_0 is the hole density in the dark

generated holes $(p - p_o)$ increases with $G^{0.5}$. This behaviour can be explained by a model containing three kinds of defects as described in /7/.

In CdTe crystals with lower dark resistivities ($\varrho \approx 10^7\ \Omega cm$) the inversion points of μ_H and $\mu_{HE\,AC}$ were observed at photoconductivities which differ more than one order of magnitude. For these samples an estimation yields trapping factors $f_r \approx 0.14$.

References

/1/ B.-P. KIETIS, R. REKSNYS, and A. SAKALAS, phys. stat. sol. (a) 39, 601 (1977).

/2/ B.-P. KIETIS, R. REKSNYS, A. SAKALAS, and P. HÖSCHL, phys. stat. sol. (a) 43, 705 (1977).

/3/ M. FINK and G. QUENTIN, phys. stat. sol. (a) 4, 397 (1971).

/4/ P. HÖSCHL, P. POLIVKA, V. PROSSER, and A. SAKALAS, Czech. J. Phys. B 25, 585 (1975).

/5/ C.A.A.J. GREEBE, Philips Res. Rep. 21, 1 (1966).

/6/ P.D. SOUTHGATE and H.N. SPECTOR, J. appl. Phys. 36, 3728 (1965).

/7/ J. ROSENZWEIG and U. BIRKHOLZ, J. Crystal Growth 59, 263 (1982).

(Received August 8, 1984)

phys. stat. sol. (a) 85, K85 (1984)
Subject classification: 14.3 and 16; 22.8

Institute of Applied Physics,
Academy of Sciences of the Moldavian SSR, Kishinev[1)]

Relaxation Peculiarities in $TlSbSe_2$ Crystals

By
I.N. GRINCHESHEN, N.S. POPOVICH, and A.A. SHTANOV

The semiconducting compounds $TlSbSe_2$ and $TlSbS_2$ form monoclinic and anortic crystal lattices with resistivities of $\approx 10^3$ and $\approx 10^9$ Ωcm, respectively, and have a strong photosensitivity /1, 2/.

The photoelectric properties of $TlSbS_2$ crystals at low and high optical excitation have been investigated /1, 3/. Deep levels (r-, s-centres) have been found in the band gap of $TlSbS_2$; they are effective recombination centres, responsible for short time ($\tau \leq 10^{-8}$ s) photoconductivity relaxation.

Investigations of lux-ampere characteristics and photoconductivity relaxation at low and high optical excitation of $TlSbSe_2$ crystals have been carried out in order to clear up the common features of the recombination mechanism in such crystals.

In experiments n-$TlSbSe_2$ crystals with resistivity $\approx 3 \times 10^2$ Ωcm have been used. Photoconductivity relaxation at low optical excitation has been investigated by means of an experimental set-up presented in /4/. At high optical excitation a Nd laser ($\hbar\omega = 1.17$ eV, $t_{imp} \leq 10^{-8}$ s, $I_{max} \approx 10^{26}$ quanta/cm^2 s) was used. The laser intensity was changed by calibrated filters. The pulse form was written on a BCI-280(DDR) integrator.

The lux-ampere characteristics of $TlSbSe_2$ crystals at low optical excitation and 80 K are shown in Fig. 1a. Two linear portions are found. The bending point moves towards higher intensities with increase of the initial intensity. In the first portion the index n appearing in the dependence $J_F = f(L^n)$ decreases from 2.2 to 0.95 with illumination intensity increase and in the second portion it increases from 0.3 to 0.45.

Such a behaviour of the lux-ampere characteristics can be explained by conversion pf trapping levels into recombination centres. With the increase of illumination intensity the electron quasi-Fermi level moves up, so that the trap-

1) 277028 Kishinev, USSR.

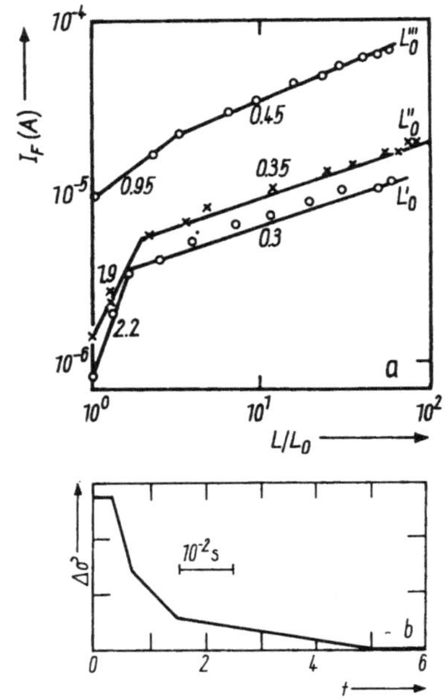

Fig. 1. a) Lux-ampere characteristics of $TlSbSe_2$ crystals at low optical excitation and 80 K (L'_0 = 0.22, L''_0 = 0.6; L'''_0 = = 10 $\mu W/cm^2$); b) photoconductivity relaxation at low optical excitation and 80 K

ping levels convert into recombination centres, which results in a sensitivity decrease.

The photoconductivity relaxation at low optical excitation is shown in Fig. 1b. There are three straight-line portions on the photoconductivity relaxation curve. In the first one $\tau_{rel} \leqq \tau_{imp}$, in the two last ones the rate of photoconductivity decrease differs by one order. This shows that within the deep levels responsible for the "photo-memory" effect, which we have found in /4/ in the band gap of $TlSbSe_2$ crystals, there are at least two trapping levels.

In order to estimate some parameters of the trapping centres of $TlSbSe_2$ crystals thermally stimulated currents have been investigated. The sample has been pre-heated up to 373 K with voltage being applied and then cooled in the darkness to 80 K. After preliminary light illumination the sample has been heated with a constant rate (β = 0.1 K/s). The TSC curve is shown in Fig. 2. Two peaks of the TSC curve show that in the $TlSbSe_2$ crystals there are two trapping levels. Using the formulas presented in /5, 6/, we estimated the depth, concentration, and the capture cross section of the trapping centres. These data are presented in Table 1.

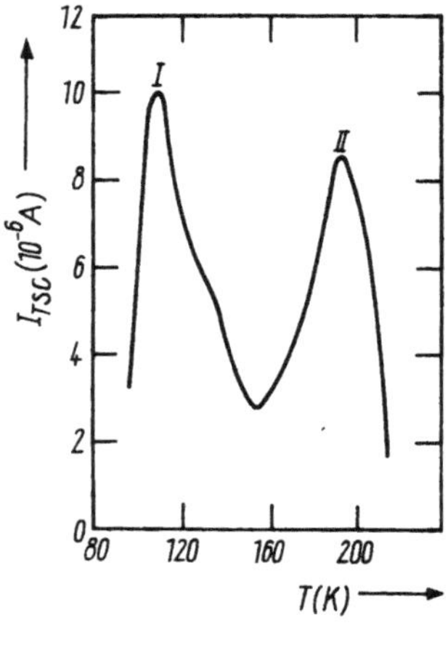

Lux-ampere characteristics (a) and the photoconductivity relaxation (b) at high optical excitation are shown in Fig. 3. As one can see, the relaxation of non-equilibrium carriers occurs during the time $\tau_{rel} \leqq \tau_{imp}$. However, a "long-time" component is observed as well.

Fig. 2. TSC curve of $TlSbSe_2$ crystals (β = 0.1 K/s; U = 20 V)

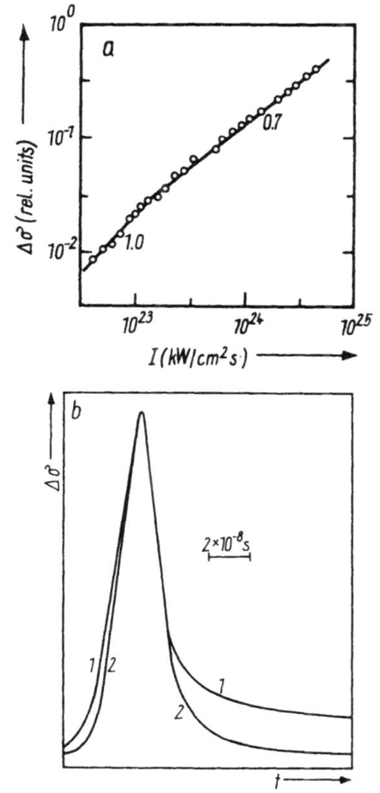

Fig. 3. a) Lux-ampere characteristic at high optical excitation and 300 K; b) photoconductivity relaxation curve of $TlSbSe_2$ crystals (1) at high optical excitation and 300 K, (2) laser pulse form

The experimental data show that the recombination of non-equilibrium carriers in $TlSbSe_2$ crystals proceeds via deep centres and trapping levels.

Unlike $TlSbS_2$, there are trapping levels in $TlSbSe_2$ crystals which prolong the photoconductivity relaxation time. Their occurrence can be considered as resulting from a structural phase transition in $TlSbSe_2$ crystals at 673 K. Therefore strong internal stress arises, which causes the appearance of extra defects in these crystals.

Table 1

parameters / peaks	E_t(eV)	N_t (10^{18} cm^{-3})	S_t (10^{-22} cm^2)
I	0.102	5	10
II	0.211	1	1

References

/1/ V.F. JITARI, N.S. POPOVICH, D.V. GITSU, and S.I. RADAUTSAN, Fiz. Tekh. Poluprov. 8, 966 (1974).

/2/ D.V. GITSU, I.N. GRINCHESHEN, and N.S. POPOVICH, phys. stat. sol. (a) 72, K113 (1982).

/3/ D.V. GITSU, N.S. POPOVICH, I.A. DAMASKIN, and S.L. PYSHKIN, in: Sbornik VIMI RIPORT, No. 20, 197 (1975).

/4/ D.V. GITSU, I.N. GRINCHESHEN, and N.S. POPOVICH, phys. stat. sol. (a) 76, K5 (1983).

/5/ J.G. SIMMONS and G.W. TAYLOR, Phys. Rev. B 5, 1619 (1972).

/6/ R.H. BUBE, Photoconductivity in Solids, Izd. in. Lit., Moscow 1962 (in Russian).

(Received May 9, 1984)

phys. stat. sol. (a) 85, K89 (1984)

Subject classification: 14.1; 21

Chemistry Division, Bhabha Atomic Research Centre, Bombay[1] (a) and Tata Institute of Fundamental Research, Bombay[2] (b)

Electrical Resistivity Studies on the Heusler Alloys $Co_2T_{1-x}Al_{1+x}$ (T = Ti or Zr)

By

J.V. YAKHMI (a), I.K. GOPALAKRISHNAN (a), and A.K. GROVER (b)

Heusler alloys have been widely studied by using magnetic and crystallographic methods /1, 2/. These alloys have the composition X_2YZ with $L2_1$ structure, where Z is generally an sp element, and Y and X are transition elements to the left of, and to the right of Fe in the periodic table, respectively. Most of the Heusler alloys show ferromagnetic order as a result of typical coordination of the transition elements. These alloys are promising systems for investigating the effects of scattering of conduction electrons by magnetic electrons, but their electrical transport properties have attracted attention only recently /3, 4/. Co_2TAl (T = Ti or Zr etc.) are interesting Heusler alloys since the entire moment is confined to the Co atoms only (viz. $0.35\mu_B$ and $0.30\mu_B$ per Co atom, respectively) /1/, unlike the much studied prototypical Heusler alloys X_2MnZ where the Mn atoms carry a highly localized moment of $\approx 4\mu_B$ irrespective of the nature of X or Z elements /2/. The magnetic behavior of Heusler alloys has been mostly studied by systematically changing one component, say Z. However, the study of related systems, in which the composition can be changed without altering the structure, i.e. where some of the Y or Z atoms occupy wrong sites is more interesting for the study of magnetic behaviour in relation to the electronic structure. It is with this objective that we report the electrical resistivity (ϱ) data on $Co_2T_{1-x}Al_{1+x}$ (T = Ti or Zr).

The alloys $Co_2Ti_{1-x}Al_{1+x}$ (x = 0, +0.25, and -0.25) and $Co_2Zr_{1-x}Al_{1+x}$ (x = 0, +0.15, and -0.15) were prepared by repeated melting together the constituents in an arc furnace under inert atmosphere. After homogenizing at 1175 K for four days, the samples for resistivity ($\approx$ 2x2x15 mm^3) were cut and annealed at 1175 K for three days in evacuated silica capsules. Accurate electrical resistivity measurements were made using a standard four-probe dc

---

1) Bombay 400085, India.

2) Bombay 400005, India.

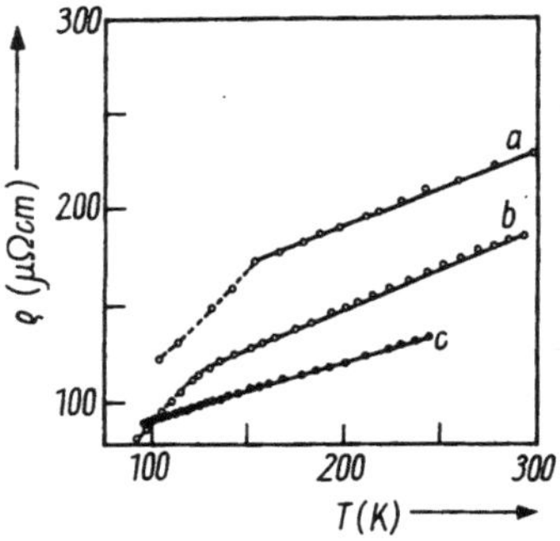

Fig. 1. Electrical resistivity of $Co_2Ti_{1-x}Al_{1+x}$ alloys: (a) $Co_2Ti_{1.25}Al_{0.75}$, (b) Co_2TiAl, and (c) $Co_2Ti_{0.75}Al_{1.25}$

set-up described elsewhere /5/. The temperature of the samples was maintained to within ± 0.2 K during each measurement. The relative values of the resistivity are estimated to be accurate to $\pm 0.02\%$. The lattice parameter values for all the alloys obtained from X-ray spectra (Debye-Scherrer) using FeK_α radiation are listed in Table 1. One of the alloys $Co_2Zr_{1.15}Al_{0.85}$ was found not to be of single phase and showed the presence of $Co_{1.14}Al_{0.6}Zr$. Lattice parameters of stoichiometric compositions (x = 0) agree well with values from literature /1/.

The resistivity (ϱ) data as a function of temperature for the alloys containing Ti are shown in Fig. 1 while that of the alloys containing Zr is shown in Fig. 2. The $\varrho(T)$ curves for all the alloys (except $Co_2Ti_{0.75}Al_{1.25}$) show a point of inflexion in the region of the expected Curie temperatures (T_C). The values of T_C and the spin-disorder resistivities (ϱ_{spd}), as determined by extrapolating the high-temperature ($T>T_C$) portion of the curves to 0 K are also listed in Table 1. For the five ferromagnetic alloys a phenomenological fit $\varrho(T) = \varrho_o + AT + BT^2$ could be given below the respective Curie temperatures, presuming that the linear and the quadratic terms arise from the electron-phonon scattering and the spin-disorder scattering, respectively. The values of ϱ_o, A, and B thus obtained are listed in Table 1, but too much significance cannot obviously be attached to this fit since we did not have $\varrho(T)$ data below 100 K, for fitting. The main conclusions are:

(i) Increasing the Ti content from

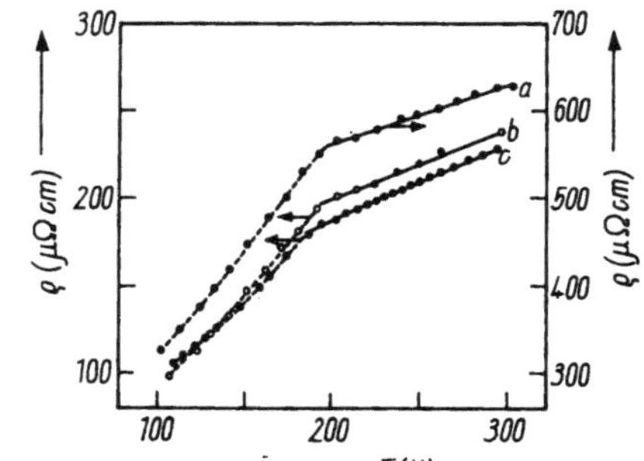

Fig. 2. Electrical resistivity of $Co_2Zr_{1-x}Al_{1+x}$ alloys: (a) $Co_2Zr_{1.15}Al_{0.85}$, (b) Co_2ZrAl, and (c) $Co_2Zr_{0.85}Al_{1.15}$. Appropriate ordinate scales are indicated by arrows

Table 1

Lattice parameter, Curie temperature, and spin-disorder resistivity values for the alloys $Co_2T_{1-x}Al_{1+x}$. Also listed are the values of the parameters ρ_0, A and B obtained from a fit $\rho(T) = \rho_0 + AT + BT^2$ given below T_C

alloy	lattice parameter (nm)	T_C (K)	ρ_{spd} ($\mu\Omega$cm)	ρ_0 ($\mu\Omega$cm)	A	B
Co_2TiAl	0.5846	126 (125) /6/	65	9.6	0.7	1.1×10^{-3}
$Co_2Ti_{1.25}Al_{0.75}$	0.5863	155	113	39	0.7	1.0×10^{-3}
$Co_2Ti_{0.75}Al_{1.25}$	0.5818	-	-	-	-	-
Co_2ZrAl	0.6082	194 (185) /1/	118	13.1	0.6	17.6×10^{-4}
$Co_2Zr_{1.15}Al_{0.85}$	0.6082	196	332(?)	39.0	1.4	3.95×10^{-3}
$Co_2Zr_{0.85}Al_{1.15}$	0.6070	188	100	9.7	0.8	5×10^{-4}

Co_2TiAl to $Co_2Ti_{1.25}Al_{0.75}$ led to a rise in T_C in agreement with Endo et al. /7/ who also reported a rise in T_C upon varying the composition from Co_2TiAl to $Co_2Ti_{1.2}Al_{0.8}$. The Ti-deficient alloy $Co_2Ti_{0.75}Al_{1.25}$ did not show any indication of ordering to the lowest temperature measured (Fig. 1), once again in agreement with /7/ where it was shown that T_C should extrapolate to zero for an approximate composition $Co_2Ti_{0.75}Al_{1.25}$. The reduction in the values of ρ_{spd} and T_C when the Zr concentration decreases from Co_2ZrAl to $Co_2Zr_{0.85}Al_{1.15}$, and the increase in the value of these parameters with increase in Ti content from Co_2TiAl to $Co_2Ti_{1.25}Al_{0.75}$ points to a definite role being played by Ti/Zr atoms occupying the Y sites in the magnetism of Co_2TAl Heusler alloys. The formation of the doubly-ordered $L2_1$ structure is a necessary but not sufficient condition for the formation of magnetic moments on Co atoms, which strongly depends on the nearest neighbour Ti/Zr atoms according to the local environment model /7/.

(ii) The departure from $L2_1$ order in $Co_2Zr_{1.15}Al_{0.85}$ implies a high defect concentration, and so a high ρ-value (Fig. 2). Our lattice parameter analysis showed that the lattice cannot accomodate 15% extra Zr atoms in the Al sites, presumably due to the difference in ionic radii. In the case of Co_2TAl it has been pointed out that the ordered $L2_1$ structure is less likely to be formed when

T = Zr or Hf, because of the size mismatch /2/. The formation in part of $Co_{1.4}Al_{0.6}Zr$ indicates that it has a very close value of enthalpy of formation compared to that of $Co_2Al_{0.85}Zr_{1.15}$.

(iii) The resistivity studies so far reported were mostly on X_2MnZ where the value of the localized magnetic moment on Mn atoms is unaffected by the conduction electron concentration, and the resistivity generally obeys $\varrho(T) = aT^n$ at very low temperatures ($T < 20$ K) and $\varrho(T) = \varrho_0 + AT + BT^2$ at higher temperatures /4/. The itinerant character of the 3d electrons in Co_2TAl alloys has been emphasized in contrast with the localized nature of Mn moments in X_2MnZ /2/. The $\varrho(T)$ plots for the ferromagnetic $Co_2T_{1-x}Al_{1+x}$ alloys, below their respective Curie temperatures, are in fair agreement with a phenomenological fit $\varrho(T) = \varrho_0 + AT + BT^2$.

(iv) The behaviour of our alloys seems to agree with Stearns' model /8/ which suggests that Ti or Zr have predominantly d_i electrons and Co has predominantly d_l electrons and a correct combination of d_i and d_l electrons gives rise to a d_i polarisation which provides the magnetic coupling.

The authors are thankful to Dr. R.M. Iyer, Head, Chemistry, Division and Prof. R. Vijayaraghavan for encouragement.

References

/1/ K.R.A. ZIEBECK and P.J. WEBSTER, J. Phys. Chem. Solids 35, 1 (1974).

/2/ K.H.J. BUSCHOW and P.G. VAN ENGEN, J. Magnetism magnetic Mater. 25, 90 (1981) and references therein.

/3/ C.M. HURD, I. SHIOZAKI, and S.P. MC ALISTER, Phys. Rev. B 26, 701 (1982).

/4/ W.H. SCHREINER, D.E. BRANDAO, F. OGIBA, and J.V. KUNZLER, J. Phys. Chem. Solids 43, 777 (1982).

/5/ J.V. YAKHMI, Ph.D. Thesis, University of Bombay, 1975.

/6/ A.K. GROVER, Ph.D. Thesis, University of Bombay, 1979.

/7/ K. ENDO, I. AN, and A. SHINOGI, J. Phys. F 7, L99 (1977).

/8/ M.B. STEARNS, Phys. Rev. B 17, 2809 (1978).

(Received January 13, 1984)

phys. stat. sol. (a) 85, K93 (1984)
Subject classification: 14.3.3; 22

Institute of Electron Technology, Warsaw[1)]

Current-Voltage Characteristic of the p-n Junction with an Exponential Impurity Distribution at Its Base Contact

By
J. PUŁTORAK

The p-n junction diode structure, given in Fig. 1, is considered. Regions I and II of that structure are homogeneous and an exponential impurity distribution in region III (of length w^*) produces an internal (built-in) field

$$E_w = -\frac{kT}{q}\frac{\ln\left[N_d(x_5)/N_d(x_4)\right]}{w^*} . \qquad (1)$$

The inhomogeneity of region III appears as a result of an unexpected diffusion of impurities from the contact into the base region during its fabrication or the use of an epitaxial layer with a not sufficiently sharp profile.

It is assumed that the length of the region I is much bigger than the corresponding excess carrier diffusion length, ohmic contacts are placed in planes $x = x_1$ and $x = x_5$, the neutrality principle within regions I, II, and III is satisfied $(n - n_o \equiv \Delta n = \Delta p \equiv p - p_o)$, the bulk recombination rate is given by $r = \Delta p/\tau$, the generation in the beyond-junction areas is neglected, processes occurring within the p-n junction depletion layer $(x_2 < x < x_3)$ are neglected, too (their contribution can be considered separately /1/), and lateral dimensions of the structure are big enough, so that the recombination on the lateral surface of the structure may be neglected /2 to 4/.

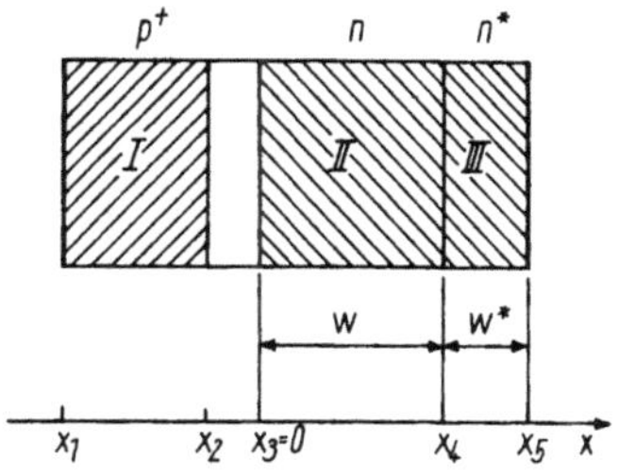

Fig. 1. The considered p-n junction diode structure

1) al. Lotników 32/46, 02-668 Warsaw, Poland.

Ambipolar transport equations derived in /5/ on the basis of /6, 7/ are used

$$\operatorname{div} J = 0 \ , \tag{2}$$

$$\frac{\partial p}{\partial t} = \operatorname{div}\left(D \operatorname{grad} \Delta p + \mu^* \Delta p \operatorname{grad} V_o\right) + \left(\frac{\sigma_{no}\sigma_p + \sigma_{po}\sigma_n}{q\sigma^2} \frac{q}{kT} \operatorname{grad} V_o - \frac{\mu^*}{\sigma} \operatorname{grad} \Delta p\right) J - \frac{\Delta p}{\tau} \ , \tag{3}$$

$$J = -\sigma \operatorname{grad} \Psi - 2q\, \mu_o \Delta p \operatorname{grad} V_o \ , \tag{4}$$

where

$$D = \left(\sigma_n/\sigma\right) D_p + \left(\sigma_p/\sigma\right) D_n \ , \tag{5}$$

$$\mu_o = (q/kT) D \ , \tag{6}$$

$$\mu^* = \left(\sigma_n/\sigma\right)\mu_p - \left(\sigma_p/\sigma\right)\mu_n \ , \tag{7}$$

and the potential, defined in /6/ and normalized in /8/, is given by

$$\Psi = V - V_o - \frac{kT}{q} \frac{b-1}{b+1} \ln \frac{\sigma}{\sigma_o} + V_{Fo} \ , \tag{8}$$

where V potential, V_o internal field potential, V_{Fo} Fermi potential in equilibrium state, σ and σ_o conductivity and conductivity in equilibrium state, respectively, $b = \mu_n/\mu_p$.

Applying the method of calculations proposed in /9/, an equation describing the I-U characteristic of the structure for currents

$$\frac{J}{J_{so}} \ll \frac{1}{b+1} \frac{w}{L_p} \frac{\sigma_{no}}{\sigma_{po}} \tag{9}$$

is obtained

$$U = \frac{kT}{q}\left\{\ln\left(1 + \frac{J}{J_{so}} \operatorname{ctgh} \frac{w}{L_p}\right) + \frac{\sigma_{po}}{\sigma_{no}} \frac{J}{J_{so}} \left[\frac{w}{L_p}\left(1 - (b+1)\frac{L_p}{w} \frac{\sigma_{po}}{\sigma_{no}} \frac{J}{J_{so}}\right) + \frac{w^*}{L_p} \frac{1}{\ln\left[\frac{N_d(x_5)}{N_d(x_4)}\right]} \times \left(1 - \frac{b+1}{3} \frac{\sigma_{po}}{\sigma_{no}} \frac{J}{J_{so}} \frac{1}{\sinh \frac{w}{L_p}}\right) + \frac{1}{\sinh \frac{w}{L_p}}\left(1 - \frac{b+1}{2} \frac{\sigma_{po}}{\sigma_{no}} \frac{J}{J_{so}} \frac{1}{\sinh \frac{w}{L_p}}\right)\right]\right\}, \tag{10}$$

where

$$J_{so} = qD_p p_o / L_p \tag{11}$$

and all physical coefficients used in the above equations refer to region II.

The first term of (10) determines the potential drop over the depletion layer of the p^+-n junction and the second one over the homogeneous part of the base region (II) including an influence of the conductivity modulation in this region. The third and fourth terms determine the potential drop over the inhomogeneous part of the base region (III) including an influence of the conductivity modulation as well as of the built-in field in this region.

The saturation current J_s of the structure, calculated from (10) for $U \to -\infty$, is

$$J_s = J_{so} \, \mathrm{tgh}(w/L_p) \ . \tag{12}$$

This equation points out (in the first approximation) that the inhomogeneity at the base contact shows similar properties as the step l-h junction of a very low surface recombination velocity.

Equation (10) can be easily simplified to the classical form $J = J_{so}\left[\exp(qU/kT) - 1\right]$ when the base is thick ($w/L_p > 1$) and made of a semiconductor fulfilling the condition $(\sigma_{po}/\sigma_{no}) \ll 1$.

For a thin base structure ($w/L_p < 1$) equation (10) may be simplified to the form

$$U = \frac{kT}{q}\left\{\ln\left(1 + \frac{J}{J_{so}} \, \mathrm{ctgh}\, \frac{w}{L_p}\right) + \frac{\sigma_{po}}{\sigma_{no}} \frac{J}{J_{so}} \frac{1}{\sinh \frac{w}{L_p}} \left(1 - \frac{b+1}{2} \frac{\sigma_{po}}{\sigma_{no}} \frac{J}{J_{so}} \frac{1}{\sinh \frac{w}{L_p}}\right)\right\} \ . \tag{13}$$

Now comparing the above result with that obtained in /9/ one can calculate that the voltage drop over region III (described by the second term of (13)) is only (b - 1) times smaller than that over the l-h junction.

The voltage drop over region III decreases with current. This leads to the conclusion that the recombination velocity in the plane $x = x_4$ should enhance due to the compensation of the built-in field created there by the current flow. Consequently minority carriers flow easier to the plane $x = x_5$ of a very high recombination velocity.

References

/1/ C.T. SAH, R.N. NOYCE, and W. SHOCKLEY, Proc. IRE 45, 1228 (1957).

/2/ J. MARCIAK-KOZŁOWSKA and J. PUŁTORAK, Electron Technology 4, 119 (1971).

/3/ J. PUŁTORAK, phys. stat. sol. (a) 73, K217 (1982).

/4/ J. PUŁTORAK, Electron Technology 15, 1/2 (1982).

/5/ S. SIKORSKI, Arch. Elektrotech. 16, 439 (1967).

/6/ W. VAN ROOSBROECK, Phys. Rev. 91, 282 (1953).

/7/ W. VAN ROOSBROECK, Bell Syst. Tech. J. 39, 515 (1960).

/8/ S. SIKORSKI, Arch. Elektrotech. 15, 703 (1966).

/9/ A. KASSUR, phys. stat. sol. (a) 23, 309 (1974).

(Received July 19, 1984)

Pre-Printed Titles

Papers to be published
in "physica status solidi (a) - applied research"
Vol. 85, No. 2

Original Papers

R.V. G. RAO and S. KUMAR PAL, Calcutta
A New Method of Evaluation of the Structure of Molten Sodium Chloride

H.C. GONZÁLEZ, Buenos Aires
Hardening of Single Crystals of Magnesium by Low Neutron Doses at 77 K

R. PICKENHAIN, K. JACOBS, W. SEIFERT, V. CHERNYI, and S. BREHME, Leipzig
Green and Yellow Light Emitting Diodes Produced from Vapour Phase Epitaxial GaP:N (I)

H. BOOYENS and J.H. BASSON, Pretoria
On the Optimal Choice of Substrate for $Hg_{0.8}Cd_{0.2}Te$ Epitaxial Layers

B.P. TYAGI and K. SEN, Dehradun
Grain Size and Temperature Dependence of Electrical Behaviour of Polysilicon

H. HESS, Graz
Luminescent Properties of Zinc Silicate Doped with Gallium

F. SCHNEIDER and S. UNTERRICKER, Freiberg
Radiation Damage and Its Annealing Behaviour in InP after Recoil Implantation of ^{118}Sb and ^{111}In Observed by TDPAC

W. HERGERT and L. PASEMANN, Halle and Leipzig
Theoretical Study of the Information Depth of the Cathodoluminescence Signal in Semiconductor Materials

G. FANTOZZI, C. ESNOUF, J. LAUZIER, and C. MINIER, Villeurbanne and Grenoble
Internal Friction at Low Temperature in a High Purity Magnesium

M. SUEZAWA and K. SUMINO, Sendai
On the Annihilation of Thermal Donors in Silicon Crystal

A.A. EL-SHARKAWY, I.H. RASHED, M.S. ZAGHLOUL, and M.H. GHONIEM, Cairo
Study of Thermal Properties of Polycrystalline KCl, KBr, and KI in the Temperature Range 300 to 600 K

A.A. KAMINSKII, S.E. SARKISOV, G.A. DENISENKO, V.V. RYABCHENKOV, V.A. LOMONOV, YU.E. PERLIN, M.G. BLAZHA, D. SCHULTZE, B. HERMONEIT, and P. REICHE, Moscow and Berlin
Growth, Spectral and Luminescence Study of Cubic $Bi_4Ge_3O_{12}:Pr^{3+}$ Crystals

A.P. TYUTNEV, V.S. SAENKO, A.V. VANNIKOV, and V.E. OSKIN, Moscow
Radiation Induced Dielectric Effect in Polymers

G. MICOCCI, A. RIZZO, and A. TEPORE, Lecce
Photoconductivity in Halogenated and Hydrogenated Amorphous Silicon Films

S. GHOSH and S. KHAN, Ujjain
Amplitude Modulation and Demodulation of Electromagnetic Waves in Magnetised Semiconducting Plasmas

K.N. JOG, R.K. SINGH, and S.P. SANYAL, Jabalpur and Trieste
The Relative Stability and Phase Transition Studies of MgO and CaO Crystals

H. KRAUSE, M. SCHELL, and H.-CHR. GOLINSKY, Ilmenau
Hopping Conduction in SiO_2 Films Thermally Grown on Phosphorus-Doped Polycrystalline Silicon

J.M. LÓPEZ and J.A. ALONSO, Valladolid
The Atomic Size-Mismatch Contribution to the Enthalpy of Formation of Concentrated Substitutional Metallic Solid Solutions

M. DAVID, W. HERREMANS, and R. GEVERS, Antwerpen
Fast-Electron Diffraction near a Translation Interface on Edge

F. WALZ, Stuttgart
Transition-Induced Magnetic After-Effect in the Amorphous Alloy $Fe_{20}Ni_{56}B_{24}$ at the Critical Temperature

N.P. ESINA, N.V. ZOTOVA, S.A. KARANDASHEV, and G.M. FILARETOVA, Leningrad
Tunneling Surface Recombination in MS Structures on p-InAs

V.V. EMTSEV, YU.N. DALUDA, P. GAWORZEWSKI, and K. SCHMALZ, Leningrad and Frankfurt/Oder
Electrical Properties of Thermal Donors Formed in Cz-Si during Heat Treatment at 450 °C

J.M. ALAMEDA, M.C. CONTRERAS, and H. RUBIO, Oviedo
Magnetic Properties of Amorphous Fe_xSi_{1-x} Films (I)

B. RAUSCHENBACH, G. BLASEK, and R. DIETSCH, Rossendorf and Dresden
Modification of Concentration Profiles in Iron and Aluminium by High-Fluence Implantation of Nitrogen and Boron Ions

S.V. SOKOLSKI, F.F. LAVRENTEV, and O.P. SALITA, Kharkov
Jump-Like Deformation and Stability of the Hardened and Structural States of Zinc Single Crystals with Forest Dislocations in the Temperature Range 293 to 4.2 K

S. NEŠPŮREK, O. ZMEŠKAL, and F. SCHAUER, Prague and Brno
An Ohmic Contact Test Using the Activation Energy of Steady-State Space-Charge-Limited Current

I.P. ZVYAGIN, Moscow
Planar Conduction in Inhomogeneous Films

A.K. ARORA, Kalpakkam
Isochronal Annealing Study of Aggregation in Doped Alkali Halide Crystals

H. SODOLSKI, Gdańsk
Anomalous Polarization Current in Dielectrics

A.V. DVURECHENSKII, B.P. KASHNIKOV, and V.V. SUPRUNCHIK, Novosibirsk
New EPR Defects in Si⟨Al⟩

V.I. GUBSKAYA, P.V. KUCHINSKII, and V.M. LOMAKO, Minsk
The Effect of the Vacancy Charge State on the Radiation Defect Formation in Silicon

P. SIRCAR and M. AUBIN, Moncton and Quebec
Pulsed UV Laser Processed Gold and Aluminum Ohmic Contacts on n^+-GaAs

R.S. VARDANIAN and A.KH. KHACHATRIAN, Erevan
Application of Invariance Principle in the Dynamic Theory of X-Ray Scattering

B.T. KOLOMIETS, V.M. LYUBIN, M.A. TAGUYRDZHANOV, and G. ZENTAI, Leningrad and Budapest
A Comparative Study of Photoconductivity in Vitreous As_2Se_3 and a-Si:H Thin Films

D.I. ALADASHVILI, L. KONSZEWICZ, and S. POROWSKI, Tbilisi and Warsaw
Studies of the Deep Levels in p-Type InSb under Pressure

C. MALGRANGE and J. GRONKOWSKI, Paris and Warsaw
Fraunhofer Diffraction of X-Ray Beams at Bragg Incidence in Distorted Crystals

P. MICHON, H. ATMANI, and C. VAUTIER, Mont Saint-Aignan
Crystallization Study of Evaporated a-Se Films by Means of DTA Technique

I. BAKER and E.M. SCHULSON, New Hamshire
On Intrinsic Stacking Faults in Polycrystalline Ni_3Al

SHU ZHEN and G.J. DAVIES, Sheffield
Calculation of Potential Energy Parameters for Some Intermetallic Compounds

V.V. KALININ and N.N. GERASIMENKO, Novosibirsk
On the Effect of Doping Impurities on the Formation of Rod-Like Defects in Silicon

P.F. LUGAKOV and V.V. SHUSHA, Minsk
A Model of Nonequilibrium Charge-Carrier Recombination in Semiconductors Containing Nonuniformities

N.I. BORGARDT, S.K. MAKSIMOV, and D.I. PISKUNOV
Phase Information in Transmission Electron-Microscope Investigationof Defects

H. TAGUCHI, M. SHIMADA, and M. KOIZUMI, Osaka and Sendai
Magnetic Properties in the System $(Sr_{1-x}Ca_x)CoO_{3-\delta}$

K. IKEZAKI, T. ISHII, and T. MIURA, Yokohama
Thermal Influence of Vacuum Deposition of Metallic Electrodes on TSC from Positively Corona-Charged Polyethylene Films

A.D. POGREBNYAK, Tomsk
Positron and Positronium States in Semiconductors Irradiated by Supercurrent Beams of Charged Particles

Papers to be published
in "physica status solidi (b) - basic research"
Vol. 125, No. 2

Original Papers

B.A. VAID and K.C. SHARMA, Simla
Ultrasonic Attenuation in Liquid Metals

G. AMBRAZEVIČIUS, G. BABONAS, and YU.V. RUD, Vilnius
Excitation Spectra of Photoluminescence Bands in $Cd_{1-x}Mn_xTe$

C. PRASAD and J.D. DUBEY, Ranchi
Electronic Structure and Properties of Cubic Boron Nitride

M.I. VLADIMIR, V.A. MOSKALENKO, and V.V. SKAFARU, Kishinev
The Distribution of the Exchange Magnetic Fields in Spin Glasses

V. ČÁPEK, Prague
To the Interplay of Vibrational Relaxation and Exciton Motion

P. PETELENZ, Cracow
Optical Detrapping in Molecular Crystals (II)

V. SURIYANARAYANAN, Y. MADHAVAN, and K. RAMACHANDRAN, Madurai
Isotope Diffusion in Cadmium Telluride

Y. MATSUOKA, S. TERAKADO, and S. TANIGAWA, Nagoya and Ibaraki
Positron Age-Momentum Correlation in Ultrafine Particles of Silver and Gold

S.W. BIERNACKI, Warsaw
Born-von Karman Cell Model for Deep Impurities

A.Z. MAKSYMOWICZ, T. STOBIECKI, E. JAROCKI, and W. KARAŚ, Cracow
Charge Transfer and Concentration Dependence of Magnetization in Amorphous Fe-B Alloys

M.A. ANDREEVA and R.N. KUZMIN, Moscow
Mössbauer Conversion Electron Spectra at Total Reflection

H.J. FISCHBECK, Berlin
The Incommensurate CDW Phase of a Quasi-One-Dimensional Peierls-Fröhlich System with a Nearly Quarter-Filled Band (I)

V.L. SAFONOV, Moscow
Interactions of Magnons in the Six-Sublattice Antiferomagnet

G. SCHUBERT and J. RICHTER, Dresden
The Electrical Resistivity of Clustering Alloys

V.V. ANTONOV-ROMANOVSKII, Moscow
On the Formation of Large Clusters of Recombining Particles of One Kind in Solids during Reaction A + B = 0 (Two- and Three-Dimensional Discrete Models)

A.R. VASCONCELLOS and R. LUZZI, Campinas
On Laser-Induced Lattice Heating in a Polar Semiconductor

B.S. TOŠIĆ and J.P. ŠETRAJČIĆ, Novi Sad
Exciton Drops in One-Dimensional Molecular Chains

P. STREITENBERGER, Magdeburg
Effective Exchange and Correlation Interaction and Dielectric Screening in the Electron Gas

B. WYNCKE, F. BREHAT, M. EL SHERIF, and G.V. KOZLOV, Nancy and Moscow
Far-Infrared and Submillimeter Dielectric Response of Sodium Nitride

A.I. MELKER and S.N. RAMANOV, Leningrad
Group Representation of the Formation Probabilities of Athermal Vacancy Clusters in Atomic Collision Cascades

V.N. ERMAKOV, D.V. KORBUTYAK, V.G. LITOVCHENKO, O.YU. MIKITYUK, and V.V. NITSOVICH, Kiev
A Study of the Reflection Spectra of Surface Exciton-Impurity Complexes

G. MANZKE, V. MAY, and K. HENNEBERGER, Güstrow
Excited States and Collective Excitations in a Dense Gas of Excitons

Y. UNUMA, Y. ABE, Y. MASUMOTO, and S. SHINOYA, Tokyo
Dynamical Behavior of the High Density Electron-Hole System in CdSe

T. IWASHITA and N. URYÛ, Kumamoto and Fukuoka
The Curie Temperature of the Ising Ferromagnet on the Honeycomb Lattice with Mixed Spins

A.N. VASILEV and V. ČÁPEK, Moscow and Prague
Vibrational Relaxation in a Localized Excited Electronic State by the GME Method

K. ŽIČKUS, A. AUDZIJONE, J. BATARŪNAS, and A. ŠILEIKA, Vilnius
The Fundamental Absorption Edge Tail of Ferroelectric SbSI

D. FRÖHLICH, A. NÖTHE, and K. REIMANN, Dortmund
Determination of Valence Band Parameters in ZnTe

M. KORKMAZ, B. AKTAŞ et M. DUPONT, Ankara et Aubière
Le spectre de R.P.E. de l'ion Mn^{2+} dans le tartrate de cadmium ($CdC_4H_4O_6 \cdot 5\,H_2O$)

S.K. PLACHKOVA and L.A. TERSIEV, Sofia
Phase Transitions and Intervalley Jahn-Teller Effect in the System $(Ge\text{-}Te)_{1-x}(AgSbTe_2)_x$

U.W. POHL, H.-E. GUMLICH, and W. BUSSE, Berlin
Zero-Phonon Lines in the Excitation Band Centred at 23 200 cm^{-1} Due to Mn^{2+} in Polymorphic ZnS

O.N. BORGES, L.A. AMARANTE RIBEIRO, and L.L. GONÇALVES, Belo Horizonte and Fortaleza
The XY Model with Nearest and Next-Nearest Neighbour Interaction in a Transverse Field

H. WERHEIT and R. FRANZ, Duisburg
States of Thermal Equilibrium in β-Rhombohedral Boron and Their Alteration by Optical Excitation

S.L. KAKANI and U.N. UPADHYAYA, Bhilwara and Udaipur
Superconducting and Magnetic Order Parameters in the Coexistence Region of Ternary Superconductors

F. ENGELKE, D. MICHEL, and F. PILLE, Leipzig
NMR Study of Dynamical Processes in Crystals of the TSCC Family

C. PISANI and E. MARCHESE, Torino
Critical Analysis and Simplified Formulation of the Travelling Cluster Approximation for Disordered Systems

V.A. KASIYAN, D.D. NEDEOGLO, A.V. SIMASHKEVICH, and I.N. TIMCHENKO, Kishinev
Hopping Conductivity and Electron Localization in Heavily Doped ZnSe

G. MILCHBERG, K. SAMINADYAR, E. MOLVA, and H.R. ZELSMANN, Grenoble
Optical Determination of the Valence Band Parameters in CdTe

A.A. ANDREEV, A.V. ZHERZDEV, A.I. KOSAREV, K.V. KOUGIA, and I.S. SHLIMAK, Leningrad
Photoluminescence and Recombination of Excess Carriers in Amorphous Hydrogenated Silicon

S.V. VOITIKOV and V.P. GRIBKOVSKII, Minsk
On the Theory of Crystal Excitation by Electric Discharges

I.G. RATISHVILI, Tbilissi
Annihilation of Frenkel Pairs in B.C.C. Metals (α-Iron) (I)

G. HEDER and J.P. SPAETH, Paderborn
ENDOR Investigation of $H_i^0(Li)$ Centres in KCl

P. STŘEDA, Prague
Universal Transport Properties of Inversion Layers in Quantizing Magnetic Field

19. Magnetic Resonance

20. Optical Properties
    - 20.1 Spectra. Optical Constants (X-Ray Spectra See 20)
    - 20.2 Lasers
    - 20.3 Luminescence (X-Ray Spectra See 20)

21. Metals. Alloys
    - 21.1 Transition Metals (Group Sc to Zn) and Their Alloys
        - 21.1.1 Fe and Fe Alloys
    - 21.2 Alkali Metals
    - 21.3 Alkali-Earth Metals
    - 21.4 Rare-Earth Metals
    - 21.5 Actinides
    - 21.6 Noble Metals
    - 21.7 Semi-Metals (Semiconducting Alloys 22.7)

22. Semiconductors and Ionic Crystals
    - 22.1 Elements
        - 22.1.1 Germanium
        - 22.1.2 Silicon
        - 22.1.3 Group VI Elements
    - 22.2 III–V Compounds
        - 22.2.1 Arsenides
        - 22.2.2 Phosphides
        - 22.2.3 Antimonides
        - 22.2.4 Mixed Crystals
    - 22.3 IV–IV, II–V, and III–VI Compounds
        - 22.3.1 Mixed Crystals
    - 22.4 II–VI Compounds
        - 22.4.1 Sulphides
        - 22.4.2 Selenides
        - 22.4.3 Tellurides
        - 22.4.4 Mixed Crystals
    - 22.5 Halides
        - 22.5.1 Silver Halides
        - 22.5.2 Alkali Halides
        - 22.5.3 Alkali-Earth Halides
        - 22.5.4 Mixed Crystals
    - 22.6 Simple Oxides (CdO, ZnO See 22.4)
        - 22.6.1 Mixed Crystals
    - 22.7 Semiconducting Intermetallic Compounds
        - 22.7.1 Mixed Crystals
    - 22.8 Tertiary and Higher Compounds
        - 22.8.1 Salts of Oxy-Acids (Carbonates, Phosphates, Silicates Including Mica etc.)
        - 22.8.2 Spinels, Garnets, Ferrites
    - 22.9 Organic Semiconductors

23. Solidified Gases

Attention

The preceding classification scheme is closely akin to the scheme used by Physics Abstracts to classify the various fields of solid state physics. A paper will either be classified under one or more of the subsections or, if none of these are relevant, it will just appear under the general section heading (i. e. sections and subsections are equivalent in this sense).

The *subject classification* given at the head of each paper begins with the main subject matter of the paper, and this is followed, if necessary, by figures referring to subsidiary subject matter and the type of solids investigated.

The papers published in this issue are listed in the order of their subject classification at the end of the table of contents.

82 914

Manuscripts and letters for physica status solidi (b) — basic research and physica status solidi (a) — applied research

should be sent to

DDR-1086 Berlin, Leipziger Straße 3–4, P. B. 1233

or

Internationaler Buchversand GmbH, 1000 Berlin 19, Eschenallee 21.

The object of "physica status solidi (b)" and "physica status solidi (a)" is to publish rapidly important and topical results concerning fundamental research in the field of solid state physics and aspects of application and experimental techniques of fundamental solid state research, respectively. The issues of "physica status solidi (b)" and "physica status solidi (a)" will be published on the 1st and the 16th day of each month, respectively. Either series will contain the following categories:

1. Review articles written by arrangement with the editors. They will be remunerated.
2. Original papers selected from those which have not been submitted for publication elsewhere, after being reviewed. (The latest date for acceptance of contributions is the 8th and the 23rd day of each month for "physica status solidi (b)" and "physica status solidi (a)", respectively, 50 days before the delivery of the journal.)
3. Short notes published in the off-set part of the journal. They are, if necessary after being reviewed, selected by editorial staff. (The latest date for acceptance of manuscripts for this part is the 12th and the 27th day of each month for "physica status solidi (b)" and "physica status solidi (a)", respectively, 16 days before the delivery of the journal.)
4. Pre-printed titles of papers to be published in the next issue of both series.

Editorial Information

Manuscripts should be submitted preferentially in English. Besides, papers in German, French, and Russian are admitted. The final decision on the incorporation of the papers in the series (b) or (a) is taken by the editorial staff after consultation of the referee. Authors are asked to observe the following rules:

1. Manuscripts have to be written as concisely as possible. As most, 20 and 4 pages (28 lines per page; abstracts, tables, references, figures, and captions of figures included, where 3 figures are counted as about one page) are admitted for original papers and short notes, respectively. All manuscripts should be type-written on one side only. Two copies are required. Badly arranged manuscripts are returned.
2. The author's name, initials, and institution should be given.
3. One abstract of about ten lines in English and one in German, French, or Russian are required for original papers.
4. Because of lack of space conclusions (if any) must not have the character of a summary (i. e. substantially a repetition of the contents of the abstract).
5. Formulae should not be written too small and not with pencil. Separate lines for formulae are desirable. Indistinct writing will lead to a return of the manuscript.
6. Drawings should be enclosed on separate sheets of paper, not exceeding the maximum and minimum sizes of 20×30 and 10×15 cm^2, respectively. Photographs should be submitted, on white paper, only for review articles and original papers. Duplicates are required.
7. Captions for the figures should be submitted on separate sheets of paper, and special instructions with regard to the size of illustrations should be given on these.
8. References should be numbered in order of appearance and should be given, on a separate sheet, in the form used in physica status solidi.
9. SI-units should be used.
10. Each paper should contain a proposal for the subject classification according to the classification scheme appearing on the second and third cover of each issue.

The authors will be sent galley proofs of review articles and original papers. Proof reading of short notes and pre-printed titles will be done by the editorial staff.

For each original paper and short note a total of 75 reprints is available free of charge. The number of reprints free of charge per review article is 25. Requests for additional reprints (against payment) should be given together with the submitted manuscripts. **The deadline for orders of additional reprints is the 15th and the 30th of each month** for "physica status solidi (b)" and "physica status solidi (a)", respectively, two weeks before the delivery of the journal.

Bei Fragen zur Produktsicherheit wenden Sie sich bitte an:
If you have any questions regarding product safety,
please contact:

Walter de Gruyter GmbH
Genthiner Straße 13
10785 Berlin
productsafety@degruyterbrill.com